CONTRIBUTIONS TO MODERN AND ANCIENT TIDAL SEDIMENTOLOGY

Proceedings of the Tidalites 2012 conference

Other publications of the International Association of Sedimentologists

Special Publication Number 47 of the International
Association of Sedimentologists

Contributions to Modern and Ancient Tidal Sedimentology

Proceedings of the Tidalites 2012 conference

Edited by

Bernadette Tessier
*CNRS - UMR 6143 M2C
University of Caen Normandie,
24 rue des Tilleuls,
14000 Caen,
France*

Jean-Yves Reynaud
*CNRS - UMR 8187 LOG,
University of Lille
Cité Scientifique,
F 59 000 Lille,
France*

SERIES EDITOR

Mark Bateman
*Department of Geography,
Winter St.,
University of Sheffield
Sheffield S10 2TN
UK*

WILEY Blackwell

Library of Congress Cataloging-in-Publication Data

Names: Tessier, Bernadette, editor. | Reynaud, Jean-Yves, 1969– editor. | International Association of Sedimentologists.
Title: Contributions to modern and ancient tidal sedimentology : proceedings of the Tidalites 2012 Conference /
    edited by Bernadette Tessier, Jean-Yves Reynaud.
Description: Chichester, West Sussex : John Wiley & Sons, Inc., 2016. | "International Association
    of Sedimentologists." | Includes bibliographical references and index.
Identifiers: LCCN 2015047530 | ISBN 9781119218371 (cloth)
Subjects: LCSH: Sedimentation and deposition–Congresses. | Marine sediments–Congresses. | Tidal flats–Congresses. |
    Sediments (Geology)
Classification: LCC QE571 .C574 2016 | DDC 551.3/6–dc23 LC record available at http://lccn.loc.gov/2015047530

A catalogue record for this book is available from the British Library.

Contents

List of contributors

Ashour Abouessa
Institut de Physique du Globe de Strasbourg
(IPGS)-UMR 7516;
Université de Strasbourg (UdS)/École et
Observatoire des Sciences de la Terre (EOST),
Centre National de la Recherche Scientifique
(CNRS), 1 rue Blessig,
Strasbourg, 67084, France

Allen W. Archer
Department of Geology, Kansas State University,
Manhattan, Kansas, 66506, USA

Andrea Baucon
UNESCO Geopark Meseta Meridional,
Geology and Paleontology Office,
6060-101-Idanha-a-Nova, Portugal

M. Isabel Benito
Departamento de Geología,
Universidad de Oviedo,
C/Jesus Arias de Velasco, s/n,
33005, Oviedo, Spain
Instituto de Geociencias IGEO (CSIC, UCM),
C/José Antonio Novais 12, 28040, Oviedo, Spain

Manuela Chamizo-Borreguero
Departamento de Estratigrafía
(UCM) Grupo de Análisis de Cuencas
Sedimentarias (UCM-CAM),
Facultad de Ciencias Geológicas,
Universidad Complutense de Madrid,
28040, Madrid, Spain

Lingling Chen
State Key Laboratory of Marine Geology,
Tongji University,
Shanghai, 200092, China

Jun Cheng
Coastal Research Laboratory,
Department of Geology,
University of South Florida,
Tampa, 33620, USA

Domenico Chiarella
Pure E&P Norway AS,
Grundingen 3,
N-0250 Oslo, Norway

Dongdong Chu
Institute of Physical Oceanography,
Ocean College, Zhejiang University,
Hangzhou, 310058, P.R. China

Robert W. Dalrymple
Department of Geological Sciences and
Geological Engineering, Queen's University,
Kingston, ON, K7L 3N6, Canada

Poppe L. de Boer
Sedimentology Group, Department of Earth
Sciences, Utrecht University, P.O. Box 80.115,
3508 TC Utrecht, The Netherlands

William A. DiMichele
Department of Paleobiology, NMNH,
Smithsonian Institution,
Washington, D.C., 20560, USA

Philippe Duringer
Institut de Physique du Globe de Strasbourg
(IPGS)-UMR 7516;
Université de Strasbourg (UdS)/École et
Observatoire des Sciences de la Terre (EOST),
Centre National de la Recherche Scientifique
(CNRS), 1 rue Blessig, Strasbourg, 67084, France

Scott Elrick
Illinois State Geological Survey,
Champaign, Illinois, 61820, USA

Daidu Fan
State Key Laboratory of Marine Geology,
Tongji University,
Shanghai, 200092, China

Fabrizio Felletti
Università di Milano,
Dipartimento di Scienze della Terra,
20133, Milano, Italy

Burghard W. Flemming
Senckenberg Institute,
Suedstrand 40, 26382 Wilhelmshaven, Germany

Lucille Furgerot
CNRS UMR 6143 M2C,
University of Caen Normandie, 24 rue des
Tilleuls, 14000, Caen, France

Joseph Hughes
U.S. Geological Survey,
Florida Water Science Centre,
Tampa, 33612, USA

Peihong Jia
The Key Laboratory of Coast & Island
Development,
School of Geographic & Oceanographic Sciences,
Nanjing University,
Hankou Rd.22, Nanjing, 210093, P. R. China
Key Laboratory of Coast and Island
Development (Nanjing University),
School of Geogarphic and Oceanographic
Sciences, Xianlin Ave. 163, Nanjing,
210023, P. R. China

Toshiyuki Kitazawa
Faculty of Geo-environmental Science,
Rissho University,
Kumagaya, 360-0194, Japan

Qing Li
The Key Laboratory of Coast & Island
Development,
School of Geographic & Oceanographic Sciences,
Nanjing University,
Hankou Rd.22, Nanjing, 210093, P. R, China
Key Laboratory of Coast and Island

Development (Nanjing University),
School of Geogarphic and Oceanographic
Sciences, Xianlin Ave. 163, Nanjing,
210023, P. R. China

Sergio G. Longhitano
Department of Sciences,
University of Basilicata, Italy

Asadollah Mahboubi
Department of Geology,
Faculty of Science,
Ferdowsi University of Mashhad, Iran

José Margotta
University Lille 1 - UMR 8187,
CNRS LOG,
Villeneuve d'Ascq, France

Ramón Mas
Departamento de Geología,
Universidad de Oviedo,
C/Jesus Arias de Velasco, s/n,
33005, Oviedo, Spain
Instituto de Geociencias IGEO (CSIC, UCM),
C/José Antonio Novais 12, 28040,
Oviedo, Spain

Nieves Meléndez
Instituto de Geociencias
(IGEO), (UCM, CSIC).

Kain J. Michaud
Petrel Robertson Consulting Ltd.,
Suite 500, 736 – 8th Avenue,
S.W. Calgary, AB, T2P 1H4, Canada

Hosien Mosaddegh
School of Earth Science,
Kharazmi University,
Tehran, Iran

Dominique Mouazé
CNRS UMR 6143 M2C,
University of Caen Normandie, 24 rue des
Tilleuls, 14000, Caen, France

Reza Moussavi-Harami
Department of Geology,
Faculty of Science,
Ferdowsi University of Mashhad, Iran

Naomi Murakoshi
Faculty of Science,
Shinshu University, Matsumoto, 390-8621,
Japan

W. John Nelson
Illinois State Geological Survey,
Champaign, Illinois, 61820, USA

Van Lap Nguyen
Ho Chi Minh City Institute of Resources
Geography,
Vietnam Academy of Science and Technology,
1 Mac Dinh Chi St., 1 Dist.,
Ho Chi Minh City, Vietnam

Jonathan Pelletier
Total, Centre Scientifique et Technique Jean
Feger, Avenue Larribau, 64000, Pau, France

I. Emma Quijada
Departamento de Geología,
Universidad de Oviedo,
C/Jesus Arias de Velasco, s/n,
33005, Oviedo, Spain
Instituto de Geociencias IGEO (CSIC, UCM),
C/José Antonio Novais 12, 28040,
Oviedo, Spain

Jean-Yves Reynaud
University of Lille - CNRS,
UMR 8187 LOG, Cité Scientifique,
F 59 000, Lille, France

Jean-Loup Rubino
Total, Centre Scientifique et Technique
Jean Feger, Avenue Larribau, 64000,
Pau, France

Yoshiki Saito
Geological Survey of Japan,
AIST, Central 7, Higashi 1-1-1,
Tsukuba, 305-8567, Japan

Mathieu Schuster
Institut de Physique du Globe de Strasbourg
(IPGS)-UMR 7516;
Université de Strasbourg (UdS)/École et
Observatoire des Sciences de la Terre (EOST),
Centre National de la Recherche Scientifique
(CNRS), 1 rue Blessig,
Strasbourg, 67084, France

Mahmoud Sharafi
Department of Geology,
Faculty of Science,
Ferdowsi University of Mashhad, Iran

Shai Shuang
State Key Laboratory of Marine Geology,
Tongji University,
Shanghai, 200092, China

Pablo Suarez-Gonzalez
Departamento de Geología,
Universidad de Oviedo,
C/Jesus Arias de Velasco, s/n,
33005, Oviedo, Spain
Instituto de Geociencias IGEO (CSIC, UCM),
C/José Antonio Novais 12, 28040, Oviedo, Spain

Thi Kim Oanh Ta
Ho Chi Minh City Institute of Resources
Geography,
Vietnam Academy of Science and Technology,
1 Mac Dinh Chi St., 1 Dist.,
Ho Chi Minh City, Vietnam

Toru Tamura
Geological Survey of Japan,
AIST, Central 7, Higashi 1-1-1,
Tsukuba, 305-8567, Japan

Akiko Tanaka
Geological Survey of Japan,
AIST, Central 7, Higashi 1-1-1,
Tsukuba, 305-8567, Japan

Bernadette Tessier
CNRS UMR 6143 M2C,
University of Caen Normandie, 24 rue des
Tilleuls, 14000, Caen, France

Alain Trentesaux
University Lille 1 - UMR 8187,
CNRS LOG, Villeneuve d'Ascq, France

Nicolas Tribovillard
University Lille 1 - UMR 8187,
CNRS LOG, Villeneuve d'Ascq, France

Junbiao Tu
State Key Laboratory of Marine Geology,
Tongji University,
Shanghai, 200092, China

Katsuto Uehara
Research Institute for Applied Mechanics,
Kyushu University,
Fukuoka, 816-8580, Japan

Ping Wang
Coastal Research Laboratory,
Department of Geology,
University of South Florida,
Tampa, 33620, USA

Pierre Weill
CNRS UMR 6143 M2C,
University of Caen Normandie, 24 rue des
Tilleuls, 14000, Caen, France

Yin Yong
The Key Laboratory of Coast & Island
Development,
School of Geographic & Oceanographic Sciences,
Nanjing University,
Hankou Rd.22, Nanjing, 210093, P. R. China

Key Laboratory of Coast and Island
Development (Nanjing University),
School of Geogarphic and Oceanographic
Sciences, Xianlin Ave. 163, Nanjing,
210023, P. R. China

Jicai Zhang
Institute of Physical Oceanography,
Ocean College, Zhejiang University,
Hangzhou, 310058, P.R. China

Yue Zhang
State Key Laboratory of Marine Geology,
Tongji University,
Shanghai, 200092, China

Contributions to Modern and Ancient Tidal Sedimentology: an introduction to the volume

BERNADETTE TESSIER[†][*] and JEAN-YVES REYNAUD[‡]

[†] *CNRS UMR 6143 M2C – University of Caen Normandie, 24 rue des Tilleuls, 14000, Caen, France*
[‡] *University of Lille - CNRS, UMR 8187 LOG, Cité Scientifique, F 59 000, Lille, France*
[*] *Corresponding author: bernadette.tessier@unicaen.fr*

HISTORY OF THE 'TIDALITES' CONFERENCE PROCEEDINGS

Besides pioneer works of the 60s, the tidal sedimentologist community really emerged in the 70s (see Klein, 1998). The first international conference on tidal sedimentology took place in 1973 in Florida (USA). It was devoted to carbonate facies, less to siliciclastic deposits and mostly to intertidal areas. The conference resulted in a book gathering case studies (Ginsburg, 1975). The fining-upward tidal flat sequence represented at this time the tidal facies model; and this was mainly applied to carbonates. The growing knowledge in siliciclastic tide-dominated environments was synthesized a few years later by Klein (1977). Following the paper of Visser (1980) demonstrating the record of tidal cycles in estuarine dunes, clastic tidal sedimentology evolved quickly towards more comprehensive and quantitative studies, both ancient and modern. A community was born.

In 1985, this community met in Utrecht (Netherlands) at the '1st Clastic Tidal Deposits symposium'. The proceeding book contains 31 papers, covering a large spectrum of topics, including facies and stratigraphic studies, from the offshore to the nearshore (de Boer *et al.*, 1988). Few articles are devoted to processes and modelling but many focus on modern shelf tidal bodies description and surveying. As noted by Davis *et al.* (1998), the concept of tidal bundles is expressed for the first time in this book.

The 2nd conference, held in 1989 in Calgary (Canada), gave rise to another book of 26 papers (Smith *et al.*, 1991). Beyond the increasing range of topics covered (e.g. the study of primary processes, such as flocculation), this book contains the pioneer paper by G. Allen establishing the estuarine tripartite facies and stratigraphic model of the Gironde estuary (SW France). The growing knowledge on modern tidal settings has been applied at the scale of petroleum reservoirs (e.g. Cretaceous Western Interior seaway).

The 3rd conference, named 'Tidal Clastics', took place in 1992 in Wilhelmshaven (Germany). The proceeding book (Flemming & Bartholomä, 1995) contains 23 papers, highlighting the increasing interest for studies dedicated to modern processes and facies in nearshore settings such as tidal inlets and tidal deltas. Wave and tide interactions are also considered. Ground penetrating radar appears as a new technique to explore ancient tidal subsurface outcrops.

In 1996, the 4th conference was held in Savannah (USA) and founded the 'Tidalites' name of the series. The proceeding book (Alexander *et al.*, 1998) contains 17 papers and three thematic sessions; one on the Wadden Sea, a second one on tidal rhythmites and a third one on stratigraphy, with study cases of reconstructions of incised valley fills (in the Holocene and the rock record).

This conference was marked by a decrease in participation and correlatively a decrease in the number of papers published in the proceedings. This probably reflects the increase in the range of topics covered by the tidal sedimentologist community and, hence, the need to publish more continuously in international journals.

This change was confirmed as the next conference, Tidalites 2000, in Seoul (South Korea), brought only 12 papers, published in a special volume of the Korean Society of Oceanography (Park & Davis, 2001) and was mostly devoted to modern tidal settings in China, Korea and Japan.

The Tidalites 2004 conference was held in Copenhagen (Denmark) and 19 papers were published in a special issue of Marine Geology (Barholdy & Kvale, 2006). Most articles are dedicated to modern processes and especially on fine-grained sediment dynamics and budgets (turbidity maximum, flocculation, tidal marsh sedimentation).

Contributions to Modern and Ancient Tidal Sedimentology: Proceedings of the Tidalites 2012 Conference,
First Edition. Edited by Bernadette Tessier and Jean-Yves Reynaud.

Only four papers deal with stratigraphy, one in the Holocene and three in the rock record.

The Tidalites 2008 conference took place in Qingdao (China) and no proceedings were published. During the conference, contributions were mostly focused on open coast tidal flats and tide-dominated deltas characteristic of Asian tidal seas, mud flats and salt marshes, as well as fluid muds in tidal channels. The conference was also marked by an increase of numerical and flume modelling of hydro-sedimentary dynamics and a rise of studies dedicated to climate and anthropogenic changes and coastal engineering.

To summarize, since the beginning, the Tidalites conference logically reflects the research made by the organiser teams rather than a general, worldwide evolution in tidal sedimentology. For instance, the North American conferences, in Calgary and Savannah, have highlighted facies and stratigraphic aspects, in relationship with a petroleum-oriented perspective, while the European meetings, in Wilhemshaven and Copenhagen, focused more on modern settings and processes. The Asian conferences, in Seoul and Qintao, put forward challenging environmental issues. At the same time, the Tidalites community has become more diverse and the pressure on young colleagues for publishing their research works in international journals has increased.

To get a more accurate idea of the tidal sedimentology production in the last years, we made a rapid overview of the articles published between 2009 and 2015 in international journals of the geosciences featuring the keywords *tide* or *tidal* in the title and *sediment* or *deposit* in the abstract. The query sent back about 400 papers mostly covering the following subjects:

- Facies and architecture in siliciclastics: IHS and fluvial-tidal transition. Tidal deltas and inlets. Wave-dominated, open-coast tidal flats. Tidal signature in open coastlines, muddy coastlines, shelves and slope systems. Carbonate peritidal flats and channels, offshore bioclastic carbonate bodies. Tidal straits.
- Biota: Benthic diatoms/foraminifera to assess tidal changes and long-term tidal flat dynamics. Ichnology of tidal environments. Tides and life: bacterial mats, Cambrian explosion.
- Processes and Modelling: Tidal bores, tidal channels and fluid muds. Tidal bars, ridges and inlets. Offshore dunes and shelf sand transport. Internal tides and deep sands, gas hydrates; tide

influenced hyperpycnal flows and turbidites. Effect of sea-level rise on tidal range, estuarine circulation. Palaeotidal reconstructions.
- Climate: Effect of storms on tidal systems. Tide-storm interplay in the evolution of offshore dunes. Rapid, climate or sea-level changes and morphodynamic evolution of coastal marshes and freshwater wetlands. Astronomical cycles and tidal rhythmites.
- Environmental studies: Carbon sequestration and geochemical tracing of tidal transport. Pollution records in tidal flats. Anthropogenic effects in tidal environments.

As a consequence of the diversification of tidal sedimentology and increase of contributors, there has been a need for more synthetic productions. Martinius & Van den Berg (2011) opened the way with their atlas of estuarine facies, partly based on the extensive lacquer peel collection of the Utrecht University. Also, the 27th IAS Meeting of Sedimentology in Alghero (Italy) in 2009 had a special session on Tidal Sedimentology, which resulted in a special issue of Sedimentary Geology providing more syntheses and fewer case studies than in the previous edited volumes (Longhitano *et al.*, 2012). During the same period, a special issue of the *Bull. Soc. Géol. France* was published on the incised-valleys around France (Chaumillon *et al.*, 2010). 6 of the 10 contributions in this volume focus on the tide-dominated to tide-influenced estuaries located along the Atlantic and Channel coasts. Finally, the textbook *Principles of Tidal Sedimentology* (Davis & Dalrymple, 2012) is the first general book dedicated to tidal sedimentology since that of Klein (1977) on clastic tidal facies and Stride (1982) on offshore tidal sands. Most authors from the steering committee of the past Tidalites conferences (except carbonate specialists) authored the chapters of this book, which provides the state of the art on typical tidal environments, including a renewed perspective on carbonates and for the first time a specific insight on the deep sea and well-known ancient tidal basins.

OUTLINE OF THE PRESENT VOLUME

The Tidalites 2012 conference was held in Caen (France) and gathered together about 100 colleagues. In addition to the 70 talks and posters covering the main fields of tidal sedimentology, the meeting offered the opportunity to visit the following sites: (i) the Arcachon basin and Gironde estuary on the Atlantic coast (Chaumillon & Féniès, 2012); (ii) the wave-dominated Somme estuary, in the Eastern Channel area (Trentesaux *et al.*, 2012); (iii) the Anjou Miocene tidal crags (André *et al.*, 2012); (iv) the Bay of Mont-Saint-Michel in the Western Channel (Tessier *et al.*, 2012). The four field trip guide-books are grouped together in a single volume (ASF, 2012).

The Caen Tidalite 2012 conference brought about 17 papers, gathered in the present volume. The book content has been organised following a progressive succession ranging from methodological papers to articles on processes and facies in modern and ancient environments and then to papers dealing with stratigraphy of tidal successions. The introductory papers highlight a diversity of tools and methodologies used in modern tidal sedimentology, such as the numerical modelling of tidal circulation in a very shallow water microtidal lagoon (Zhang *et al.*), the satellite monitoring of deltaic mouthbars using SAR data (Tanaka *et al.*) or the GIS database setup for microtidal flat ichnofacies (Baucon & Felletti). The next three papers reflect the relatively recent interest for tidal bore research. Two of them are process-oriented: Furgerot *et al.* document resuspension processes due to the tidal bore in the Mont-Saint-Michel estuary, whilst Fan *et al.* considered the morphodynamic impact of the tidal bore in the Qiantang river. The third paper links tidal bores to sediment supply in a Cretaceous fluvio-estuarine system (Chamizo *et al.*). The recognition of tidal facies is still a matter of discoveries and debate. Fluvial to lacustrine floodplains can be misinterpreted as tidal flats (Flemming), as they share many similar features (Quijada *et al.*). The imprint of tides on the growth of stromatolites is also questioned (Suarez-Gonzalez *et al.*). The geometric analysis of crossbeds is used to locate bedforms within a larger-scale tidal landscape (Chiarella *et al.*). Tidal rhythmite deposition and preservation are discussed with respect to rapid increase in accommodation, either due to tidal channel migration at a local scale (Pelletier *et al.*) or melt-water pulses at a basin scale (Archer *et al.*). The final group of papers illustrates the continued interest in replacing the tidal facies in a high-resolution sequence stratigraphic framework. The multiplicity of tidal ravinement surfaces within a tide-dominated Pleistocene estuarine fill is exemplified (Kitazawa & Murakoshi), while the

estuarine to shoreface transition is documented within the infilling of a Holocene coastal plain (Margotta *et al.*). The tide-to-wave, estuarine-to-marine transition is also addressed in an example from the Devonian of Iran (Sharafi *et al.*). Finally, the transgressive reworking of lowstand deltas into headland-attached, tide-dominated sandbodies is documented from the classic example of the Roda sandstones in Northern Spain (Michaud & Dalrymple).

ACKNOWLEDGEMENTS

We are very grateful to the Tidalites community for the opportunity given to organise the Caen 2012 conference and then to publish this volume. Bernadette Tessier is particularly grateful to all her colleagues of the M2C lab for their assistance in the Conference organisation, with special thanks to Olivier Dugué. Reviewing, gathering and organising the articles of the present volume, as well as writing this editorial, was a stimulating experience that helped to clarify our own view of the scientific production of our tidal community. We would like to thank warmly the authors for their contributions to the volume and for their patience. We are very grateful to the reviewers, as well as to the editorial board of the IAS, Thomas Stevens and Mark Bateman, the series editors, and Adam Corres the editorial manager, for their continued assistance during this long editorial story. At last, we wish great success to the next Tidalites Conference (Tidalites 2015) that is going to be held in Puerto Madryn, Argentina, in November 2015.

Bernadette Tessier
Caen, France

Jean-Yves Reynaud
Lille, France

REFERENCES

Alexander, **C.R.**, **Davis**, **R.A.** and **Henry**, **V.J.**, Eds (1998) Tidalites: processes and products. *SEPM Spec. Publ.*, **61**, 171 p.

André, **J.-P.**, **Redois**, **F.**, **Gagnaison**, **C.** and **Reynaud**, **J.-Y.** (2012) The Miocene Tidal Shelly Sands of Anjou-Touraine, France. In: Tidalites 2012, the 8th International Conference on Tidal Environments. Field trip booklet. *Editions ASF*, **72**, 65–102.

ASF (2012) Tidalites 2012, the 8th International Conference on Tidal Environments. Field trip booklet. *Editions ASF*, **72**, 200 p.

Bartholdy, **J.** and **Kvale**, **E.P.**, Eds (2006) Proceedings of the 6th international congress on Tidal Sedimentology (Tidalites 2004). *Marine Geology*, 235, 271 p.

Chaumillon, **E.** and **Féniès**, **H.** (2012) The Incised-Valleys of SW France: Marennes-Oléron Bay, Gironde Estuary and Arcachon Lagoon. In: Tidalites 2012, the 8th International Conference on Tidal Environments. Field trip booklet. *Editions ASF*, **72**, 3–63.

Chaumillon, **E.**, **Tessier**, **B.** and **Reynaud**, **J.-Y.**, Eds (2010) French incised valleys and estuaries, *Bull. Soc. Géol France*, **181**, 224 p.

Davis, **R.A.**, **Alexander**, **C.R.** and **Henry**, **V.J.** (1998) Tidal sedimentology: historical background and current contributions. In: Tidalites: processes and products (Eds **C.R. Alexander**, **R.A. Davis** and **V.J. Henry**). *SEPM Spec. Publ.* **61**, 1–4.

Davis, **R.A.** and **Dalrymple**, **R.W.**, Eds (2012) *Principles of tidal sedimentology*. Springer, 621 p.

De Boer, **P.L.**, **Van Gelder**, **A.** and **Nio**, **S.D.**, Eds (1988) Tide-Influenced Sedimentary Environments and Facies. D Reidel Publishing Company, Dordrecht. 530 p.

Flemming, **B.W.** and **Bartholomä**, **A.**, Eds (1995) Tidal Signatures in Modern and Ancient Sediments. *Int. Assoc. Sedimentol. Spec. Publ.*, **24**, 358 p.

Ginsburg, **R.N.**, Ed. (1975) Tidal deposits. A casebook of recent examples and fossil counterparts, Springer-Verlag, NY, 428 p.

Klein, **G. de V.** (1977) Clastic tidal facies. CEPCO, Champaign, Illinois, 149 p.

Klein, **G. de V.** (1998) Clastic Tidalites: a partial retrospective view. In: Tidalites: processes and products (Eds C.R. Alexander, R.A. Davis and V.J. Henry). *SEPM Spec. Publ.* 61, 1–4.

Longhitano, **S.**, **Mellere**, **D.** and **Ainsworth**, **B.**, Eds (2012) Modern and ancient tidal depositional systems: perspectives, models and signatures. *Sed. Geol.*, **279**, 186 p.

Martinius, **A.W.** and **Van den Berg**, **J.H.** (2011) Atlas of sedimentary structures in estuarine and tidally-influenced river deposits of the Holocene Rhine-Meuse-Scheldt system: Their application to the interpretation of analogous outcrop and subsurface depositional systems. EAGE Publication, 298 p.

Park, **Y.A.** and **Davis**, **R.A.**, Eds (2001) Proceedings of Tidalites 2000, *The Korean Society of Oceanography, Special publications*, 103 p.

Smith, **D.G.**, **Reinson**, **G.E.**, **Zaitlin**, **B.A.** and **Rahmani**, **R.A.**, Eds (1991) Clastic Tidal Sedimentology. *Mem. Can. Soc. Petrol. Geol.*, **16**, 387 p.

Stride, **A.H.**, Ed. (1982) Offshore tidal sands: processes and deposits. Chapman & Hall, London, 222 p.

Tessier, **B.**, **Bonnot-Courtois**, **C.**, **Billeaud**, **I.**, **Weill**, **P.**, **Caline**, **B.** and **Furgerot**, **L.** (2012) The Mt St Michel bay, NW France: Facies, sequences and evolution of a macrotidal embayment and estuarine environment. In: Tidalites 2012, the 8th International Conference on Tidal Environments. Field trip booklet. *Editions ASF*, **72**, 149–195.

Trentesaux, **A.**, **Margotta**, **J.** and **Le Bot**, **S.** (2012) The Somme bay, NW France: a wave-dominated macro tidal estuary. In: Tidalites 2012, the 8th International Conference on Tidal Environments. Field trip booklet. *Editions ASF*, **72**, 103–147.

Visser M.J. (1980) Neap-spring cycles relected in Holocene subtidal large scale bedforms deposits: a preliminary note. *Geology*, **8**, 543–546.

Hydrodynamic modelling of salinity variations in a semi-engineered mangrove wetland: The microtidal Frog Creek System, Florida

JICAI ZHANG[†*], DONGDONG CHU[†], PING WANG[‡], JOSEPH HUGHES[§] and JUN CHENG[‡]

[†] Institute of Physical Oceanography, Ocean College, Zhejiang University, Hangzhou, 310058, P.R. China
[‡] Coastal Research Laboratory, Department of Geology, University of South Florida, Tampa, 33620, USA
[§] U.S. Geological Survey, Florida Water Science Centre, Tampa, 33612, USA
* Corresponding Address: 866 Yu-Hang-Tang Road, Ocean College, Zi-Jin-Gang Campus, Zhejiang University, Hangzhou, 310058, P.R. China; E-mail: Jicai_Zhang@163.com

ABSTRACT

As components of a large-scale ecosystem restoration project, three intertidal lagoons are proposed offline of the Frog Creek and Terra Ceia River (Frog Creek System, Florida), which are mangrove-covered and micro-tidal estuaries. A three-dimensional hydrodynamic model has been developed based on EFDC (Environmental Fluid Dynamics Code) and the effects of proposed lagoons on short-time-scale salinity variations have been evaluated. High resolution airborne LiDAR data is employed to depict the bathymetry of mangrove areas. The model has been calibrated and verified by using water level and salinity observations. Due to the proposed engineered lagoons, the tidal prism will be changed and the following conclusions have been obtained from the numerical experiments. (1) The effect of three engineered lagoons is insignificant under low, moderate and super high inflow conditions and the high inflow condition has the most significant effect on salinity regime. (2) In upstream areas, the salinity is increased because the lagoons will import more saline water. In downstream areas, the salinities with and without lagoons are almost the same during flood tide. However, the surface salinity with lagoons is larger than that without lagoons during ebb tide. (3) In downstream areas, the absolute differences between surface salinities with and without lagoons are larger than those of bottom salinities. On the contrary, the absolute differences of bottom salinities are larger than those of surface salinities in upstream areas. It is of great importance to evaluate reasonably the influence of human activities or natural changes on surrounding environments; and this model can serve as a powerful tool in wetland analysis.

Keywords: Frog Creek System, EFDC, Salinity, Microtidal wetlands, Ecosystem Restoration, Numerical prediction.

INTRODUCTION

Wetland systems are becoming increasingly important for ecological, hydrological and recreational purposes. A better understanding of the functional dynamics of these systems requires a good understanding of the hydrodynamics. The hydrodynamics in estuarine wetlands are highly complex, characterized by tidal influence currents, rough bathymetry, energetic turbulence and steep density gradients caused by the interaction between ocean water and fresh water discharges (MacCready & Geyer, 2010). For coastal environments, complexities can also arise because the intertidal zones may become dry and blocked during low tides (Yang & Khangaonkar, 2009). As a result, in the past decades numerical models have acted as a powerful tool in the study and prediction of estuarine hydrodynamics.

Contributions to Modern and Ancient Tidal Sedimentology: Proceedings of the Tidalites 2012 Conference,
First Edition. Edited by Bernadette Tessier and Jean-Yves Reynaud.
© 2016 International Association of Sedimentologists. Published 2016 by John Wiley & Sons, Ltd.

One of the most difficult aspects is that the numerical models for wetlands have to cope with shallow water depths and complex bottom topography. For estuarine wetland systems, the wetting and drying processes due to the changes of surface water elevation are essential (Ji *et al.*, 2001). Consequently, in order to simulate the estuarine hydrodynamics accurately, high-resolution bathymetric data are necessary, not only for deep river channels but also for intertidal zones. Elevations and geometry details of intertidal zones with subtidal channels have been shown to play an important role in transport and exchange processes in estuaries (Ralston & Stacey, 2005). Airborne LiDAR (Light Detection And Ranging) is a method of detecting distant objects and determining their position and other characteristics by analysis of pulsed laser light reflected from their surfaces. Airborne LiDAR is now being applied in coastal environments to produce accurate, high resolution, cost-efficient bathymetric and topographic datasets (Schmid *et al.*, 2011). Traditional techniques and satellite remote sensing are generally unable to penetrate forest canopies and are not at a sufficiently high level of resolution to depict the micro-topography of mangrove communities. Therefore, LiDAR data can be especially useful for mangrove covered areas even under dense canopies (Knight *et al.*, 2009). With the help of LiDAR data, the accuracy of model bathymetry in the tidal flats can be improved significantly and features of multiple tidal channels can be better represented (Yang & Khangaonkar, 2009).

Located in Tampa Bay area, the Terra Ceia Aquatic Preserve (TCAP) is characterized by inlets and embayments of a drowned shoreline. With increasing development, recreation and economic pressures, the aquatic resources have the potential to be significantly impacted. The TCAP area is composed of open water, inlet bays and tidally influenced creeks. The Terra Ceia River and Frog Creek provide fresh water to the wetland system. A better understanding of the hydrodynamics, such as water level, salinity, stratification, destratification, flushing time and residence time, is urgently needed to provide suggestions for resource management and protection. A large-scale ecosystem restoration project has been undertaken in the wetlands associated with Terra Ceia Bay. As components of a wetland restoration project, three intertidal lagoons have been proposed offline of the Frog Creek System. It is unknown whether the proposed intertidal lagoons will have a significant effect on the existing salinity regime of Frog Creek System. Temperature, salinity and tidal fluctuation are all important physical factors influencing the estuarine environments. For instance, mangroves require an annual average water temperature of about 19°C to survive and mangroves have adapted to the saltwater environment by excluding salt from plant tissues. Although they can survive in fresh water, salt water is a key element in reducing competition from other plants, thus allowing mangroves to flourish. Consequently, understanding the structure and variability of the salinity regime in estuaries is critical to ecological and engineering management decisions. The objective of this work, therefore, is to develop a three-dimensional hydrodynamic model to evaluate the effect of the proposed lagoons on the salinity regime and provide suggestions to ecosystem management. Airborne LiDAR data will be employed to depict the micro-structure of the topography in mangrove covered areas.

DATASETS AND STUDY AREA

Study area

Adjacent to the Gulf of Mexico, TCAP is located along mid-peninsula Florida and is characterized by a humid subtropical climate. The average low air temperature for the area is 16°C; and this generally occurs in January. The average high temperature for the area is 28°C; occurring between July and August. The climate of this area is significantly influenced by the Gulf of Mexico. The annual average rainfall is approximately 1100 mm and occurs primarily during a distinct wet season (June to September) with frequent convective summer thunderstorms. According to Meyers *et al.* (2007), the typical values of evaporation rates for the Tampa bay area range from near zero to about 0.60 cm/day and the long-term average evaporation is 0.28 cm/day.

With the mouth located at the northern end of Terra Ceia Bay, Terra Ceia River and Frog Creek extends in a north and north-east direction for approximately 3.5 km, then continues east for about 8 km (Fig. 1; Zhang *et al.*, 2012). Both Terra Ceia River and Frog Creek are shallow with reduced tidal action and are covered by mangroves. As there is no clear difference between Terra Ceia River and Frog Creek, they are usually considered a single entity and are collectively referred to as the Frog Creek System in this paper. The tidal creek connecting the Frog Creek System

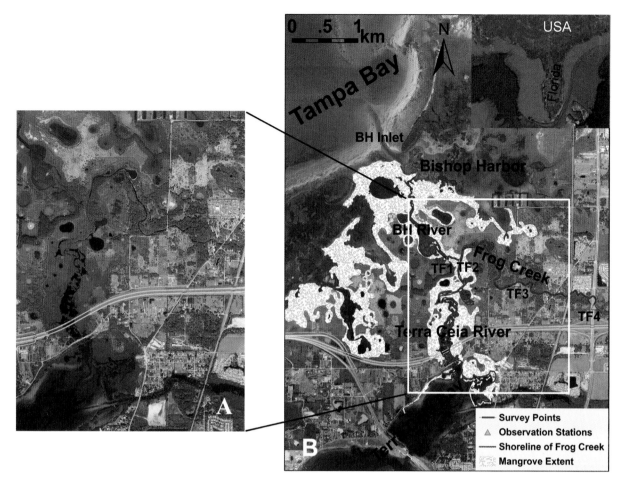

Fig. 1. Study area, showing: (A) The satellite image of the Frog Creek System; (B) Detailed information of the Frog Creek System, where red lines denote the river contours, blue lines indicate the bathymetry survey points, green triangles are the locations of observation stations in the channel and the mangrove covered areas are indicated by the green stippled regions.

to Bishop Harbor is a distinct and unnamed creek, called Bishop Harbor (BH) River in this work. An analysis of sea-level at St. Petersburg shows that about 24% of the variance is associated with the semi-diurnal tidal component, 42% with the diurnal tidal component; and 31% with longer time scales, mostly of non-tidal origin by weather and steric effects (Weisberg & Zheng, 2006). The tidal range is small, with an average value around 0.3 m. No measurements are available but flow velocities associated with tidal dynamics are also weak.

As shown in Fig. 1, the tidally influenced portions of the Frog Creek System are covered by mangrove communities (mangrove forests, mangrove swamps and mangrove islands). There are also some natural lagoons with karstic features which are connected to the Frog Creek System. Water depths range from 0.3 to 1.0 m for most of the study area. The average depth is less than

1.0 m and the deepest depth occurs in the eastern portion of the Frog Creek System, about 1.5 m to 2.3 m. Based on observations over more than four years, the monthly average values of the river discharge of the Frog Creek System are $0.26\,m^3\,s^{-1}$ for June, $0.80\,m^3\,s^{-1}$ for July, $0.95\,m^3\,s^{-1}$ for August, $1.32\,m^3\,s^{-1}$ for September and around $0.10\pm0.03\,m^3\,s^{-1}$ for other months. Storm-induced maximum inflows can be as large as $20.00\,m^3\,s^{-1}$ and usually occur in August and September. In the eastern part of the Frog Creek system, these storm-induced inflows can lead to high current velocities with a value larger than $1.0\ m\,s^{-1}$.

Data sources

The USGS LiDAR data for Frog Creek System, with a horizontal resolution of 1.5 m by 1.5 m are available. It is especially useful to depict the micro-topography of mangrove covered areas. With the help of LiDAR, the grid steps for the numerical model in this work can achieve a minimum resolution of around 4 m. In order to obtain the accurate depth of the channels and natural karstic lagoons, several surveys were carried out during the favourable high tide using RTK; and the survey lines are shown in Fig. 1B (blue lines).

The locations of observations used in this work are shown in Fig. 1B. Hourly water level and wind data for Port Manatee Station and hourly atmosphere pressure data for St. Petersburg Station were obtained from the National Oceanic and Atmospheric Administration-National Ocean Service (NOAA-NOS). The hourly water level data for Manatee River Station, located in Terra Ceia Bay, were provided by the U.S. Geological Survey (USGS). Supported by the TCAP water quality monitoring project, the 15 minutes water level data of TF1, TF2 and TF3, located in the channel of the Frog Creek System, were measured by the USGS. For the same time period, the 15 minutes surface and bottom salinity data of Manatee River Station, TF1, TF2 and TF3, were also obtained from the USGS. Hourly precipitation data for the Frog Creek System were provided by South-west Florida Water Management District (SWFWMD). The hourly inflow data for station TF4, the most upstream station, were obtained from a USGS stream gage located at the eastern end of Frog Creek. All data were quality controlled and gap-filled.

Proposed engineered ponds

As indicated by Fig. 1B, the mangrove communities have been degenerated in the northern and north-eastern parts of the Frog Creek System. As part of the Surface Water Improvement and Management (SWIM) Program, three intertidal ponds, A, B and C, shown in Fig. 2, have been proposed in order to recover the wetland environments for marine species. Station TF3 is located in the upstream areas of Frog Creek, upstream of the three ponds. At this station, the high bottom salinities indicate that the saline water can persistently intrude here as a result of favourable bathymetry for upstream transport of saline water, especially under moderate and low inflow conditions. According to the bathymetry survey results, the values of bottom elevation are around −0.7 m near TF1, −1.0 m near TF2 and −2.0 m near TF3; all values refer to the North American Vertical Datum

of 1988 (NAVD88). This persistent salt intrusion near TF3 will benefit the purposes of proposed lagoons. The lagoons will be connected to the main waterway of the Frog Creek System through canals which will be deeper than the lagoons to allow for sediment deposition.

MODEL DEVELOPMENT

Model description

A three-dimensional hydrodynamic model, EFDC (Environmental Fluid Dynamics Code) has been modified and used in the present study. EFDC has been applied successfully in many water bodies such as estuaries, lakes, rivers and coastal bays (Ji *et al.*, 2001; Shen & Lin, 2006; Xu *et al.*, 2008; Gong *et al.*, 2009; Shi *et al.*, 2009). EFDC solves the Navier-Stokes equations with free surface, which can simulate density, and topographically-induced circulation, tidal and wind-driven flows, spatial and temporal distributions of salinity, temperature and conservative/non-conservative tracers. It employs stretched (namely sigma) vertical coordinates and curvilinear, orthogonal horizontal coordinates. Another important reason for selecting the EFDC model is that it includes sediment and water quality modules, which will be suitable for future studies of the Frog Creek System.

The Mellor-Yamada's 2.5-level turbulence closure sub-model is implemented in the EFDC model (Mellor & Yamada, 1982). The turbulence sub-model calculates vertical eddy viscosity and diffusivity through simulation of turbulence energy and length scale. Vertical boundary conditions for the solution of the momentum equations are based on the specification of kinematic shear stresses. The bottom friction is described by the quadratic law with the drag coefficient determined by the logarithmic bottom layer as a function of bottom roughness height. Wind stress is specified at the water surface.

Model setup

The bathymetric measurements from *in-situ* RTK surveys and USGS LiDAR datasets are interpolated to the centre of model grids by using an inverse distance weighting method. Specifically, the values for the grids in the river channel are calculated from *in-situ* measurements and the values for the grids in mangrove areas are deduced

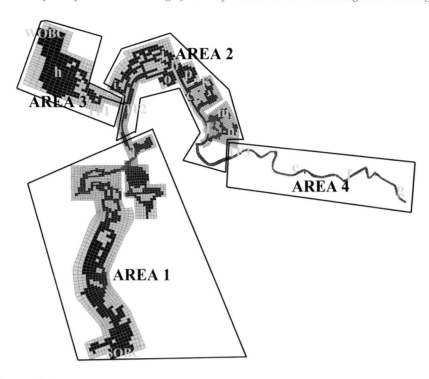

Fig. 2. The wet (blue) and dry (grey) grids for the Terra Ceia River and Frog Creek hydrodynamic model. The grid points selected for discussing the differences between simulated salinities with and without lagoons are indicated by a, b, c, d, e, f, g, h, i, j, k, m, n, o, p, TF1, TF2 and TF3. SOBC and WOBC mean south and west open boundary conditions, respectively. Area 1 contains the grid points located south of Point a. The grid points located west of TF1 belong to Area 3. The eastern part of Frog Creek, from TF3 to the eastern end, constitutes Area 4. The rest, mainly the western part of Frog Creek, belongs to Area 2, which includes the three proposed lagoons.

from USGS LiDAR datasets. Fig. 2 gives the wet and dry grids for the present model. There are a total of 3762 horizontal grids in the computing area. The horizontal grid resolution ranges from 3.8 m to 56.1 m and the time step is set to 1.5 seconds to satisfy the CFL condition. The size of model grids varies with relatively smaller cells for the channel of Frog Creek and the northern part of Terra Ceia River and larger cells for mangrove areas and the channel of the southern part of the Terra Ceia River. The water column is divided into 8 layers in the vertical direction.

The model is driven by the water level elevations specified along open boundaries, river discharge at the eastern headwater, winds and atmospheric pressures. Hourly wind data from Port Manatee station and hourly atmospheric pressure data from St. Petersburg station are applied uniformly to the water surface of entire model domain. The hydrodynamics of the Frog Creek System are co-dominated by the tidal waves propagating from Terra Ceia Bay and Bishop Harbor (Fig. 1). Consequently, the south open boundaries for the present model are set at the

southern end of Terra Ceia River and the west open boundaries are prescribed in the middle of BH River. The hourly water level observations at Manatee River and TF1 are used as incoming tidal waves. The salinity along the open boundaries for EFDC can specify either observed salinity or a maximum incoming salinity boundary value and a recovery time from the outflow salinity to the maximum incoming salinity. In the present work, the hourly salinity observations at Manatee River Station and TF1 are taken as the incoming salinities. At the eastern headwater, hourly fresh water discharges measured at TF4 are utilized (Fig. 3A).

Model calibration

The model's initial condition was obtained by running the model iteratively until the modelled salinity distribution reached the quasi-equilibrium state, which needed 30 days as the spin-up time. Wetting and drying processes in mangrove areas were simulated in the model and a water depth of 5 cm was used as the dry cell criterion. Model results were compared with water level and salinity

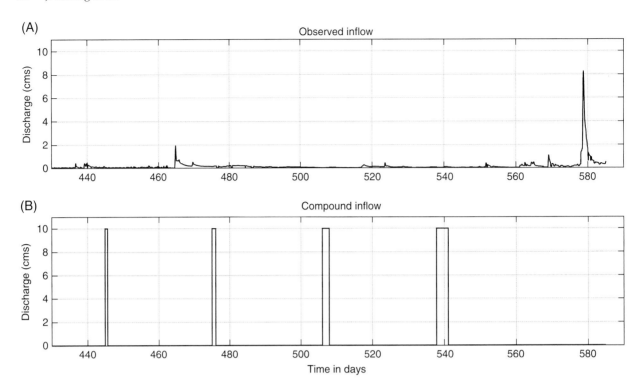

Fig. 3. (A) The time series of river discharge observed at station TF4 (east Frog Creek) from days 430 to 590; (B) The fifth inflow condition (compound inflow). The date starts from 01 January 2006.

observations to calibrate the model. Model calibration on water level and salinity was conducted from March 7 to August 9, 2007 (155 days). The water elevation was calibrated by adjusting the bottom roughness height and open boundary forcing to make the simulated values agree well with the observations. The bottom roughness height was finally set to 0.002 m (Yand & Khangaonkar, 2009; Shi *et al.*, 2009).

The simulated and observed values of water level at TF1, TF2 and TF3 have been shown in Fig. 4A, Fig. 5A and Fig. 6A, respectively. It can be seen that the modelled water level elevation compares favourably with the observations, which indicates the characteristics of tidal propagation from open boundaries to upstream areas have been well reproduced by the model. For TF2 and TF3, relatively large discrepancy occurred around day 578, which might be caused by the unresolved, storm-induced, extreme inflow and rainfall. The average absolute differences between observed and simulated water levels for TF1, TF2 and TF3 are 1.1 cm, 1.6 cm and 2.0 cm, respectively.

Comparisons of observed and modelled surface and bottom salinities for TF1, TF2 and TF3 are plotted in the middle and bottom panels of Fig. 4,

Fig. 5 and Fig. 6, respectively. The model results matched the observations reasonably well. The average absolute differences for the surface salinities at TF1, TF2 and TF3 are 3.37, 3.12 and 2.77, respectively; and 2.50, 2.72 and 1.66 for bottom salinities. In the study area, the tidal dynamics are weak and the salinity in the river channel is very sensitive to river discharge. The spectrum analysis results of observations have indicated that the processes with subtidal frequencies introduced by physical processes with longer periods, such as spring-neap tidal variability and seasonal freshwater river discharge variability, played a very important role in the salinity variations of the Frog Creek System (Zhang *et al.*, 2012). As shown by the figures, the present model reasonably replicated the subtidal salinity variations. In contrast, it was apparently deficient in modelling the variations of salinities with diurnal or semidiurnal tidal frequencies. Most probably, the reasons should be attributed to the unresolved micro-bathymetry and the effect of vegetation resistance which was not considered in the present model.

As shown by Fig. 3A, around day 465, the river discharge increased to about 2.0 m³/s. The observations of salinities at TF1, TF2 and TF3 indicated

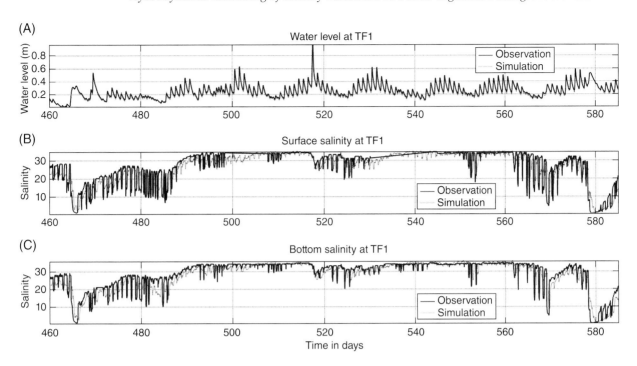

Fig. 4. The calibration of water level (A), surface salinity (B) and bottom salinity (C) at station TF1 (western end of Frog Creek). The date starts from 01 January 2006.

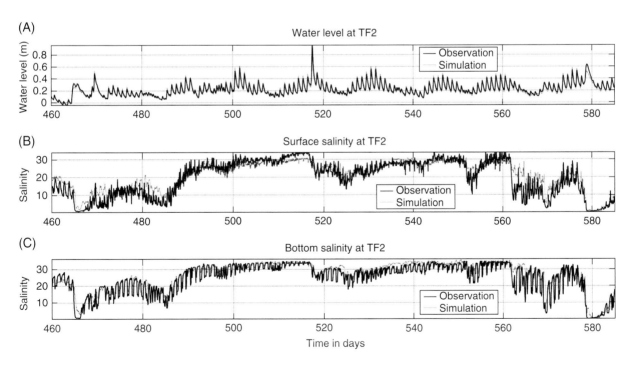

Fig. 5. The calibration of water level (A), surface salinity (B) and bottom salinity (C) at station TF2 (western end of Frog Creek, about 2 km upstream of TF1). The date starts from 01 January 2006.

(A)

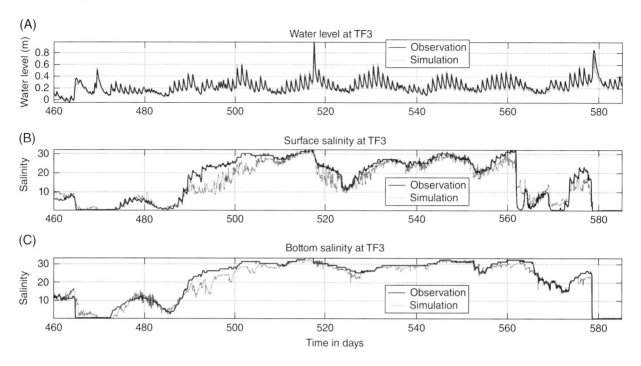

Fig. 6. The calibration of water level (A), surface salinity (B) and bottom salinity (C) at station TF3 (middle Frog Creek). The date starts from 01 January 2006.

that the saline water was flushed out of the river channel and then recovered after the inflow decreased. The present model has reasonably replicated the salinity variation caused by this event. At around day 580 the river discharge exceeded $8.0\,m^3\,s^{-1}$, which was caused by storm-induced precipitation. At TF3, observations have shown that the saline water was totally flushed without recovery from about day 578 to day 585. In contrast, at TF1 and TF2 the recovery process after flushing was very rapid. The different effects of this extreme inflow on the salinity variations were also reproduced accurately by the numerical model. Overall, the simulation results of bottom salinity were better than those of surface salinity. The authors think that the reason was that the surface salinity was more sensitive to river inflow. Consequently, it would introduce larger errors to the simulation of surface salinities if the observed river discharges were not very accurate.

RESULTS AND INTERPRETATION

Scenarios of numerical experiments

The major purpose of this work is to discuss the response of salinity regime to three proposed engineered lagoons for the Frog Creek System.

The salinity distribution of estuaries is governed by a balance between downstream advection of salt by river flow and upstream transport of salt by tidal induced processes (MacCready & Geyer, 2010). For the present research, the engineered lagoons will change the tidal prism of the total system and then influence the salinity regime. Meanwhile, the variations of fresh water discharge from the headwater will also generate different spatial and temporal distributions of salinity. Consequently, in this section, experiments have been designed to discuss the effects of these two factors. According to the design, the depth of lagoons is set to 1 m and 3 m, respectively. The salinities with and without lagoons are then simulated and compared under different inflow (fresh water discharge) conditions and water depth of lagoons.

The responses of salinity regime under 5 inflow conditions are studied. The first four correspond to low, moderate, high and super high inflow conditions, respectively. The exceedance probability used in rainfall and flood statistics is introduced to determine the values of 4 kinds of inflow conditions (Liu *et al.*, 2007). To calculate the exceedance probability (p), the hourly observations of river discharge are first rearranged from the largest to the smallest. Assuming the

Table 1. Setup of model scenarios for the production run.

Exp.	Inflow condition	Selection of Lagoons	Designed Depth	Incoming salinities	Simulation period
E11	$Q_{0.8}$ (0.04 m³ s⁻¹)	Without	---	34	60 days
E12	$Q_{0.8}$ (0.04 m³ s⁻¹)	A, B and C	1 m	34	60 days
E21	$Q_{0.5}$ (0.10 m³ s⁻¹)	Without	---	34	60 days
E22	$Q_{0.5}$ (0.10 m³ s⁻¹)	A, B and C	1 m	34	60 days
E31	$Q_{0.2}$ (0.30 m³ s⁻¹)	Without	---	34	60 days
E32	$Q_{0.2}$ (0.30 m³ s⁻¹)	A, B and C	1 m	34	60 days
E41	$Q_{0.05}$ (1.40 m³ s⁻¹)	Without	---	34	60 days
E42	$Q_{0.05}$ (1.40 m³ s⁻¹)	A, B and C	1 m	34	60 days
E51	Compound	Without	---	34	60 days
E52	Compound	A, B and C	1 m	34	60 days
E53	Compound	A, B and C	3 m	34	60 days

total number of river discharge observations is m and the index is i ($1 \leq i \leq m$ and $i = 1$ for the sampling time with the largest value of discharge), then p can be given by

$$p = 100 \times \frac{i}{m+1}.$$

where $0 < p < 1$. Note that smaller values of p correspond to larger river discharge. Suppose Q_{ep} is the value of discharge with an exceedance probability of ep. In this section the low, moderate, high and super high inflow conditions are figured out by $Q_{0.8}$, $Q_{0.5}$, $Q_{0.2}$ and $Q_{0.05}$ respectively. The values of $Q_{0.8}$, $Q_{0.5}$, $Q_{0.2}$ and $Q_{0.05}$ were calculated based on more than 4 years of observations obtained from station TF4. This obtained $Q_{0.8} = 0.04\,m^3/s$, $Q_{0.5} = 0.10\,m^3/s$, $Q_{0.2} = 0.3\,m^3/s$ and $Q_{0.05} = 1.4\,m^3/s$. The fifth inflow condition (compound inflow), plotted in Fig. 3B, is designed to discuss the response of salinity to extreme inflow, which is often caused by the summer storm. For this case, the base inflow is $Q_{0.8}$ and the extreme inflow with a value of $10.0\,m^3/s$ is triggered every 30 days (see the 4 peaks in Fig. 3B). The duration time for the extreme inflow is set to 12 hours, 1 day, 2 days and 3 days, respectively. By doing this, we can discuss the response of recovery time of salinity to proposed lagoons under different strength of extreme inflow.

All the scenarios of the numerical experiments are described in Table 1. These experiments are numbered by Emn, where m is the code for the inflow conditions and n is the code for the different choice of lagoons or designed values of water depth. The first five series of experiments employ idealized inflow conditions and constant incoming salinities (with a value of 34) to discuss the response of salinity regime to different type of

inflow. Eleven grid points (h, a, TF1, TF2, b, c, d, TF3, e, f and g. Location in Fig. 2) are selected to analyse the simulation results. The authors have divided the whole study area into four parts (Fig. 2). In order to evaluate the differences of salinity with and without the engineered lagoons, the absolute differences were calculated. Suppose S_0^i and S_1^i are the simulated salinities without and with lagoons, i is the index of time and $1 \leq i \leq N$. The time varying absolute difference Δ_0^i is simply defined by

$$\Delta_0^i = \left| S_1^i - S_0^i \right|.$$

The average absolute difference Δ_1 is given by

$$\Delta_1 = \frac{\sum_{i=1}^{N} \left| S_1^i - S_0^i \right|}{N}.$$

For all the experiments, there are eight vertical layers for the present model. In order to analyse the differences clearly, we calculate the surface, middle, bottom and depth-averaged salinities from the original eight-layer results. Specifically, the surface salinity is defined as the average value of the first two layers, the bottom salinity is defined as the average of the last two layers and the middle salinity is given by the average of the middle four layers.

Response under different inflow conditions

The differences between simulated salinities with and without proposed lagoons for selected points and subareas under low ($Q_{0.8}$), moderate ($Q_{0.5}$), high ($Q_{0.2}$), super high ($Q_{0.05}$) and compound inflow conditions are shown in Table 2.

Table 2. Differences between simulated salinities with and without proposed lagoons for selected points and subareas under low (E11 *vs.* E12), moderate (E21 *vs.* E22), high (E31 *vs.* E32), super high (E41 *vs.* E42) and compound (E51 *vs.* E52 and E51 *vs.* E53) inflow conditions.

Exp.	Location	Points											Areas				
		h	a	TF1	TF2	b	c	d	TF3	e	f	g	Area 1	Area 2	Area 3	Area 4	Whole
E11&	Bottom	0.09	0.26	0.15	0.18	0.58	1.27	1.25	1.36	1.33	1.27	1.06	0.18	0.80	0.09	1.26	0.57
E12	Middle	0.11	0.34	0.18	0.33	0.98	1.38	1.39	1.37	1.23	1.09	0.96	0.19	0.96	0.11	1.17	0.58
	Surface	0.14	0.47	0.26	0.52	1.38	1.51	1.48	1.08	0.97	0.94	0.86	0.22	1.15	0.15	0.99	0.57
	Averaged	0.10	0.32	0.18	0.29	0.84	1.38	1.37	1.29	1.19	1.10	0.96	0.18	0.92	0.11	1.15	0.56
E21&	Bottom	0.10	0.46	0.22	0.29	0.87	2.55	2.44	2.74	2.43	2.12	1.47	0.30	1.52	0.11	2.23	1.01
E22	Middle	0.15	0.61	0.27	0.59	1.88	2.55	2.61	2.45	2.06	1.56	1.14	0.30	1.77	0.16	1.87	0.95
	Surface	0.22	0.84	0.5	0.93	2.47	2.53	2.35	1.55	1.27	1.16	0.96	0.39	2.00	0.26	1.30	0.87
	Averaged	0.14	0.53	0.28	0.48	1.60	2.54	2.50	2.3	1.95	1.60	1.18	0.29	1.67	0.15	1.82	0.92
E31&	Bottom	0.13	1.1	0.36	0.74	1.79	4.68	4.83	5.44	3.10	1.38	0.18	0.56	3.00	0.16	2.65	1.48
E32	Middle	0.22	1.21	0.5	1.29	3.33	4.14	4.37	3.27	1.35	0.52	0.05	0.54	3.05	0.27	1.55	1.17
	Surface	0.44	1.53	1	1.7	3.54	3.08	2.41	1.15	0.49	0.29	0.04	0.73	2.77	0.55	0.59	0.96
	Averaged	0.22	1.00	0.5	0.93	2.85	4.01	4.00	3.29	1.57	0.68	0.08	0.53	2.79	0.27	1.58	1.13
E41&	Bottom	0.29	1.6	1.48	1.70	1.17	0.63	0.52	0.03	0.00	0.00	0.00	0.65	1.24	0.53	0.02	0.54
E42	Middle	0.69	1.01	1.21	1.18	0.76	0.34	0.25	0.01	0.00	0.00	0.00	0.55	0.78	0.77	0.01	0.45
	Surface	0.93	0.76	1.1	0.77	0.51	0.13	0.04	0.00	0.00	0.00	0.00	0.59	0.47	0.92	0.00	0.43
	Averaged	0.59	1.03	1.17	1.09	0.79	0.36	0.27	0.01	0.00	0.00	0.00	0.55	0.78	0.69	0.01	0.44
E51&	Bottom	0.22	0.77	0.36	0.5	1.24	1	0.96	1.02	1	1.01	0.88	0.59	0.9	0.25	0.99	0.7
E52	Middle	0.27	1.04	0.43	1	1.03	1.06	1.07	1.08	0.96	0.88	0.79	0.66	1.06	0.29	0.94	0.74
	Surface	0.31	1.14	0.58	1.21	1.15	1.2	1.2	0.9	0.81	0.77	0.72	0.77	1.18	0.38	0.82	0.76
	Averaged	0.24	0.98	0.42	0.9	1.01	1.05	1.04	1	0.93	0.88	0.8	0.66	1	0.28	0.92	0.72
E51 &	Bottom	0.35	2.32	0.67	1.51	3.15	2	2.3	2.35	2.23	2.11	1.82	1.46	2.23	0.41	2.17	1.59
E53	Middle	0.45	2.48	0.82	2.49	1.9	2.12	2.37	2.16	1.99	1.81	1.62	1.58	2.29	0.52	1.93	1.59
	Surface	0.58	2.18	1.04	2.25	2.06	2.19	2.09	1.87	1.7	1.61	1.47	1.75	2.14	0.7	1.67	1.59
	Averaged	0.43	2.33	0.81	2.14	2.01	2.06	2.21	2.09	1.97	1.83	1.63	1.58	2.15	0.51	1.91	1.59

The effect of proposed lagoons is insignificant under low inflow condition ($Q_{0.8}$). For the whole area, the average absolute differences of bottom, middle, surface and depth-averaged salinities are 0.57, 0.58, 0.57 and 0.56, respectively. It has been found that Area 4 (the eastern part of Frog Creek) is the most significantly influenced area. For Area 4, the average absolute differences of bottom, middle, surface and depth-averaged salinities are 1.26, 1.77, 0.99 and 1.15, respectively. This maximum influence can also be proved by the calculated differences at Points TF3, e, f and g (Table 2).

The proposed lagoons under moderate inflow conditions ($Q_{0.5}$) have similar but amplified effects on the salinity regime. For the whole area, the average absolute differences of bottom, middle, surface and depth-averaged salinities are 1.01, 0.95, 0.87 and 0.92, respectively. Similar to the results under low inflow condition, Area 4 will still be the most significantly influenced area; and the next most significantly influenced is Area 2 (the area including the three lagoons). The average absolute differences of bottom, middle, surface and depth-averaged salinities are 2.23, 1.87, 1.30 and 1.82 respectively for Area 4 and 1.52, 1.77, 2.00 and 1.67 respectively for Area 2. The time series of simulated salinities for E21 and E22 at TF3 clearly show that the salinity will increase (Fig. 7), which is similar to the low inflow condition. Based on the results of Table 2, we can conclude that the proposed lagoons would import more saline water to Area 4 and Area 2, which will increase the salinity of these areas under low or moderate inflow conditions. However, in downstream areas the effect of lagoons is different. Time series of simulated salinity in E21 and E22 at TF1 demonstrates that the salinities with and without lagoons are almost the same during flood tide (Fig. 8). The authors' calculations showed, on the contrary, that during ebb tide the surface salinity was larger with lagoons than without. The reason is that part of the fresh water will flow into the lagoons and therefore the volume of fresh water to downstream areas will be reduced, especially during ebb tide. As a result, if the lagoons are considered, during ebb tide the surface salinity of downstream areas will be increased because the volume of fresh water for mixing is decreased. Similar changes can be found in bottom and middle salinities, but not as obvious as in surface salinity (Fig. 8B and C).

Among the four inflow conditions in this section, the effect of lagoons under high inflow condition ($Q_{0.2}$) is the most significant. For the whole area, the average absolute differences of bottom, middle, surface and depth-averaged salinities are 1.48, 1.17, 0.96 and 1.13, respectively (Table 2). Comparing the results under low and moderate inflow conditions, Area 2, instead of Area 4, is the most significantly affected area during high flow incoming conditions. The average absolute differences of bottom, middle, surface and depth-averaged salinities are 3.00, 3.05, 2.77 and 2.79 respectively for Area 2 and 2.65, 1.55, 0.59 and 1.58 respectively for Area 4. The time series of simulated salinity for E31 and E32 at Point e (within Area 4) are plotted in Fig. 9. The absolute differences at Points c, d and TF3 are the largest, especially for bottom salinities (around 5). The reason is also that the lagoons will introduce more saline water to the upstream areas and therefore the bottom salinity is significantly increased (Fig. 9C). The absolute difference for the surface salinity is smaller than the bottom salinity in the upstream area. Contrarily, for the downstream areas (such as Points h, a, b, TF1 and TF2), the absolute difference of the surface salinity is larger than that of the bottom salinity, as demonstrated by the simulated salinity for E31 and E32 at TF2 (Fig. 10).

Under the super high inflow condition ($Q_{0.05}$), the saline water in the middle and eastern part of the Frog Creek System is flushed, no matter whether the lagoons are considered. It has been found that there is almost no difference in salinity in the whole of Area 4 (Tab. 2). In the whole system, including the four areas, the average absolute differences of bottom, middle, surface and depth-averaged salinities are 0.54, 0.45, 0.43 and 0.44, respectively. The largest depth-averaged difference of salinity between E41 and E42, only about 1, occurs at points a, TF1 and TF2 (Table 2). It can thus be concluded that the effect of lagoons is insignificant under super high inflow conditions ($Q_{0.05}$).

Response of salinity recovery time

The fifth inflow condition is the compound inflow (Fig. 3B), which is designed to discuss the response of salinity to extreme inflow induced by summer storm-induced rainfall. By doing this, we can discuss the response of recovery time of salinity to proposed lagoons under different strengths of extreme inflow. The depth of the proposed lagoons is set to 1 m (E52) and 3 m (E53), respectively.

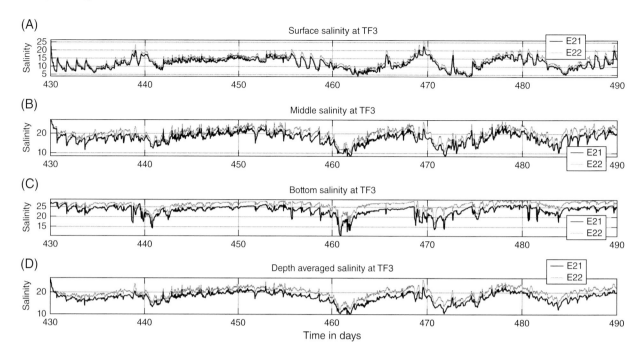

Fig. 7. The simulated surface (A), middle (B), bottom (C) and depth averaged (D) salinities in the water column at TF3 (middle Frog Creek, Area 4) for moderate inflow conditions, without (E21) and with (E22) proposed lagoons (of 1 m water depth). The date starts from 01 January 2006.

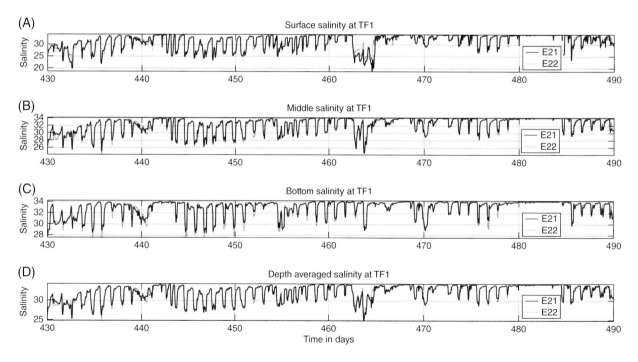

Fig. 8. The simulated surface (A), middle (B), bottom (C) and depth averaged (D) salinities in the water column at TF1 (western end of Frog Creek, Area 3) for moderate inflow conditions, without (E21) and with (E22) proposed lagoons (of 1 m water depth). The date starts from 01 January 2006.

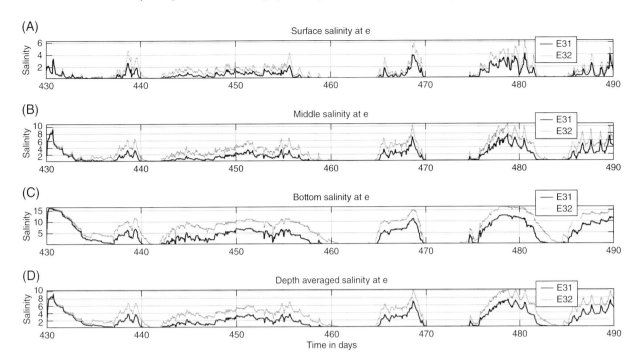

Fig. 9. The simulated surface (A), middle (B), bottom (C) and depth averaged (D) salinities in the water column at Point e (eastern part of Frog Creek, Area 4) for high inflow conditions, without (E31) and with (E32) proposed lagoons (of 1 m water depth). The date starts from 01 January 2006.

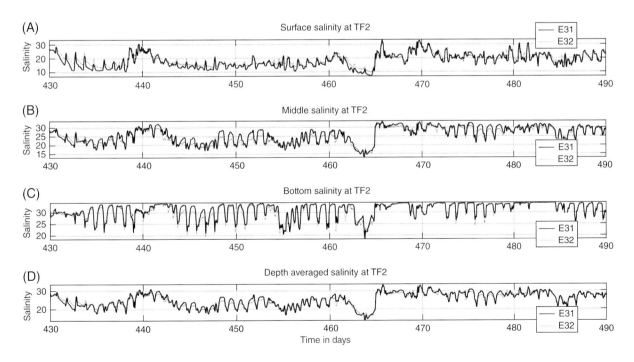

Fig. 10. The simulated surface (A), middle (B), bottom (C) and depth averaged (D) salinities in the water column at TF2 (western end of Frog Creek, Area 2) for high inflow conditions, without (E31) and with (E32) proposed lagoons (of 1 m water depth). The date starts from 01 January 2006.

(A)

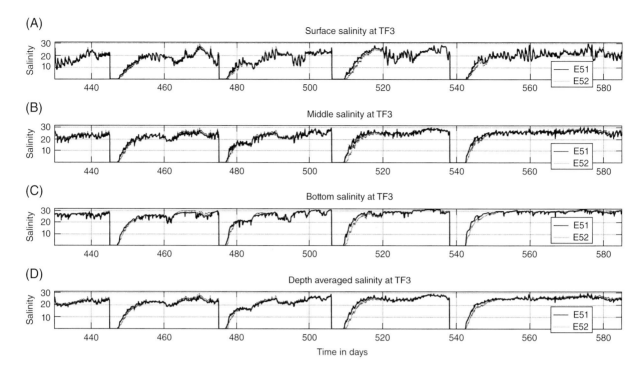

Fig. 11. The simulated surface (A), middle (B), bottom (C) and depth averaged (D) salinities in the water column at TF3 (middle Frog Creek, Area 4) for compound inflow conditions, without (E51) and with (E52) proposed lagoons (of 1 m water depth). The date starts from 01 January 2006.

The differences between E51 and E52, E51 and E53 are shown in Table 2. The differences of depth-averaged salinity between E51 and E52, E51 and E53 are 0.72 and 1.59 respectively for the whole area and 1.00 and 2.15 respectively for Area 2. The time series of simulated salinity for E51 and E52 at Point TF3 is plotted in Fig. 11. It is shown that the salinity will require slightly more time (a few hours) to recover from flushing status when the lagoons are taken into account. The longer the extreme inflow lasts, the more time needed to recover the salinity regime.

CONCLUSIONS

A large-scale ecosystem restoration project has begun in the wetlands associated with Terra Ceia Bay. As components of wetland restoration, three intertidal lagoons are proposed offline of the northern loop of Frog Creek before the creek bends to the south and becomes the Terra Ceia River. In this work, a three-dimensional hydrodynamic model (EFDC) was developed in order to evaluate and the effect of the proposed lagoons on the salinity regime. LIDAR data was employed to depict the bathymetry of mangrove covered areas. The model

was calibrated by using water level and salinity observations. The responses of salinity regime under different inflow conditions were studied and the conclusions will provide appropriate suggestions for wetland management. This paper is one of the initial modelling works for the Frog Creek systems. In the future, a better understanding of the hydrodynamics, such as water level, salinity, stratification, destratification, flushing time and residence time, is needed to provide suggestions for resource management and protection. Based on preliminary results, the following questions might be worthy of being further studied using the model:

1. Observations indicate that there are great differences between the water level variations in Tampa Bay and in the river channel, the latter being characterized by reduced tidal energy and increased subtidal regime. The resistance effect of vegetation (mainly mangroves) and the complex topography should be the most probable reasons. It will be a great challenge for the numerical models to replicate the interaction between flow and vegetation. Also, the wetting and drying technique is especially important to resolve the effect of topography on the hydrodynamics.

2. The salinity structure in Frog Creek Systems is very complex and influenced by tides, subtidal water motions and river discharges. The observations of salinity at TF1, TF2 and TF3 have shown very different characteristics. Discrepancies also occur between the surface and the bottom salinities. The EFDC model has reasonably reproduced the subtidal variations of salinity. However, the model was apparently deficient in modelling the variations of salinities with diurnal or semidiurnal tidal frequencies. The ability of numerical models to further improve the accuracy of salinity simulation is another challenging issue of this project.

ACKNOWLEDGMENTS

This work was supported by the National Natural Science Foundation of China through grant 41206001, the Natural Science Foundation of Zhejiang Province through Grant LY15D060001, Natural Science Foundation of Jiangsu Province through grant BK2012315 and by the United States Geological Survey and the University of South Florida.

REFERENCES

Gong, W.P., Shen, J. and Hong, B. (2009) The influence of wind on the water age in the tidal Rappahannock River. *Mar. Environ. Res.*, **68**, 203–216.

Ji, Z.G., Morton, M.R. and Hamrick, J.M. (2001) Wetting and drying simulation of estuarine processes. *Estuar. Coast. Shelf. Sci.*, **53**, 683–700.

Knight, J.M., Dale, P.E.R., Spencer, J. and Griffin, L. (2009) Exploring LiDAR data for mapping the micro-topography and tidal hydro-dynamics of mangrove systems: An example from southeast Queensland, Australia. *Estuar. Coast. Shelf. Sci.*, **85**, 593–600.

Liu, W.C., Chen, W.B., Cheng, R.T., Hsu, M.H. and Kuo, A.Y. (2007) Modeling the influence of river discharge on salt intrusion and residual circulation in Danshuei River estuary, Taiwan. *Cont. Shelf. Res.*, **27**, 900–921.

MacCready, P. and Geyer, W.R. (2010) Advances in Estuarine Physics. *Annu. Rev. Marine. Sci.*, **2**, 35–58.

Mellor, G.L. and Yamada, T. (1982) Development of a turbulence closure model for geophysical fluid problems. *Rev. Geophys. Space. Phys.*, **20**, 851–875.

Meyers, S.D., Luther, M.E., Wilson, M., Havens, H., Linville, A. and Sopkin, K. (2007) A numerical simulation of residual circulation in Tampa Bay. Part I: Low-frequency temporal variations. Estuar. *Coast.*, **30**, 679–697.

Ralston, D.K. and Stacey, M.T. (2005) Stratification and turbulence in subtidal channels through intertidal tidal flats. *J. Geophys. Res.*, **110**, C08009.

Schmid, K.A., Hadley, B.C. and Wijekoon, N. (2011) Vertical accuracy and use of topographic LIDAR data in coastal marshes. *J. Coast. Res.*, **27**, 116–132.

Shen, J. and Lin, J. (2006) Modeling study of the influences of tide and stratification on age of water in the tidal Jams River. *Estuar. Coast. Shelf. Sci.*, **68**, 101–112.

Shi, J.H, Li, G.X. and Wang, P. (2009) Anthropogenic Influences on the Tidal Prism and Water Exchanges in Jiaozhou Bay, Qingdao, China. *J. Coast. Res.*, **21**, 835–842.

Weisberg, R.H. and Zheng, L. (2006) The circulation of Tampa Bay driven by buoyancy, tides, and winds, as simulated using a finite volume coastal ocean model. *J. Geophys. Res.*, **111**, C01005.

Xu, H.Z., Lin, J. and Wang, D.X. (2008) Numerical study on salinity stratification in the Pamlico River Estuary. *Estuar. Coast. Shelf. Sci.*, **80**, 74–84.

Yang, Z.Q. and Khangaonkar, T. (2009) Modeling tidal circulation and stratification in Skagit River estuary using an unstructured grid ocean model. *Ocean. Model.*, **28**, 34–49.

Zhang, J.C., Wang, P. and Hughes, J. (2012) EOF Analysis of Water Level Variations for Microtidal and Mangrove-Covered Frog Creek System, West-Central Florida. *J. Coast. Res.*, **28**(5), 1279–1288.

Temporal changes in river-mouth bars from L-band SAR images: A case study in the Mekong River delta, South Vietnam

AKIKO TANAKA*§, KATSUTO UEHARA†, TORU TAMURA*, YOSHIKI SAITO*, VAN LAP NGUYEN‡ and THI KIM OANH TA‡

* *Geological Survey of Japan, AIST, Central 7, Higashi 1-1-1, Tsukuba, 305-8567, Japan*
† *Research Institute for Applied Mechanics, Kyushu University, Fukuoka, 816-8580, Japan*
‡ *Ho Chi Minh City Institute of Resources Geography, Vietnam Academy of Science and Technology, 1 Mac Dinh Chi St., 1 Dist., Ho Chi Minh City, Vietnam*
§ *Corresponding author: akiko-tanaka@aist.go.jp*

ABSTRACT

ALOS (Advanced Land Observing Satellite) PALSAR (Phased Array type L-band SAR) data acquired between December 2006 and January 2011 were used to investigate changes in the areal extent of three river-mouth bars in distributaries of the Mekong River delta. In the SAR data, river-mouth bars are characterized by strong backscatter, whereas the adjoining river water is characterized by weak backscatter. The areal extents of the river-mouth bars, extracted from the SAR data by using a histogram thresholding algorithm, increased gradually on an annual time scale. Tidal heights at the time of SAR data acquisition clearly correlated with the areal extent of the river-mouth bars. Seasonal variations in the areal extent of river-mouth bars were also recognized; these were controlled mainly by monsoon-influenced water-level variations. This study demonstrates that SAR data can be used to quantify long-term changes in the area and shape of river-mouth bars.

Keywords: synthetic aperture radar (SAR), Mekong River delta, river-mouth bar, tidal height, inter-annual and intra-annual changes.

INTRODUCTION

Global changes of sea-level and climate are expected to have a major impact on river and coastal systems (IPCC, 2007). To properly assess possible future changes of coastal environments it is necessary to understand past and present coastal systems and their morphodynamic processes. Ground-based surveys and aerial photography provide good spatial resolution but the temporal coverage of these methods is limited and very site-specific. Furthermore, many variables and parameters that are difficult to estimate are needed for understanding the morphodynamic processes; consequently, little of the world's coastline morphology has been studied quantitatively (e.g. Mills *et al.*, 2005).

Satellite data provide coverage of large areas at regular intervals and the increasing variety of passive and active spaceborne sensors available in the visible, infrared and microwave ranges has benefited the monitoring of large river systems (e.g. Lillesand *et al.*, 2008). Much of the use of satellite data has focused on optical sensors of visible and near-infrared reflectance, such as Landsat Thematic Mapper and SPOT data (e.g. Jensen, 2007). In contrast, synthetic aperture radar (SAR) sensors provide their own source of illumination and return information about surface roughness and dielectric properties, with the added advantage of cloud penetration (e.g. Elachi, 1988). This all-weather capability makes SAR attractive for use in regions affected by cloud cover. Radar backscatter intensity depends on surface roughness

Contributions to Modern and Ancient Tidal Sedimentology: Proceedings of the Tidalites 2012 Conference, First Edition. Edited by Bernadette Tessier and Jean-Yves Reynaud.

elements that scale with the radar wavelength projected onto the scattering surface. Therefore, mesoscale oceanic phenomena, such as internal waves and shallow underwater bottom topography, can be shown in SAR images of the ocean surface because they are associated with a variable sea surface roughness (e.g. Alpers & Hennings, 1984). Hence, the difference in surface roughness makes it possible to discriminate land from sea (e.g. Alpers, 1983; McCandless & Jackson, 2004).

Numerous studies have demonstrated the value of SAR amplitude (radar backscatter) data to depict coastal systems (e.g. Smith, 1997). For example, Brakenridge *et al.* (1994) used ERS-1 images to delineate flood inundation area during the 1993 Mississippi floods. Smith (1997) suggested that the spaceborne SAR imagery was useful to detect changes in water surface area in floodplains and large rivers. Kushwaha *et al.* (2000) used ERS-1 SAR data for discrimination of mangrove wetlands in the Sundarban delta in West Bengal state, India, and found that the use of multi-temporal SAR images and the integration of SAR data with the multi-temporal optical sensor data could improve the information on wetlands. However, SAR images have coherent speckle noise that arises from the coherent interference of signals scattered from different parts of the radar resolution element (e.g. Elachi, 1988). This noise is suppressed by the spatial multi-looking at the expense of reduced spatial resolution. Therefore, the resolution of these data is often too coarse for shoreline mapping and speckle noise often inhibits the detection and mapping of shorelines (e.g. Morang & Gorman, 2005).

Deltaic systems are major elements to be considered for coastal monitoring. The wide, low-lying delta plains, especially in Asian mega-deltas, are dynamic regions that receive large amounts of sediment and are vulnerable to sea-level rise. However, limited amounts of available data and poor accessibility to these deltas have prevented the establishment of systematic monitoring. One of many processes involved in the evolution of delta is the deposition of river-mouth bars because they are active places in deposition and erosion of sediments. In addition, positions of the network bifurcations might be past locations of river-mouth bars that formed in front of old distributary mouths (Edmonds & Slingerland, 2007). However, the formation of river-mouth bars is a dynamic and complex process which is not yet fully understood, partially due to the interactions and combinations amongst rivers, buoyancy, waves and tides (Wright, 1977) over various time scales.

In this paper, the use of SAR backscatter intensity data to detect temporal changes of the areal extents was examined and applied to three river-mouth bars of the Mekong River delta. Detected temporal variations of river-mouth bars are discussed in their relationships with other contributing factors such as tidal level and sediment discharge. This analysis, that may also apply to other coastal environments, is rather simple and robust.

REGIONAL SETTING

The Mekong River rises in the highlands of the Himalayas and the Tibetan Plateau and flows 4600 km southward before debouching into the South China Sea, at the southern tip of Vietnam (Fig. 1A). It traverses a drainage basin with an area of about $8.0 \times 10^5 \, km^2$ (Milliman & Syvitski, 1992), which provides high volumes of both water and sediment. The Mekong River has been ranked tenth in the world in terms of water discharge and ninth in terms of sediment discharge (Milliman & Syvitski, 1992). The large volumes of sediment transported to the coast have formed an extensive delta plain, since 8000 years ago (Tamura *et al.*, 2009). The modern Mekong River delta, which is usually classified as a tide-dominated and wave-dominated delta (Ta *et al.*, 2002), has formed over the last ~3500 years (Tamura *et al.*, 2012).

The hydrodynamic environment and climate of the delta is dominated by the wet and dry seasons of the tropical monsoon (Hu *et al.*, 2000). The rainy season usually extends from late May or early June until October. The water discharge of the Mekong River varies seasonally with the monsoon, reaching a maximum in September and October (Tamura *et al.*, 2010).

The South China Sea coast of the Mekong River delta is affected by semi-diurnal tides with ranges of about 1 m and 2 m at neap and spring tides respectively. Tidal range increases northward along the coast to reach about 2 m at neap tide and 4.4 m at spring tide at the mouth of the Saigon River (Kubicki, 2008) and to about 20 km north at Vam Kenh (Fig. 1B). Tide gauge data from Vung Tau (Fig. 1B) between 1979 and 2001 (Permanent Service for Mean Sea Level [PSMSL]) show a relative sea-level rise of 6 mm yr^{-1} (Syvitski *et al.*, 2009), together with annual trends and irregular changes. Monthly mean sea-level data during the period 1979 to 2001 show higher levels in winter and lower levels in summer, with a range of about 40 cm. Average monthly wind direction and speed

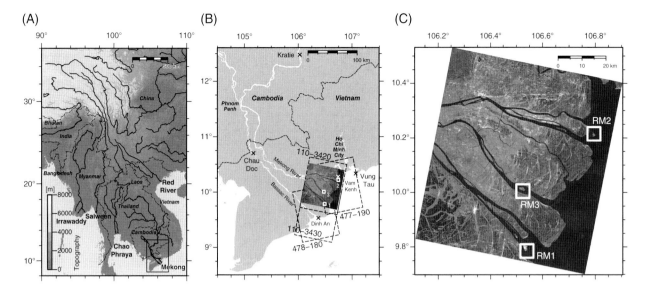

Fig. 1. (A) Map of the whole Mekong River. The study area is marked by the black square. (B and C) Maps of the study area showing locations of the three river-mouth bars (small white squares) investigated in this study. On (B) dashed rectangles show the approximate areas of coverage of the ALOS PALSAR data used (path and frame numbers are annotated).

in South East Asia observed by QuickScat (http:// www.ssmi.com/qscat/) show that, in the Mekong delta region, south-westerly winds from the land are dominant in the rainy season, whilst strong north-easterly winds from the ocean are dominant in winter.

It is generally known that river-mouth bars (distributary mouth bars) are formed as a consequence of the decrease in discharge and bed shear stress associated with the loss of flow competence at the mouth of the distributary channel (Wright, 1977; Bhattacharya, 2003). The river-mouth bars are rapidly evolving features and their morphologies depend on the balance between hydrodynamic processes and sediment budgets. Tamura *et al.* (2012) suggested that the Mekong delta plain propagated laterally seaward during the late Holocene and it was formed by the successive deposition of longshore river-mouth bars. Also, they mentioned that the geometry of the Mekong River delta has been controlled by contributions of the river, tides and monsoon waves.

Study area

Phased Array type L-band SAR (PALSAR) images from the Advanced Land Observing Satellite (ALOS) were acquired over the Mekong River delta near its entry to the South China Sea. For the present study, three isolated river-mouth bars of almost similar size were chosen in contrasting morphodynamic environments of the delta (Fig. 1C). River-mouth

bar 3 (RM3 hereafter) is an island (Cu Lao Dat) in the stream of the Song Ham Luong (Lower Mekong River), about 15 km upstream from the open sea. It is shown on a 1:153,000 scale map (Naval Oceanographic Office, United States, 1966) based on French Chart 5903 (1:50,000, Ed. 1942, Rev. 1946) and a 1:200,000-scale map based on surveys in 1973 (GUNIO, 1989). River-mouth bars 1 and 2 (RM1 and RM2 hereafter) were described as inter-tidal areas on these two maps (Naval Oceanographic Office, United States, 1966; GUNIO, 1989). RM1 (known as Con Ngheu) in the mouth of the Co Chien River was identified as salt marshes on a 1:100,000-scale map published by the Ministry of Natural Resources and Environment, Vietnam (2001). These three kinds of historical maps suggest that RM1 and RM2 were not exposed above the sea-level before at least 1973. Also, Landsat archive data for January 16, 1989 (http://landsat.org/) show no emergent bars at the present-day locations of RM1 and RM2. However, RM1 and RM2 are shown to be subaerially exposed in a Landsat image acquired on November 16, 2000. Thus, RM1 and RM2 became emergent between January 1989 and November 2000. RM1 looks a typical tide-dominated river mouth bar, with a V-shape pointing landward. From Google Earth image (copyright 2013 Cnes/ Spot Image) as shown in Fig. 2M, the bar seems to consist of a relatively fine-grained tidal flat partly covered on its most elevated parts by salt marshes. RM2 has a typical shape of a wave-dominated bar, as shown in Fig. 2N, with a V-shape pointing seaward. It is made of two symmetric recurved sand spits. Landward of these spits are large, protected tidal flats with salt marshes developed between them. RM2 is not located at a river mouth but between two river mouths.

SAR DATA AND ANALYSIS

ALOS PALSAR data acquired between December 6, 2006 and January 3, 2011 at a minimum interval of 46 days were used. All images used in this study were from both ascending tracks (south to north; path 478, frame 180 and path 477, frame 190) and descending tracks (north to south, path 110, frames 3420 and 3430). The beam mode was either FBS (fine beam single polarization) or FBD (fine beam dual HH-polarization and HV-polarization) with nominal resolution of 4.7 m and 9.4 m, respectively; and the off-nadir angle was 34.3°. A range over-sampling by a factor two is applied to the FBD scene using only the HH polarization data to transform it to the same sample spacing as the FBS scene. For RM2 and RM3, 17 ascending (path 477, frame 190) and 4 descending (path 110, frame 3420) images were used. In addition to these data, 25 ascending (path 478, frame 180) and 4 descending (path 110, frame 3430) images were also used for RM1. During 49 months, 50, 21 and 21 images were used for RM1, RM2 and RM3 respectively. The limited number of used images results from the availability of acquired data.

SAR processing was conducted using Gamma software (Wegmüller & Werner, 1997). Level 1.0 PALSAR data were processed to the single look complex (SLC). In this study, multi-look factors 2 and 4 in range and azimuth, respectively, are used because the range/azimuth ratio pixel size of PALSAR is about 2. The multi-look image has in both directions approximately 15 m in pixel size. A set of SLCs were co-registered at sub-pixel accuracy. Co-registration consists of the computation of the offsets in range and azimuth among a set of SLCs to a common reference and resampling of one image to match with the other images. Then, co-registered SAR images were geocoded to produce SAR backscatter intensity images (Fig. 2A, E and I). During the geocoding process, the images were resampled to a pixel size of one arc-seconds square. NIH ImageJ software, a public domain program developed at the U.S. National Institutes of Health (available at http://rsb.info.nih.gov/ij/), was used to remove noisy pixels by applying a 3×3 neigh-

bourhood median filter that preserves small objects with high contrast and ensures minimum distortion of the original data (Fig. 2B, F and J).

Radar backscatter depends mainly on surface roughness and the dielectric properties of the target (e.g. Elachi, 1988). Histogram thresholding is widely used for image segmentation (e.g. Gonzalez & Woods, 2002). The iterative self-organizing data analysis technique algorithm (ISODATA), which chooses the threshold that is equidistant from the average intensity of pixels below and above it, was used to extract the target river-mouth bars from the SAR data. This algorithm is commonly used for unsupervised classification of image pixels into spectral clusters (e.g. Ridler & Calvard, 1978).

As shown in Fig. 2C, G and K, isolated river-mouth bars with strong backscatter (radar bright) that are surrounded by water (weak backscatter) can be extracted from SAR data. These show good correspondence to satellite images downloaded from Google Earth as shown in Fig. 2 M, N and Q. Note that since SAR images such as ALOS PALSAR and optical sensor images on Google Earth deploy different physical principles for the data acquisition, each image may contain different types of land cover information. Moreover, each image was acquired on a different time and under different conditions. The areal extents of the three river-mouth bars, which are defined by the condition that areas are not connected to an outer frame, were measured from geocoded binary SAR images by counting the number of pixels as shown in Fig. 2D, H and L.

RESULTS AND DISCUSSION

SAR data analyses allow us to highlight changes in area extent of the three selected bars at different time scales in relation to fluctuating parameters such as tidal heights (high frequency sea-level fluctuations), climate (river discharge and coastal dynamic changes) and mean sea-level (global increase).

Tidal heights and bar area extents

Bar area extents are related to tidal heights at a daily time scale. This can be illustrated with the case of the RM1. RM1 was sometimes imaged twice within about 12 hours, firstly by a descending orbit (~03:10 AM at local time) and later by an ascending orbit (~03:35 PM). Changes of the areal extent of

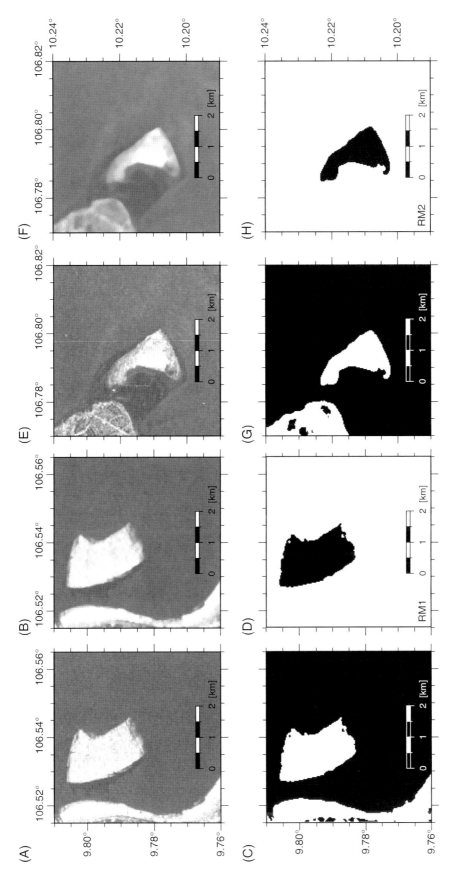

Fig. 2. (A), (E) and (I) ALOS PALSAR HH-polarized FBS backscatter intensity image of RM1, RM2 and RM3 respectively acquired on January 3, 2011. (B). (F) and (J) Images after noise reduction with a 3 × 3 median filter. (C). (G) and (K) Result of segmentation with threshold selection using the ISODATA algorithm.

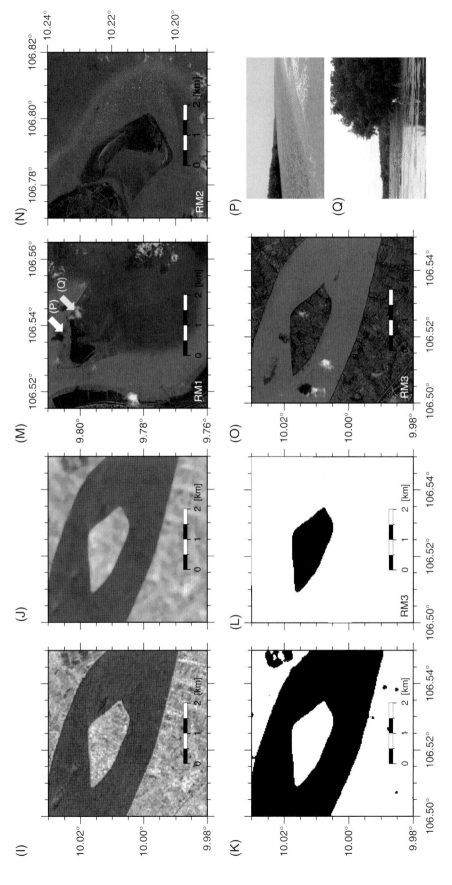

Fig. 2. (Continued) (D), (H) and (L) Images used for estimation of the area of RM1, RM2 and RM3, respectively. (M), (N) and (O) Satellite image, reproduced from Google Earth (copyright 2013 Cnes/Spot Image) of RM1, RM2 and RM3. White and yellow arrows on (M) indicate the locations of photographs shown on (P) and (Q).

RM1 plotted versus tidal height at SAR data acquisition times on December 28, 2008 and January 3, 2011 show RM1 to be larger at low tide (Fig. 3). When differences in tidal heights are small at the times of SAR data acquisitions, differences of bar areal extents are negligible (this is the case for examples with data on August 12, 2008 and September 27, 2008). This primary result suggests that the method, based on the analysis of SAR data, has a sufficient temporal resolution to capture the diurnal fluctuations of sea-level due to tides.

To delineate more clearly the relationship between the areal extent of the river-mouth bars and tidal heights, area residuals computed from linear annual trends (discussed in the next section) were plotted against tidal heights at the times of SAR data acquisition, which were estimated on the basis of hourly tidal data at Vung Tau, obtained from the

University of Hawaii Sea Level Center [UHSLC] website. For RM1 and RM2 (Fig. 4A and B), the larger residual areas correspond to lower tidal heights. The variations in the areal extents associated with the tidal heights can thus be pointed out although the annual changes in areal extents are significant. For RM3 a similar relationship with tidal heights seems to exist (Fig. 4C) although this bar shows almost constant area residuals throughout the study period (see next section).

Annual changes of river-mouth bars

The time series of SAR backscatter intensity images from December 2006 to January 2011 (Fig. 5) shows systematic changes of the areal extents of the three river-mouth bars. Each data is plotted within an envelope representing a 95%

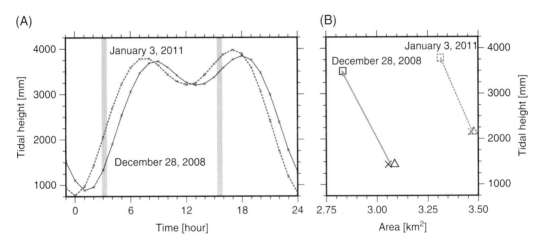

Fig. 3. (A) Hourly variations of tidal height on December 28, 2008 (solid line) and January 3, 2011 (dashed line). Grey vertical bars indicate time of acquisition of ALOS PALSAR data. (B) Areal extent of RM1 versus tidal height at time of data acquisition. Squares indicate data from an ascending track (path 477, frame 180) and crosses and triangles indicate data from a descending track (path 110, frames 3420 and 3430).

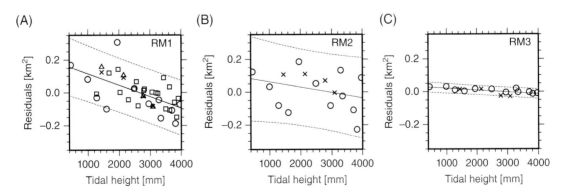

Fig. 4. Relationship between residual areas after removal of linear annual trends and estimated tidal heights for (A) RM1, (B) RM2 and (C) RM3 with 95% confidence intervals (dashed lines).

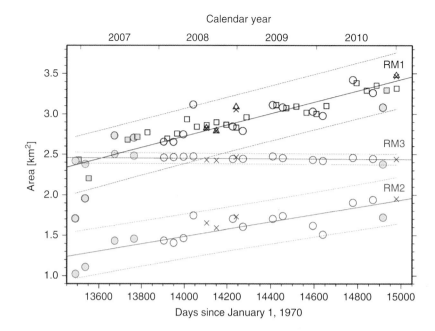

Fig. 5. Temporal changes of areal extent for RM1 (black), RM2 (blue) and RM3 (red). Linear least-squares fits (line) with 95% confidence intervals (dashed lines) are also shown. Circles and squares indicate data from an ascending track (path 478, frame 180; path 477, frame 190); crosses and triangles indicate data from a descending track (path 110, frames 3420 and 3430). Symbols with grey fill indicate area data without corresponding hourly tidal height data.

confidence interval. The most internal bar, RM3, located in the middle of the river, not at the river mouth, showed negligible change in size. Between 2006 and 2011 its areal extent was similar to that shown by the map published by the U.S. Naval Oceanographic Office in 1966. RM3 has remained stable and has been cultivated, as inferred from published maps referred to here and a Google Earth satellite image (Fig. 2O). In contrast, RM1 and RM2, located at or close to a river mouth, showed significant increases in size. The average annual increases for RM1 and RM2 between December 2006 and January 2011 were 0.2 and $0.1 \, \text{km}^2 \, \text{yr}^{-1}$, respectively, although seasonal fluctuations about these averages clearly occurred (Fig. 5). Note that the SAR images are taken at different times in the tidal cycle so that the areal extent is estimated at different tidal height. This proves that the annual changes of areal extents are significant compared to diurnal changes of areal extent due to tidal height fluctuations at the time of the data acquisition. The processed SAR data as well as previously published maps and archived Landsat images all indicate that RM1 and RM2 have increased in size in recent years. If constant annual increases of 0.2 and $0.1 \, \text{km}^2 \, \text{yr}^{-1}$ are assumed for RM1 and RM2, respectively, both may have been emergent since the mid-1990s,

which is consistent with archived Landsat data that show no evidence of RM1 and RM2 in 1989 but show both subaerially exposed in 2000. Thus, the gradual growth of RM1 and RM2 shown clearly over the period 2006 to 2011 by SAR data is supported by other data sources.

RM1 was imaged from different paths and frames and from both ascending and descending orbits. The areal extents measured for RM1 from images from ascending tracks (path 478, frame 180; path 477, frame 190) from the far and near ranges of the swath, show consistent annual trends (Fig. 5, squares and circles). Similarly, the areal extents measured from images from different descending tracks (path 110, frames 3420 and 3430) show almost the same values (Fig. 5, crosses and triangles). These results verify the validity of the method.

Seasonal changes of river-mouth bars

Seasonal variations of the areal extents of the two river-mouth bars, RM1 and RM2, are also apparent in the SAR data (Fig. 5). As mentioned before, the annual changes of the areal extents are high enough no matter when the data is acquired. Therefore, to evidence the seasonal changes more precisely, using all the SAR data from 2006 to

2011, the annual linear trends were removed to provide monthly residuals, which were then averaged to show annual seasonal patterns (Fig. 6A and B). Positive residuals in summer and negative residuals in winter for RM1 and RM2 are highlighted, although the averaged value has a substantial error partially due to the limited number of data available. On the other hand, the areal extent of RM3 shows no seasonal variations (Fig. 6B).

Fig. 6C shows monthly variations of mean sea-level for 2008 and 2009, based on hourly tidal data from Vung Tau (UHSLC) and the monthly tidal height data averaged over 1992 to 2001, based on data from PSMSL. This seasonal pattern of summer sea-levels, about 40 cm lower than winter sea-levels, indicates similar trends during 1992 to 2001, 2008 and 2009. The ~40 cm difference reflects monsoonal wind setup in the South China Sea (Fang *et al.*, 2006). The relationship of the areal extent of the river-mouth bars (Fig. 6A and B) to monsoon-driven sea-level fluctuations (Fig. 6C) is clear.

Wind and wave regime might have a seasonal pattern and it is very probable that they have an influence on the seasonal variations of the areal extents of the river-mouth bars. However, there are no available data and no evidence to support this suggestion. The sediment discharges may be estimated from an empirical relation between water discharge and sediment discharge at a site, referred to as a sediment-transport curve, when water discharge data are available but sediment data are not (e.g. Colby 1956). This empirical relationship is not always present; however, it might be used as a rough estimate. Considering this condition, sediment discharge from the Mekong delta to the sea might occur mostly in summer as a result of high water discharge from July to October (Fig. 6D). During the period of increasing discharge of water and sediment from June to September, the area residuals for RM1 and RM2 (Fig. 6A and B) decrease as wind setup and mean relative sea-level increase (Fig. 6C). This indicates that sediment discharge is not the main factor controlling seasonal changes in the emerged areal extent of the river-mouth bars. However, the annual increase in size clearly shows that sediment has been deposited in and around the river-mouth bars. Therefore, it is suggested that the sediment supplied from the river mainly from June to November is deposited in and around mouth bars as the water level rises after June

(Tamura *et al.*, 2012; Xue *et al.*, 2012), although there is unfortunately no direct evidence such as simultaneous *in-situ* measurements to support this hypothesis

In order to highlight temporal changes and erosion/accumulation patterns of the areal extent of RM1 and RM2, colour additive analysis were used (Fig. 7). For this analysis, older binary SAR data were shown in cyan and newer data in red. The colour additive process results in areas coloured in black, red and cyan, respectively, representing those parts of the river-mouth bars that were unchanged, those where the area of the bar increased and those where the area of the bar decreased. The left and centre panels of Fig. 7 show changes of the areal extent of RM1 and RM2 over two intervals of about six months. From January to June 2009, which is mostly during the dry season, both bars grew outward in all directions. In contrast, from June 2009 to January 2010, which includes the wet season, the areal extent of both bars decreased, mainly on the seaward side. These semi-annual changes in areal extents are possibly caused by alternate accumulation and erosion phases. During the rainy season of the relatively weak south-westerly summer monsoon, large volumes of sediment are discharged from the Mekong River (Tamura *et al.*, 2010). During the winter dry season, longshore drift driven by strong north-easterly winds and waves removed some of the sediment deposited during the preceding summer monsoon. This semi-annual pattern of deposition and erosion is the opposite of the pattern that drives the changes of the size of RM1 and RM2. There is no evidence to support that accumulation processes at river-mouth bars that face the open sea may be determined by the balance between the semi-annual differences of sediment supply and tidal heights.

The panels on the right of Fig. 7 show changes to RM1 and RM2 over a period of about two years (from January 2008 to January 2010). For RM1, red areas are dominant, indicating that this bar tends to grow almost in all directions but with a dominant seaward evolution. This may support the V-shape with a vegetated apex landward of the RM1 based on the Google Earth satellite image (Fig. 2M). This also seems to correspond to historical maps published in 1966 and 1989 and satellite images acquired in 1989 and 2000 from the Landsat archives; and is consistent with rapid growth with downstream trails (Fig. 5). Photos on Fig. 2P and Q were taken on January 15, 2005 and December 27,

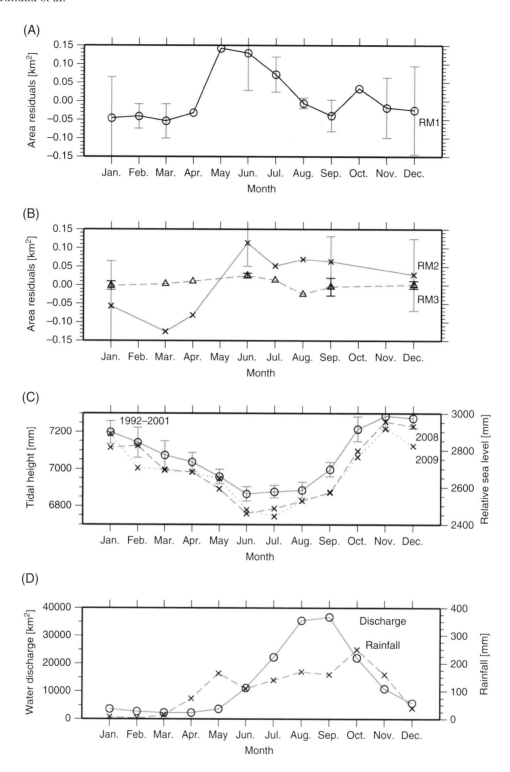

Fig. 6. (A) Monthly changes in residuals after removal of linear annual trends for areas of RM1, with one-standard-deviation error bars. (B) Same as (A) except for RM2 and RM3. (C) Monthly variations of mean monthly water level at Vung Tau (from tide gauge data obtained from the PSMSL database). Mean monthly data averaged over period from 1992 to 2001 with error bars (solid line). Dashed and dotted lines show variations of mean monthly sea-level for 2008 and 2009, respectively, calculated from hourly tidal level data from Vung Tau (UHSLC website). (D) Mean monthly discharge of lower Mekong River mainstream averaged over period from 1960 to 2004 at Kratie, eastern Cambodia (Mekong River Commission, 2009) and average monthly rainfall at Chau Doc on the Bassac River (Mekong River Commission, 2009). See Fig. 1B for locations of Kratie and Chau Doc.

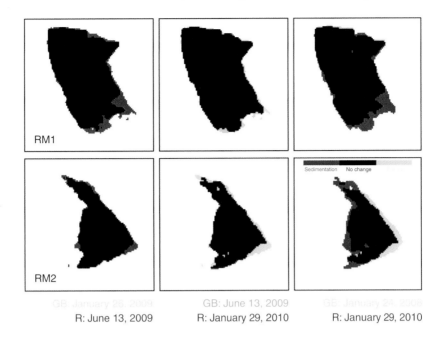

Fig. 7. Changes of areal extents of RM1 (upper panels) and RM2 (lower panels) derived by colour additive analysis over three periods in this study. Left panels, January 26, 2009 to June 13, 2009; centre panels, June 13, 2009 to January 29, 2010; right panels, January 24, 2008 to January 29, 2010.

2002, respectively, and show gently sloping downstream faces. According to the photo on Fig. 2Q, the mangrove is expected to migrate seaward, supporting that RM1 is migrating seaward through a seaward accretion of intertidal flats.

For RM2, accumulation occurred mainly on the landward side whereas erosion took place on its seaward side, indicating that as a whole the bar moved landward (Fig. 7). RM2 is assumed to be a wave-dominated bar, according to its V-shape pointing seaward (Fig. 2N). This strongly suggests that the erosion/deposition processes highlighted with the SAR data are rather related with wave dynamics. Since RM2 has a tentency to increase in area during the study period (Fig. 5), seaward erosion is overbalanced by landward accretion.

Our analysis succeeds in monitoring long-term (over a few years) changes of emerged surfaces of the river-mouth bars. These changes, to be detected, should be greater than those linked to water level fluctuations, noise in the data and uncertainties caused from the algorithm. As discussed above, the spatial growth rate of river-mouth bars, which should be corrected of the changes due to sea-level and/or tidal heights, can be obtained from this analysis, although *in-situ* ground measurements are required for verification and for quantifying the processes involved in

the river-mouth bar evolution. However, SAR data analysis can be considered as a valuable monitoring approach when the target area has poor accessibility and limited amounts of available data. There is no guarantee that the areas extracted by our method correspond exactly to the areal extents of river-mouth bars. Moreover, many other time-dependent factors, such as the state of the surface of river-mouth bars, beach slope and wave height, could also influence SAR data. However, the area extracted from our analysis of SAR data seems to provide a good approximation of river mouth-bar long-term evolution.

CONCLUSIONS

Time-series analyses of ALOS PALSAR data proposed here show that radar amplitude (backscatter) images can be used to delineate changes quantitatively, spanning several years, of the areal extent of river-mouth bars. Changes on this time scale have not been reported previously. The areal extents of three river-mouth bars on the Mekong River delta were measured from geocoded SAR images and annual and seasonal variations were revealed. The two river-mouth bars that face the South China Sea on the delta coast show nearly constant annual growth rates in the spatial extent. Changes of the size of these two river-mouth bars also show clear seasonal variations, i.e. an increase during the dry season and decrease during the wet season. Sequential SAR images of these river-mouth bars also provided an overview of patterns of accumulation and erosion at a wide range of spatial and temporal scales. A third river-mouth bar, located about 15 km upstream from the river mouth, showed no significant annual change of size during the study period from December 2006 to January 2011. However, the data presented here revealed a relationship between the areal extent of the bar and tidal height, which implies that emergence and submergence of river-mouth bars is controlled largely by tidal heights.

The method described in this paper provides a simple and robust way of delineating the spatial evolution of river-mouth bars on different time scales. At present, there has been no ground verification of the interpretation of SAR data from the Mekong River delta. However, the analysis presented here may provide an efficient tool for coastal monitoring and for understanding the behaviour of coastal environments over time. Nonetheless, further improvements to the method should be sought and opportunities to combine its use with other data investigated. One promising analysis method is the differential interferometric SAR (InSAR), although the delta is a difficult area to monitor using InSAR, due to the signal scattering properties as it is affected by variable surface conditions. However, Erban *et al.* (2013) were the first to estimate land subsidence rates at the Mekong delta area using InSAR, demontrating the high potential of the method in surveying the behaviour of this deltaic area.

ACKNOWLEDGEMENTS

The authors wish to thank the editor Dr. Bernadette Tessier and the reviewers for their constructive comments, which helped in improving the quality of our paper. The Ministry of Economy, Trade and Industry (METI) and the Japan Aerospace eXploration Agency (JAXA) retain joint ownership of the original ALOS PALSAR data, which were distributed by Japan Space Systems. Some of figures in this manuscript were prepared using GMT (Wessel & Smith, 1998). This study was partially supported by the Asia delta project of the Ministry of Environment, Japan and by NAFOSTED (The Vietnam National Foundation for Science and Technology Development; Project 105.01-2012.24).

REFERENCES

Alpers, **W.** (1983) Imaging ocean surface waves by synthetic aperture radar - a review. In: *Satellite Microwave Remote Sensing* (Ed. **T.D. Allan**), Chapter 6. Ellis Horwood Ltd.

Alpers, **W.** and **Hennings**, **I.** (1984) A theory of the imaging mechanism of underwater bottom topography by real and synthetic aperture radar. *J. Geophys. Res.*, **89**, 10529–10546.

Bhattacharya, **J.** (2003) Deltas and estuaries. In: *Encyclopaedia of Sediments and Sedimentary Rocks* (Ed. **G.V. Middleton**), pp. 195–203. Springer.

Brakenridge, **G.R.**, **Knox**, **J.C.**, **Paylor** II, **E.D.** and **Magilligan**, **F.J.** (1994) Radar remote sensing aids study of the Great Flood of 1993, *EOS Trans. Am. Geophys. Union*, **75**(45), 521-527, doi:10.1029/EO075i045p00521.

Colby, **B.R.** (1956) Relationship of sediment discharge to streamflow. *US Geol. Surv. Open-File Report*, **56–27**, 170 pp.

Edmonds, **D.A.** and **Slingerland**, **R.L.** (2007) Mechanics of river mouth bar formation: Implications for the morphodynamics of delta distributary net- works. *J. Geophys. Res.*, **112**, F02034, doi: 10.1029/ 2006JF000574.

Elachi, **C.** (1988) *Spaceborne Radar Remote Sensing: Applications and Techniques.* IEEE Press, NY, 255 pp.

Erban, **L.E.**, **Gorelick**, **S.M.**, **Zebker**, **H.A.** and **Fendorf S.** (2013) Release of arsenic to deep groundwater in the Mekong Delta, Vietnam, linked to pumping-induced land subsidence, *Proc. Natl Acad. Sci. USA*, **110**(34), 13751–13756, doi: 10.1073/pnas.1300503110.

Fang, **G.**, **Chen**, **H.**, **Wei**, **Z.**, **Wang**, **Y.**, **Wang**, **X.** and **Li**, **C.** (2006) Trends and interannual variability of the South China Sea surface winds, surface height and surface temperature in the recent decades. *J. Geophys. Res.*, **111**, C11S16, doi:10.1029/2005JC003276.

Gonzalez, **R.C.** and **Woods**, **R.E.** (2002) *Digital Image Processing.* 2nd edn, Prentice-Hall, Upper Saddle River, 793 pp.

GUNIO (Glavnoe upravlenie navigatsii i okeanografii Ministerstva oborony SSSR) (1989) Russian nautical chart 62510 (Level 3, scale 1:200,000).

Hu, **J.**, **Kawamura**, **H.**, **Hong**, **H.** and **Qi**, **Y.** (2000) A review on the currents in the South China Sea: seasonal circulation, South China Sea Warm Current and Kuroshio Intrusion. *J. Oceanogr.*, **56**, 607–624.

IPCC (2007) Climate change 2007: impacts, adaptation and vulnerability. In: *Contribution of Working Group II to the Fourth Assessment Report of the IPCC* (Eds **M.** Parry, **O.** Cansiani, **J.** Palutikof, **P.** Van der **Linden** and **C.** Hanson), Cambridge University Press, Cambridge, 996 pp.

Jensen, **J.R.** (2007) *Remote Sensing of the Environment: An Earth Resource Perspective.* 2nd edn, Prentice-Hall, Upper Saddle River, 592 pp.

Kubicki, **A.** (2008) Large and very large subaqueous dunes on the continental shelf off southern Vietnam, South China Sea. *Geo-Mar. Lett.*, **28**, 229–238.

Kushwaha, S.P., Dwivedi, R.S. and Rao, B.R. (2000) Evaluation of various digital image processing techniques for detection of coastal wetlands using ERS-1 SAR data. *Int. J. Remote Sens.*, **21**, 565–579.

Lillesand, T., Kiefer, R.W. and Chipman, J. (2008) *Remote Sensing and Image Interpretation.* 6th edn, John Wiley and Sons, 756 pp.

McCandless, S.W. and Jackson, C.R. (2004) Principles of synthetic aperture radar. In: SAR Marine User's Manual (Ed. **C.R. Jackson** and **J.R. Apel**). *National Oceanic and Atmospheric Administration*, **2004**. 1–23.

Mekong River Commission (2009) MRC Management Information Booklet Series No.2, The Flow of the Mekong, Mekong River Commission (MRC), Vientiane, Lao PDR, 12 pp., 2009. Overview of the Hydrology of the Mekong Basin, *MRC (Mekong River Commission) ISSN:* **1728** 3248.

Milliman, J.D. and Syvitski, J.P.M. (1992) Geomorphic/tectonic control of sediment discharge to the oceans: the importance of small mountain rivers. *J. Geol.*, **100**, 525–544.

Mills, J.P., Buckley, S.J., Mitchell, H.L., Clarke, P.J. and Edwards, S.J. (2005) A geomatics data integration technique for coastal change monitoring. *Earth Surf. Proc. Land.*, **30**, 651–664.

Ministry of Natural Resources and Environment (Vietnam) (2001) Vietnam 1:100,000 Scale Topographic Maps, *Thanh Phu* **C48–58** (6328).

Morang, A. and Gorman, L.T. (2005) Monitoring coastal geomorphology. In: *Encyclopaedia of Coastal Science* (Ed. **M.L. Schwartz**), pp. 663–674. Springer, The Netherlands.

Naval Oceanographic Office, United States (1966) Mouths of the Mekong River to Mui Ba Kiem, Vietnam, South China Sea, Asia (Scale 1:153,000), 2nd edn.

Ridler, T.W. and Calvard, S. (1978) Picture thresholding using an iterative selection method. *IEEE Transactions on Systems, Man and Cybernetics*, **8**, 630–632.

Smith, L.C. (1997) Satellite remote sensing of river inundation area, stage and discharge: a review. *Hydrol. Process.*, **11**, 1427–1439.

Syvitski, J.P.M., Kettner, A.J., Overeem, I., Hutton, E.W.H., Hannon, M.T., Brakenridge, G.R., Day, J., Vorosmarty, C., Saito, Y., Giosan, L. and Nicholls, R.J. (2009) Sinking deltas due to human activities. *Nature Geoscience*, **2**, 681–686.

Ta, T.K.O., Nguyen, V.L., Tateishi, M., Iwao, K. and Saito, Y. (2002) Sediment facies and late Holocene progradation of the Mekong River Delta in Bentre Province, southern Vietnam: an example of a tide- and wave-dominated delta. *Sed. Geol.*, **152**, 313–325.

Tamura, T., Horaguchi, K., Saito, Y., Nguyen, V.L., Tateishi, M., Ta, T.K.O., Nanayama, F. and Watanabe, K. (2010) Monsoon-influenced variations in morphology and sediment of a mesotidal beach on the Mekong River delta coast. *Geomorphology*, **116**, 11–23.

Tamura, T., Saito, Y., Nguyen, V.L., Ta, T.K.O., Bateman, M.D., Matsumoto, D. and Yamashita, S. (2012) Origin and evolution of inter-distributary delta plains, insights from Mekong River delta. *Geology*, **40**, 303–306.

Tamura, T., Saito, Y., Sieng, S., Ben, B., Kong, M., Sim, I., Choup, S. and Akiba, F. (2009) Initiation of the Mekong River delta at 8 ka: evidence from the sedimentary succession in the Cambodian lowland. *Quatern. Sci. Rev.*, **28**, 327–344.

Ulaby, F.T., Moore, R.K. and Fung. A. K. (1986) *Microwave Remote Sensing: Active and Passive*, Vol. III - Volume Scattering and Emission Theory, Advanced Systems and Applications, Artech House, Inc., 1065–2162.

Wegmüller, U. and Werner, C.L. (1997) Gamma SAR processor and interferometry software, *Proc. 3rd ERS Symposium, ESA*, **SP-414**, 1686–1692.

Wessel, P. and Smith, W.H.F. (1998) New, improved version of the Generic Mapping Tools released, *EOS Trans. Am. Geophys. Union*, **79** (47), p. 579.

Wright, L.D. (1977) Sediment transport and deposition at river mouths: a synthesis. *Geol.Soc. Am. Bull.*, **88**, 857–868.

Xue, Z., He, R., Liu, J.P. and Warner, J.C. (2012) Modeling Transport and Deposition of the Mekong River Sediment. *Cont. Shelf Res.*, **37**, 66–78.

Does the Ichnogis method work? A test of prediction performance in a microtidal environment: The Mula di Muggia (Northern Adriatic, Italy)

ANDREA BAUCON*† and FABRIZIO FELLETTI*

* Università di Milano, Dipartimento di Scienze della Terra, 20133, Milano, Italy
(andrea@tracemaker.com)
† UNESCO Geopark Meseta Meridional, Geology and Paleontology Office, 6060-101,
Idanha-a-Nova, Portugal

ABSTRACT

The purpose of the present study is to evaluate the performance of the IchnoGIS method, which is a high-resolution framework to capture, manage, analyse and display geographically referenced ichnological data. By surveying neoichnological, sedimentological and environmental variables, the IchnoGIS method allows users to identify the environmental factors controlling the distribution of modern ichnoassociations. If the IchnoGIS method holds what it promises, then it should be possible to predict environmental parameters directly from ichnoassociations. Based on this hypothesis, the prediction performance of an ichnological model derived from the IchnoGIS method (Grado Model) was tested against a real-world situation: the Banco della Mula di Muggia (Northern Adriatic, Italy). The study site is a shallow-water microtidal system consisting of a sheltered zone protected by elongated sand ridges (bars, barrier islands). The performance test proceeded by first assessing the potential distribution of environmental parameters on the basis of both the ichnoassociations and the Grado model. Thereafter, the performance of the model was evaluated successively from its success at predicting environmental parameters. The results of the performance test show that the model is able to predict modern environmental parameters from neoichnological data. It is suggested that the IchnoGIS method may have great potential for the interpretation of ancient sedimentary sequences.

Keywords: Ichnology, Adriatic Sea, network analysis, geostatistics, barrier islands.

INTRODUCTION

Biogenic sedimentary structures generated in tide-influenced environments commonly have a good preservation potential in the geological rock record (Gingras & MacEachern, 2012; Gingras et al., 2012). As traces are manifestations of biologic behaviour and this behaviour, in turn, is a function of the environment, they can be used to assess the environmental conditions concurrent with their formation (Seilacher, 2007; see Baucon et al., 2012 for a historical review of the concept). In modern tidal environments, traces and associated behaviour can be observed directly together with the local environmental variables which, as a

consequence, provide a solid basis for palaeoenvironmental analysis.

In light of these assumptions, a new method for ichnological analysis (IchnoGIS), based on geostatistics and network analysis, has recently been developed (Baucon & Felletti, 2012, 2013a, 2013b). Its application to modern intertidal flats revealed a predictive capacity which suggested that environmental variables could be derived from ichnological features. The question therefore arose as to what extent this observation could be generalised, i.e. whether the IchnoGIS method produced realistic results in all kinds of environmental settings. Thus, if a predicted scenario is consistent with observations then the method can, in principle, be regarded

Contributions to Modern and Ancient Tidal Sedimentology: Proceedings of the Tidalites 2012 Conference,
First Edition. Edited by Bernadette Tessier and Jean-Yves Reynaud.

to work. Based on this hypothesis, the purpose of the present study was to test the validity of the IchnoGIS method by evaluating its prediction performance in a shallow, subtidal to intertidal environment, the Mula di Muggia system (Adriatic Sea, Italy). A successful outcome of the test would have important implications not only for the analysis and understanding of modern tidal systems but potentially also for the reconstruction of palaeoenvironmental parameters from trace fossils.

GEOGRAPHICAL SETTING

The study area is located in the Northern Adriatic Sea (Italy), on the seaward margin of the Grado barrier island/lagoon system (Fig. 1). More precisely, the predictive model was derived from observations on the intertidal flats lining the coast

between Grado and Grado Pineta (model area) adjacent to a shallow subtidal zone surrounding the intertidal Mula di Muggia sandbank (test area). In correspondence, the model has been named 'the Grado model' (Baucon & Felletti, 2013b).

Both the test and the model area are part of the same marginal-marine depositional system, where intertidal and shallow subtidal flats are protected from the open sea by elongated sand ridges. With respect to sea-level, three types of ridges are distinguished: supratidal (barrier islands *sensu* Otvos, 2012), intertidal and subtidal ridges (bars *sensu* Otvos, 2012). The largest sand bar in the study area is the Mula di Muggia sandbank, the shallowest parts of which reach up to a few centimetres below the high tide level; the barrier-island of the model area represents the western prolongation of the Mula di Muggia sandbank (Fig. 1). The sandbank is commonly assumed to

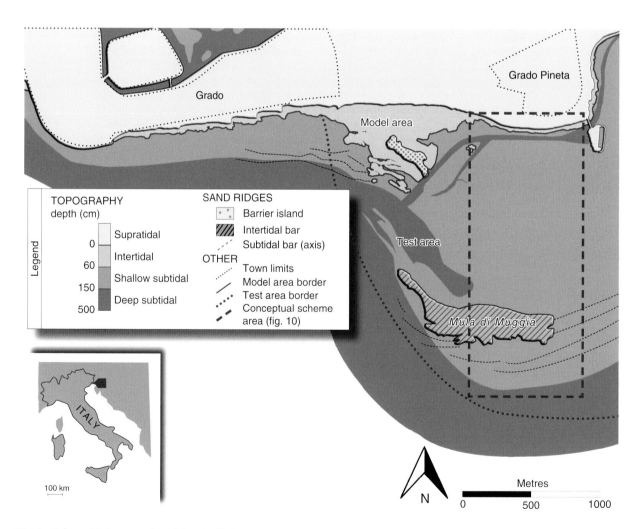

Fig. 1. Schematic topography of the studied area.

represent the remnants of the former Isonzo river delta (Venturini, 2003 and references therein), having formed during the Middle Ages. Today, muddy sands and microbial mats prevail in its shelter, whereas rippled sands dominate the shallow waters seaward of the sandbank (Baucon & Felletti, 2013a, b). The two regions, i.e. the areas landward and seaward of the Mula di Muggia sandbank (Fig. 2A) are here referred to as the back-bar and the fore-bar respectively.

The study area experiences semidiurnal tides that are unusually high for the Mediterranean Sea, ranging from 22 cm (neap tidal range) to 105 cm (spring tidal range); mean tidal range is 65 cm (Dorigo, 1965; Sconfietti *et al.*, 2003; Covelli *et al.*, 2008; Fontolan *et al.*, 2012; Fig. 3).

Concomitant with spring tides, seiches generated by the Scirocco wind and the passage of atmospheric low pressure systems are able to amplify tidal water levels up to 160 cm ('acqua alta'; Fontolan *et al.*, 2007). The prevailing winds blow from ENE (Bora wind) and SE (Scirocco wind), the latter being less intense but with a longer fetch (Fontolan *et al.*, 2007; Pervesler & Hohenegger, 2006). Wave heights conform to the general conditions of the Northern Adriatic, fair-weather waves being smaller than 0.5 m (Fontolan *et al.*, 2007). Within the confines of the sheltered Adriatic Sea, these conditions conform to the definition of fetch-limited barrier island systems (Pilkey *et al.*, 2009).

METHODOLOGY

Reliability test of the Grado model

To test the reliability of the Grado model (Fig. 4), the actual distribution of environmental variables was compared to that predicted by the model itself. For this purpose, the test area was crossed in several directions on foot or by kayak, thus drawing sampling tracks along which the distribution of ichnoassociations and environmental indicators was mapped (Fig. 2B). Trackpoints were positioned by a GPS unit (Garmin GPSmap 62 s; positioning error less than 10 m); each trackpoint refers to an observation area of approximately 5 m².

In most cases, environmental indicators and ichnoassociations were observable from a subaerial position or by snorkelling at the surface. In rarer cases, poor visibility of the sea floor required underwater inspection; the GPS positions of diving locations being recorded prior to each dive.

Consequently, sampling points to be inspected had to be positioned by the GPS unit and, successively, they were ichnologically and sedimentologically analysed.

Overall, the sampling tracks consisted of 9794 points, which allowed description of the observed and the predicted environmental scenarios:

1. Observed environmental scenario. In line with the Grado model, the mapped attributes were emersion time, hydrodynamics, sediment firmness and degree of microbial binding, as derived from field indicators ('observed environmental indicators' in Fig. 2C). More specifically, each sampling point was classified environmentally based on direct and indirect field diagnostic indicators (Table 1). The test area was also analysed from selected observation zones ('viewing areas' in Fig. 2B), offering optimal viewpoints on the surrounding intertidal and subtidal flats. Observations at trackpoints and viewing areas were compiled into a map synthesizing the observed environmental scenario. Satellite imagery was used to confirm and complement the interpretation of field data.
2. Predicted environmental scenario. Each sampling point was classified ichnologically by identifying the dominant ichnoassociation in conformity with the classes previously distinguished by Baucon & Felletti (2013b). This information, together with viewing areas, allowed the generation of an ichnoassociation map of the study area. Satellite imagery was used to confirm and complement the interpretation of field data. As the Grado model establishes a relationship between ichnoassociations and the environment, it can be applied to the observed ichnoassociations in order to derive a predicted environmental scenario (Fig. 2C), represented as a map.

Finally, comparison of the predicted with the observed scenario enabled an assessment of the reliability of the Grado model (Fig. 2C).

Additional data

In addition to the test of the Grado model, ichnological data (bioturbation intensity, ichnofaunal composition, burrow morphology) were observed. A camera with built-in GPS was used to take georeferenced pictures, whilst bathymetry was measured by a metre stick in both the fore-bar (10 measurements) and the back-bar (10 measurements).

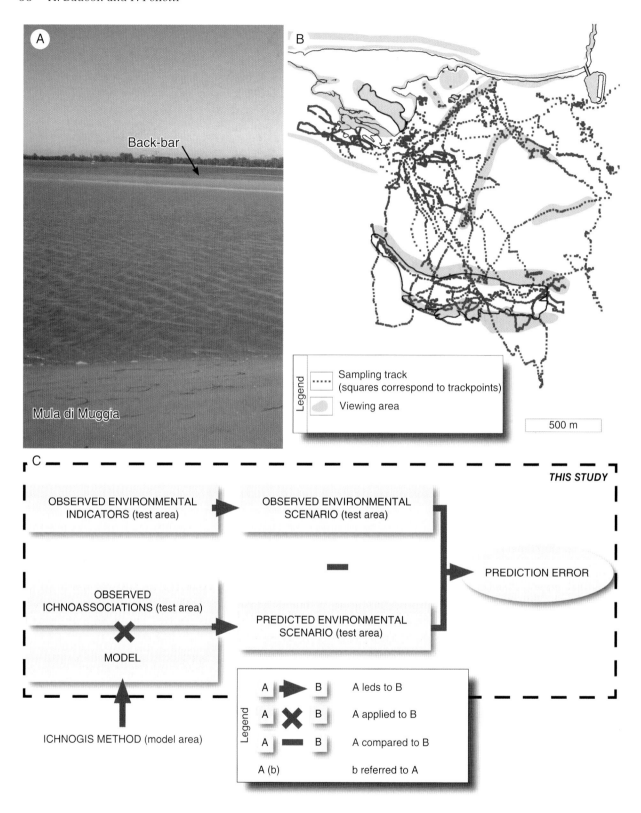

Fig. 2. Geological and methodological setting. (A) View of the back-bar area as seen from the Mula di Muggia at low tide. (B) The data sources for testing the Grado model are sampling routes, along which environmental and ichnological parameters were noted and control routes, along which robustness of data interpretation was checked. (C) Synthetic illustration of the testing method.

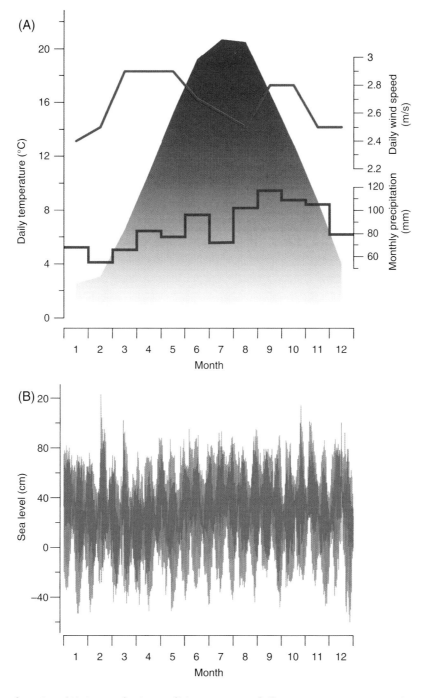

Fig. 3. Environmental setting. (A) Atmospheric conditions: average daily temperature at 180 cm (average 1999 to 2012); monthly precipitation (average 1961 to 2000); daily wind speed (average 1999 to 2012). Data from OSMER FVG, Grado station. (B) Sea-level for 2011 (interval: 10 minutes). Data from ISPRA, Grado station.

Development of the Grado model

This paper aims to test the Grado model, which was developed during previous research (Baucon & Felletti, 2013b) by applying the IchnoGIS method.

The IchnoGIS approach integrates hardware, software and data for capturing, managing, analysing and displaying geographically referenced ichnological data. IchnoGIS is based on four survey tasks relying on relatively low-cost tools (GPS unit,

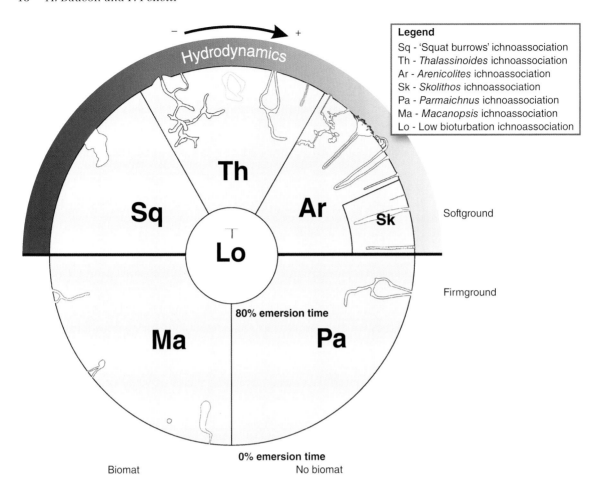

Fig. 4. Graphic depiction of the environmental significance of ichnoassociations in the Grado model in the form of specific positions in a circular diagram. More specifically, the upper hemisphere refers to softgrounds, the lower one to firmgrounds; the left hemisphere of the scheme is related to matgrounds. Decreasing emersion times are mapped with the distance from the centre of the scheme, while hydrodynamics is shown through a speedometer-like representation.

water quality instruments) and free/open-source software (Libre Office, PAST, Gephi):

1. Survey design. The first stage is the definition of the traces of interest and their correspondence with existing ichnotaxa.
2. Ichnological and sedimentological sampling. Two sampling approaches were followed to collect ichnological and sedimentological data, which were georeferenced by a GPS unit;
   a) Quadrat sampling. At each sampling site, a quadrat of a set size was placed on the substrate; facies type and abundance of each ichnotaxon were recorded in conjunction with the spatial coordinates of the sampling site.
   b) Trench sampling. Quadrat sampling was complemented by the study of vertical trenches, excavated at regularly spaced sites. At each sampling site the facies type and presence/absence of each ichnotaxon was recorded in conjunction with the spatial coordinates of the sampling site.

3. Environmental and topographical sampling. Water properties (e.g. pH, Eh, nutrients, salinity) and sediment features (e.g. depth of the redox potential discontinuity, substrate firmness, emersion time) were recorded at representative sampling sites. Major topographic features (e.g. intertidal/subtidal limit, landform boundaries) were mapped using a GPS unit.
4. Data visualization and analysis. Finally, the relationship between traces and the environment was deduced from both the spatial distribution of ichnotaxa and their association patterns:
   a) Spatial organization. Trench sampling data are displayed by point maps showing presence/absence of each ichnotaxon. A geostatistical approach based on kriging (Matheron, 1962) was adopted to interpolate the number of traces at unsampled positions, thus producing density maps of each ichnotaxon. Both point and density maps allow comparison of the spatial distribution of traces with

Table 1. Mapped attributes, their physical significance and the indicators used to recognize them. Small ripples are characterized by a crest-to-crest distance ranging between 10 and 25 cm. Emersion time refers to the amount of time during which the sediment is above water level; expressed as percentage of a complete tidal cycle. Sediment texture is determined by visual comparison and manipulation. Current velocity is intended as the average current velocity near the sea bottom, induced either by unidirectional or oscillatory flows. Physical significance of attribute classes is deduced by field measurements (Baucon & Felletti, 2013a, b).

Attribute	Value class	Environmental indicators		Inferred physical significance
		Direct	Indirect	
HYDRO-DYNAMICS	High	• Wave height >15 cm • Constant wave action • Small breakers • Capillary waves never dominant	• Sandy sediment (sand >50%) • Shell coquina (diameter >1 cm) • Small, straight to sinuous ripples • Fore-bar position	Current velocity: 20 to 50 cm s^{-1}
	Moderate-low	• Wave height <15 cm	• Sandy sediment (sand >50%) • Back-bar position	
	Low	• Wave height <15 cm • Frequent calm conditions (glassy sea)	• Conspicuous muddy fraction	Current velocity <20 cm s^{-1}
EMERSION TIME	Supratidal	• Constant emersion		Emersion time = 100%
	Intertidal	• Emersion at low tide		100% >Emersion time >0%
	Subtidal	• Submersion at high and low tide		Emersion time = 0%
MICROBIAL BINDING	Biomat	• Dominance of laminated or filamentous mats (coverage >90%)		
	No biomat	• Absence of evident biomat features (coverage <90%)		
FIRMNESS	Firmground	• Relative movement of sediment particles not possible	• Brinell test	Firmness ≥10^6 Pa
	Softground	• Relative movement of sediment particles possible	• Brinell test	Firmness <10^6 Pa

environmental and topographical data, including satellite imagery.

b) Patterns of association. Ichnological data, collected during trench sampling, are described by a network graph (ichnonetwork) which maps ichnotaxa as nodes and their association relationships as links. Consequently, network theory (Brandes & Erlebach, 2005; Fortunato, 2010) can be used to explore the ichnological system on the basis of the association relationships between ichnotaxa. Ichnonetwork analysis allows determination of how individual ichnotaxa are embedded in the studied system and discern patterns of organization (ichnoassociations). Regularities are successively searched for in the environmental properties associated with each ichnoassociation in order to identify the major structuring processes of ichnoassociation composition and distribution.

RESULTS

Ichnoassociations

In conformity with Baucon & Felletti (2013a, b), traces were abundant and morphologically diverse in the study area, tending to be organized in distinct groups (ichnoassociations). This section describes the ichnoassociations characterizing the study area, based on the classes previously distinguished by Baucon & Felletti (2013b). In particular, each ichnoassociation is addressed with respect to three major aspects:

1. Ichnological features. Major ichnological attributes include characteristic ichnotaxa, which have been used as key recognition criteria during sampling. The recognized ichnotaxa, to be considered incipient (see Bromley, 1996), are described in Table 2; the reader is referred to Baucon & Felletti (2013a) for a

Table 2. Incipient ichnotaxa of the Grado ichnosite (based on Baucon & Felletti, 2013a). Different morphotypes are indicated by the suffixes XL (very large), L (large), M (medium), S (small), XS (very small).

	Incipient ichnotaxon	Amended description	Tiering depth	Dominant behaviour	Tracemaker
Branched structures	*Thalassinoides* XL	Burrow network with multiple openings, one of which corresponds to a sediment mound.	40 cm	Deposit feeding	*Pestarella candida* (Crustacea: Decapoda)
	Thalassinoides L	Burrow network with chambers filled by seagrass	20 to 40 cm	Deposit feeding	*Pestarella tyrrhena* (Crustacea: Decapoda)
	Parmaichnus	Y shaped burrow with swellings	20 to 40 cm	Suspension feeding, ?deposit feeding	*Upogebia pusilla* (Crustacea: Decapoda)
	Polykladichnus	I-shaped or U-shaped burrow with Y-shaped bifurcations	5 to 10 cm	Deposit feeding, suspension feeding	*Nereis diversicolor* (Annellida: Polychaeta)
U-burrows	*Arenicolites* XL	U-burrow with single faecal cast	20 to 40 cm	Deposit feeding	*Sipunculus nudus* (Sipuncula: Sipunculidae)
	Arenicolites L	U-burrow with radial faecal casts	20 to 40 cm	Deposit feeding	*Sipunculus nudus* (Sipuncula: Sipunculidae)
	Arenicolites S	U-burrow	3 to 10 cm	Deposit feeding	*Corophium volutator* (Crustacea: Amphipoda)
Chambered burrows	'Squat burrow'	Squat burrow with terminal disc-shaped chamber	5 to 15 cm	Mating	*Carcinus maenas* (Crustacea: Decapoda)
	Macanopsis	Clavate burrow	2 to 7 cm	Feeding, sheltering, reproducing	*Heterocerus flexuosus* (Insecta: Coleoptera)
Plug-shaped structures	*Bergaueria*	Rounded, plug-shaped depressions	3 to 10 cm	'omnivourous' suspension feeding	*Cereus* (Cnidaria: Sagartiidae), *Condylactis* and *Anemonia* (Cnidaria: Actiniidae)
Winding structures	*Helminthoidichnites*	Unbranched horizontal burrow	<1 mm	Undermat mining	Dipteran larvae (Insecta: Diptera)
Simple structures	*Skolithos* L	Vertical burrow with 8-shaped opening	>15 cm	Suspension feeding	*Solen marginatus, Ensis ensis, Ensis minor* (Mollusca: Bivalvia)
	Skolithos M	Vertical burrow with constructional lining	10 to 15 cm	Suspension feeding	*Megalomma sp.* (Annelida: Polychaeta)
	Skolithos S	Vertical burrow	3 to 7 cm	?Suspension feeding	Worm-like organisms
	Skolithos XS	Vertical unlined burrow	1 to 5 cm	Sheltering	*Talitrus saltator* (Crustacea: Amphipoda)
	Monocraterion	Vertical burrow with funnel-shaped opening	10 cm	?Suspension feeding	Worm-like organisms
Trackways and footprints	*Avipeda-/Ardeipeda*-like	Footprints with three digits directed forward	0.1 to 1 cm	Locomotion / Feeding	*Ardea cinerea, Egretta garzetta* (Aves: Ardeidae), *Laurus michahellis* (Aves: Laridae)
	Canipeda	Tetradactyl footprints with heel pad	0.3 to 2 cm	Locomotion	*Canis lupus familiaris* (Mammalia: Canidae)
	Parallel furrows	Sets of parallel, elongated furrows	0.2 cm	Locomotion	*Carcinus maenas* (Decapoda: Brachyura)

	Incipient ichnotaxon	Amended description	Tiering depth	Dominant behaviour	Tracemaker
Trails	*Archaeonassa*	Median furrow flanked by two lateral ridges	0.1 cm	Locomotion	*Bolinus brandaris*, *Hexaplex trunculus* (Gastropoda: Muricoidea), *Cerithium vulgatum* (Gastropoda: Cerithiidae), *Pirenella conica* (Gastropoda: Potamidi-dae), *Sphaeronassa mutabilis*, *Nassarius nitidus* (Gastropoda: Nassariidae)
	Nereites biserialis	A furrow flanked on both sides by lobes	0.1 cm	Locomotion and feeding	hermit crabs, probably *Pagurus* (Decapoda: Paguridae)
	Nereites uniserialis	Furrow flanked by a single row of lobes	0.1 cm	Locomotion and feeding	hermit crabs, probably *Pagurus* (Decapoda: Paguridae)
Miscellaneous group	'Diverging shafts'	V-shaped tunnels with circular cross-section	0.3 to 5 cm	Suspension feeding	*Abra alba*, *Donax trunculus*, *Solecurtus sp.*, *Venus sp.* (Bivalvia: Veneroida)
	Lockeia S	Almond-shaped burrow	3 to 5 cm	Suspension feeding	*Mactra corallina*, *Venus* sp. (Bivalvia: Venroida)
	Lockeia XS	Almond-shaped burrow	<3 cm	Suspension feeding	*Abra alba*, *Donax trunculus* (Bivalvia: Veneroida)
	Mottling	Indistinct bioturbation	1 to 40 cm	Rooting Locomotion	*Zostera marina*, *Zostera noltii*, *Cymodocea nodosa*, *Posidonia oceanica* (Magno-liophyta: Najadales), *Nereis diversicolor* (Annellida: Polychaeta)

Table 3. Descriptors of bioturbation intensity used in this study. Ichnofabric index (II) quantifies bioturbation as seen in section, while bedding plane bioturbation index (BPBI) refers to the sea floor surface as seen from top-view.

Index	% bioturbation	Ichnofabric index (Droser & Bottjer, 1986)	Bedding plane bioturbation index (Miller & Smail, 1987)
1	0	No bioturbation recorded; all original sedimentary structures preserved	No bioturbation. If the surface is not flat and featureless, the only disruption is that caused by physical or chemical processes
2	0 to 10	Discrete, isolated trace fossils	Bioturbation may be represented by zones of generalized disruption or by discrete trace fossils. Most discrete structures are isolated but some intersect
3	10 to 40	Burrows are generally isolated but locally overlap	Bioturbation represented by discrete traces, zones of generalized disruption, or by both
4	40 to 60	Last vestiges of bedding discernible. Burrows overlap and are not always well-defined	Bioturbation represented by discrete traces, zones of generalized disruption, or by both. Interpenetration of discrete structures is more common than in less bioturbated surfaces
5	60 to 100	Bedding completely disturbed but burrows are still discrete in places and the fabric is not mixed	Surface has been disrupted by biological activity

more comprehensive treatment of individual traces. In addition, ichnofabric index (II; Droser & Bottjer, 1986) and bedding plane bioturbation index (BPBI; Miller & Smail, 1997) are used to describe typical bioturbation intensity as expressed in vertical profile and on the sea floor surface, respectively (Table 3). It should be noted that Miller & Smail (1997) use the categories 1–5 of Droser & Bottjer (1986) (II 6 denotes nearly or totally homogenized bedding). Ichnodiversity focuses on the number of ichnotaxa present at the observation scale (see Bromley, 1996).

2. Occurrence. The presence/absence of a given ichnoassociation with respect to the test and the model area (Fig. 1) is highlighted.

3. Environmental significance. This aspect concerns the environmental significance of a given ichnoassociation, as expressed by the Grado model (Fig. 4). In this regard, substrate firmness is described with the terms 'softground' (unconsolidated sediment) and firmground (partially consolidated sediment; see Bromley, 1996). Emersion time refers to the percentage of time during which a given area is subaerially exposed. For instance, 100% emersion time indicates subaerial conditions for the entire tidal cycle. Finally, the Grado model provides hydrodynamic information in relative terms ('high' to 'low') as specific studies are lacking in the study area. Nevertheless, an approximate estimate of average current velocity near the sea bottom (Table 1) can be deduced by considering sedimentary features (Baucon & Felletti, 2013a, b in conjunction with Nichols, 2009 and Stow *et al.*, 2008) and general hydrographic features

(Dorigo, 1965; Sconfietti *et al.*, 2003; Covelli *et al.*, 2008; Fontolan *et al.*, 2012).

Low bioturbation ichnoassociation (Fig. 5A)

Ichnological features

This ichnoassociation is characterized by no bioturbation or low bioturbation intensity (II 1–2; BPBI 1-2; Fig. 5A), as evidenced by the frequent preservation of parallel lamination. Occasionally, vertebrate trampling may produce higher bioturbation intensities. In this context, *Skolithos* XS marks the landward limit of a wide low-bioturbation belt, whilst tetrapod footprints are rarer.

Occurrence

This ichnoassociation is documented from the model area only.

Environmental significance

According to the Grado model, the low bioturbation ichnoassociation characterizes areas undergoing prolonged emersion (emersion time longer than 80% of tidal cycle duration).

Thalassinoides ichnoassociation (Fig. 5B)

Ichnological features

High densities of decapod burrows (*Thalassinoides, Parmaichnus*) and intense root-related mottling are common (Fig. 5B, C). Small U-burrows (*Arenicolites* S) are locally abundant. Typical bioturbation intensity is moderate to high (II 3–5; BPBI 3–5). Based

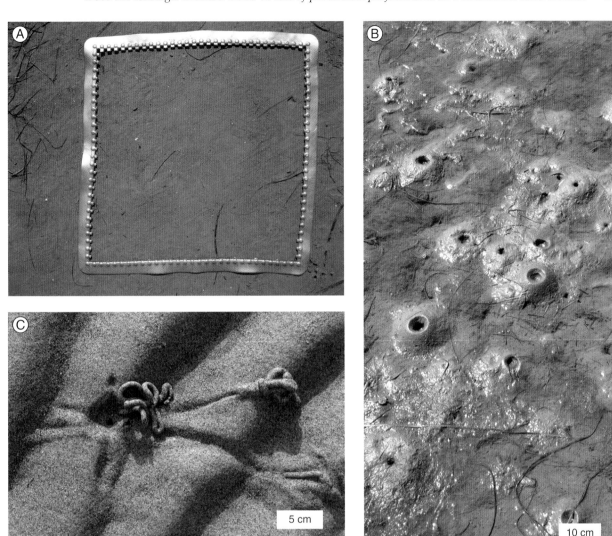

Fig. 5. Ichnoassociations of softgrounds without microbial binding. (A) Low bioturbation ichnoassociation. Quadrat side (0.5 m) for scale; top view. (B) The *Thalassinoides* ichnoassociation is characterized by intense bioturbation. Picture shows numerous openings and mounds produced by decapod crustaceans. (C) The pictured paired openings correspond to the U-shaped burrow *Arenicolites*. Note faecal casts. Top view.

on field observations of the test area, two subunits are distinguished:

2.1 A mottling-dominated subunit is character-ized by intense sediment reworking, which is usually produced by roots and rhizomes of seagrasses (*Cymodocea, Zostera, Posidonia*). For mapping purposes, dominance of seagrass (percent cover>50%) has been considered to be the major diagnostic criterion, together with the absence of sediment mounds. In fact, sediment mounds correspond to the ichnoge-nus *Thalassinoides* XL (Table 2), which is the characteristic ichnotaxon of the *Thalassinoides*-dominated subunit.

2.2 A *Thalassinoides*-dominated subunit is char-acterized by the diagnostic presence of large sediment mounds, corresponding to the ichnotaxon *Thalassinoides* XL. Sediment mounds are commonly found within dense seagrass meadows, which are accompanied by extensive mottling but occurrences in unvegetated substrates are also common. *Parmaichnus* and *Thalassinoides* L may be present as accessory components.

Occurrence

The *Thalassinoides* ichnoassociation is docu-mented from both the model and the test area.

Environmental significance

According to the Grado model, the *Thalassinoides* ichnoassociation characterizes softgrounds with moderate hydrodynamics. The substrate may be colonized by biomats, while emersion time is shorter than 80% of the tidal cycle duration.

Arenicolites/Skolithos ichnoassociation (Figs 5C and 6)

Ichnological features

Large U-burrows (*Arenicolites* XL, L), vertical burrows (*Skolithos* M, L, S, *Monocraterion*) and horizontal trails (*Nereites, Archaeonassa*) are the most distinctive ichnotaxa, while biodeformational structures, *Lockeia*, 'diverging shafts' and *Thalassinoides* may also be present. Bioturbation intensity is low (II 1–2; BPBI 2), although more intensely bioturbated patches may be present.

Based on field observations, two subunits are distinguished:

3.1 A low ichnodiversity subunit is characterized by large *Arenicolites* (types XL and L, Table 2), *Skolithos* M, L, S and *Nereites.* Ichnodiversity and bioturbation intensity fluctuate in space but are generally very low (BPBI 2; II 1–2). In fact, traces of epibenthic activity (*Nereites, Archaeonassa*) are frequently the only distinct structures, being separated with metre-size patches of virtually unbioturbated sediment. Nevertheless, higher bioturbation intensities

are also found especially at landward sites, with densities of *Arenicolites* up to 20 individuals per m². *Thalassinoides* and 'diverging burrows' are also documented. Baucon & Felletti (2013b) subdivided the *Arenicolites/Skolithos* ichnoassociation into two subclasses, which are characterized respectively by prevailing vertical burrows (*Skolithos* ichnoassociation) and the dominance of U-shaped burrows (*Arenicolites* ichnoassociation). The present subunit has more affinity with the *Skolithos* ichnoassociation in that it has similar ichnotaxa (*Skolithos* M, L, S) and bioturbation intensities, although it covers a wide ichnological spectrum including elements of the *Arenicolites* ichnoassociation (i.e. *Arenicolites*) as well.

3.2 A moderate ichnodiversity subunit shares most of the ichnotaxa with the previously described one but it differs in that it has higher ichnodiversity and bioturbation intensity (II 2; BPBI 2–5). In fact, the uppermost tier appears frequently reworked by the activity of epibenthic organisms, resulting in distinct trails (*Nereites, Archaeonassa*) being frequently superimposed on an indistinctly bioturbated background. Distinct burrows are sparsely distributed, with denser assemblages at the transition with the *Thalassinoides* ichnoassociation. *Skolithos* M is one of the most typical components, although not ubiquitous. Small 'diverging shafts' have been documented frequently (opening width: 0.3 to 0.4 cm), while larger specimens produced by *Solecurtus*

Fig. 6. *Arenicolites/Skolithos* ichnoassociation. The suspension-feeding polychaete *Megalomma* extends its crown of tentacles from its vertical burrow, which corresponds to the ichnogenus *Skolithos*. Other individuals are marked by arrows; 0.5 m water depth.

(opening width: 0.3 to 1.5 cm) seem to be diagnostic of this subichnoassociation. *Thalassinoides, Skolithos* L, *Lockeia* are common.

With regard to the subclasses distinguished by Baucon & Felletti (2013), this ichnoassociation is more affined to the *Arenicolites* ichnoassociation although it incorporates features of both *Thalassinoides* and *Skolithos* ichnoassociations (i.e. *Thalassinoides,* moderate to high bioturbation; *Skolithos*).

Occurrence

The *Arenicolites/Skolithos* ichnoassociation is recognized both in the model and the test areas.

Environmental significance

According to the Grado model, the *Arenicolites/ Skolithos* ichnoassociation corresponds to softgrounds without biomat cover. Hydrodynamics is moderate-high, while emersion time is shorter than 80% of the tidal cycle duration. Given the ichnological affinities of the low and moderate ichnodiversity subunits, they are referred to as having low and moderate energy levels, respectively.

Macanopsis ichnoassociation (Fig. 7A)

Ichnological features

Insect burrows (*Macanopsis, Helminthoidichnites*) are the most typical component of this ichnoassociation,

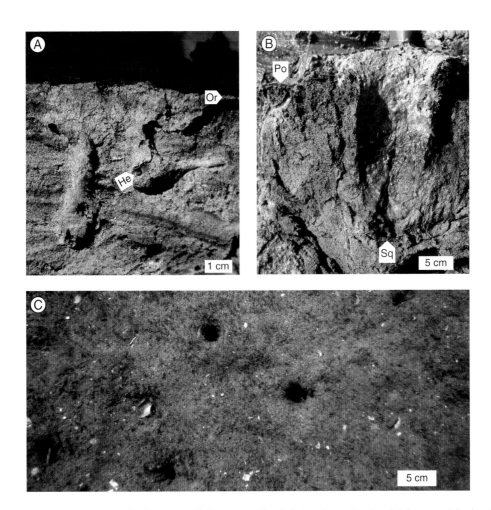

Fig. 7. Ichnoassociations of microbial mats and firmgrounds. (A) Laminated microbial mat with the *Macanopsis* ichnoassociation. Two specimens of *Macanopsis* are visible, one of which presents the trace maker (heterocerid beetle: He). Note the organic-rich horizon of the microbial mat (Or). Profile view. (B) Filamentous microbial mat with the 'Squat burrows' ichnoassociation. *Polykladichnus* (Po) and a 'Squat burrow' (Sq) are visible. Profile view; dashed line indicates the water/ sediment interface. (C) Firmground with the *Parmaichnus* ichnoassociation. Numerous openings produced by upogebiid crustaceans are visible. Top view, ca. 1 m water depth.

which is often accompanied by *Polykladichnus*. Bioturbation intensity is low (II 2; BPBI 1–2).

Occurrence

This ichnoassociation occurs in the model area only.

Environmental significance

According to the Grado model, the *Macanopsis* ichnoassociation is restricted to biomat-dominated firmgrounds experiencing emersion times shorter than the 80% of the tidal cycle duration. Although the Grado model focuses on intertidal settings (0%<emersion time<100%), Baucon & Felletti (2013 a, b) suggested that, in the study area, microbial mats are restricted to intertidal settings of the study area.

'Squat burrows' (Fig. 7B)

Ichnological features

Crab mating burrows ('squat burrows') are the diagnostic element of this ichnoassociation, which is related to filamentous mats. Small individuals of *Arenicolites* and *Polykladichnus* may be abundant locally. Bioturbation intensity is moderate to high (II 2–5; BPBI 2–5).

Occurrence

This ichnoassociation is found in the model area.

Environmental significance

According to the Grado model, this ichnoassociation characterizes biomat-dominated softgrounds emerged for less of the 80% of the tidal cycle duration.

Parmaichnus ichnoassociation (Fig. 7C)

Ichnological features

Monoichnospecific assemblages of *Parmaichnus* characterize this ichnoassociation. Bioturbation intensity is low to high (II 2–5; BPBI 2–4).

Occurrence

Both the model and the test area display this ichnoassociation.

Environmental significance

According to the Grado model, the *Parmaichnus* ichnoassociation is found in firmgrounds without microbial mats. Emersion time is shorter than 80% of the entire tidal cycle duration.

The spatial organization of the aforementioned ichnoassociations is revealed by the ichnoassociation map (Fig. 8), in which the spatial distribution of the major ichnological features of the test area is represented. The map shows a marked dichotomy between *Thalassinoides* and *Arenicolites/Skolithos* ichnoassociations, the latter being distributed in the more distal fore-bar environment. Here, ichnodiversity and bioturbation intensity are lower, with U-burrows (*Arenicolites* L, XL), vertical traces (*Skolithos* M) and trails (*Nereites, Archaeonassa*) as the dominant biogenic structures. Isolated patches with the *Arenicolites/Skolithos* ichnoassociation are also found in the back-bar area but the ichnological features are different from the fore-bar manifestations (i.e. moderate ichnodiversity subunit of the *Arenicolites/Skolithos* ichnoassociation).

The map also shows a correlation between the occurrence of *Parmaichnus* and the position of a dredged channel, although observations were more difficult in this area because of the frequent presence of boats and poor visibility. It should also be noted that the *Macanopsis* and 'squat burrows' ichnoassociations are not represented in the test area, despite their widespread presence in the intertidal model area (Baucon & Felletti, 2013b).

Predicted *vs.* observed scenario

The Grado model (Fig. 4) allows the derivation of the potential distribution of environmental variables from the previously discussed ichnoassociation map (Fig. 8). The resulting map (Fig. 9A) predicts a marked hydrodynamic contrast between the back-bar area and the fore-bar, both of which would be dominated by softgrounds without microbial binding. According to the prediction map, firmgrounds would be widespread in the dredged channel and adjacent areas.

The observed distribution of environmental parameters (Fig. 9B), as deduced from process-based indicators (Table 1), is consistent with the predicted distribution of environmental parameters. In fact, the back-bar/fore-bar dichotomy is evident and finds full correspondence with the

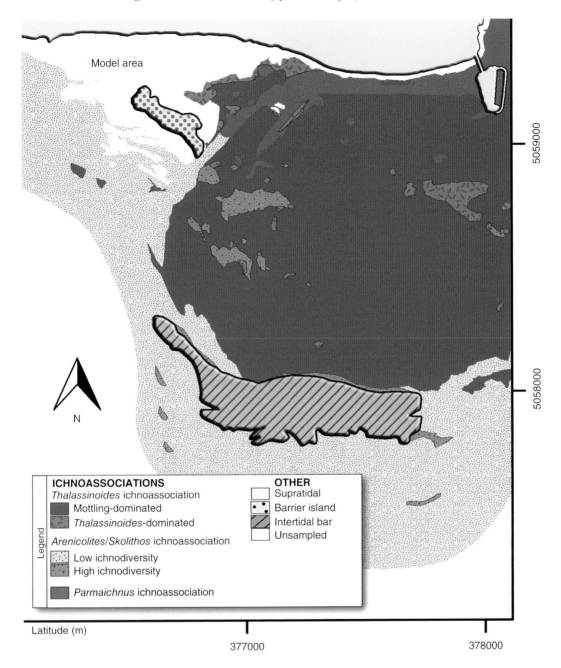

Fig. 8. Ichnoassociation map of the test area.

predicted data. Both the predicted and the observed maps show areas of moderate hydrodynamics within the sheltered back-bar. These areas are identified by the moderate ichnodiversity subunit of the *Arenicolites/Skolithos* ichnoassociation, despite not fitting exactly with the *Arenicolites* subclass distinguished by Baucon & Felletti (2013b),

Minor differences are found in small patches, some of which are indicated by arrows in Fig. 9.

Major discrepancies are found in the distribution of firmgrounds, which is deduced from the presence of the *Parmaichnus* ichnoassociation. In fact, although the *Parmaichnus* ichnoassociation is commonly related to firmgrounds (Fig. 7C), it is also documented from softgrounds. More specifically, softground occurrences of *Parmaichnus* are found in the artificial channel and its surroundings, where firmgrounds are also common but distributed patchily (Fig. 8).

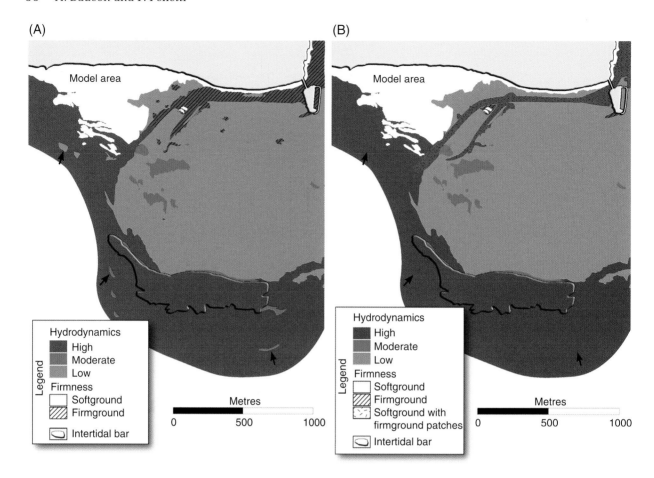

Fig. 9. Environmental maps of the test area. (A) Predicted scenario. Arrows show selected differences with the observed scenario. (B) Observed scenario. Arrows show selected differences with the predicted scenario.

DISCUSSION

The model and the test areas are located adjacent to each other and therefore they are exposed to similar environmental parameters, despite their slightly different bathymetric setting. Although this similarity prevents assessment of the global validity of the Grado model, it allows better evaluation of the prediction performance of the IchnoGIS method. In fact, the similarity of environmental factors actually restricts the interpretation of the results.

In terms of the overall pattern of ichnoassociation distribution, the predicted scenario is significantly consistent with the observed distribution of environmental parameters. Quantitatively, the predicted and observed maps overlap 97.031%, implying that the Grado model describes the trace-environment relationships of the study area in a consistent way.

Although the aforementioned result is central to this research, the role played by control factors (hydrodynamics, biomat binding, emersion time, substrate firmness) is important as well. In fact, understanding how control factors are embedded in the studied system is crucial not only for evaluating coherence of the model but also for a better comprehension of the ichnosite itself. More specifically, this issue is synthesized by two complementary questions: (a) How do control factors influence other biologically-significant parameters? (b) How are control factors modulated by general physiographic features?

In this regard, hydrodynamics is controlled strongly by the Mula di Muggia sandbank, which represents one of the major geomorphological features of the studied site. It functions as a barrier against waves entering, the influence of which is less intense in the back-bar than in the fore-bar (Fig. 10). A similar environmental dichotomy is

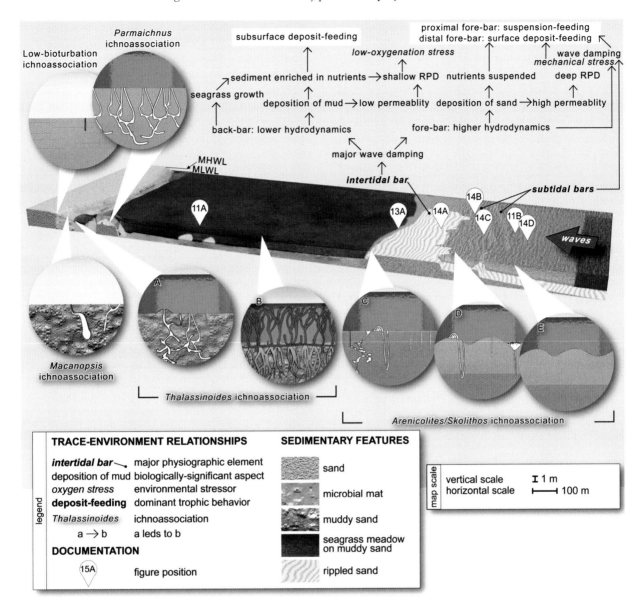

Fig. 10. Conceptual diagram of the major trace-environment relationships. The area in the figure corresponds to the eastern sector of the study site (see Fig. 1 for exact location). The represented area is 815 m wide and 2255 m long; vertical scale is exaggerated (maximum depth: 2.5 m) while sedimentary structures (i.e. ripples) are not to scale. The sea-level in the figure corresponds to mean low water level (MLWL), whereas mean high water level (MHWL) is indicated by a dashed line. To better show the bathymetry, water has not been represented in the narrow westernmost belt. The flow-chart shows how environmental features control trophic strategies and ichnoassociations. Position of georeferenced photos is indicated by placeholders (number refers to figure number). The icons present ichnoassociations and their typical ichnotaxa: low-bioturbation ichnoassociation (*Skolithos* XS), *Parmaichnus* ichnoassociation (*Parmaichnus*), *Macanopsis* ichnoassociation (from left to right: *Helminthoidichnites, Macanopsis, Polykladichnus*), *Thalassinoides* ichnoassociation (A – *Thalassinoides*-dominated subunit, from left to right: *Thalassinoides, Parmaichnus Arenicolites;* B – mottling-dominated subunit, mottling), *Arenicolites/Skolithos* ichnoassociation (C – moderate ichnodiversity subunit, from left to right: *Thalassinoides* L, *Arenicolites, Skolithos* M; D – low ichnodiversity subunit, from left to right: *Arenicolites* XL, *Nereites, Skolithos* M, *Arenicolites* L; E – low ichnodiversity subunit, *Nereites*). Burrow morphology based on field observations and literature (Dworschak, 1987, 2002; Atkinson & Froglia, 2000).

widely regarded as amongst the major features of barrier island systems, which are typically characterized by sediment barriers separating a shallow lagoon from the open sea (Otvos, 2012; see also Baucon & Felletti, 2013a). However, being periodically submersed, the Mula di Muggia sandbank does not strictly fit the definition of a barrier island (Otvos, 2012) but it does represent an intertidal bar. Rather, it may be in the early process of evolving into a barrier island system, as exemplified by the supratidal ridge which constitutes its western prolongation ('barrier island' in Fig. 1). It should also be noted that, in the context of the test of the Grado model, both the predicted and the observed scenarios denote a marked hydrodynamic contrast between the fore-bar and the back-bar regions.

Hydrodynamics controls sedimentation and, consequently, influences nutrient availability. In conformity with the 'food resource paradigm' (Pemberton *et al.*, 2001), sheltered, low-energy areas favour the deposition of particulate organic matter, which tends to remain suspended in more turbulent areas. For this reason, deposit-feeding strategies (e.g. *Thalassinoides*) dominate the back-bar, whilst suspension-feeding burrows (e.g. *Skolithos* M) are commonly found in the fore-bar and in the more energetic sectors of the back-bar.

These interpretations are confirmed by georeferenced photographic documentation (Fig. 11), showing the marked ichnological and sedimentological dichotomy of the studied area.

In line with the intertidal model area, hydrodynamics exerts a significant control on interstitial oxygenation as well (Fig. 10). In fact, sediment permeability controls the depth of the redox potential discontinuity (RPD); that is, the horizon where oxidizing processes become replaced by reducing processes (Fenchel & Riedl, 1970). Since permeability increases with grain size, which is ultimately controlled by hydrodynamics, the RPD is located at greater depth in coarser sediments (Raffaelli & Hawkins, 1996). Consequently, the RPD is found at greater depth in the turbulent fore-bar and a shallower depth in the sheltered back-bar (Fig. 12).

The highest bioturbation intensities are found in the back-bar, in spite of the fact that the muddy sands display widespread dysoxic conditions. This is explained by the greater abundance of organic matter which has deposited in the back-bar as a consequence of its quieter hydrodynamic regime (see Baucon & Felletti, 2013a). For this reason, lagoonal animals commonly develop adaptations for survival in low-oxygen and sulphide-rich environments (Fenchel & Riedl, 1970; Vismann, 1990 and references therein). Marine seagrass are favoured by low-energy conditions and, at the same time, they attenuate current speed and wave energy (Koch, 2001). Wave action, in turn, represents a source of mechanical stress in the fore-bar, which is well oxygenated but relatively nutrient-depleted (Fig. 12C). Georeferenced documentation shows clearly that tracemaking organisms are more abundant in the back-bar (Figs 11, 13 and 14), whereas the lower bioturbation intensity of the fore-bar is explained by inherently low density of tracemaking organisms.

This interpretation is in line with previous studies, as muddy sands and sandy muds are known to support higher population densities than either sandier or muddier environments because of inherently higher concentrations of organic matter in the sediment (Flemming & Delafontaine, 2000; Flemming, 2012).

These observations show that the fore-bar and the back-bar are characterized by a marked environmental and ichnological dichotomy (Fig. 9), which is also recognized for the back-barrier and fore-barrier regions of the model area (Fig. 15; Baucon & Felletti, 2013 a, b). This aspect is particularly manifest in the spatial distribution of *Arenicolites,* as revealed by the geostatistical module of the IchnoGIS method, which mapped the density of large-sized *Arenicolites* (types L and XL; Fig. 15A, B) from quadrat samples. The interpolation of such data resulted in a density map of large *Arenicolites* (Fig. 15A), showing its preferential distribution in the fore-barrier. Symmetrically, Baucon & Felletti (2013b) demonstrated that *Thalassinoides* is preferentially distributed in the back-barrier. The same trends are recognized in the test area, as shown by the contrasting ichnoassociations of the fore-bar and back-bar (Fig. 9). A similar scenario is portrayed by the ichnonetwork of the model area (Fig. 15C), representing ichnotaxa (nodes) and their association relationships (edges; Baucon & Felletti, 2013b). According to the ichnonetwork approach, a pair of traces is connected if they co-occur at least in one trench sample, whereas the strength of the connection (represented as edge thickness) depends on the number of samples in which two ichnotaxa co-occur (Fig. 15C). Consequently, the existence of different structural areas – manifested as sets of densely connected nodes – reveals

(A)

(B)

Fig. 11. Ichnological and sedimentological dichotomy between the fore-bar and the back-bar. Picture locations are given in Fig. 10; for difficult underwater conditions, scale bar dimensions are approximate. (A) – Dense seagrass meadows characterize the back-bar area. (B) – Rippled sand characterizes the fore-bar.

Fig. 12. Redox potential discontinuity (RPD). (A) Redox potential discontinuity (RPD, dashed line) in the rippled sands of the fore-bar. Profile view; note various but thin vertical burrows. (B) Redox potential discontinuity (RPD, dashed line) in the muddy sands of the back-bar area. Profile view; note a small U-shaped burrow (U: *Arenicolites*) and a decapod burrow (D: *Parmaichnus* or *Thalassinoides*). Profile view.

inhomogeneities in the association patterns, suggesting a certain degree of organization of the ichnological system. In the specific case of the ichnonetwork of the model area, major inhomogeneities mirror the fact that the fore-barrier and the back-barrier are characterized by specific ichnoassociations (Fig. 15C; Baucon & Felletti, 2013b). The ichnonetwork approach has recently been applied to the palaeoenvironmental reconstruction of Paleozoic and Mesozoic ecosystems (Baucon *et al.*, 2014; Baucon *et al.*, 2015).

The structuring role of hydrodynamics is demonstrated by georeferenced photographs and qualitative observations realized at the Mula di Muggia sandbank (Figs 10 to 14). In fact, within fore-bar settings, water turbulence decreases with decreasing distance from the intertidal Mula di Muggia sandbank. This aspect is explained by the sheltering function of sand bars, as energy of entering waves is progressively dissipated over them. This energy gradient is accompanied by a bioturbation gradient: within the fore-bar, higher bioturbation indexes (BPBI 2; II 2) are found on the intertidal sandbank and its immediate surroundings, where 'diverging shafts', *Arenicolites* L, *Arenicolites* XL, *Skolithos* M, L, *Nereites* are distributed sparsely. In these settings, hydrodynamic energy may be used by deposit-feeding burrowers for ventilating their

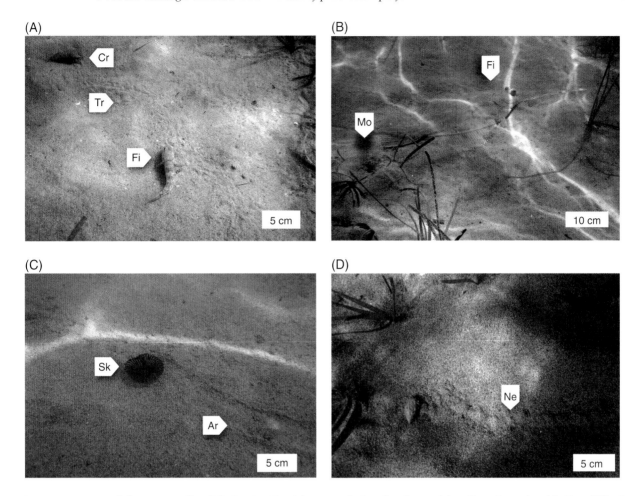

Fig. 13. Features of the *Arenicolites/Thalassinoides* ichnoassociation (moderate ichnodiversity subunit). For difficult underwater conditions, scale bar dimensions are approximate. (A) – Shallower tiers are reworked by epifauna, such as the figured gobiid fish (Fi) and hermit crab (Cr). Note the track (Tr) behind the hermit crab. (B) – Occasionally, vertical burrows with mounds (Mo) are associated to biodeformational structures produced by epifauna, such as the figured flatfish (*?Platichthys flesus,* Fl). Note also seagrass, which is responsible for root-related mottling. (C) – A sabellarid worm protrudes from its burrow, corresponding to the ichnogenus *Skolithos* M (Sk). Note also the *Archaeonassa* trail (Ar). (D) – Hermit crab with its trail (*Nereites,* Ne) behind.

burrows, in order to exploit nutritious layers at greater depth (Fig. 14A, B). Bioturbation intensity and ichnodiversity rapidly decrease towards the open sea, where *Nereites* and biodeformational structures are largely prevailing over vertical burrows, reworking only the uppermost layers of the sediment (BPBI 1–2; II 1; Fig. 14C, D). Bioturbation intensity seems to increase at greater depths (≈ 5 m), as wave influence on the sea floor is weaker.

Wave action is reduced notably within back-bar environments, which are sheltered from waves entering by subtidal and intertidal bars.

The aforementioned aspects support the validity of the Grado model but a problem of resolution arises. In fact, the Grado model predicts the back-bar/fore-bar dichotomy but the fore-bar gradient is not depicted coherently. In addition, the effect of wave turbulence and tidal currents on ichnofauna needs to be investigated better, whereas trails (i.e. *Nereites, Archaeonassa*) have to be implemented in the model, as their distribution appears to be controlled environmentally. Consequently, the general validity of the Grado model is demonstrated but further application of the IchnoGIS method in subtidal settings is needed to achieve greater accuracy and resolution.

Although the aforementioned hydrodynamic dichotomy is depicted conveniently by the predicted scenario, minor discrepancies exist with

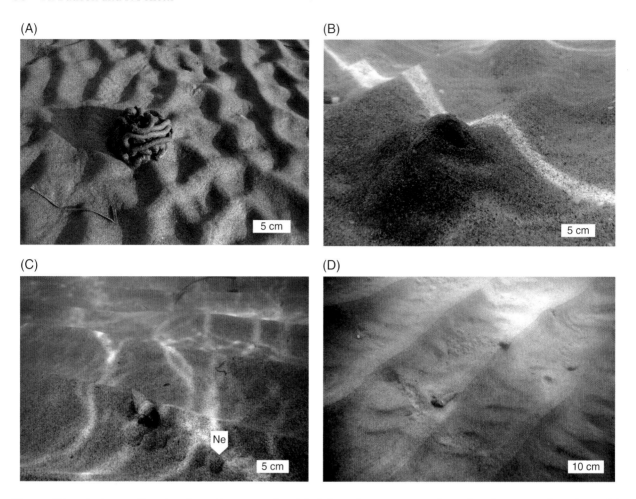

Fig. 14. Bioturbation gradient involving the *Arenicolites/Skolithos* ichnoassociation (low ichnodiversity subunit). For difficult underwater conditions, scale bar dimensions are approximate in B, C, D. (A) – Faecal casts, revealing the presence of *Arenicolites* XL, are common on the Mula di Muggia sandbank. (B) – Faecal casts, here partially washed-out by waves, are found in the more proximal sectors of the fore-bar. (C) – *Nereites* (Ne), produced by hermit crabs, is a dominant ichnotaxon in distal fore-bar settings. (D) – Trails dominate over vertical burrows in the distal fore-bar.

the observed conditions (e.g. Fig. 9, arrow). In fact, small differing patches are present, which can represent (a) stochastic fluctuations in ichnoassociation distribution (from which the predicted map is derived) or (b) areas in which physical indicators (Table 1) are less precise. However, neither case seriously challenges the general validity of the predicted map.

Alongside hydrodynamics, substrate firmness is a major driver of ichnoassociation composition. In this regard, firm substrates are characteristically of interest to the *Parmaichnus* ichnoassociation, which is represented by nearly monoichnospecific assemblages of Y-shaped burrows (Fig. 7C). The absence of other ichnotaxa is explained by the mechanical requirements necessary to burrow into partially consolidated

substrates; firm substrates are also difficult to exploit for deposit-feeders, for which reason firmgrounds are dominated by suspension-feeding strategies (i.e. *Parmaichnus* itself).

Nevertheless, a major discrepancy concerns the distribution of firm substrates, as derived from the *Parmaichnus* ichnoassociation. In fact, the affinity of the *Parmaichnus* ichnoassociation with softgrounds appears to question its utility as a firmground indicator. However, in the present case, *Parmaichnus*-bearing softgrounds are always close to firmgrounds (Figs 8 and 9B), supporting a proximity relationship with cohesive substrates and/or current-swept erosive settings. This assumption is supported by numerous reports of *Parmaichnus* in fossil and modern firmgrounds (i.e. Asgaard *et al.*, 1997; Pervesler & Hohenegger, 2006).

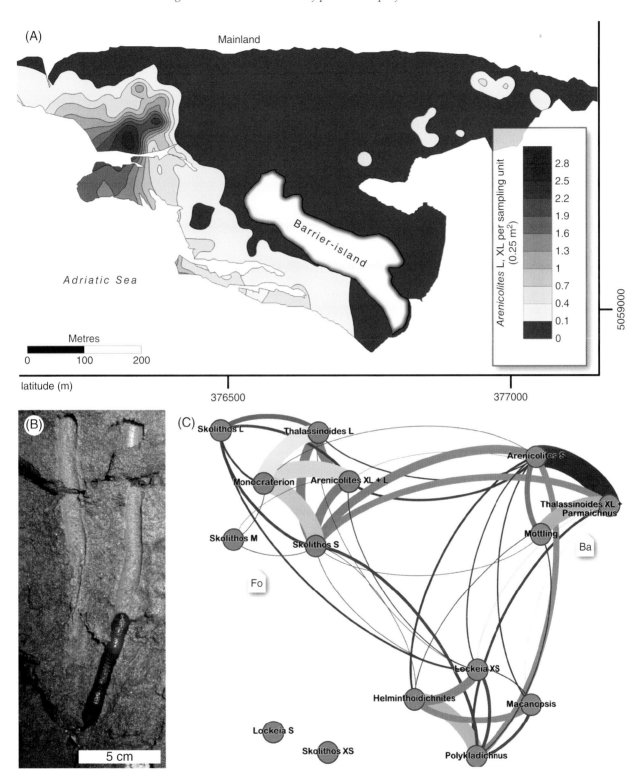

Fig. 15. Data from the model area. (A) Density of *Arenicolites* (large forms: XL and L) in the intertidal model area. (B) A sipunculan worm inside its U-shaped burrow (*Arenicolites*). Profile view; sediment-water interface not pictured. (C) The fore-bar/back-bar dichotomy is mirrored in the Grado ichnonetwork, which displays a set of densely connected ichnotaxa (Ba, *Thalassinoides* ichnoassociation: *Arenicolites* S, mottling, *Thalassinoides* XL and *Parmaichnus*) corresponding to back-barrier conditions. In symmetric view it is possible to distinguish fore-barrier ichnotaxa (Fo, *Arenicolites/Skolithos* ichnoassociation: *Skolithos* S, M, L, *Monocraterion, Thalassinoides* L, *Arenicolites* XL, L). Edge thickness and colour measure the degree of association between nodes representing ichnotaxa. See text for more detailed discussion.

Besides hydrodynamics and firmness, microbial binding and emersion time are the major control factors of ichnoassociation distribution (see Baucon & Felletti, 2013b). In the light of the bathymetric characteristics of the test area (Fig. 1), which is largely subtidal, the effect of emersion time is less evident than in the intertidal model area. Nevertheless, the widespread presence of large suspension-feeding structures (i.e. *Skolithos* M) reflects the short or zero emersion times characterizing the test area.

In fact, the feeding time of suspension feeders increases with the increasing submergence time because a water column is needed to support nutritious particle suspensions.

Despite their abundance in the intertidal model area, microbial mats do not occur in the test area. Biomat-related ichnoassociations are therefore absent, which is an aspect to be considered when extending the Grado model to the subtidal realm.

The studied site is characterized by a notable abundance and diversity of biogenic sedimentary structures, but to what degree will its ichnological heritage be preserved in the rock record? The answer is difficult due to the complexity of the ichnological system, concomitant with the diversity of the physical phenomena influencing the sedimentary dynamics at the Mula di Muggia. In this context, several tiering levels are present, with the deeper ones having a higher preservation potential (Bromley, 1996; Savrda, 2007). For this reason, large specimens of *Thalassinoides* and *Arenicolites* have good preservation potential. Furthermore, burrow overprinting may obliterate previously emplaced structures (Bromley, 1996; Savrda, 2007); therefore, lower tiers of high-bioturbation ichnoassociations (i.e. *Thalassinoides* ichnoassociation) will probably have been completely reworked. This scenario is complicated by the interplay of sedimentary dynamics and high-energy events, as evidenced by previous research on modern and fossil systems (van Straaten, 1954; Reineck & Singh, 1980; Dott, 1983, 1988; van der Spek, 1996; Bromley, 1996; Chang *et al.*, 2006; Davis & Flemming, 1995; Flemming, 2012). In this regard, lined burrows, such as from large specimens of *Skolithos* (Fig. 6) may resist the mechanical stress induced by storm events. In conclusion, the taphonomic future of the ichnological heritage of the Mula di Muggia is difficult to assess with confidence, although large specimens of *Thalassinoides, Arenicolites* and *Skolithos* probably became fossilized.

CONCLUSIONS

This study has identified the spatial distribution of ichnoassociations as a function of environmental factors, thus confirming concordance with the Grado model. The prediction performance of the model is excellent, opening new avenues for its extension to the subtidal realm. The results confirm that hydrodynamics, sediment firmness, microbial binding and emersion time are the major control factors of the tidal ichnoassociations in the wider study area. Studies on biological zonation of other tidal systems (Dahl, 1952; Dittmann, 2000; Jaramillo *et al.*, 1993; Raffaelli & Hawkins, 1996; Salvat, 1964) support this idea and suggest a global adaptability of the model, although further studies are required to confirm corresponding trends at a global scale, in particular with regard to the *Parmaichnus* ichnoassociation.

The excellent results of the performance test confirm the reliability of the IchnoGIS method, which has been able to describe and analyse the studied ichnological system. Nevertheless, finer patterns of ichnological variation are found within the fore-bar, therefore further application of the IchnoGIS method is required to achieve optimal accuracy and resolution of prediction.

Finally, this study encourages future application of the IchnoGIS method on other tidal systems. It has demonstrated that knowledge of the mechanisms regulating the distribution of biogenic traces is important for understanding present and past environments.

ACKNOWLEDGMENTS

We thank the participants of Tidalites 2012 for providing helpful discussions on traces. Burghard Flemming (Senckenberg Research Institute) and an anonymous reviewer are thanked for improving the manuscript. Peter Dworschak is gratefully acknowledged for his help on decapod burrows. We thank Murray Gingras for his precious help. Orrin Pilkey and Andrew Cooper are acknowledged for help on fetch-limited barriers. Ezio Fonda is acknowledged for historical information on the Pineta area. Mark Aidan Shields is thanked for discussions on sipunculans.

REFERENCES

Asgaard, U., **Bromley**, R.G. and **Hanken**, N.-M. (1997) Recent firmground burrows produced by a upogebiid crustacean; paleontological implications. *Courier Forschungsinstitut Senckenberg*, **2001**, 23–28.

Atkinson, R. and **Froglia**, C. (2000) Burrow structures and eco-ethology of burrowing fauna in the Adriatic sea. Impact of trawl fishing on benthic communities. *Proceedings ICRAM 2000*, Rome, 79–94.

Baucon A., **Bordy E.**, **Brustur T.**, **Buatois L.**, **Cunningham T.**, **De C.**, **Duffin C.**, **Felletti F.**, **Gaillard C.**, **Hu B.**, **Hu L.**, **Jensen S.**, **Knaust D.**, **Lockley M.**, **Lowe P.**, **Mayor A.**, **Mayoral E.**, **Mikulas R.**, **Muttoni G.**, **Neto de Carvalho C.**, **Pemberton S.**, **Pollard J.**, **Rindsberg A.**, **Santos A.**, **Seike K.**, **Song H.**, **Turner S.**, **Uchman A.**, **Wang Y.**, **Yi-ming G.**, **Zhang L.**, **Zhang W.** (2012) A history of ideas in ichnology. In: Trace Fossils as Indicators of Sedimentary Environments (Eds **D. Bromley** and **D. Knaust**). *Dev. Sedimentol.*, **64**, 3–43.

Baucon, A. and **Felletti**, F. (2012) A quantitative tool for the ichnological analysis of tidal environments: the IchnoGIS method. *Tidalites 2012, 8th Conference on tidal environments,* pp. 1–2, Caen.

Baucon, A. and **Felletti**, F. (2013a) Neoichnology of a barrier-island system: the Mula di Muggia (Grado lagoon, Italy). *Palaeogeogr. Palaeoclimatol. Palaeoecol.*, **375**, 112–124.

Baucon, A. and **Felletti**, F. (2013b) The IchnoGIS method: network science and geostatistics in ichnology. Theory and application (Grado lagoon, Italy). *Palaeogeogr. Palaeoclimatol. Palaeoecol.*, **375**, 83–111.

Baucon, A., **Ronchi**, A., **Felletti**, F. and **Neto de Carvalho**, C. (2014) Evolution of Crustaceans at the edge of the end-Permian crisis: ichnonetwork analysis of the fluvial succession of Nurra (Permian-Triassic, Sardinia, Italy): *Palaeogeogr., Palaeoclimatol., Palaeoecol.*, **410**, 74–103.

Baucon, A., **Venturini**, C., **Neto de Carvalho**, C. and **Felletti**, F. (2015) Behaviours mapped by new geographies: ichnonetwork analysis of the Val Dolce Formation (lower Permian; Italy-Austria): *Geosphere*, **11**, 744–776.

Brandes, U. and **Erlebach**, T. (Eds) (2005) *Network Analysis: Methodological Foundations.* Berlin: Springer, p. 472.

Bromley, R.G. (1996) *Trace fossils: biology, taphonomy and applications* (Second Edition). London: Chapman and Hall, p. 361.

Chang, T.S., **Flemming**, B.W., **Tilch**, E., **Bartholomä**, A. and **Wöstmann**, R. (2006) Late Holocene stratigraphic evolution of a back-barrier tidal basin in the East Frisian Wadden Sea, southern North Sea: transgressive deposition and its preservation potential. *Facies*, **52**, 329–340.

Covelli, S., **Faganeli**, J., **Devittor**, C., **Predonzani**, S., **Acquavita**, A and **Horvat**, M. (2008) Benthic fluxes of mercury species in a lagoon environment (Grado Lagoon, Northern Adriatic Sea, Italy). *Appl. Geochem.*, **23**(3), 529–546. doi:10.1016/j.apgeochem.2007.12.011

Dahl, E. (1952) Some aspects of the ecology and zonation of the fauna on sandy beaches. *Oikos*, 1–27.

Davis, R.A. Jr. and **Flemming**, B.W. (1995) Stratigraphy of a combined wave- and tide-dominated intertidal sand body: Martens Plate, East Frisian Wadden Sea, Germany. *Int. Assoc. Sedimentol. Spec. Publ.*, **24**, 121–132.

Dittmann, S. (2000) Zonation of benthic communities in a tropical tidal flat of north-east Australia. *J. Sea Res.*, **43**(1), 33–51. doi:10.1016/S1385-1101(00)00004-6

Dorigo, L. (1965) La Laguna di Grado e le sue foci e Ricerche e rilievi idrografici, **155**, Magistrato delle Acque - Ufficio Idrografico, 231 pp.

Dott, R.H. Jr. (1988) An episodic view of shallow marine clastic sedimentation. In: *Tide-influenced sedimentary environments and facies* (Eds **P.L. de Boer**, **A. van Gelder** and **S.D. Nio**). D Reidel Publ Co, Dordrecht, pp 3–12.

Dott, R.H. Jr. (1983) Episodic sedimentation – how normal is average? How rare is rare? Does it matter? *J. Sed. Petrol.*, **53**, 5–23.

Droser, M.L. and **Bottjer**, D.J. (1986) A semiquantitative field classification of ichnofabric. *J. Sed. Petrol.*, **56**, 588–559.

Dworschak, P.C. (1987) Feeding behaviour of *Upogebia pusilla* and *Callianassa tyrrhena* (Crustacea, Decapoda, Thalassinidea). *Investigación Pesquera*, **51**, 421–429.

Dworschak, P.C. (2002) The burrows of *Callianassa candida* (olivi 1972) and *C. whitei* Sakai 1999 (Crustacea: Decapoda: Thalassinidea). In: *The Vienna School of Marine Biology: A Tribute to Jörg Ott* (Eds **M. Bright**, **P.C. Dworschak** and **M. Stachowitsch**), Facultas Universitätsverlag, Wien, 63–71.

Fenchel, T.M. and **Riedl**, R.J. (1970) The sulfide system: a new biotic community underneath the oxidized layer of marine sand bottoms. *Mar. Biol.*, **7**, 255–268.

Flemming, B.W. and **Delafontaine**, M.T. (2000) Mass physical properties of muddy intertidal sediments: some applications, misapplications and non-applications. *Cont. Shelf Res.*, **20**, 1179–1197.

Flemming, B.W. (2012) Siliclastic back-barrier tidal flats. In: *Principles of Tidal Sedimentology*, (Eds **R.A. Davis** Jr. and **R.W. Dalrymple**), Springer, Dordrecht, pp. 231–267.

Fontolan G., **Pillon S.**, **Bezzi A.**, **Villalta R.**, **Lipizer M.**, **Triches A.** and **D'Aietti A.** (2012) Human impact and the historical transformation of saltmarshes in the Marano and Grado Lagoon, northern Adriatic Sea. *Estuar. Coast. Shelf Sci.*, **113**, 41–56.

Fontolan, G., **Pillon**, S., **Quadri**, F.D. and **Bezzi**, A. (2007) Sediment storage at tidal inlets in northern Adriatic lagoons: Ebb-tidal delta morphodynamics, conservation and sand use strategies. *Estuar. Coast. Shelf Sci.*, **75**, 261–277. doi:10.1016/j.ecss.2007.02.029

Fortunato, S. (2010) Community detection in graphs. *Physics Reports*, **486**, 75–174.

Gingras, M.K. and **MacEachern**, J.A. (2012) Tidal Ichnology of Shallow-Water Clastic Settings. In: *Principles of Tidal Sedimentology* (Eds **R.A. Davis** Jr. and **R.W. Darlymple**), Springer, pp. 57–77.

Gingras, M.K., **MacEachern**, J.A. and **Dashtgard**, S.E. (2012) The potential of trace fossils as tidal indicators in bays and estuaries. *Sed. Geol.*, **279**, doi:10.1016/j.sedgeo.2011.05.007

Jaramillo, E., **McLachlan**, A. and **Coetzee**, P. (1993) Intertidal zonation patterns of macroinfauna over a range of exposed sandy beaches in south-central Chile. *Mar. Ecol. Prog. Ser.*, **101**, 105–118.

Koch, E.W. (2001) Beyond Light: Physical, Geological and Geochemical Parameters as Possible Submersed Aquatic Vegetation Habitat Requirements. *Estuaries*, **24**, 1–17.

Matheron, G. (1962) Traité de géostatistique appliquée. *Bur. Rech. Géol. Min. Mem.*, p. 333.

Miller, M. and **Smail, S.E.** (1997) A semiquantitative field method for evaluating bioturbation on bedding planes. *Palaios*, **12**, 391–396.

Nichols, G. (2009) *Sedimentology and stratigraphy.* Chichester: Wiley-Blackwell, p. 419.

Otvos, E.G. (2012) Coastal barriers — Nomenclature, processes and classification issues. *Geomorphology*, **139-140**, 39–52. doi:10.1016/j.geomorph.2011.10.037

Pemberton, S.G., Spila, M., Pulham, A.J., Saunders, T., MacEachern, J.A., Robbins, D. and **Sinclair, I.K.** (2001) Ichnology & Sedimentology of Shallow to Marginal Marine Systems. *Geol. Assoc. Canada, Short Course Notes*, **15**, 343, AGMV Marquis, St. John's.

Pervesler, P. and **Hohenegger, J.** (2006) Orientation of crustacean burrows in the Bay of Panzano (Gulf of Trieste, Northern Adriatic Sea). *Lethaia*, **39**(2), 173–186. doi:10.1080/00241160600715297

Pilkey, H.O., Cooper, J.A.G. and **Lewis, D.A.** (2009) Global Distribution and Geomorphology of Fetch-Limited Barrier Islands. *J. Coastal Res.*, **25**(4).

Raffaelli, D.G. and **Hawkins, S.J.** (1996) *Intertidal ecology.* Dordecht: Kluwer, p. 356.

Reineck, H.-E. and **Singh, I.B.** (1980) *Depositional sedimentary environments.* 2ⁿᵈ edition. Springer-Verlag, Berlin.

Salvat, B. (1964) Les conditions hydrodynamiques interstitielles des sediments meubles intertidaux et la repartition verticale de la faune endogene. *CR Acad. Sci. Paris*, **259**, 1576–1579.

Savrda, C.E. (2007) Taphonomy of trace fossils: In: *Trace Fossils: Concepts, Problems, Prospects* (Ed. **W. Miller** III). Elsevier, p. 92–109.

Sconfietti, R., Marchini, A., Ambrogi, A.O. and **Sacchi, C.F.** (2003) The sessile benthic community patterns on hard bottoms in response to continental vs. marine influence in northern Adriatic lagoons. *Oceanol. Acta*, **26**, 47–56.

Seilacher, A. (2007) *Trace fossil analysis.* Berlin, Heidelberg: Springer, p. 226.

Stow, D.A., Hernández-Molina F.J., Llave, E., Sygo-Gil M., del **Río, V.D.** and **Branson, A.** (2008) Bedform-velocity matrix: The estimation of bottom current velocity from bedform observations. *Geology*, **37**(4), 327–330.

van der Spek, A.J.F. (1996) Holocene depositional sequences in the Dutch Wadden Sea south of Ameland. In: Coastal studies on the Holocene of the Netherlands (Eds **D.J. Beets, M.M. Fischer** and **W. de Gans**). *Mededel. Rijks Geol. Dienst*, **57**, 41–69.

van **Straaten, L.M.J.U.** (1954) Composition and structure of recent marine sediments in the Netherlands. *Leidse Geol. Mededel.*, **19**, 1–110.

Venturini, C. (2003) Il Friuli nel Quaternario: l'evoluzione del territorio. In: *Glacies* (Ed. **G. Muscio**), Udine: Comune di Udine/Museo Friulano di Storia Naturale, pp. 23–106.

Vismann, B. (1990) Sulfide detoxification and tolerance in *Nereis* (*Hediste*) *diversicolor* and *Nereis* (*Neanthes*) *virens* (Annelida: Polychaeta). *Mar. Ecol. Prog. Ser.*, **59**, 229–238. doi:10.3354/meps059229

Suspended sediment dynamics induced by the passage of a tidal bore in an upper estuary

LUCILLE FURGEROT[†*], PIERRE WEILL[†], DOMINIQUE MOUAZÉ[†] and BERNADETTE TESSIER[†]

[†] CNRS UMR 6143 M2C, University of Caen Normandie, 24 rue des Tilleuls, 14000, Caen, France
[*] Corresponding author: lucille.furgerot@unicaen.fr

ABSTRACT

Tidal bores are complex but popular phenomena that occur in estuarine rivers during rising tides. Until recently, few studies were devoted to the understanding of this process, although it should have a significant influence on sediment erosion and deposition in estuarine systems. Herein, we propose to reconstruct the evolution of sediment concentration close to the channel bed and in the water column during the passage of several tidal bores. Field data were collected in the Sée River channel, approximately 15 km upstream of the Mont-Saint-Michel outer estuary (NW France). An Acoustic Doppler Velocimeter (ADV) was used to measure longitudinal and vertical flow velocities. Suspended Sediment Concentration (SSC) evolution was monitored using an Argus Surface Meter (144 OBS sensors) as well as direct water sampling at different elevations above the channel bed. The highest sediment concentration was recorded at the channel bed a few seconds after the passage of the bore front. Values reached up to $53.5\,g\,L^{-1}$. Then, in the few minutes after the bore's passage, an upward advection of suspended sediment was observed which resulted in the homogenization of the water column concentration to approximately 10 to $15\,g\,L^{-1}$. These results demonstrate the high potential of this tidal process to induce very large upstream-directed sediment fluxes in upper estuaries.

Keywords: Tidal bore, Suspended sediment concentration, Optical measurements, Vertical concentration profile, Mont Saint Michel Bay.

INTRODUCTION

Sedimentary processes in estuaries have for a long time been the matter of numerous studies, dealing especially with fine-grained sediment transport (e.g. Dyer, 1995; Lesourd et al., 2003), turbidity maximum evolution in time and space (e.g. Avoine et al., 1981; Castaing, 1981; Deloffre et al., 2004), tidal current patterns and sediment balance (Li & Zhang, 1998; van der Spek, 1997). These studies are critical since many estuaries experience rapid infilling, compromising their high biological and socio-economical interests (Allen, 2000; Green & MacDonald, 2001; Tessier et al., 2012). In this general frame of estuarine research, tidal bores, in comparison with other common estuarine processes, have been poorly studied until recent years. A tidal bore is a hydraulic jump that propagates upstream in estuarine rivers as the tide is rising. Depending mostly on tidal ranges, fluvial conditions and channel morphologies, tidal bores display variable amplitudes and shapes but in all cases they are highly sheared flows associated with high celerity (commonly 3 to $4\,m\,s^{-1}$). Thus, tidal bores are assumed to provoke heavy sediment reworking along channel bed and banks as they propagate (Wolanski et al., 2004). Although measurements of hydrodynamic parameters and induced sediment transport are commonly performed in estuarine rivers, measurements carried out during the passage of a tidal bore are much less frequent in the literature.

However, during the last few years several field studies have been purposely conducted with the aim of better characterizing fluid and sediment motion involved during tidal bore propagation. Wolanski *et al.* (2004) monitored a tidal bore in the Daly River, Northern Australia, and found that the sediment concentration was not directly correlated with the current speed. They suggested that patches of high turbidity waters might be advected from a fluidified bed, although they did not observe vertical velocity close the the channel bottom. Simpson *et al.* (2004) investigated the production, fluxes and dissipation of turbulent kinetic energy in the tidal bore of the Dee estuary (North Wales). Uncles *et al.* (2006) reported extensive sediment concentration measurements in the Humber estuary but did not found any evidence of sediment resuspension by the small-size tidal bore. Chanson *et al.* (2011), Furgerot *et al.* (2012a) and Fan *et al.* (2012) investigated flow velocity and sediment concentration evolution associated to tidal bores, respectively in the Garonne and Sée rivers (France) and in the Qiantang River (China). However, contradictory conclusions on how tidal bores impact sediment transport arose from these papers, mainly because the resuspension phenomenon is difficult to quantify. Thus, it appears that there is a need to strengthen our knowledge on the processes of sediment resuspension on the channel bed in relation to the tidal bore hydrodynamics by performing mesurements at higher frequency. In the scope of this analysis, ten field campaigns were conducted in the Mont-Saint-Michel Bay between January 2011 and May 2012, which resulted in the monitoring of 35 tidal bore events. The objective of these campaigns was to collect a large set of hydro-sedimentary parameters to describe the physical processes associated to these tidal bores' passage. The results presented in this paper are principally extracted from one of these campaigns (7th May 2012) that was especially dedicated to the measurement of the SSC from the channel bed to the free surface in relation with the passage of a tidal bore in an estuarine river. These SSC measurements were performed using indirect (optical sensors) and direct (pumps) means, highlighting the specificities of SSC dynamics during a rising tide accompanied by a tidal bore.

ENVIRONMENTAL SETTING

The Mont Saint-Michel estuary

The measurement site is located in the Mont-Saint-Michel Bay (English Channel, NW France), which forms a $500 \, km^2$ depression developed in the south of the Normandy-Brittany Gulf (Fig. 1). In this area of the English Channel, the reflection along the Cotentin peninsula of the incoming tidal wave propagating from the Atlantic Ocean induces an amplified standing wave (Larsonneur, 1989). As a result, a tidal range up to 14 m is recorded in the Mont-Saint-Michel Bay, which can therefore be classified among the few type C hypertidal coastal systems around the world (Archer, 2013). The tidal regime is semi-diurnal with an insignificant diurnal inequality.

According to its general morphology with respect to tidal, wave and fluvial dynamics, the Mont-Saint-Michel Bay is commonly divided into three morphosedimentary environments (Billeaud *et al.*, 2007): 1) an embayment in the west, characterized by extensive tidal flats; 2) a sandy to muddy estuarine system in the eastern corner of the Bay, at the entrance of three rivers (The Sée, Sélune and Couesnon rivers) (Fig. 1); and 3) an elongated coastal barrier composed of sandy beaches and aeolian dunes, in the north-east, at the transition between the estuarine mouth and the open marine entrance.

In relation with the hypertidal range of the bay, high-energy alternative tidal currents penetrate into the estuary where they control sedimentary processes and channel migration. In agreement with the model of Dalrymple *et al.* (1992), this tide-dominated estuary comprises seaward a high-energy braided system made of a complex channel-and-shoal network, followed upstream by a straight-meandering-straight transitional fluvio-tidal channel in connection with each river (Lanier & Tessier, 1998). The measurement site is precisely located in the upper estuary, on the right bank of the transitional channel connected to the Sée River (cf. Fig. 1 for location). Tidal range is maximum at the seaward entrance of the braided domain and reaches 14 m. It then decreases as the tide propagates upstream, so that the tidal range is no more than 1.5 m at the measurement site (Fig. 2).

Despite this microtidal range, tidal current velocities commonly reach up to $2 \, m \, s^{-1}$ during high spring tide periods. In addition, the Sée River, like the Sélune and Couesnon rivers, is a

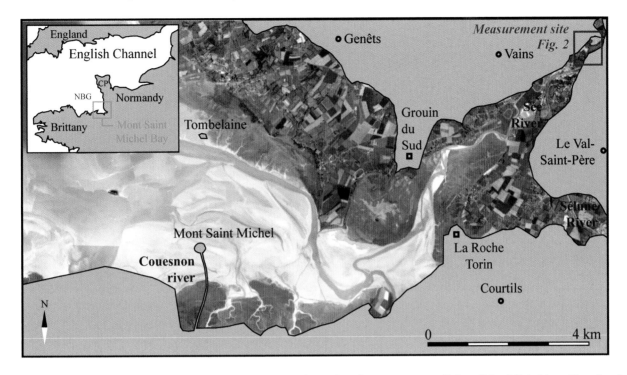

Fig. 1. Satellite image (Spot 2007) of the estuarine system formed in the eastern part of Mont-Saint-Michel bay. The site of tidal bore measurements ('Le Bateau') is located upstream on the Sée River (red square, cf. Fig. 2). CP: Cotentin Peninsula; NBG: Normandy Brittany Gulf.

minor river with very low water discharge rates. The annual water discharge is less than $10\,m^3\,s^{-1}$ and never exceeds a few tens of cubic metres per second during the wet season (Bonnot-Courtois *et al.*, 2002). These average tidal and fluvial conditions, combined with the general morphology of the Sée River channel, are favourable to tidal bore development.

The Sée River channel morphology at the measurement site

At the measurement site 'Le Bateau', located some 15 km upstream of the estuary mouth (i.e. Tombelaine, Fig. 1), the Sée River features a 570 m-long straight stretch between elbow meanders, well channelized in the salt marsh (Fig. 2A). Bathymetric profiles were performed along and across the channel using a tacheometer, with a measurement every metre. The channel width, fairly constant along the stretch, is around 30 m. The channel width did not change during the survey. The channel cross-section (Fig. 2C) reveals steep banks (~30°) and a flat 20 m-wide channel bed. At some field campaigns the Sée River was empty of sediment and the bedrock was outcropping on the channel bed.

In these conditions, from the top of the banks, the channel maximum depth was around 2.5 m (Fig. 2C). This value decreased when sediment deposition occured on the channel bed. The longitudinal channel slope was 0.16% on average, with relatively smooth bed morphology.

Water depth in the channel depends on the river discharge and tidal conditions. It can be shallow at low slack water (less than 1 m) and can increase to bankfull conditions at high tide and high fluvial discharge. Both undular (Fig. 2B) and breaking tidal bores have been observed on this section of the Sée channel during different field campaigns. Generally, the tidal bore was undular over the whole rectilinear section, although sometimes it evolved progressively toward a breaking bore at the end of the section.

Sediment characteristics

The characterization of the sediment at the measurement site is critical since the objective herein is to define the role of tidal bore dynamics in sediment reworking.

In addition to very low water discharge, the solid charge of the three rivers entering the bay is

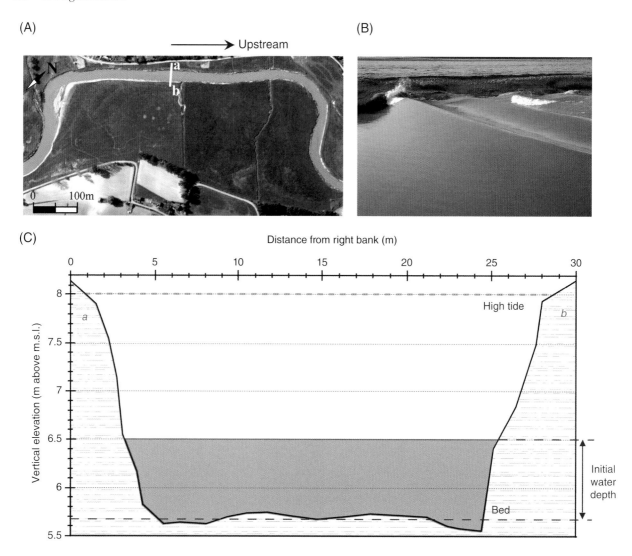

Fig. 2. The measurement site 'Le Bateau' (cf. Fig. 1 for location). A: Aerial photograph of the surveyed stretch of the Sée River; B: Typical undular tidal bore that commonly develops at 'Le Bateau'; C: Transverse bathymetric profile (a-b) of the Sée River channel at the measurement point (profile located on Fig. 2A) (m.s.l.: mean sea-level).

almost negligible (Larsonneur, 1989; Migniot, 1997; Bonnot-Courtois *et al.*, 2002). This implies that the sediments that infill the Mont-Saint-Michel Bay and its estuary are almost exclusively of marine origin. They are mixed sediments composed of siliciclastic material reworked from the English Channel sea floor by tidal currents and waves; and of biogenic carbonates (principally shell clasts and red algae). Mean sediment grain-size decreases progressively from offshore to the most internal parts of the bay. In the middle to upper estuary, including the measurement site, sediments are defined as silty sands to sandy silts, the local name of which is the 'tangue' (Bourcard

& Charlier, 1959; Larsonneur, 1989). In detail, the grain-size of the tangue ranges between fine silt (min ~3.9 μm) and fine sand (max ~200 μm). At the measurement site, the median grain-size of the tangue collected on the channel bank and channel bottom is 72 μm and 80 μm, respectively (Fig. 3B). It contains an average of 50% of carbonates, mainly composed of shell fragments and foraminifera tests (Fig. 3A) and almost no clay and organic matter.

The tangue presents some mechanical cohesion, despite the low percentage of clay. This significant mechanical cohesion is mainly due to the shape of bioclastic particles. The tangue is thus a

(A)

(B)

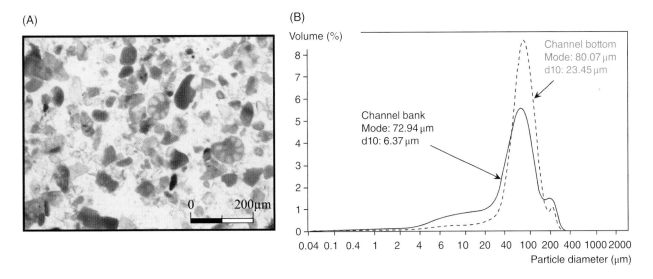

Fig. 3. Characteristics of the sediment from the measurement site. A: Sediment sample under binocular displaying the typical mixed siliciclastic and carbonated components of the tangue (local name of this mixed siliciclastic-carbonated silt-dominated sediment); B: Grain-size distribution curves (analysis using a Laser Particle Size Analyser Beckman-Coulter LS230).

particulate sediment (Bourcard & Charlier, 1959) but only few studies have investigated its physical properties and hydrodynamic behaviour.

Experimental measurements of settling velocity in calm water were conducted by Migniot (1997), revealing, amongst other results, that the tangue undergoes compaction quickly after deposition. For the present study, additional tests in the laboratory were performed.

First, the tangue compaction was quantified. Settling velocity experiments were conducted in a tank after sediment, sampled in the field, was resuspended. The density of the deposits was then measured regularly over several days. Sediment density evolved from $700\,g\,L^{-1}$ a few hours after sedimentation to more than $1300\,g\,L^{-1}$ after 3 days.

The second test consisted of measuring the evolution of the critical shear stress (τ_e) for tangue erosion as a function of its compaction. This test was achieved by performing vertical velocity profile measurements in a small recirculating flume (see Weill *et al.*, 2010 for flume description). The result was that τ_e increases non-linearly as a function of duration after deposition (the details of the experimental setup, data aquisition and results may be found in Furgerot, 2014). One hour after sediment deposition, τ_e values ranged between 0.05 and $0.1\,N\,m^{-2}$. Twelve hours after deposition, τ_e value reached $0.29\,N\,m^{-2}$. It then slowly increased up to $0.55\,N\,m^{-2}$ after 30 days of compaction.

Shear stress resistance measurements were also performed *in-situ* at the 'Le Bateau' site using a RocTest H-60 field vane tester. By this means, the mechanical rigidity (Cu) of the sediment at different depths was obtained. According to Migniot (1982), the friction velocity (U*) and the critical shear stress (τ_e) can be related to Cu. The measured mechanical rigidity was in the order of $2\,kPa$ for surficial sediment on the channel bed. This value increased to $25\,kPa$ at a depth of 0.3 to $1\,m$ below the channel bed. Our *in-situ* measurements and observations confirmed the conclusions of previous authors (Bourcard & Charlier, 1959; Larsonneur, 1989), underlining that the tangue approaches a thixotropic behaviour: a minimum shear stress had to be applied for a certain time before the material started to deform and liquefy.

MATERIAL AND METHODS

The evolution of the suspended sediment concentration (SSC) associated to the passage of the tidal bore was monitored thanks to an Argus Surface Meter (ASM) and a direct water sampling system. The Argus Surface Meter is an optical apparatus allowing vertical concentration profile measurements. Flow velocity and pressure data were collected thanks to an Acoustic Doppler Velocimeter (ADV). The *in-situ* experimental set-up is detailed in Fig. 4.

Distance from right bank (m)

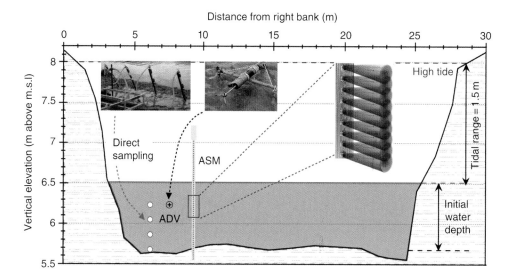

Fig. 4. Schematic diagram and photos illustrating the arrangement of the different instruments deployed in the Sée River to measure SSC and current velocities into a tidal bore (ASM: Argus Surface Meter; ADV: Acoustic Doppler Velocimeter; m.s.l.: mean sea-level).

The ASM consisted of 144 sensors embedded every centimetre in a 1.5 metre stainless steel rod. Each sensor included an infrared laser transmitter and a receiver. Its principle relies on the measurement of the quantity of backscattered light received by the sensor, which is a function of the suspended matter concentration. Numerous attempts have been made to correlate backscattered signal from optical or acoustic devices to absolute SSC values, by means of laboratory calibration prior to *in-situ* deployment (e.g. Hosseini *et al.*, 2006; Sottolichio *et al.*, 2011). For this study, the ASM has been calibrated in laboratory using a recirculating water column with natural sediment (tangue from the measurement site "Le Bateau"). The calibration was carried out from 0 to 30 g.L^{-1}. Known masses of sediment were introduced in the column and the ASM signal was recorded after water homogenization. A calibration curve between backscattered intensity signal and absolute suspended matter concentration was thus obtained. As it will be discuss later, ASM data displayed some discrepancies compared to direct sampling values. The choice of the calibration curve probably explains these differences (Furgerot *et al.*, 2012b). Nevertheless, the ASM provided at least accurate information of relative evolution of the SSC in the water column, with a temporal and spatial resolution which is difficult to achieve with usual systems.

A water sampling system was deployed in association with the ASM and provided absolute references of the SSC to the relative measurements of the optical system. It consisted of 4 flexible plastic tubes mounted on a rigid metal stick fixed vertically in the channel bed (Fig. 4). The sampling tubes were positioned at 0, 20, 40 and 60 cm above the channel bed with their extremity facing downstream. Each tube was connected to a manual pump located on the channel bank. Water was sampled simultaneously at the four elevations every 2 seconds. The sampling started a few seconds before the passage of the tidal bore and lasted for at least 40 minutes. The time delay between the entrance and output of the tubes had been considered (i.e. ~3 seconds). Finally, 800 samples per tidal bore were collected and processed at the laboratory. The SSC value was calculated after measuring the weights of (1) the water/sediment sample and its container, (2) the dry sediment (dried in an oven at 40°C) and its container; and (3) the clean, dry and empty sampling container.

The ADV probe (Vector Nortek 64 Hz) was deployed 60 cm above the channel bed (6.20 m above m.s.l.), to measure the flow velocity and pressure evolution during the passage of the tidal bore. Due to the sampling tubes generating flow disturbances, the ADV was positioned 1.5 m away from the latter, toward the channel centre (Fig. 4).

The ADV head was placed perpendicular to the current direction, so that Vx positive component pointed downstream. The ADV programming was carried out according to flow conditions and suspended matter concentration (ADV technical characteristics can be found in Hosseini *et al.*, 2006). The first parameter to set, according to Meuret *et al.* (2003), was the Nominal Velocity Range (NVR). A value of $2\,\mathrm{m\,s^{-1}}$ was chosen. As SSC was high during the survey, the transmission power was set to a minimum value in order to minimise noise on the recording. The size of the sampling volume had to be chosen according to the velocity gradient. A strong vertical velocity gradient was expected in the tidal bore and, in order to reduce the Signal to Noise Ratio (SNR), a small sampling volume size of $145\,\mathrm{mm^2}$ was set ($6.6\,\mathrm{mm}$ in height and $7\,\mathrm{mm}$ of invariant radius). Correlation coefficients and SNR were used to assess the quality and reliability of the measurements. The free surface evolution was tracked thanks to the ADV pressure sensor.

RESULTS

The results are presented at two time scales: i) At the scale of a few tenths of minutes, which includes the end of ebb, the tidal bore, the flood, the high slack water stage and the beginning of ebb; and ii) At the scale of the tidal bore (the first minutes of flood). Due to the upstream location of the measurement site, the flood stage was very short in comparison to the ebb.

Current velocity and water depth

The hydrodynamics of the channel flow before, during and after the tidal bore has been described in terms of longitudinal velocity, vertical velocity and water depth (Fig. 5A).

The water depth in the channel before the passage of the tidal bore was $0.9\,\mathrm{m}$ (Fig. 5A). The elevation of the free surface was $6.5\,\mathrm{m}$ above m.s.l. (Fig. 4), which corresponds to the normal river level at low tide. The longitudinal velocities were directed downstream (positive values), with a mean value of $0.3\,\mathrm{m\,s^{-1}}$. Small fluctuations of vertical velocities (Vz) around the value $0.05\,\mathrm{m\,s^{-1}}$ indicate a low level of turbulence.

The beginning of the flood was marked by the front of the tidal bore, corresponding to the propagation front of the dynamical tide. The bore was around $0.5\,\mathrm{m}$-high and created a quasi-instantaneous increase (in the order of one second) of the water depth, up to $1.4\,\mathrm{m}$. Synchronously, longitudinal velocities reversed upstream to a value of

almost $1.5\,\mathrm{m\,s^{-1}}$ under the crest of the bore. A peak of upward-directed velocity with a value of $0.5\,\mathrm{m\,s^{-1}}$ was recorded half a second before the tidal bore crest (Fig. 5B).

After the front passage, whelps created oscillations of the free surface with amplitude of between 10 and 15 cm; and wavelength of ca. 6 m. Fluctuations of the longitudinal and vertical velocities were associated to these whelps . The longitudinal velocity signal is in phase with the water depth signal. The longitudinal velocity (Vx), directed upstream, oscillates around $1\,\mathrm{m\,s^{-1}}$, with maximum and minimum values corresponding to the whelp crests and troughs, respectively. The vertical velocity (Vz) and the water depth signals were out of phase. Vz maxima occurred approximately $0.25\,\mathrm{s}$ before whelp crests. Between crests and troughs, Vz direction reversed from downward to upward (Fig. 4B). After a few tenths of seconds, the whelps dampened. The longitudinal velocities tended to stabilise at $1.3\,\mathrm{m\,s^{-1}}$ for several minutes (Fig. 4A). The mean vertical velocity was positive ($\sim 0.2\,\mathrm{m\,s^{-1}}$) and occasionally instantaneous values were negative (Fig. 4A). The free surface steadily increased up to a maximum value of $2.4\,\mathrm{m}$ at high water slack stage (cf. Fig. 6A). This water depth corresponded to an elevation of 8 m above m.s.l. No overbank occurred during this tide, as the altitude of the channel banks was $8.1\,\mathrm{m}$ above m.s.l. (Fig. 2C).

Evolution of suspended sediment concentration

SSC evolution was monitored for 40 minutes from the beginning of the flood (Fig. 6). As mentioned previously, this duration included the whole flood, the high water slack stage and the beginning of the ebb. SSC values obtained from *in-situ* water sampling are presented for the 4 different elevations above the bed (0, 20, 40 and 60 cm), together with the water depth evolution (Fig. 6A and B). ASM measurements have been presented on a spatio-temporal graph (Fig. 6C and D). The saturated signal before the tidal bore (0 s) and below the channel bed (0 m) was the response of the sensors at the base of the ASM stick buried in the sediment bed. The noisy signal recorded above 1 m is the response of emerged sensors at the top of the stick. At low tide slack water, before the passage of the tidal bore, SSC was very close to $0\,\mathrm{g\,L^{-1}}$ (Fig. 6). At the passage of the tidal bore, due to the sudden water level elevation, all the sensors of the ASM stick were submerged. Immediately

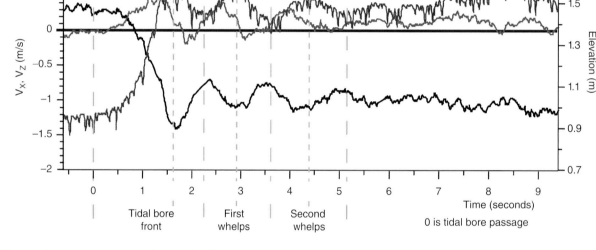

Fig. 5. ADV measurements of water depth and current velocity (Vx: longitudinal component; Vz: vertical component). A) 4 minutes of recording after the passage of the tidal bore; B) time window on the first 9 seconds after the tidal bore.

after the bore's passage, the SSC on the channel bed increased dramatically. The ASM data shows a saturated signal on the first 10 cm above channel bed, with a burst indicating SSC values of at least $30\,g\,L^{-1}$ up to 30 cm above channel bed. Data from direct sampling at the channel bed actually revealed concentration values up to $53.5\,g\,L^{-1}$ (Fig. 6A and B). This peak of SSC at the channel bed was followed by a slower decrease, until the SSC signal stabilised at $15\,g\,L^{-1}$ one minute after the tidal bore's passage. The SSC at 20 cm above the channel bed started to increase rapidly a few seconds after the peak of SSC at the channel bed,

up to a value of ca. $5\,g\,L^{-1}$. SSC values at 40 and 60 cm above the channel bed started to increase 30 seconds after the bore's passage, with a more gradual trend. One minute after the bore's passage, SSC values at 20, 40 and 60 cm homogenised around $5\,g\,L^{-1}$. This upward advection of suspended sediment is highlighted clearly within the ASM data: although the reddish and greenish colours in Fig. 6D indicate that high concentration values near the channel bed attenuated after the first 30 seconds; the limit between dark blue $(0\,g\,L^{-1})$ and light blue $(5\,g\,L^{-1})$ colours migrated upward in the water column with time. One minute after the bore's passage,

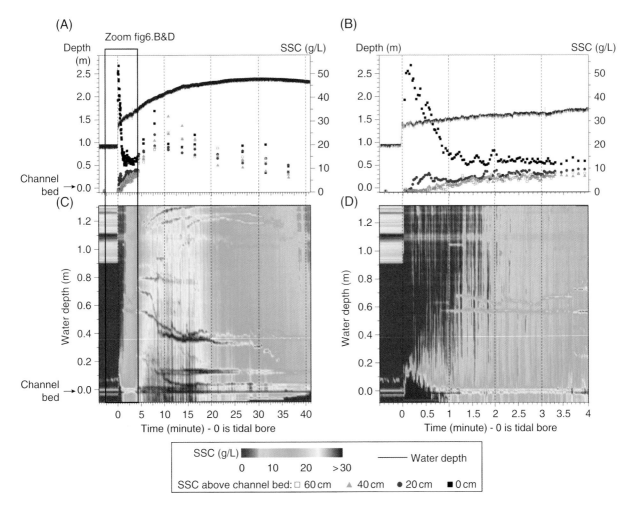

Fig. 6. Evolution of suspended sediment concentration (SSC); A: SSC measured by direct sampling at 4 elevations in the water column (during 40 min); B: Zoom on direct sampling results during 4 minutes; C: SSC measured with ASM probe (during 40 min); D: Zoom on ASM probe results focussing on the tidal bore passage and a few minutes after. Note that the yellow to red pseudo-horizontal lines on the ASM concentration plots result from debris (algae, leaves or branches) trapped in the ASM stick, producing noise in the signal.

this limit was between 60 and 80 cm above channel bed. This pattern of ASM data provides good evidence that the increase of turbidity in the water column resulted from the upward advection of sediment reworked from the channel bottom.

Between 1 and 2 minutes after the bore's passage, regular fluctuations of SSC were clearly recorded on the water sampling data at 0 and 20 cm above channel bed with a period of 10 to 20 seconds (Fig. 6B). These oscillations do not appear clearly on the ADV data of longitudinal and vertical velocities (Fig. 5A), which were more sensitive to fluctuations created by the whelps of the bore, with a period of a few seconds (Fig. 5B). However, these oscillations of the SSC might be attributed to convective cells of turbid and turbulent water, as observed on the field, that periodically outburst toward the water surface. Sensors at the base of the ASM stick that were buried in the sediment at the beginning of the recording started to show lower concentration values of between 10 and 20 g.L^{-1} one minute after the tidal bore's passage (Fig. 6D). This indicates channel bed erosion. Upward sediment advection in the water column from the channel bed was observed up to 13 minutes after the passage of the tidal bore. ASM measurements reveal a quite homogeneous SSC in the water column; around 20 g L^{-1}. Sediment concentration near the channel bed was always greater, reaching up to 30 g L^{-1}.

Ten minutes after the bore's passage, SSC in the water column started to decrease gradually

(Fig. 6A and C). This corresponds to the end of the flood and the beginning of the high tide slack water, as indicated by the pressure sensor (t~30 min). As revealed by the gradual saturation of the sensors at the base of the ASM stick, the settling of suspended sediment resulted in channel bed accretion. At 40 minutes after the bore's passage, the water level started to decrease. SSC was comprised of between 5 and 10 g L^{-1} respectively at the top and bottom of the water column. Due to sediment deposition on the channel bed, the first 10 cm of the ASM sensors were buried again.

DISCUSSION

The shape and height of a tidal bore in a river depends upon the tidal range, the water depth and especially the overall morphological context and local bathymetry. In the Sée River, tidal bores are most of the time undular, with an average height of 50 cm. This height is relatively small when compared to other undular tidal bores in the world, despite the fact that the Mont-Saint-Michel bay experiences one of the largest tidal ranges in the world. The bathymetry and geometry of the estuary control the way the tidal wave propagates and is amplified, which in turn determines the tidal bore height and shape. For example, in the Garonne River in France (Bonneton *et al.*, 2012) and in the Qiantang River in China (Fan *et al.*, 2012), where tidal bores are higher (1.3 and 2 m-high respectively), channels are deeper and wider. Channel depths at low tide are 3 and 9 m respectively and channel width values range from a few hundred metres to several kilometres. Such channel sizes, in addition to the general funnel shape of these two estuaries, induce an amplification of the tidal range upstream. In the Garonne estuary, the tidal range evolves from about 5 m at the mouth (Gironde entrance) to more than 6 m 127 km upstream (Bonneton *et al.*, 2012). In the Qiantang River (Fan *et al.*, 2012) the tidal range is 2 m at the mouth and increases up to 5.5 m some 80 km upstream. As a result, high mesotidal to macrotidal ranges still occur in the innermost portions of these hypersynchronous estuaries, contributing to the generation of high amplitude tidal bores. The Mont-Saint-Michel estuary is rather synchronous to hyposynchronous. As mentioned previously the tidal range decreases rapidly as soon as the tide propagates into the upper estuary, evolving from about 14 m at the outer mouth to 1.5 m at the

measurement site, i.e. 15 km upstream only. Due to this microtidal range, tidal bore heights are limited into the upper estuary, as compared to other regions. Nevertheless, these small amplitude tidal bores generate significant sediment resuspension and transport since they propagate into shallow and narrow channels.

Detailed information on the hydrodynamics and SSC evolution in the water column during the flood tide, especially during the passage of tidal bores, is provided thanks to the deployment of instruments in the upper Sée estuarine river, where well developed, mainly undular, tidal bores occur commonly. The data collected allows the reconstruction of a general model of SSC evolution (Fig. 7).

The sudden rise in water depth associated to a tidal bore is accompanied almost simultaneously by a reversal of the longitudinal velocity component (Vx). Initially directed downstream with the river flow, it decreases down to 0 m s^{-1} and then increases in the opposite upstream direction with the flood tide (Vx~1.5 m s^{-1} at the bore front). At the same time a peak of upward directed velocity of 0.5 m s^{-1} (positive vertical velocity component Vz) occurs. Direct sampling and ASM data have revealed that a dramatic sediment resuspension at the channel bed, with a resulting erosion of ca. 10 cm, occurs at this instant. Hence, this study demonstrates first of all that both the sudden reversal of longitudinal velocities and the significant upward directed velocity related to the tidal bore's passage result in the formation of a sudden highly concentrated suspended sediment layer (up to 53.5 g L^{-1}) on the channel bed. This result is critical in terms of sediment transport since this combination of sudden horizontal current reversing and upward vertical current is specific to the tidal bore's passage only, i.e. not to all ebb-to-flood reversing tidal phases.

Vane tests performed on the channel bed revealed a superficial layer of soft sediment, 10 cm-thick on average, with a very low critical shear strength value. This thickness is consistent with the erosion depth observed at the channel bed thanks to the ASM data. It is probable that this low resistance top layer of sediment will be easily reworked due to the tidal bore's passage.

After the passage, it is assumed that a part of the reworked sediment was redeposit. Then, this sediment was firstly advected upward by means of positive vertical velocities and was also diffused in the water column by the turbulence few minutes

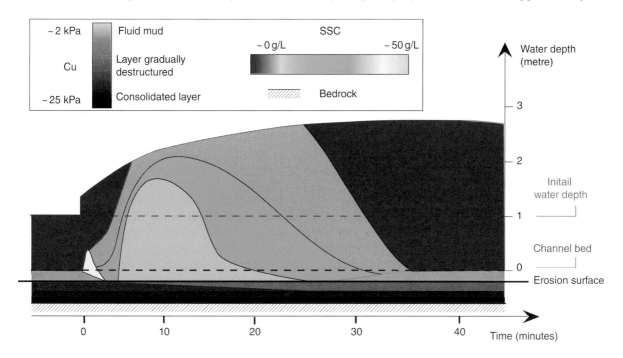

Fig. 7. Evolution of suspended sediment concentration during the passage of a tidal bore modelled on the basis of field data (ASM, direct water sampling) collected in the inner estuary of the Sée River. Time axis starts at the passage of the tidal bore front (0 min).

after the front. As the flood velocity progressively increased in time, SSC in the water column increased as well, reaching a maximum value up to 10 minutes after the tidal bore's passage. At this time, the upstream-directed longitudinal velocity ranged between 1.5 and 2 m s^{-1}, with a turbulence level sufficient to maintain sediment in suspension. The water column was almost homogeneous in terms of SSC, with a mean value around 15 g L^{-1}. SSC values close to the channel bed were slightly higher.

As soon as the flood current decelerated, SSC values in the water column decreased and sedimentation occurred at the channel bed. The deposited sediment was fluid (Cu~2 kPa) and therefore the shear stress required for erosion was low. During high tide slack water, which lasts for about 30 minutes, the remaining suspended sediment in the water column can settle. The thickness of the deposited sediment is comparable with the thickness of the sediment layer eroded by the tidal bore's passage, in the order of 10 cm. This fluid mud layer is then partly resuspended with the ebb current in the downstream direction.

The monitoring of 4 consecutive tidal bores (from 6th to 8th May) revealed that the thickness of eroded sediment during the passage of the tidal bore increases with the number of consecutive

tides. It was observed that the first well-developed tidal bore that reached the measurement site during a high spring tide period did not trigger much erosion on the channel bed, highlighting that the shear strength of the surficial sediment layer exceeded the bottom shear stress exerted by the tidal bore. This suggests that during periods without a tidal bore, i.e. neap tide and low spring tide periods, processes of sediment reworking and deposition above the channel bed are reduced, allowing the sediment layer deposited during the previous high spring tide period to compact. This was confirmed by vane tests performed on the channel bed during a high spring tide period when tidal bores occurred. The tests showed that the shear strength of the 'tangue' sediment increases, logically, with depth. Below the surficial layer of fluid mud, sediment layers of higher shear strength were found (Fig. 7). These represent compacted remains of deposits that occurred during previous high spring tide periods.

The increasing thickness of eroded sediment during consecutive tidal bores might be related to the thixotropic behaviour of the tangue sediment (Migniot, 1997) and to a cumulative effect of sediment bed destructuration due to high fluctuating pressures and strong accelerations during the passage of successive tidal bore fronts.

This behaviour, due to the tidal bore effect, is confirmed by the occurrence of convolute beddings preserved into estuarine tidal successions and attributed to tidal bore impact (e.g. Tessier & Terwindt, 1994 in the Mont-Saint-Michel estuary; Greb & Archer, 2007 in the Turnagain Arm estuary, Alaska; Fan *et al.*, 2012, 2014 in the Qiantang River). Thus we believe that the passage of the tidal bore, rather than eroding them, weakens the underlying compacted layers, becoming more erodible by the following tidal bores. Further laboratory experiments are in progress on the tangue (P' Institute, Poitiers, France) to better quantify the rheological properties of this particular carbonate sediment. During equinoctial tides (i.e. a very high spring tide period), up to 10 successive tidal bores can occur. This can trigger important sediment destructuration over significant thicknesses, which will facilitate sediment erosion by tidal bore-associated flood currents and increase sediment fluxes in the estuary. On a longer time scale, it appears that: i) little sediment is moved during periods without tidal bore because tidal currents are too slow and velocity reversals too progressive to erode the sediment bed, or to erode and transport sediment downstream if river water discharge is high; and ii) during periods of high spring tides when tidal bores develop, large quantities of sediment are eroded and transported upstream, toward the innermost parts of the estuary. However, this statement requires stronger field evidence, especially a longer monitoring of sediment erosion and deposition on the channel bed over several spring-neap tidal cycles.

CONCLUSION

This paper reports for the first time *in-situ* measurements with high spatial and temporal resolutions of hydro-sedimentary processes associated to a tidal bore. Current velocity, water level and SSC evolutions were monitored at a fixed station in an upper estuary during undular tidal bore passages (Sée River, Mont-Saint-Michel bay, NW France). On the basis of the results obtained during several successive tidal bores, a model of SSC evolution in the water column during and after the tidal bore's passage is proposed, highlighting the following main processes:

- important sediment resuspension due to highly sheared flow during the bore's passage, resulting in a 10 cm-thick fluid mud layer with concentration values up to 53.5 g/L, created on the channel bed as the bore is propagating;
- upward advection from this high concentration layer in the water column by positive vertical velocities (Vz ~0.5 m/s) and turbulence;
- a homogenization of SSC in the water column. The sediment is maintained in suspension due to sufficient horizontal velocity of flow and transported upstream.

The two first processes at least should be considered as the typical signature of a tidal bore's passage, occurring probably in most estuaries, since ebb-to-flood reversal phases without bores, being more progressive, cannot produce very high concentrated sediment layer at the channel bed.

In addition, the authors believe that the highly fluctuating pressures and strong accelerations that typify the passage of a tidal bore generate important bed destructuration, enhancing resuspension by subsequent tidal bores. This phenomenon will be particularly critical in the case of thixotropic sediment, such as in the Mont-Saint-Michel estuary.

Finally, the results of this study emphasise the high potential of tidal bores to rework sediment and to enhance significatively sediment flux towards upper estuaries. It demonstrates that this short but intense tidal phenomenon mobilises large quantities of sediment, as compared to the modal flood current processes occurring most of the year.

ACKNOWLEDGEMENTS

This study was part of Lucille Furgerot's PhD work, supported by the ANR project 'Mascaret' (ANR-2010-BLAN-0911), the coordinator of which, Prof. Pierre Lubin (I2M Bordeaux), is warmly thanked, and by the regional council of Basse Normandie. We also thank Dr. Alain Crave (Géosciences Rennes) for the calibration and loan of the ASM probe. We are grateful to all the colleagues of the 'Mascaret' team and of the M2C research lab, as well as friends who provided helpful assistance for field surveys. Special thanks go to Sylvain Haquin (M2C lab) for designing the water pumping system and to Laurent Perez (M2C lab) for developing the scan bar code assistance software. We thank warmly Jean-Yves Cocaign, the Director of 'la Maison de la Baie' in the Mont-Saint-Michel, for the logistical assistance he kindly supplied during the surveys.

REFERENCES

Allen, J.R.L. (2000) Morphodynamics of Holocene salt marshes: a review sketch from the Atlantic and Southern North Sea coasts of Europe. *Quatern. Sci. Rev.*, **19**, 1155–1231.

Archer, A.W. (2013) World's highest tides: Hypertidal coastal systems in North America, South America and Europe. *Sed. Geol.*, 284–285, 1–25.

Avoine, J., Allen, G.P., Nichols, M., Salomon, J.C. and Larsonneur, C. (1981) Suspended sediment transport in the Seine estuary, France: effect of man-made modifications on estuary-shelf sedimentology. *Mar. Geol.*, **40**, 119–137.

Billeaud, I., Tessier, B., Lesueur P. and Caline, B. (2007) Preservation potential of highstand coastal sedimentary bodies in a macrotidal basin: Example from the Bay of Mont-Saint-Michel, NW France. *Sed. Geol.*, **202**, 754–775.

Bonneton, N., Bonneton, P., Parisot, J.P., Sottolichio, A. and Detandt, G. (2012) Ressaut de marée et Mascaret - Exemples de la Garonne et de la Seine, *C.R. Geosci.*, **344**, 508–515.

Bonnot-Courtois, C., Caline, B., L'Homer, A. and Le Vot, M. (2002) *The Bay of Mont-Saint-Michel and the Rance Estuary: Recent development and evolution of depositional environments.* Vol. **26**. CNRS, EPHE & Total-Fina-Elf.

Bourcart, J. and Charlier, R. (1959) The tangue: a "nonconforming" sediment. *Geol. Soc. Am. Bull.*, **70**, 565–568.

Castaing, P. (1981) Le transfert à l'océan des suspensions estuariennes. Cas de la Gironde. Thèse d'Etat, Université Bordeaux 1 (France), 701.

Chanson, H., Reungoat, D., Simon, B. and Lubin, P. (2011) High-frequency turbulence and suspended sediment concentration measurements in the Garonne River tidal bore. *Estuar. Coast. Shelf. Sci.*, **95**, 298–306.

Dalrymple, R.W., Zaitlin, B.A. and Boyd, R. (1992) Estuarine facies models: conceptual basis and stratigraphic implication. *J. Sed. Petrol.*, **62**, 1130–1146.

Deloffre, J., Lafite, R., Lesueur, P., Verney, R., Lesourd, S., Cuvilliez, A. and Taylor, J. (2004) Controlling factors of rhythmic sedimentation processes on an intertidal estuarine mudflat – Role of the turbidity maximum in the macrotidal Seine estuary, *France. Mar. Geol.*, **235**, 151–164.

Dyer, K.R. (1995) Sediment transport processes in estuaries. In: *Geomorphology and sedimentology of estuaries* (Ed. G.M.E. Perillo). *Dev. Sedimentol.*, **53**, 423–449.

Fan, D., Cai, G., Shang, S., Wu, Y., Zhang, Y. and Gao, L. (2012) Sedimentation processes and sedimentary characteristics of tidal bores along the north bank of the Qiantang Estuary. *Chinese Sci. Bull.*, **57**, 1478–1589.

Fan, D.D., Tu, J.B., Shang, S. and Cai, G.F. (2014) Characteristics of tidal-bore deposits and facies associations in the Qiantang Estuary, *China. Mar. Geol.*, **348**, 1–14.

Furgerot, L. (2014) Propriétés hydrodynamiques du mascaret et de son influence sur la dynamique sédimentaire – Une approche couplée en canal et in situ (estuaire de la Sée, Baie du Mont-Saint-Michel). PhD Thesis. Université de Caen Basse-Normandie. 319pp (available on http://tel.archives-ouvertes.fr/tel-01061118).

Furgerot, L., Mouazé, D., Tessier, B. and Brun-Cottan, J.-C. (2012a) Tidal bore: eulerian velocities and suspended sediment concentration measurements. *Proc. Int. Conf. of River Flow, San José, Costa Rica*, 399–406.

Furgerot, L., Mouazé, D., Tessier, B. and Haquin, S. (2012b) Acoustic Doppler Velocimeter (ADV) measurements in a tidal bore results from field experiments. *Proc. 33rd Int. Conf. on Coast. Eng. (ICCE), Sandanter, Spain.*

Greb, S.F. and Archer, A.W. (2007) Soft-sediment deformation produced by tides in a meizoseismic area, Turnagain Arm, Alaska. *Geology*, **35**, 435–438.

Green, M.O. and MacDonald, I.T. (2001) Processes driving estuary infilling by marine sands on an embayed coast. *Mar. Geol.*, **178**, 11–37.

Hosseini, S., Shamsai, A. and Ataie-Ashtiani, B. (2006) Synchronous measurements of the velocity and concentration in low density turbidity currents using an Acoustic Doppler Velocimeter. *Flow Meas. Instrum.*, **17**, 59–68.

Lanier, W.P. and Tessier, B. (1998) Climbing ripple bedding in fluvio-estuarine system; a common feature associated with tidal dynamics. Modern and ancient analogues. In: *Tidalites: Processes and Products* (Eds C. Alexander, R.A. Davis Jr. and V.J. Henry) *SEPM Spec. Pub.*, **61**, 109–117.

Larsonneur, C. (1989) La baie du Mont-Saint-Michel. *Bulletin de l'Institut Géologique du Bassin d'Aquitaine*, **46**, 1–75.

Lesourd, S., Lesueur, P., Brun-Cottan, J.C., Garnaud, S. and Poupinet, N. (2003) Seasonal variations of the superficial sediments in a macrotidal estuary: the Seine inlet, *France. Estuar. Coast. Shelf. Sci.*, **58**, 3–16.

Li, J. and Zhang, C. (1998) Sediment resuspension and implications for turbidity maximum in the Changjiang estuary. *Mar. Geol.*, **148**, 117–124.

Meuret, A., Drevard, D., Piazzola, J. and Rey, V. (2003) Caracterisation technique du velocimetre doppler vector et applications à la mesure de la houle. *Proc. of 9th Journées de l'Hydrodynamique, Poitiers,* France.

Migniot C. (1982) *Étude de la dynamique sédimentaire marine, fluviale et estuarienne*, Thèse de Doctorat d'Etat. Université Paris Sud – Orsay, 488 pp.

Migniot, C. (1997) Mission Mont Saint Michel; Rétablissement du caractère maritime du Mont Saint Michel; Synthèse générale des connaissances sur les problèmes hydro-sédimentaire. DDE Manche.

Simpson, J.H., Fisher, N.R. and Wiles, P. (2004) Reynolds stress and TKE production in an estuary with a tidal bore. *Estuar. Coast. Shelf. Sci.*, **60**, 619–627.

Sottolichio, A., Hurther, D., Gratiot, N. and Bretel, P. (2011) Acoustic turbulence measurements of near-bed suspended sediment dynamics in highly turbid waters of a macrotidal estuary. *Cont. Shelf Res.*, **31**, 36–49.

Tessier, B., Billeaud, I., Sorrel, P., Delsinne, N. and Lesueur, P. (2012) Infilling stratigraphy of macrotidal tide-dominated estuaries. Controlling mechanisms: Sea-level fluctuations, bedrock morphology, sediment supply and climate changes (The examples of the Seine estuary and the Mont-Saint-Michel Bay, English Channel, NW France). *Sed. Geol.*, **279**, 62–73.

Tessier, B. and Terwindt, J.H.J. (1994) An example of soft sediment deformations in an intertidal environment: the effect of a tidal bore. *C.R. Acad. Sci. II*, **319**, 217–223.

Uncles, **R.**, **Stephens**, **J.** and **Law**, **D.** (2006) Turbidity maximum in the macrotidal, highly turbid Humber Estuary, UK: Flocs, fluid mud, stationary suspensions and tidal bores. *Estuar. Coast. Shelf. Sci.*, **67**, 30–52.

van der Spek, **A.J.F.** (1997) Tidal asymmetry and long-term evolution of Holocene tidal basins in The Netherlands: simulation of palaeo-tides in the Schelde estuary. *Mar. Geol.*, **141**, 71–90.

Weill P., **Mouazé D.**, **Tessier B.** and **Brun-Cottan J.-C.** (2010) Hydrodynamic behaviour of coarse bioclastic sand from shelly cheniers. *Earth Surf. Proc. Land.*, **35**, 1642–1654.

Wolanski, **E.**, **Williams**, **D.**, **Spagnol**, **S.** and **Chanson H.** (2004) Undular tidal bore dynamics in the Daly estuary, northern Australia. *Estuar. Coast. Shelf. Sci.*, **60**, 629–636.

Morphodynamics and sedimentary facies in a tidal-fluvial transition with tidal bores (the middle Qiantang Estuary, China)

DAIDU FAN*, JUNBIAO TU*, SHUAI SHANG*, LINGLING CHEN* and YUE ZHANG*

*State Key Laboratory of Marine Geology, Tongji University, Shanghai, 200092, China;
E-mail: ddfan@tongji.edu.cn (D. Fan).

ABSTRACT

The middle Qiantang Estuary breeds the world's biggest tidal bores. It has unique morphodynamics and sediment facies which have been studied using a time series of field photographs, satellite images and by using the textural and structural composition of 23 short cores from two tidal flats along the north and south banks. The results show that the channel morphology is extremely mobile in terms of rapid growing and shifting of intertidal banks over both short-term and large spatial scales because channel sediments, dominated by fine sand and coarse silt, are easily resuspended and dispersed by shooting flood flows due to energetic tidal bores and by strengthening river runoff during rainy seasons. This generally favours sediment dispersal upstream, accumulation along the north bank and the consequent development of extensive intertidal flats during dry seasons, while sediment dispersal downstream develops extensive intertidal flats along the south bank during rainy seasons over multiyear periods. Four sedimentary facies are identified across the channel: (1) tidal-bore deposits (TBDs) in the main channel and on the lower tidal flat, (2) hybrid deposits (HDs) near the mean low-water neaps, (3) tidal rhythmites (TRs) with (incomplete) spring-neap tidal cycles on the middle to upper tidal flats and (4) annual rhythmites (ARs) on the upper tidal flat and marshland. A typical TBD consists of at least two of the following sedimentary features: (1) erosion surface, (2) massive bedding, (3) graded bedding, (4) parallel lamination and (5) soft-sediment deformation structures. TBDs also differ from tidal sand deposits (TSDs) in having a coarser grain-size and poorer sorting because of rapid deposition. The well-developed TRs with (incomplete) spring-neap tidal cycles are attributed to three favourable conditions: (1) abundant sediment supply from the main channel resuspended and dispersed by tidal bores and rapid flood flows, (2) a relative protected setting with regard to wave attack and (3) very rapid accretion of intertidal flats over a short-term scale. Consequently, a stacked facies succession of TBDs, HDs and TRs is assumed as the best indicator for a tidal-fluvial transition with tidal bores.

Keywords: Morphodynamics, tidal-bore deposits, tidal rhythmites, facies associations, tidal-fluvial transition, Qiantang Estuary.

INTRODUCTION

The fluvial-marine transition zone encompasses the most prosperous waterways and harbours, extremely diversified ecosystems and crucial connections along the source-to-sink journey of sediments and pollutants. This is typically true for tide-dominated fluvial-marine transitions, where tides may intrude inland for tens to hundreds of kilometres. Tidal-fluvial transitions are important today and have also been important in the geological past, in terms of paleogeography and the prediction of oil/gas reservoirs. Several recent papers summarise diagnostic criteria for the tidal-fluvial transition zone based on specific research (Lanier & Tessier, 1998; Greb & Martino, 2005; van den Berg et al., 2007) and reviews (Dalrymple & Choi, 2007; Longhitano et al., 2012).

Contributions to Modern and Ancient Tidal Sedimentology: Proceedings of the Tidalites 2012 Conference, First Edition. Edited by Bernadette Tessier and Jean-Yves Reynaud.

Undular bores occur in tens of estuaries (Bartsch-Winkler & Lynch, 1988; Chanson, 2012), including the Salmon River (Canada), the Severn River (UK), the Sélune and the Sée (France), together with the Daly River and the Ord River (Australia). In certain hypertidal-fluvial transitions, undular bores develop into breaking bores; in the Qiantang Estuary (China), the Amazon Delta (Brazil) and the Turnagain Arm of Cook Inlet (Alaska). Hydrodynamic surveying and modelling results show that tidal bores are highly capable of sediment reworking and transport, exerting great impact on estuarine ecosystems, waterways and infrastructures (Wolanski *et al.*, 2001; Simpson & Wiles, 2004; Pan *et al.*, 2007; Chanson, 2012). However, channel morphology and stratigraphy in the context of tidal bores have so far been little studied (Dalrymple *et al.*, 2012; Tessier, 2012). This hampers the unsuspicious interpretation of fossil analogues. For instance, Martinius & Gowland (2011) pondered over the bore-genesis interpretation of a typical sedimentary package with an undular erosion surface overlain by homogeneous sand and bi-directional cross-bedded units in the Late Jurassic Lourinhã Formation (the Lusitanian Basin in Portugal).

The Qiantang Estuary breeds the world's biggest tidal bores with rolling and breaking fronts during spring tides (Lin, 2008; Fan *et al.*, 2012). The mechanism for tidal-bore development and decay has been simulated numerically on the basis of observations of the hydraulic processes (Han *et al.*, 2003; Pan *et al.*, 2007; Lin, 2008). Hydrodynamics and sedimentary characteristics of tidal bores and related processes have been preliminarily discussed using ADCP and OBS data over two semidiurnal tidal cycles and by examining 16 short cores in terms of textural and structural compositions, respectively (Fan *et al.*, 2012, 2014). In this paper, time series based on field photographs and satellite images are compared to understand the highly mobile channel morphology and its potential linkage with the tidal-bore action. Grain-size data and sedimentary structures in 23 short cores have been analysed for vertical and lateral variations in tidal deposits in relation to the tidal and tidal-bore activity. A schematic stacked facies model is then presented of the channel-flat complex in the bore-affected Qiantang estuarine reach, providing a basis for comparison with other modern and ancient analogues.

STUDY AREA

The Qiantang River is, in total, about 600 km-long and has a catchment of 4.9×10^4 km². The average river runoff is $952\,m^3\,s^{-1}$ and the annual average sediment discharge (majorly suspended load) is 6.6×10^6 ton (Han *et al.*, 2003). The river debouches into the ~120 km-long Hangzhou Bay, which has a funnel shape with a bay-mouth and bay-head width of ~100 km and <20 km (Fig. 1A). The strong decrease in bay width greatly deforms advancing tidal waves, with a mean tidal range from <2 m at the mouth to ~5.5 m near Ganpu, where mean and maximum spring tidal ranges reach 6.44 m and 9.00 m, respectively (Han *et al.*, 2003; Lin, 2008; Fan, 2012). From there, the tidal range decreases upstream quickly because of increasing friction by the ascending and narrowing river channel (Fig. 1A). The tidal limit is at Lucibu, approximately 190 km upstream of Ganpu. Multiyear average salinity at ZK (Zha-Kou) is 0.4‰.

A tripartite model of the Qiantang Estuary usually sets the boundaries at WY (Wen-Yan) and Ganpu (Chen *et al.*, 1990; Zhang & Li, 1996; Fan *et al.*, 2014). The upper estuary between Lucibu and WY is a tide-influenced freshwater river with thalweg sediments mostly composed of gravely and coarse sand (Fig. 1A). The middle estuary, between WY and Ganpu, features a highly meandering channel with thalweg sediments predominantly consisting of fine sand and coarse silt. The outer estuary between Ganpu and Jinshan, coincident with the inner part of Hangzhou Bay, is characterised by well-developed linear erosion troughs and accretion ridges on the silty bed of basin plain. The troughs and ridges are composed of coarse sand with some gravel and medium-fine sand with silt, respectively. The outer part of Hangzhou Bay is a shallow and smooth plain with an average depth of 8 to 10 m during low tide. It is covered predominantly with mud, which is sourced from the Yangtze (Changjiang) River plume. Consequently, the bay is considered to be part of the Yangtze subaqueous delta (Chen *et al.*, 1990; Zhang & Li, 1996; Han *et al.*, 2003; Fan *et al.*, 2012, 2014).

There is a gigantic sand bar occupying the middle estuary (Fig. 1B). Its highest part is located between CQ (Cang-Qian) and QB (Qi-Bao), with a relatively steep upstream slope and a very gentle downstream-sloping gradient of 0.02% (Chen

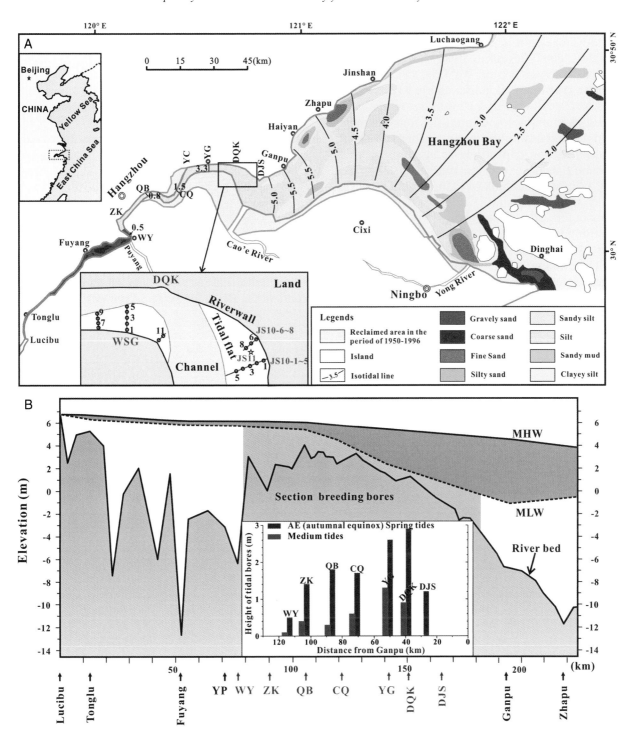

Fig. 1. (A) Surface sediment distribution map for the macro-to-hypertidal Hangzhou Bay - Qiantang Estuary system with co-range lines and short-core sampling locations (after Feng *et al.*, 1990; Fan *et al.*, 2012); (B) difference in tidal levels and bore-head heights along the Qiantang Estuary (after Lin, 2008).

et al., 1990; Hang *et al.*, 2003; Lin 2008; Yu *et al.*, 2012). The elevated riverbed further deforms the incoming tidal wave and its front develops into undular surges during spring tides between

Ganpu and DJS (Da-Jian-Shan), where the mean water depth is generally less than 5 m during low tide (Fig. 1B). The undular bores grow upstream into breaking bores with a peak between DQK

78   *D. Fan* et al.

(Da-Que-Kou) and YG (Yan-Guan), where the mean water depth is usually less than 1 m during low tide. The bores decay upstream from YG and finally disappear behind WY (Fig. 1B). The reach of these tidal bores thus is roughly 110 km, more or less coinciding with the middle estuary.

Mid-channel bars and intertidal flats are developed extensively in the middle estuary with significant erosion/deposition cycles. Most of these bars and flats have been reclaimed since the 1960s (Fig. 1A). DJS tidal flats were the largest intertidal banks along the winding channel, with growing bores before 2010, but have largely been eroded by strengthening river runoff during two successive rainy seasons (2010 and 2011). WSG (Wei-Shi-Gongduang) tidal flats are presently the largest intertidal area along the straight channel with peak bores.

METHODS

Regular visits to watch tidal bores at YG have been carried out since 2005. DJS and WSG tidal flats were visited several times between 2010 and 2012 for short-core collection and monitoring of channel morphology. Time-series of fieldtrip photos show rapid erosion and deposition cycles of tidal banks over the seasons. Four Landsat satellite images (from the USGS official website) with roughly 5 year intervals were chosen to study multiyear variations of channel morphology in the lower middle estuary from YG to Ganpu (Figs 2, 3 and 4). The focus was on images taken at low tide on sunny days, so that the lower waterline could be easily defined.

Eight short cores were collected along two transects on the DJS tidal flats in April 2010, numbered

Fig. 2. Rapid shifts of intertidal banks and main channels at the lower middle Qiantang Estuary from Ganpu to YG. There was a broad (hundreds of metres) intertidal flat along the north bank near YG in 2007 (A), a medium year during a dry period. In 2006 and 2008 there was a shallow channel at the same location. When the main channel shifted back to the north bank after 2007, extensive intertidal flats near WSG continued growing, with an extreme width of ~1200 m in 2011 (B). DJS intertidal flats were extensive with a maximum width of 3000 m in 2010 (C); most of these were eroded with the main channel approaching the foot of river walls in 2011 after two continuous rainy seasons (D).

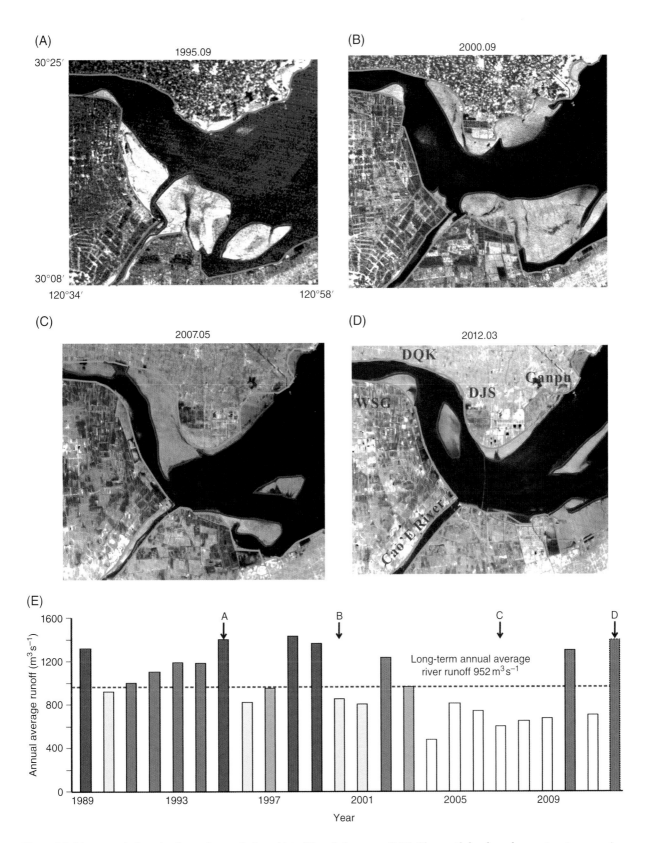

Fig. 3. Multiyear variations in channel morphology (A to D) and river runoff (E). The north bank underwent serious erosion with significant accretion at the south bank in 1995, a medium year of a rainy period from 1989 to 1999 (A). In 2000, at the transition between a rainy and a dry period the accretion zones along the south bank shifted downstream and the north bank began accreting (B). Development of intertidal flats was extreme along the north bank with serious erosion along the south bank in 2007, a medium year of the dry period (C). Serious erosion occurred along the north bank with initiate growth of intertidal flats along the south bank in 2012 (D). Long-term variations in river runoff exhibit roughly decadal alternations of dry and rainy periods (E). Cyan and red lines mark the locations of river walls and low water line, respectively.

Fig. 4. Rapid erosion and accretion of the DJS tidal flats during the period 2010 to 2012. Black and blue arrows are two reference houses for comparison and the red arrow points to the coring locations JS11-1 to 4.

JS10-1 to 5 and JS10-6 to 8 (Fig. 1A). An approximately 4 m-high erosion cliff occurred in August 2011 at the protected corner of the DJS tidal flats and sediments were sampled continuously from the top to the lower waterline in four cores, each roughly 1 m-long and numbered JS11-1 to 4. Eleven cores were sampled along three transects on the WSG tidal flats in August 2011, numbered WSG1 to 11.

Sediment cores were split into halves in the laboratory. On the working halves, the following analyses were performed: facies description, XRF scanning, laser-diffraction grain-size analyses (refer to Fan *et al.*, 2012, 2014 for a detailed description).

MORPHODYNAMICS OF THE MIDDLE ESTUARY WITH TIDAL BORES

The river channel between YC (Yan-Cang) and DQK extends over 20 km in roughly an east-west direction with a width of 2.5 to 3.5 km. The bended

part grades into a gigantic meander loop (Fig. 1A). The convex south bank usually experiences accretion and has well-developed intertidal flats. The concave north bank often experiences erosion with the main channel approaching the cutbank (Fig. 3). Tidal bores are consequently very strong along the north bank and the bore fronts are usually high and regular with a straight or slight sigmoidal form while propagating through the long, straight channel. A reversal sometimes occurs when the main channel shifts to the south bank; a significant accretion may then occur along the north bank. This happened in 2007, in a period with small bores. Extensive intertidal flats were then developed along the north bank (Figs 2A and 3C). The flats were several hundred metres wide and over ten kilometres long in 2007, but were not seen during fieldtrips in 2006 and 2008 except for a shallower than normal channel, denoting a very rapid variation in channel morphology. The accretion-erosion pattern seems to be linked with multiyear variations in river runoff and weather variability in the catchment. The year of 2007 was in the midst of a 10-year dry period (Fig. 3E). Weak fluvial runoff promoted siltation at the north bank with sediments dispersed upstream by tidal bores and rapid flood flows. The main channel consequently shifted toward the south bank, leading to erosion there. The main channel shifted back to the north bank after 2007, when river runoff increased again. The river thalweg has been relocated close to the river walls along the north bank, while extensive tidal flats have been growing along the south bank since 2010 (Fig. 2B), when a new multiyear rainy period began.

The river channel widens seaward from ~4 km near DQK to ~20 km near Ganpu, occupied by highly mobile mid-channel bars and intertidal flats (Fig. 3). This looks like a big meander loop but behaves differently to fluvial meanders. The north bank usually undergoes accretion (erosion) by the weakening (strengthening) fluvial current during the dry (rainy) seasons and periods (Fig. 3). DJS tidal flats grew toward an extreme extent during a sustained dry period from 2004 to 2009, with an area of over 25 km² and a maximum width of 3 km. The extensive tidal flats were still observed during a fieldtrip on May 2nd, 2010, before the rainy season started (Fig. 2C). Most of the tidal flats were rapidly eroded during the following two rainy seasons (Figs 2D and 3D), except for a few higher tidal flats and marshlands at protected settings (Fig. 4B). On the contrary, erosion (accretion) took place along the south bank during the dry (rainy) periods (Fig. 3). Intertidal flats and mid-channel bars were extensively developed near the Cao'E River outlet in 1995 (Fig. 3A). In 2000 they shifted downstream with significant erosion at the upstream part of the river outlet (Fig. 3B). The years 1995 and 2000 were in the midst of, and one year after, a rainy period from 1989 to 1999, respectively (Fig. 3E). The downstream translation of intertidal flats and mid-channel bars along the south bank (Fig. 3A and B) is attributed to continuous reworking and seaward dispersal of eroded sediments by ebb flows, which were significantly strengthened by increasing river runoff during rainy seasons (periods). At the same time, intertidal flats started growing along the north bank in 2000, attesting to the rule that weakening fluvial discharge during dry periods favours sedimentation along the north bank.

The intertidal flats along the straight channel between YC and DQK are usually devoid of higher intertidal zones, potentially related to a rapid erosion and deposition cycle over a few years. Their topography is ordinarily very gentle and flat with some small creeks and small-scale ripples highly abundant on the surface (Fig. 2). The intertidal flats and shoals between DQK and Ganpu usually have a broader extent and a more complex topography, with well-developed tidal creeks and flood barbs (Figs 2C, 3A and B, Fan *et al.*, 2014). Higher intertidal flats and marshlands potentially develop near the river margins; and some relatively protected settings survive severe erosion by strong ebb flows during rainy seasons (Figs 4A and B). Seasonal variations in erosion and deposition are also highly evident in the relatively protected settings, where tidal creeks and scouring features are quite abundant on the vegetated and bare muddy higher flats (Fig. 4).

In short, the lower middle estuary is highly mobile in terms of growing and decaying intertidal flats and mid-channel bars over large areas under the interaction of tidal bores, tidal flows and river runoff (Fig. 5). The north bank is usually eroded by ebb flows, which are greatly strengthened by increased river runoff during rainy seasons/periods and the eroded sediments tend to be dispersed seaward and to accumulate along the south bank (Fig. 5B). As rainy periods continue, sediments are carried further downstream and the channel becomes deeper, typically between DQK and YC, favouring the formation of stronger bores. During dry seasons and periods sediments tend to be redistributed upstream by flood-dominated tides, resulting in the growth of intertidal flats along the north bank and significant sedimentation in the main channel, which in turn suppresses the growth of tidal bores (Fig. 5A). Intertidal flats are usually less mature with the lower flats showing rapid erosion-deposition patterns, especially along the narrow straight channel from DQK to YC. The broad lower estuary tends to develop more extensive intertidal flats and detached shoals with a complex topography, including flood barbs, tidal creeks and scarped higher muddy flats and marshlands.

SEDIMENTARY FACIES OF THE MIDDLE ESTUARY WITH TIDAL BORES

Sedimentary features of the WSG tidal flats

The eleven short cores collected on the WSG tidal flats (Figs 1 and 6) show significantly uniform sedimentary features in terms of types and

82   *D. Fan* et al.

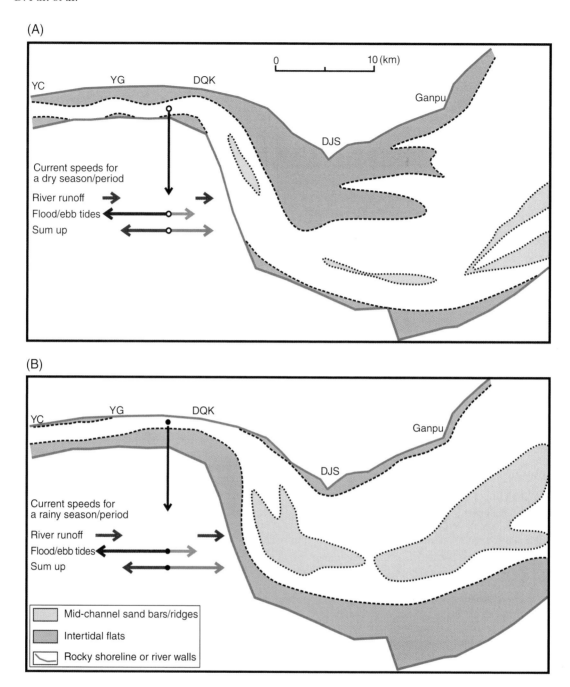

Fig. 5. Schematic maps to show alternative patterns of erosion and deposition at the lower reach of middle Qiantang Estuary in response to transition from a dry (A) to a rainy season/period (B).

abundance of both primary and secondary sedimentary structures (Figs 6 and 7). The most abundant primary structures are erosion surface (ES), massive bedding (MB) and parallel lamination (PL). Erosion surfaces vary from straight to undular. Thick sand layers usually display fining-upward sequences or graded bedding (GB), typically along the western transect (WSG6 to 9,

Fig. 6). Mud pebbles (MP) and bidirectional cross stratification (CS) are rare and only present at the bottom of WSG11 and in the upper part of WSG10, respectively.

Secondary sedimentary structures are as outstanding as primary structures in these short cores. The most common secondary structure is convolute bedding (CB) with different forms

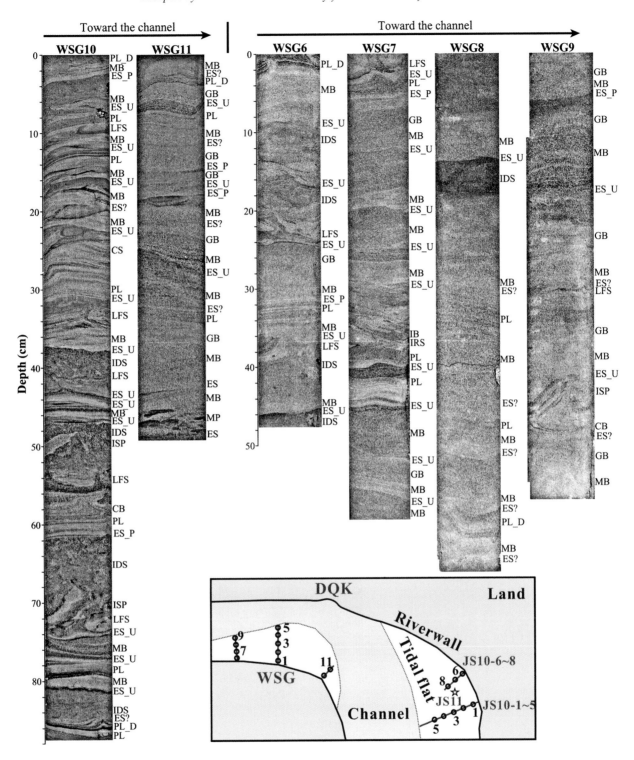

Fig. 6. Core photos of WSG6 to 11 showing typical sedimentary structures produced by tidal bores and associated processes at the eastern and western parts of the WSG intertidal flats. ES: erosion surface, ES_P/U: planar/undular ES; CS: cross stratification, GB: graded bedding, MB: massive bedding, PL: parallel lamination, PL_D: slight deformation PL; CB: convolute bedding, IDS: irregular deformation structures, ISP: invasive sand patches, LFS: load-flame structures, PS: pipe-like structures, SD: sand dikes. Note: colour contrasts of core photos have been adjusted to show clearly the internal structures.

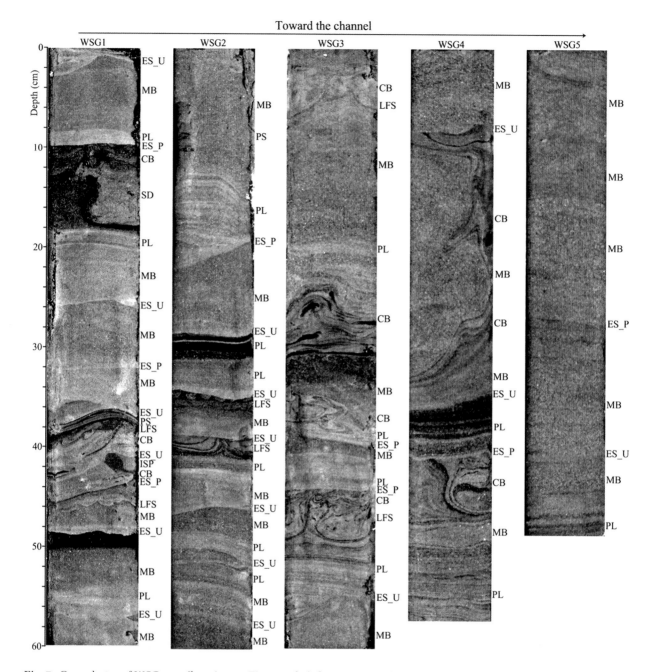

Fig. 7. Core photos of WSG1 to 5 (location on Figs 1 and 6) showing typical sedimentary structures produced by tidal bores and associated processes in the central WSG intertidal flats (after Fan *et al.*, 2014). The encoded structures by letter abbreviations are similar to Fig. 6.

(Figs 6 and 7). Load and flame structures (LFS) are also common. Scattered secondary structures include pipe-like structures (PS), invasive sand patches (ISP), sandy dikes (SD) and irregular deformation structures (IDS). Convoluted beds and other deformation structures are typically developed in layers rich in dark, platy minerals, like WSG 1 to 4, 6, 7 and 10 (Figs 6 and 7).

Packages of a few cm-thick often consist of an erosion surface at the base, covered by massive bedding with parallel lamination on top (Figs 6 and 7). The parallel-laminated layers on the top are usually deformed into convolute bedding with distinguishable original laminations or highly deformed sand patches (typically evident in SWG10, Fig. 6). Sand dikes and patches are also present in massive sand beds and may not result from bioturbation in such a low-salinity (<2‰) and mobile-bed setting. These features together suggest a link with highly energetic tidal bores and associated processes. In each package, the erosion surface at the base is produced by tidal bores and the rapid flood flows that follow them, with a huge amount of sediment being resuspended. Massive beds and graded beds are laid down during the rapidly decelerating flow phase with firstly deposition of poorly sorted coarse sand and then a fining-upward sequence of better sorted sand and silt. Parallel laminations rich in platy minerals are assumed to form at the transition flow from supercritical to subcritical regimes. Over a semidiurnal tidal cycle, the ebb flow may not have created its own deposits or they did not survive erosion by subsequent tidal bores. This assumption is strengthened by the mere presence of only one unit with an opposite direction of inclined bedding to the others in WSG10. All secondary structures basically result from soft-sediment deformation due to tidal bores. The passage of tidal bores undoubtedly increases the interstitial water pressure of the newly deposited water-saturated sand layers; while parallel laminations at the top serve as a seal to keep the interstitial pressure increasing until breaking through, consequently, with the strong deformations. The same mechanism has been proposed to explain the extensive occurrence of soft-sediment deformation structures in intertidal-flat deposits flanking the main channels with tidal bores in the Mont-Saint-Michel Bay (Tessier & Terwindt, 1994) and the Turnagain Arm (Greb & Archer, 2007). Moreover, bottom shear stress by shooting flood flows also has the potential to significantly increase the interstitial water pressure, resulting in liquefaction and deformation structures (Fan *et al.*, 2012).

Lateral variations in sedimentary features are remarkable along three transects (Figs 6 and 7). The cores at the low waterline usually have coarser sediments with well-developed massive bedding, parallel to inclined bedding but lacking distinct erosion surfaces and deformed structures (WSG5, 9 and 11). The latter is ascribed to the coarseness of these sediments and the absence of sealing layers (like platy-mineral rich laminations) on the top so that no large interstitial water pressure could build up. Sediments generally become finer shoreward with the abundance of platy minerals and erosion surfaces and deformation structures.

The differences between sediment characteristics of the western transect WSG6 to 9 and the central transect WSG1 to 5 are minor because of their similar hydrodynamic and morphodynamic setting. However, the eastern transect, WSG10 to 11, is a little different from the other two transects in terms of having finer sediments, an increased abundance of platy minerals and the presence of mud pebbles and small-scale cross lamination by the ebb flow. This may be attributed to the relatively weak influence of small bores along the eastern transect (Fig. 1).

Sedimentary features of the DJS tidal flats

Twelve short cores were examined in detail for lateral variations in sedimentary facies on the DJS tidal flats (Fig. 8). Five short cores collected along transect JS10-1 to 5, from the upper to the lower tidal flats (Fig. 1), are characterised by a gradual shoreward transition from massive sand to heterolithic rhythmites (Fig. 8A). JS10-5 is mainly composed of massive sand with a few distorted sand-mud interlayers at the top. Close examination showed the abundance of irregular deformation structures (IDS) in massive sand. The lower and middle-upper parts of JS10-4 consist of sandy beds with massive bedding and irregular deformation structures, respectively. The lower part of JS10-3 is composed of sandy beds with massive bedding, parallel laminations and an erosion surface. Its middle and upper parts consist of two layers of tidal rhythmites (TRs) and a few massive sandy layers with well-developed deformation structures. JS10-1 and 2 feature alternations of sand-dominated and mud-dominated packages, their thicknesses decreasing upward. Regular alternations of thickening and thinning sand-mud-couplet packages in JS10-1 denote (incomplete) spring-neap tidal cycles. In some spring-tide (thicker) depositional packages, two consecutive sandy laminae with different thicknesses denote the diurnal inequality of two successive semidiurnal tides. Sedimentary features in JS10-6 to 8 are analogous to those of JS10-2 and 3 because of their similar coring settings (Figs 1, 8A and B).

Four cores of JS11-1 to 4 were taken along an erosion cliff from the vegetated surface down to the lower waterline (Fig. 4B). They disclose distinctly the vertical association of intertidal-flat facies and marsh deposits (Fig. 8C). The lower section (2.53 to 3.80 m-deep, including JS11-4 and the lower part of JS11-3) is mostly composed of

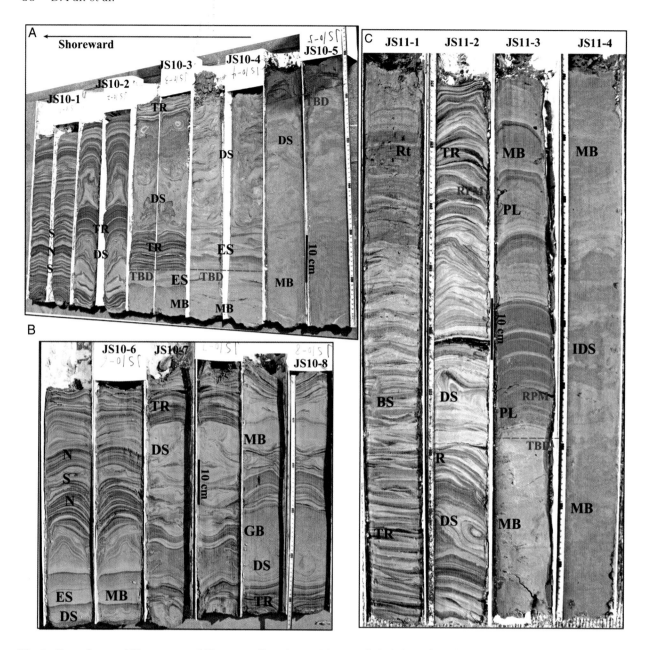

Fig. 8. Core photos of JS10-1 to 8 and JS11-1 to 4 (location on Figs 1 and 6). (A) Landward change in sedimentary structures from massive beds of JS10-5 to tidal rhythmites of JS10-1; (B) Landward change in sedimentary structures from alternations of massive and rhythmic beds (JS10-8) to predominance of tidal rhythmites (JS10-6); (C) Upward change in sedimentary structures from massive beds in JS11-4 to annual rhythmites in JS11-1. BS: burrow structures, DS: deformed structures, ES: erosion surface, GB: graded bedding, MB: massive bedding, N: neap tides, PL: parallel lamination, RPM: rich in platy minerals, Rt: plant roots, S: spring tides, TBD: tidal-bore deposits, TR: tidal rhythmites.

massive sands with some deformation structures, which look exactly like those of JS10-5 from the lower tidal flat. The intermediate section (1.30 to 2.53 m-deep) consists of alternations of sand-dominated and mud-dominated packages with highly deformed laminations, which highly mimic JS10-2, 3 and JS10-6 to 8 taken from the

middle tidal flat. The upper section (0.57 to 1.30 m-deep, including the lower part of JS11-1 and the upper part of JS11-2) shows heterolithic rhythmites, highly similar to JS10-1 from the upper tidal flat, denoting incomplete spring-neap cycles. The topmost section (0 to 0.57 m-deep) is also characterised by thin heterolithic rhythmites but with some bioturbation structures, plant roots and iron-manganese concretions, which is therefore interpreted as marsh deposits with seasonal cycles (Fig. 8C; Tessier, 1998).

In summary, a fining-upward sequence exposed in an erosion cliff is a vertical stack from lower to upper tidal-flat deposits (Fig. 8). The lower tidal flat is exposed to energetic tidal-bore impacts, so the rapid deposition after tidal bores and rapid flood flows produce massive sand beds, which are subject to deformation by the subsequent tidal bores through liquefaction. The upper tidal flat is not affected by tidal bores; and heterolithic rhythmites with spring-neap tidal cycles are deposited there. Consequently, the alternating packages of thin heterolithic rhythmites and thick massive

sand beds with deformation structures reflect the switching off and on of tidal-bore influence on the middle tidal flat during neap and spring tides, respectively.

Grain-size of DJS intertidal deposits

Grain-size analyses of 197 and 192 samples taken respectively from JS10-1 to 5 and from JS11-1 to 4 showed that these largely consist of fine sand and coarse silt with a mean size of 4.61 to 6.79 φ (Fig. 9). Standard deviation varies from 1.33 to 1.86 φ, reflecting a poor sorting (Cai *et al.*, 2014). Skewness and kurtosis vary from 0.29 to 2.44 (fine skewed to very fine skewed) and 2.31 to 9.50 (mesokurtic to very leptokurtic). A general fining-shoreward of

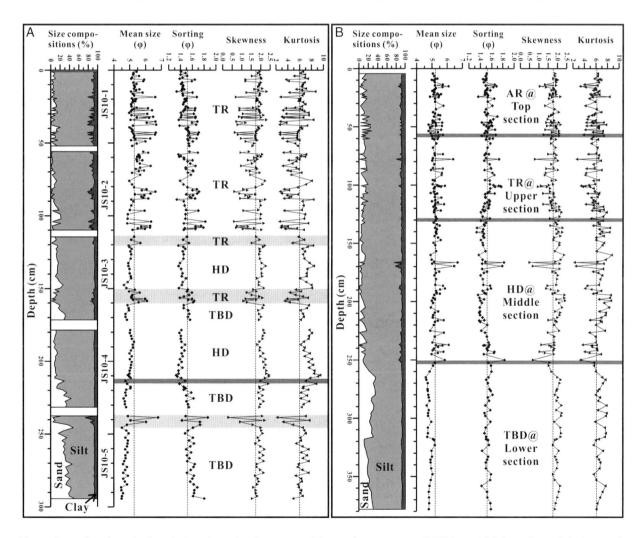

Fig. 9. Lateral and vertical variations in grain-size composition and parameters of DJS intertidal deposits and their genetic interpretations. JS10-5 to 1 are used to reconstruct shoreward lateral evolution (A) and JS11-4 to 1 upward vertical evolution (B). ARs: annual rhythmites, TRs: tidal rhythmites with (incomplete) spring-neap tidal cycles, TBDs: tidal-bore deposits, HDs: hybrid deposits having similar textural composition as regular tidal deposits and their structures analogous to TBDs.

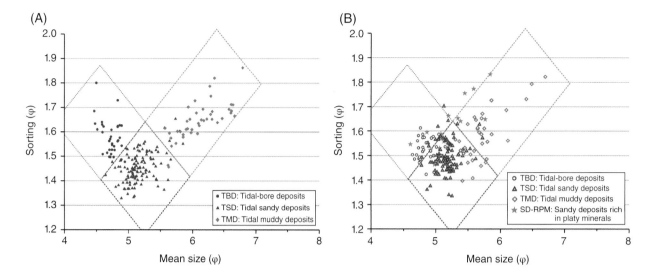

Fig. 10. Bivariate plots of sorting *vs.* mean size for 197 and 192 samples from JS10-1 to 5 (A) and JS11-1 to 4 (B). Both show a similar distribution pattern and allow discriminating between tidal-bore deposits (TBDs), tidal sandy deposits (TSDs) and tidal muddy deposits (TMDs).

sand composition is observed along the transect JS10-1 to 5 with a fining-upward in each core. Size-parametric curves of JS10-1 to 5 and JS11-1 to 4 are comparable (Fig. 9), denoting similar controlling factors and mechanisms.

Sandy and muddy layers were sampled and analysed separately, with respect of their different genetic mechanisms. Sandy layers have generally lower sorting values and a greater skewness and kurtosis than muddy layers, so they can easily be discriminated by scatter plotting of any two parameters (Fig. 10; Fan *et al.*, 2012, 2014). Seesaw curves with strongly changing amplitudes significantly reflect the parametric contrast between muddy and sandy layers in the upper parts of JS10-1 to 5 and JS11-1 to 4, while the curves have a more smooth shape in the lower massive sandy beds (Fig. 9).

Sandy layers can be ascribed to various deposition processes with different structural and textural composition (Figs 8 and 10). Tidal-bore deposits (TBDs) are generally coarser and poorer sorted than tidal-sand deposits (TSDs) because TBDs are formed by rapid deposition during and immediately after tidal bores and shooting flood flows in the main channel and on the lower tidal flat. TSDs are formed by selective transportation and deposition of regular tidal flows over the higher tidal flats (Fan *et al.*, 2012, 2014). These textural contrasts are especially clear for thick massive sandy beds on the lower tidal flat with direct bore impacts and tidal (annual) rhythmites

(TRs/ARs) on the higher tidal flats. On the intermediate flat, sandy layers have similar structures with typical TBDs in terms of massive bedding and deformation structures but mimic typical TSDs with respect to the textural composition. They are consequently named hybrid deposits (HDs) and typically occur in the middle section of JS11-1 to 4 and in the middle and upper parts of JS10-3 and 4 (Fig. 9). HD formation is ascribed to rapid deposition by the deceleration of shooting flood flows over the intermediate flat without direct bore impacts (see fig. 8 in Fan *et al.*, 2014 for more discussion), so the sorting is improved shoreward with finer sizes after selective export from the lower tidal flat with energetic bores.

In the main channel fine sand and coarse silt are highly mobile and usually transported as saltation and suspension load. This is attested by the C-M diagram of JS10-1 to 5 and JS11-1 to 4 in terms of a lack of the segments denoting bedload transportation (Fig. 11). The TMD (tidal mud deposits) points scatter over a wide range but most of them fall in the SR segment, parallel to the X-coordinate (M value), representing sediments transported as a uniform suspension (Passega, 1964). The TSD (including HD) and TBD points are mostly distributed at the lower and upper part of the RQ segment, respectively. The RQ segment, parallel to the C = M line, denotes sediments transported as a graded suspension. Some TSD points above the RQ segment with larger C values are attributed to a sampling bias in the two segments rich in platy

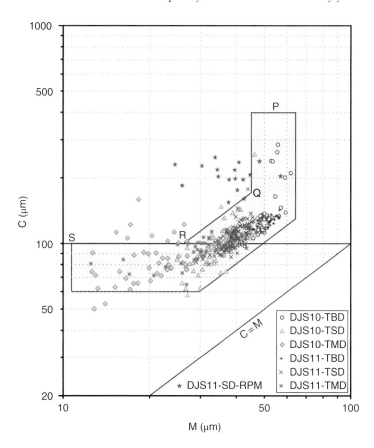

Fig. 11. C-M diagram of DJS intertidal deposits indicating that sediments in the middle Qiantang Estuary with tidal bores are largely transported as suspension load with little bedload. The mode of sediment transport is inferred to be uniform suspension in the SR segment, graded suspension in the RQ segment and graded suspension with some rolling components in the PQ segment. C and M values are the 1th and 50th percentile diameters on the grain-size cumulative curve, respectively.

minerals in JS11-1 to 4 (RPM, Fig. 8). A few TBD points scattering around the PQ segment, parallel to the Y-axis (C value), represent sediments transported mainly in a graded suspension with some rolling components. It is therefore concluded that most of sediment particles in the bore-affected estuarine section were transported as suspended load, with a small fraction, if any, transported as bed load.

DIAGNOSTIC CRITERIA FOR RECOGNIZING TIDAL-BORE DEPOSITS

A tidal bore is a high-energy destructive agent eroding and dispersing with no or little deposition. However, as sediments which have been eroded somewhere should be deposited elsewhere, most of the sediments in the middle Qiantang Estuary have a linkage with tidal bores. Surface sediments in the main channel and on the lower tidal flat are subject to erosion and resuspension by energetic tidal bores. The resuspended particles are potentially maintained in a graded suspension during highly turbulent supercritical flows (Fig. 11). Some of them settle back to the main channel floor and the lower tidal flat after deceleration of the flood flow, resulting in rapid deposition with well-developed massive bedding, graded bedding and (or) parallel lamination, which have a high potential to be deformed by liquefaction triggered by latter tidal bores (Fig. 12). Others are dispersed and deposited by shooting and regular tidal flows on the middle and upper tidal flats, producing HD, TR and AR deposits (Figs 9 and 12). Tidal-bore deposits (TBDs) are here strictly referred to as the depositional packages produced and deformed directly by tidal bores in the main channel, on the adjacent lower tidal flat and in secondary channels. They obviously differ from other intertidal deposits in terms of textural and structural composition (Figs 6 to 9 and 12).

A typical TBD at least comprises two of the following sedimentary features: (1) erosion surface, (2) massive bedding, (3) graded bedding, (4) parallel lamination and (5) soft-sediment deformation structures; although, none of these structures is exclusive to other geological processes, like river floods and turbidity currents. Additional criteria are therefore needed to constrain the depositional environment in a tidal-fluvial transition zone.

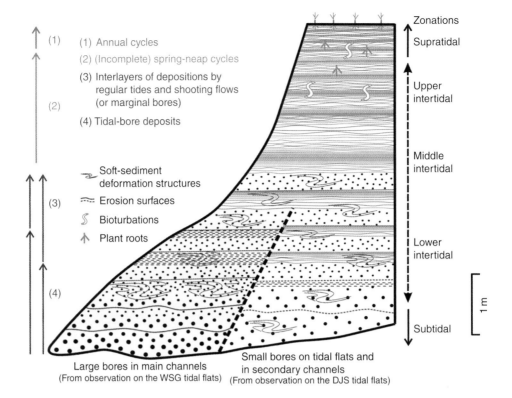

Fig. 12. A schematic facies model of the channel-flat complex in the middle Qiantang Estuary with tidal bores.

Bi-directional cross-stratifications are a good indicator of tidal settings but they are little recognised in core sediments with TBDs. Thalweg sediments in the middle estuary are generally finer and better sorted than in the two ends because of bedload convergence (BLC, Dalrymple *et al.*, 2012; Tessier, 2012), but they should be traceable over a very great distance in a macro-to-hyper-tidal estuary. For example, the middle Qiantang Estuary extends roughly 115 km from Ganpu to WY (Figs 1 and 2). So the target is hardly achieved to trace fossil sediment variations along the river valley by examining outcrops of their limited extent. Lateral (perpendicular to the river valley) variations in sedimentary facies are usually obvious over a limited distance (a small thickness), so they have been part of routine procedures to seek out clues for a tidal-setting interpretation.

Shoreward facies transitions from TBDs, through to HDs and TRs, to ARs are clearly shown along transect JS10-1 to 5 and profile JS11-1 to 4 on the DJS tidal flats (Figs 8 and 9). TRs with (incomplete) spring-neap tidal cycles are generally exclusive to the middle Qiantang Estuary due to sheltering against direct impacts of big waves and to abundant sediment supply resuspended from the main

channel by tidal bores and dispersed by tidal flows. This is completely different from open-coast tidal flats along the outer Qiantang Estuary and Hangzhou Bay, where rhythmic bedding generally reflects seasonal alternations of rough and calm seas (Fan, 2012, 2013). Other modern tidal rhythmites have so far been reported to develop principally in middle estuaries with tidal bores, like the Cobequid Bay-Salmon River Estuary (Dalrymple *et al.*, 1991), Mont-Saint-Michel Bay (Tessier, 1993) and Turnagain Arm (Greb *et al.*, 2011). Furthermore, some typical structures linked to tidal-bore activity are also well-developed in these estuaries, like parallel sandy lamination on the UFR (upper-flow regime) braided sand flats in the Cobequid Bay-Salmon River Estuary (Dalrymple *et al.*, 1991), soft-sediment deformation structures on the intertidal flats flanking the main channel in Mont-Saint-Michel Bay (Tessier & Terwindt, 1994) and at Turnagain Arm (Greb & Archer, 2007). Thus, the typical stacked facies sequence of TBDs, HDs, TRs and ARs (Fig. 12) is assumed to be a good indicator of a middle estuary with tidal bores (also see Tessier, 1998).

There are some differences between DJS and WSG tidal-flat deposits. WSG tidal flats have coarser sediments and a greater abundance and diversity of bore-genetic sedimentary structures than DJS tidal flats (Figs 6 and 7 *vs*. Fig. 8) because tidal bores are relative smaller at DJS and grow to a maximum near WSG, opposite to DQK (Fig. 1B). WSG tidal flats are at present devoid of higher tidal flats and marshlands, differing from DJS tidal flats which have all subenvironments. However, tidal flats were much more developed with higher tidal flats before large-scale land reclamation during the 1950s to 1990s. Therefore, the schematic facies model of stacked channel-flat complexes reflecting these subentities is potentially useful in comparative studies of modern and fossil analogues in the middle estuary with small or large tidal bores (Fig. 12). Applications of this facies model should also pay attention to sediment availability, size composition and river runoff which exert significant impact on sediment dynamics, morphodynamics and strata formation.

CONCLUSIONS

The middle Qiantang Estuary is roughly 115 km in length, in a hypertidal-fluvial transition, breeding the world's largest tidal bores. It features a highly meandering channel with thalweg sediments predominantly consisting of fine sand and coarse silt, which are easily resuspended by energetic tidal bores and shooting flood flows. So the channel morphology is highly mobile in terms of short-term growing and shifting of intertidal banks over large spatial extents under complex interactions of tidal bores, tidal flows and river runoff. During dry seasons and periods huge amounts of sediment are resuspended and transported upstream by tidal bores and shooting flood flows, leading to the extensive development of intertidal flats along the north bank and significant siltation in the main channel, which in turn suppress the growth of tidal bores. This trend is reversed during rainy seasons and periods with obvious erosion along the north bank and in the main channel, whilst extensive intertidal flats develop along the south bank and tend to shift downstream as the rainy period continues. Intertidal banks have a more complex morphology in the downstream broad channel than in the upstream narrow channel and the former tend to preserve full-developed channel-flat sequences with various facies subdivisions.

Lateral variations in sedimentary facies are obvious in the middle Qiantang Estuary. The facies succession starts with tidal-bore deposits (TBDs) in the main channel and on the lower tidal flat, through to hybrid deposits (HDs) near the mean low-water neaps (MLWN) and tidal rhythmites (TRs) with (incomplete) spring-neap tidal cycles on the middle to upper tidal flats, to annual rhythmites (ARs) on the upper tidal flats and marshlands. A typical TBD is here strictly referred to as a depositional package produced and deformed directly by tidal bores which at least consists of two of the following sedimentary features: (1) erosion surface, (2) massive bedding, (3) graded bedding, (4) parallel lamination and (5) soft-sediment deformation structures. TBDs differ from tidal sandy deposits (TSDs) in having coarser sizes and poorer sorting because of their rapid deposition. HDs refer to certain sand-dominated packages having similar structures as TBDs but their textural composition is analogous to that of TSDs within the spring-neap tidal cycles. Their generation is linked to activities of shooting tidal flows near MLWN, with imprints of selective sediment export from the main channel before rapid deposition and later deformation by liquefaction, resulting from intense bottom shear stress induced by shooting flood flows. The well-developed TRs with (incomplete) spring-neap tidal cycles are attributed to three favourable conditions: (1) abundant sediment supply from the channel resuspended and dispersed by tidal bores and rapid flood flows, (2) relatively protected settings with regard to direct wave impact and (3) very rapid accretion of intertidal flats.

TBDs have a few distinct sedimentary structures but none of them is exclusive. Additional criteria are therefore needed to constrain the depositional environment in a tidal-fluvial transition. The good criterion is an ideal stacked facies succession of the channel-flat complex with massive sandy beds rich in erosion surfaces and deformation structures at the base and heterolithic rhythmites with (incomplete) spring-neap tidal cycles at the top.

ACKNOWLEDGEMENTS

This work was supported by the National Natural Science Foundation of China (NSFC) under Grant Numbers 41276045 and 41076016, the Fundamental Research Funds for the Central University and the China Geological Survey (GZH201100203). We wish to thank Yijing Wu, Guofu Cai and Mengying Zhang for their assistance in fieldtrips and laboratory work. Great appreciation goes to Profs. B. Tessier and P.L. de Boer for constructive suggestions on previous versions and language polishing.

REFERENCES

Bartsch-Winkler, S. and Lynch, D.K. (1988) *Catalog of Worldwide Tidal Bore Occurrences and Characteristics.* US Geol. Surv. Circ., No. **1022**, 17 pp.

Cai G., Fan, D., Shang, S., Wu, Y. and Shao, L. (2014) Difference in grain-size parameters of tidal deposits derived from the graphic and moment methods and its potential causes. *Mar. Geol. Quat. Geol.*, **34**, 195–204.

Chanson, H. (2012) *Tidal Bores, Aegir, Eagre, Mascaret, Pororoca: Theory and Observations.* World Scientific, Singapore, 200 pp.

Chen, J., Liu, C., Zhang, C. and Walker, H.J. (1990) Geomorphological development and sedimentation in Qiantang Estuary and Hangzhou Bay. *J. Coastal Res.*, **6**, 559–572.

Dalrymple, R.W. and Choi, K. (2007) Morphologic and facies trends through the fluvial-marine transition in tide-dominated depositional systems: A schematic framework for environmental and sequence-stratigraphic interpretation. *Earth-Sci. Rev.*, **81**, 135–174.

Dalrymple, R.W., Mackay, D.A., Ichaso, A.A. and Choi, K.S. (2012) Processes, morphodynamics and facies of tide-dominated estuaries. In: *Principles of Tidal Sedimentology* (Eds R.A. Davis, Jr. and R.W. Dalrymple), 79–107. Springer, London.

Dalrymple, R.W., Makino, Y. and Zaitlin, B.A. (1991) Temporal and spatial patterns of rhythmite deposition on mud flats in the macrotidal Cobequid Bay-Salmon River estuary, Bay of Fundy, Canada. In: *Clastic Tidal Sedimentology* (Eds D.G. Smith, G.E. Reinson, B.A. Zaitlin and R.A. Rahmani). *Can. Soc. Petrol. Geol. Mem.*, **16**, 137–160.

Fan, D. (2013) Classification, sedimentary features and facies associations of tidal flats. *J. Palaeogeogr.*, **2(1)**, 66–80.

Fan, D. (2012) Open-Coast Tidal Flats. In: *Principles of Tidal Sedimentology* (Eds R.A. Davis, Jr. and R.W. Dalrymple), 187–229. Springer, London.

Fan, D., Cai, G., Shang, S., Wu, Y., Zhang, Y. and Gao, L. (2012) Sedimentation processes and sedimentary characteristics of tidal bores along the north bank of the Qiantang Estuary. *Chin. Sci. Bull.*, **57**, 1157–1167.

Fan, D., Tu, J., Shang, S. and Cai, G. (2014) Characteristics of tidal-bore deposits and facies associations in the Qiantang Estuary, China. *Mar. Geol.*, DOI: 10.1016/j.margeo.2013.11.012.

Feng, Y., Li, Y., Xie, Q. and Zhang, L. (1990) Morphology and activity of sedimentary interfaces of the Hangzhou Bay. *Acta Oceanol. Sinica*, **12**, 213–223.

Greb, S.F. and Archer, A.W. (2007) Soft-sediment deformation produced by tides in a meizoseismic area, Turnagain Arm, Alaska. *Geology*, **35**, 425–438.

Greb, S.F., Archer, A.W. and de Bore, D.G. (2011) Apogean-perigean signals encoded in tidal flats at the fluvial-estuarine transition of Glacier Creek, Turnagain Arm,

Alaska: implications for ancient tidal rhythmites. *Sedimentology*, **58**, 1434–1452.

Greb, S.F. and Martino, R. (2005) Fluvial-estuarine transitions in fluvial-dominated successions: examples from the Lower Pennsylvanian of the Central Appalachian Basin. In: *Fluvial Sedimentology VII* (Eds M.D. Blum, S.B. Marriott and S.F. Leclair). *Int. Assoc. Sedimentol Spec. Publ.*, **35**, 425–451.

Han, Z., Dai, Z. and Li, G. (2003) *Regulation and Exploitation of Qiantang River Estuary.* China Water Publication, Beijing, 554 pp.

Lanier, W.P. and Tessier, B. (1998) Climbing-ripple bedding in the fluvio-estuarine transition: a common feature associated with tidal dynamics (modern and ancient analogues). In: *Tidalites: Processes and Products* (Eds C.R. Alexander, R.A. Davis, Jr. and V.J. Henry). *SEPM Spec. Publ.*, **61**, 110–117.

Lin, B. (2008) *Characters of Qiantang Bore.* Chinese Ocean Press, Beijing, 212 pp.

Longhitano, S.G., Mellere, D., Steel, R.J. and Ainsworth, R.B. (2012) Tidal depositional systems in the rock record: a review and new insights. *Sed. Geol.*, **279**, 2–22.

Martinius, A.W. and Gowland, S. (2011) Tide-influenced fluvial bedforms and tidal bore deposits (Late Jurassic Lourinhã Formation, Lusitanian Basin, Western Portugal). *Sedimentology*, **58**, 285–324.

Pan, C., Lin, B. and Mao, X. (2007) Case study: Numerical modeling of the tidal bore on the Qiantang River. *China. J. Hydraul. Eng.*, **133(2)**, 130–138.

Passega, R. (1964) Grain size representation by CM patterns as a geological tool. *J. Sed. Petrol.*, **3**, 830–847.

Simpson, J.H., Fisher, N.R. and Wiles, P. (2004) Reynolds stress and TKE production in an estuary with a tidal bore. *Estuar. Coast. Shelf Sci.*, **60**, 619–627.

Tessier, B. (2012) Stratigraphy of tide-dominated estuaries. In: *Principles of Tidal Sedimentology* (Eds R.A. Davis, Jr. and R.W. Dalrymple), 109–128. Springer, London.

Tessier, B. (1998) Tidal cycles: annual versus semi-lunar records. In: Tidalites: Processes and Products (Eds: Alexander, C., Davis Jr., R.A. & Henry, V.J.), SEPM Special Publication, **61**, 69–74.

Tessier, B. (1993) Upper intertidal rhythmites in the Mont-Saint-Michel Bay (NW France): perspectives for paleo-reconstruction. *Mar. Geol.*, **110**, 355–367.

Tessier, B. and Terwindt, J.H.H. (1994) An example of soft-sediment deformations in an intertidal environment: the effect of tidal bores. *CR Acad. Sci. Paris*, **319**, 217–223.

Van Den Berg, J.H., Boersma, J.R. and Van Gelder, A. (2007) Diagnostic sedimentary structures of the fluvial-tidal transition zone – evidence from deposits of the Rhine and Meuse. *Neth. J. Geosci.*, **86(3)**, 287–306.

Wolanski, E., Moore, K., Spagnol, S., D'Adamo, D. and Pattiaratchi, C. (2001) Rapid, human-induced siltation of the macro-tidal Ord river estuary, western Australia. *Estuar. Coast. Shelf Sci.*, **53**, 717–732.

Yu, Q., Wang, Y., Gao, S. and Flemming, B. (2012) Modeling the formation of a sand bar within a large funnel-shaped, tide-dominated estuary: Qiantangjiang Estuary, China. *Mar. Geol.* **299–302**, 63–76.

Zhang, G. and Li, C. (1996) The fills and stratigraphic sequences in the Qiantangjiang incised paleovalley, China. *J. Sed. Res.* **66**, 406–414.

Tidal-bore deposits in incised valleys, Albian, SW Iberian Ranges, Spain

MANUELA CHAMIZO-BORREGUERO*, NIEVES MELÉNDEZ*† and POPPE L. DE BOER‡

* Departamento de Estratigrafía (UCM) Grupo de Análisis de Cuencas Sedimentarias (UCM-CAM), Facultad de Ciencias Geológicas, Universidad Complutense de Madrid, 28040, Madrid, Spain
† Instituto de Geociencias (IGEO), (UCM, CSIC)
‡ Sedimentology Group, Department of Earth Sciences, Utrecht University, P.O. Box 80.115, 3508 TC Utrecht, The Netherlands

ABSTRACT

Well-preserved Albian tidal-bore deposits in the south-western Iberian Basin, bordering the Tethys, are interbedded with ephemeral alluvial, tidal and aeolian sediments. In five sections, five sedimentary facies associations are distinguished, corresponding to tidal flat (TF), tidal bore (TB), aeolian (A), ephemeral alluvial (EA) and overbank (OB) deposits. Ephemeral alluvial discharge and tidal reworking alternated with varying dominance and the continuous presence of ventifacts, feldspar and fine windblown sand reflects an arid source area and continued aeolian activity. Three main units are separated by laterally continuous, deep and sharp incisions, attributed to drops of relative sea-level. Unit 1 records the evolution from a tidally reworked, arid, ephemeral alluvial system to tidal flats and aeolian dunes. The most remarkable characteristics of Units 2 and 3 are valley incisions at their bases with a fill of up to 9m-thick high-energy flood-tide-supplied deposits interpreted as the product of tidal bores. After filling up of the incisions, Units 2 and 3 show a reactivation of alluvial discharge, although weaker than in Unit 1. Thus, erosion and the subsequent fill of these valleys indicate a relative sea-level fall leading to valley incision, followed by continued deposition by episodic, long-lasting tidal bore activity in the gradually increasing accommodation space provided by a very slow, gradual sea-level rise. As sea-level changes may have been 2 or 3 orders of magnitude slower than in glacial periods, such fill may have lasted thousands of years. The up to 9m-thick tidal bore deposits consist of well-sorted sands with upper-stage plane-bed low-angle planar lamination and cm to dm-scale very low-angle tangential planar cross bedding with internal erosion surfaces. The sands are carbonate-cemented and contain dinoflagellates. Where observable, the palaeocurrent pattern is persistently inland. Tidal-bore deposits were generated by high-energy flood tides while alluvial discharge was low or absent. The tide was amplified due to basin resonance in the adjacent Tethys and in the funnel-shaped estuary.

Keywords: tidal bore, incised valleys, tide amplification, tide influence, estuary.

INTRODUCTION

A tidal bore is a flood-current surge, commonly driven by a large tidal range in the adjacent sea, entering a shallow funnel-shaped river mouth with a gently sloping bottom (Lynch, 1982; Kjerfve & Ferreira, 1993; Chanson, 2001; 2011). Tidal bores have been described in detail for 67 river mouths and/or estuaries around the world (Bartsh-Winkler & Lynch, 1988; Kjerfve & Ferreira, 1993) and worldwide over 450 estuaries are affected by a tidal bore, on all continents except Antarctica (Chanson, 2011). Tidal bores depend on a delicate balance between features such as topography, tidal amplitude, fluvial runoff and the morphology of the river mouth (Lynch, 1982; Chanson, 2011).

Contributions to Modern and Ancient Tidal Sedimentology: Proceedings of the Tidalites 2012 Conference, First Edition. Edited by Bernadette Tessier and Jean-Yves Reynaud.
© 2016 International Association of Sedimentologists. Published 2016 by John Wiley & Sons, Ltd.

Although any high tide can generate a bore, about half of the known bores are associated with tidal amplification due to resonance in the adjacent tidal basin (Lynch, 1982; Chanson, 2011). Due to the critical conditions, some of the known examples have been greatly reduced or have disappeared due to human activity in recent history (Bartsh-Winkler & Lynch, 1988).

Tidal bores have been studied for their hydrodynamical and physical properties (Lynch, 1982; Kjerfve & Ferreira, 1993; Chanson, 2001; Donnelly & Chanson, 2005; Chanson *et al.*, 2011, Fan *et al.*, 2012), from an ecological point of view (Donnelly & Chanson, 2005; Chanson, 2005, 2011) and experiments have been carried out in flumes (Chanson, 2001; Donnelly & Chanson, 2005;

Chanson & Docherty, 2012; Kherzi & Chanson, 2012). The effects of tidal bores on sedimentary processes in the Qiantang Estuary were studied by Chen *et al.* (1990) and Fan *et al.* (2012, 2014) and tidal bore-induced soft sediment deformation has been reported (Tessier & Terwindt, 1994; Greb & Archer, 2007; Fan *et al.*, 2012, 2014). Examples from the fossil sedimentary record are rare. Martinius & Gowland (2011) reported on tide-influenced fluvial deposits in the Late Jurassic Lourinhã Fm (Western Portugal) with signals of tidal bore activity. This rarity may be due to geologically short-lived conditions favouring tidal bores and/or because criteria for their identification are not well established, so that tidal bore deposits were not recognised as such.

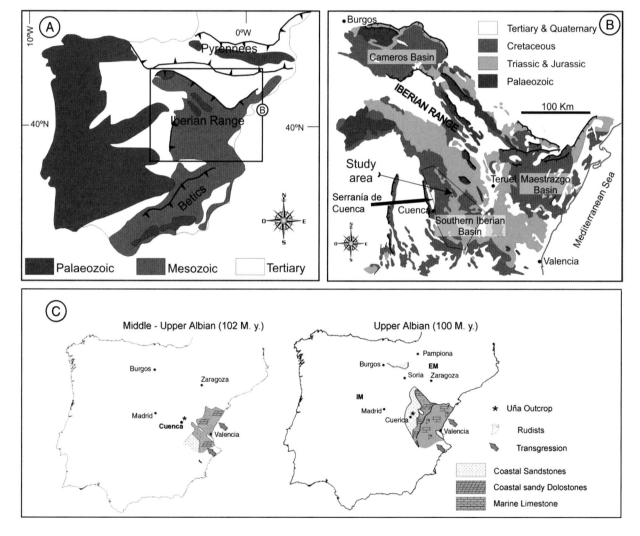

Fig. 1. (A) The Iberian Range in Eastern Spain. (B) Simplified geological map and subdivision of the Iberian Range with the location of the study area within the Serranía de Cuenca sub-basin (Southern Iberian Basin). (C) Albian palaeogeography of the Iberian Basin (modified after Segura *et al.* 2004). IM: Iberian Massif; EM: Ebro Massif.

A well-exposed outcrop of Albian siliciclastic deposits near Uña (SW Iberian Ranges, Spain) (Fig. 1) shows an ephemeral alluvial system with an arid source area and sediment redistribution by high-energy tidal currents. Two intervals, up to 9 m-thick, of stacked sandstone beds with upper-stage plane-bed lamination and persistent flood-dominated transport directions, separated by internal erosion surfaces, filling incised valleys, are described and interpreted to be the result of repeated accumulation by tidal bores. This fits to the palaeogeographical configuration with an adjacent funnel-shaped shallow marginal seaway, bordering the Tethys Ocean, that allowed tidal amplification and the generation of tidal bores (cf. Sztanó & de Boer, 1995).

Taking advantage of this well-exposed outcrop, the main objectives of this paper are: (i) to describe and interpret deposits related to tidal bore action. These are unusual facies, poorly described and recognised in the sedimentary record; (ii) to discuss the occurrence of these tidal bore deposits infilling depressions interpreted as incised valleys; and (iii) to discuss the relation between preserved tidal bore deposits and relative sea-level rise.

GEOLOGICAL SETTING

The Iberian Range (Fig. 1) forms a NW-SE striking intraplate fold belt and represents an uplifted (inverted) Mesozoic basin. The Iberian Basin had a similar outline as the present Iberian Range and was formed by crustal thinning during the Late Permo–Triassic and Late Jurassic–Early Cretaceous rift stages (Salas & Casas, 1993; Capote *et al.*, 2002; Martín-Chivelet *et al.*, 2002).

The Iberian Basin was fragmented and four major fault-controlled palaeogeographic domains evolved, with several sub-basins controlled by reactivated NW-SE and NE-SW late-Variscan and post-Variscan and/or Triassic faults (Soria *et al.*, 2000; Salas *et al.*, 2001). Compression during the Tertiary led to deformation and uplift (Capote *et al.*, 2002).

The study area is located within the southwestern domain of the Iberian Basin (Fig. 1B) in the Serranía de Cuenca sub-basin. This NW-SE elongated sub-basin was controlled by NW-SE and NNW-SSE extension faults and bordered by the remnants of uplifted Palaeozoic bedrock, consisting mainly of quartzites and slates (Gutiérrez-Marco, 2004). Internally minor NNW-SSE faults divided the sub-basin and several stepped SSW-NNE extension faults produced palaeo-thresholds during the Upper Cretaceous sea-level rise. The synrift sedimentary record of this sub-basin (Fig. 2) includes the Barremian Weald Facies (La Huérguina Fm) and a scarce record of shallow marine deposits, time equivalent to the Urgonian Platforms (Aptian) towards the open marine Tethys in the SE (Meléndez, 1983; Meléndez *et al.*, 1994).

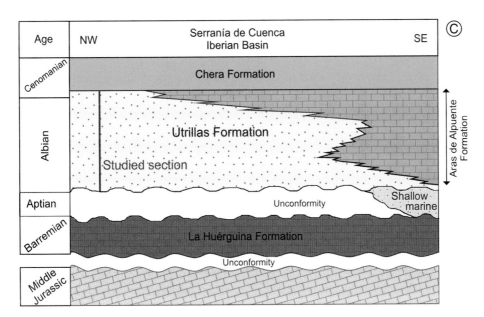

Fig. 2. Chronostratigraphic chart of the Middle Jurassic to the middle Cretaceous in the Serranía de Cuenca. The stratigraphic position of studied sections in the Utrillas Formation is indicated.

The synrift deposits in the Serranía de Cuenca sub-basin are covered by the siliciclastic Albian to Early Cenomanian Utrillas Sandstone Formation, the lateral equivalent of the mid-Cretaceous erg system between the Variscan Massif and the western Tethyan margin (Iberian Basin) (Rodríguez-López *et al.*, 2008) to the NW and transgressive shallow-marine carbonates to the SE (Meléndez, 1983; Segura *et al.*, 2004; García & Mas, 2004). The erg system is up to 300 m-thick and extends over more than 18,000 km² throughout the Iberian Basin (Rodríguez-López *et al.*, 2008; 2012). This paper focuses on a laterally equivalent outcrop near Uña, which records the interaction of alluvial, aeolian and shallow marine processes. The NNE-SSW running Albian Tethys palaeocoastline (Fig. 1C) was located ~30 km from the Uña outcrop towards the SE (Meléndez, 1983; Segura *et al.,* 2004; García & Mas, 2004).

FACIES ANALYSIS

Based on body geometry, lithology, internal structures, bioturbation and palaeocurrent characteristics, 14 sedimentary facies are distinguished (Table 1), grouped into: tidal flat (TF), tidal bore (TB), aeolian (A), ephemeral alluvial (EA) and overbank deposits (OB) facies associations (Figs 3 to 10).

Tidal Flat Deposits (TF)

Description

This facies association consists of facies TF1 to TF4 (Table 1; Fig. 3). TF1 is characterised by decimetre-scale, planar, cross-bedded coarse sandstones and pebbles (Fig. 3A). Sets are usually fining upwards and occasionally foresets have been deformed. Facies geometry is tabular and laterally continuous over more than 70 m. Boundary surfaces are planar and erosive. Maximum stacking thickness is 6 metres. Cross bedding points to palaeocurrent directions mainly towards the NE. This facies grades laterally into facies TF2.

TF2 is characterised by white fine-grained sandstones. Trough cross bedding (Fig. 3B) shows a predominant NW palaeocurrent direction and is intensively bioturbated by 'Arenicola-type' organisms, *Skolithos/Arenicolites Ichnofacies* (Fig. 3D).

TF3 consists of medium-grained to coarse-grained sandstones with NE-oriented climbing ripples. Ripple height varies from 5 to 20 cm and ripple sets stack up to 1.2 m vertically. Sets are continuous over up to 2 m and cover overbank/floodplain deposits with a slightly erosive, sharp surface. The upper boundary surface is sharply erosive (Fig. 3C, E and F). It is an uncommon facies in the area and only appears in section III as the lowermost sediment within a depression (Fig. 3C).

Facies TF4 consists of laterally continuous, up to one metre thick, massive grey clays with scarce reddish mottling. It is interbedded by facies TF1 and TF2.

Interpretation

The tabular and laterally continuous, decimetre-scale, planar, cross-bedded sandstones of facies TF1 are interpreted as tidal sand flat deposits and the fine-grained, trough-cross-bedded sandstones of facies TF2, covering facies TF1 with an erosive contact, as tidal channel deposits.

Considering the Albian palaeogeography (Fig. 1; Meléndez, 1983; Segura *et al.* 2004), NE to NW palaeocurrent directions can be assigned to inland currents. Thus the TF1 and TF2 palaeocurrent pattern is clearly inland and ascribed to flood dominance in the upper part of the intertidal zone and a consequent residual sand transport in the flood direction (cf. de Boer *et al.*, 1989; de Boer, 1998). *Arenicola*-type bioturbation confirms the intertidal conditions (cf. Ekdale *et al.*, 1984).

The climbing ripples in facies TF3 also reflect inland palaeocurrents with a high sediment supply in a flood-dominated setting. Wunderlich (1969) reported on such structures from tidal flats with high sedimentation rates which indeed are found in point bars of sinuous tidal channels (De Mowbray & Visser, 1984; Yokokawa *et al.*, 1995; Choi, 2010). Lanier & Tessier (1998) reported on climbing-ripple bedding as a common feature in the fluvio-estuarine transition zone. In our case this facies association occurs in one unique stratigraphic interval above the floodplain facies, indicating incidental large-scale sand supply.

TF4 is ascribed to mudflat deposition with scarce influence of vegetation.

Table 1. Facies discussed grouped according to the generating processes.

Facies associations	Facies code	Lithology and grain size	Components	Structures	Geometry	Bioturbation	Transport directions	Process	Interpretation
Tidal flats	TF1	Ochre heterometric sandstones	Feldspars and scarce pebbles	Dm-scale planar cross bedding; deformed foresets	Laterally continuous tabular bodies	Moderate	NE	Bedload transport	Sand flats
	TF2	White heterometric sandstones	Feldspars and scarce pebbles	Cm-scale trough cross-bedding	Lenticular	Intensively bioturbated by Arenicola-type organisms	NW	Bedload transport	Sand flats
	TF3	Medium to very coarse sandstones		Climbing ripples	?		NE	High sedimentation rates	Tidal channel
	TF4	Grey Clays		Massive	Tabular; few decimetres to 1.5 m-thick			Settling from suspension	Low-energy environment (mudflat)
Tidal bore deposits	TB	Alternating medium and fine ochre sandstones	Carbonate-cemented	Dm-scale high energy low angle planar lamination and trough cross-bedding	Laterally continuous tabular bodies		NW	Very high-energy bedload transport	Tidal bore infill
Aeolian deposits	A1	Very fine well-sorted white and reddish arkosic sandstones		Cm to dm-scale planar and trough cross-bedding	Laterally continuous tabular bodies			Aeolian transport	Aeolian dunes
	A2	Very fine well-sorted white sandstones		Massive	Lenticular			Aeolian transport	Windblown sandstones/ aeolian pods
	A3	Linings of a single quartz pebble and or ventifacts			One pebble-thick, flat-sharp based, laterally continuous			Deflation	Deflation lags
Alluvial ephemeral deposits	EA1	White heterometric arkosic siliciclastic sandstones to micro-conglomerates	Disperse and reworked ventifacts, polished pebbles (up to 9cm), mud clasts and feldspar grains (occasionally up to 40% plant remains)	Dm-scale planar and trough cross-bedding	m-scale lenticular bodies	Skolithos/ Arenicolites Ichnofacies	Dominantly NW and SE	Bedload transport	Ephemeral channel with occasional tidal influences
	EA2	Clast supported white heterometric arkosic sandstones	Disperse and reworked ventifacts and polished pebbles. Pebbles up to 8 cm Ø. Feldspar grains	Massive, locally planar cross-bedding	Flat based dm-scale tabular layers		SE and NW	High energy bedload transport	Sheet flow events
Overbank deposits	OB1	Red sandy clays	Disperse sands	Massive; intense reddish mottling	Tabular, bed thickness up to 2 m	Root traces		Settling from suspension	Low-energy oxidized shallow environment with soils
	OB2	Green and grey sandy clays	Disperse sands	Massive. Moderate reddish mottling	Tabular. Thickness range from few dm to 1.5 m	Root traces		Settling from suspension	Low-energy, reducing environment with soils
	OB3	Siliciclastic fine to coarse sandstones.	Occasionally pebbles	Massive and nodular aspect. Occasional trough cross-bedding	Lenticular bodies. Maximum thickness 0.8 m	Root traces		Low-energy Bedload transport	Crevasse channels/ Proximal crevasse splay

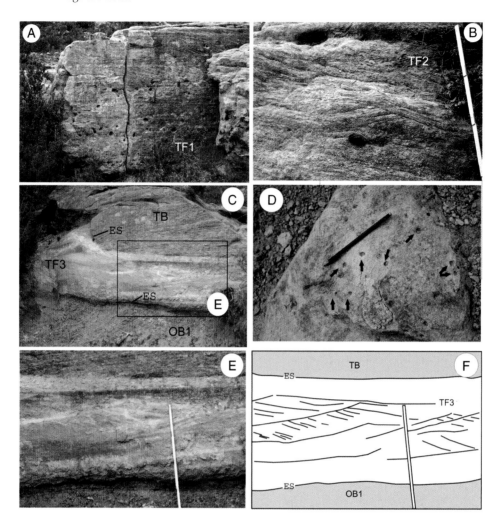

Fig. 3. Tide-influenced deposits. (A) Tabular body of decimetre-scale planar cross-bedding within coarse sandstones with dispersed pebbles (Facies TF1: tidal sand flat). (B) Well-preserved trough cross-bedding in medium to fine-grained white sandstone (Facies TF2: tidal channel). The upper surface of this facies is intensively bioturbated (Fig. 3D). (C) Detail of section III (see correlation panel in Fig. 11 for position) with vertical staking of facies OB1 (overbank deposits with palaeosols), facies TF3 (tidal flat deposits) and facies TB (tidal bore deposits). Facies TF3 is confined in between two erosive surfaces (ES). (D) Detail of *Arenicola*-type burrows on top of facies TF2. (E) Close up of facies TF3 (visible part of measuring rod is 80 cm); (F) Line drawing of climbing ripples in E.

Tidal Bore Deposits (TB)

In this facies association (TB) (Figs 4, 5, 6 and 7) alluvial ephemeral deposits (facies EA1) are occasionally interbedded as a minor facies (Table 1).

Description

Facies association TB consists of alternating medium and coarse to very coarse ochre sandstone, with a minor proportion of feldspars, in gently tangential tabular beds with low-angle planar lamination (Fig. 6B and C) and flat bases and tops, 10 to 20 cm-thick and laterally continuous along the up to 54 m-wide outcrops. Occasionally faint internal reactivation surfaces (Fig. 6D and E) occur. Locally, carbonate cement and dinoflagellates are present (Fig. 7). Bed contacts are erosive and characterised by slightly undulating erosive surfaces that continue through the outcrop. The very low-angle foresets (Fig. 6B and C) are continuous over tens of metres and reflect a persistent NW palaeocurrent.

Eighty percent of the sand grains in TB facies are between 1Φ (0.5 mm) and 3Φ (0.125 mm) and they are very well sorted ($\sigma = 0.21\,\Phi$; sorting value (σ) of very well sorted sand is $<0.35\,\Phi$). In SEM analysis (Fig. 7) grain surfaces are clean and polished and show V-shaped scars.

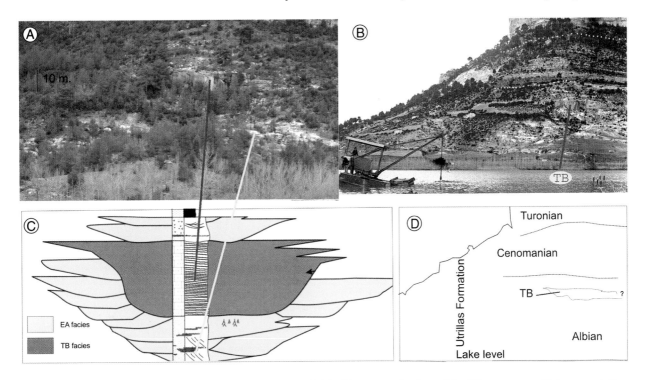

Fig. 4. Tidal bore deposits. (A) Current panoramic view and (B) old panoramic view of the Uña outcrop with less vegetation than today. In both A and B, incision and isolated body geometry with facies TB is indicated. (C) Sketch showing the deep incision in facies EA and the geometry of one isolated body with facies TB. (D) Sketch of the old panoramic view (B) showing the stratigraphical position of the studied deposits.

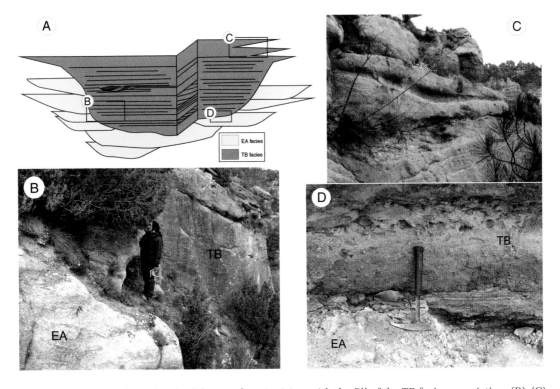

Fig. 5. Tidal bore deposits. (A) 3D sketch of the 9 m-deep incision with the fill of the TB facies association. (B), (C) and (D) show details (see A for location) of the incision fill. (B) View of the westernmost part of the deep and sharp incision into facies EA. (C) Detail of the upper eastern part of the TB deposits, showing a winged geometry. (D) Lowermost part of the TB body showing the erosive contact with facies EA.

Fig. 6. Tidal bore deposits. (A) 3D sketch of the incision and TB fill with location of the other pictures indicated. (B) General view (W-E-oriented) of the fill showing the homogeneity of facies TB. (B') Close-up of B with details of facies TB. (C) General view (SSW-NNE oriented) of the body fill. Lower and upper limits of facies TB are marked with dotted lines: The lower one corresponds to the basis of the deep incision and the upper one is due to erosion by the reactivation of alluvial discharge. (C') Closer view of facies TB, showing laminae with grain-size variations. (D and D') Closer view and sketch of facies TB (W-E oriented) showing a slight internal erosive surface (IES) and reactivation surfaces (RS). (E and E') Perpendicular to picture D: Closer view and sketch of TB deposits (S-N oriented), showing persistent northward palaeocurrent direction and internal erosion surfaces (IES). The resultant palaeocurrent from D & E is towards the NW, i.e. inland.

The TB facies association only fills several steep erosive incisions (up to 9 m-deep and 64 m-wide) and appears in two intervals (Fig. 11). Small reworked ventifacts and polished pebbles are strikingly absent. Alluvial ephemeral deposits, facies EA1, cover the upper part of these incision fills (Figs 4 and 11; Table 1).

Interpretation

The grain-size distribution between 1 and 3 Φ, the good sorting and V-shaped microtextures seen on the quartz grains in SEM pictures are considered evidence of aeolian preselection (see aeolian deposits, below) in combination with coarser arkosic sands brought in by ephemeral rivers. Well-sorted aeolian sands in desert dunes display a mean grain-size of about 2 to 3 Φ (Livingstone *et al.*, 1999). The presence of dinoflagellates in these sands (Fig. 7) and the clean and polished grain surfaces with V-shaped scars indicate that facies TB partly represents marine reworked aeolian sands with admixtures from other sources (Fig. 7; cf. Rodríguez-López, 2006; Trabucho-Alexandre *et al.*, 2011). In addition, marine reworking will have destroyed some of the feldspar from the aeolian and alluvial sands.

Low-angle planar lamination and parallel lamination, interpreted as upper stage plane bed deposits, are the unique structures preserved in these fills. The rare occurrence of small ventifacts and polished pebbles in the well-sorted, fine to medium, quartz-dominated sand and the persistent inland (NW) orientation of very low-angle

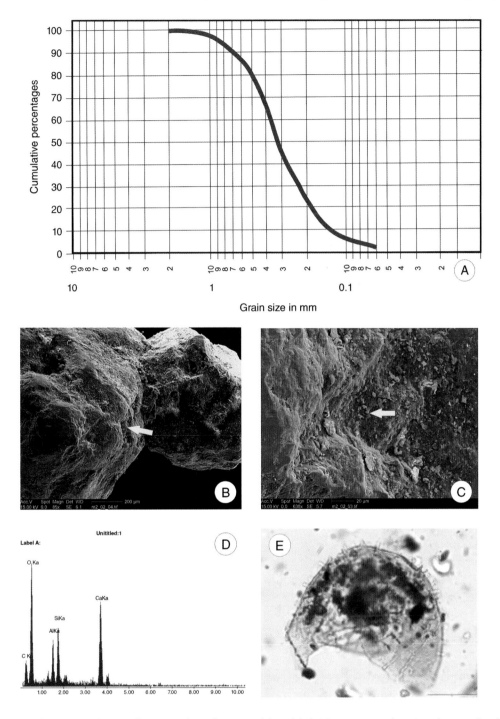

Fig. 7. Tidal bore deposits. (A) Granulometry of TB deposits. (B) and (C) SEM pictures showing clean and polished grain surfaces and V-shaped scars (yellow arrows).(D) Mineralogical composition of carbonate cement in facies TB. (E) Detail of marine dinoflagellate within facies TB; bar atthe bottom right of the picture is 20 μm-wide.

foresets suggest upper-flow regime conditions transporting sediment inland through a narrow incised valley.

Kherzi & Chanson (2012) demonstrated that a tidal bore can transport sediment upstream. Experiments with a movable gravel bed with natural granite gravels of 4.75 mm to 6.70 mm showed that the longitudinal pressure gradient is the dominant agent, de-stabilising the particles and inducing upstream sediment transport.

Fig. 8. Aeolian deposits. (A) Cross-bedded aeolian dune sandstone (facies A1) and (B) line drawing of (A). Superimposition of dunes generates interdune surfaces (IS). Aeolian dune foresets display reactivation surfaces (R). (C) Cross-section through deflation lag (Facies A3, black arrow) in ephemeral alluvial deposits (Facies EA). (D) SEM photograph of an aeolian quartz grain showing typical dish-shaped depressions (white arrow) and elongate shape with rounded edges (cf. Rodríguez-López *et al.*, 2006).

In our case, the TB facies association is interpreted as the result of many tidal bores (multiple events) which contributed to the fill. The slightly undulating surfaces at the contacts between TB layers are interpreted as erosion surfaces and may span numerous bore events. Continuous tidal bore activity during an ongoing and regular sea-level rise thus would have gradually filled the incised valley, over periods of hundreds to thousands of years. Such a prolonged period of infill of the incised valleys is related to the rate of sea-level change in ice-cap-free periods being 2 to 3 orders of magnitude slower than during the Quaternary (cf. Schlager, 1981).

The tidal bore was powerful enough to import medium to coarse-grained sand and to generate low-angle inland-sloping planar lamination. The backflow was more spread in time and thus less

energetic, so that it could not transport the coarser fraction back to the sea. Moreover, some of the tidal bore waters may have spilled over higher up in the valley and may have returned to the sea along a different route, over the surface and as groundwater.

Finally, it is important to consider that all this occurred in a desert system, with very rare precipitation so that rivers supplied fresh water only (very) incidentally and, from time to time, catastrophically. In such cases part of the tidal bore succession may have been eroded, after which erosion and deposition by tidal bores resumed.

Inland sediment transport by modern tidal bores has been documented from the Ord River (Wolanski *et al.*, 2001), the Rio Mearim (Kjerfve & Ferreira, 1993), the Turnagain Arm (Bartsch-Winkler

Fig. 9. Ephemeral alluvial deposits. (A) Ephemeral alluvial deposits (facies EA) erosively overlying overbank deposits (OB). (B) Vertical stacking of facies EA. (C) Line drawing of B (IES: Internal erosion surface) Successive alluvial pulses partly eroded deposits of the previous ones leading to vertical amalgamation of facies EA. (D) Close up of facies EA with feldspars. (E) Interbedded facies EA1 and EA2. (F) Detail of the picture B. (G) Subfacies EA1b. (H) Closer view of (G) showing rhythmic grain-size alternation foresets.

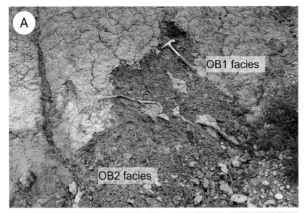

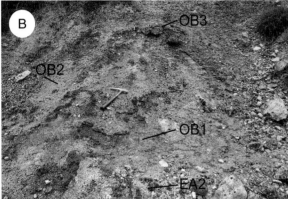

Fig. 10. Overbank deposits. (A) Vertical transition from green-greyish sandy clays (facies OB2) to reddish sandy clays (OB1). (B) Vertical stacking of root traces and deep reddish mottled levels on ephemeral alluvial deposits of facies EA2 and OB1. Lenticular sandstone body on top corresponds to crevasse splay deposits (OB3). (C) Root traces in ephemeral alluvial deposits (EA1).

et al., 1985), the Sée and Sélune (Tessier & Terwindt, 1994; Furgerot *et al.*, 2013), Hangzhou Bay (Chen *et al.*, 1990) and the Qiantang Estuary (Fan *et al.*, 2012; 2014).

Reported examples from the fossil sedimentary record are rare, even though tidal bores should have been active in certain fossil macrotidal embayments/estuaries and produced distinctive deposits during sea-level rise. Macrotidal embayments and estuaries may have been more prevalent earlier in geological time, particularly during periods in the evolution of Pangea and Panthalassa when nearshore and offshore tidal ranges may have been potentially larger than at present (Fan, 2011). However, a lack of preservation potential may have reduced the number of fossil examples and also such deposits may have been misinterpreted because of the absence of an appropriate facies model (Fan *et al.*, 2012).

Martinius & Gowland (2011) reported on tidal bore deposits in the Late Jurassic Lourinhã Fm (Western Portugal). These deposits are characterised by sandy bedforms about 30 cm-thick, with a very homogeneous grain-size distribution and with low-angle landward dipping laminae. The Lourinhã bore deposits were found interbedded within tidally modulated fluvial bedforms and are interpreted to be the result of strong flood tides during low river discharge (Martinius & Gowland 2011). The authors of the present article ascribe the absence of ebb-flow sedimentary structures within tidal bore deposits in Uña to tidal asymmetry, as in the present-day Qiantang Estuary, where the ebb flow is generally weak and has a limited influence (Fan *et al.*, 2012). In addition, the general absence of fluvial water supply in the arid climate of the Iberian Basin may have contributed to the absence of preserved ebb-flow sedimentary structures.

Aeolian Deposits (A)

Description

Aeolian deposits in the study area are represented by facies A1 to A3 (Fig. 8 and Table 1).

Facies A1 is characterised by stacked, laterally continuous beds of cm-scale to dm-scale (very low-angle) planar and trough-cross-bedded, well-sorted very fine sandstones with internal bounding surfaces that truncate cross strata (Fig. 8A and B). These sandstones largely consist of quartz grains, with a variable proportion of feldspars; in addition kaolinite cement is very abundant. SEM photographs (Fig. 8D) show V-shaped scars and bulbous edges on the very well rounded quartz grains. Mean grain-size is in 2 to 3 Φ (0.250 to 0.125 mm).

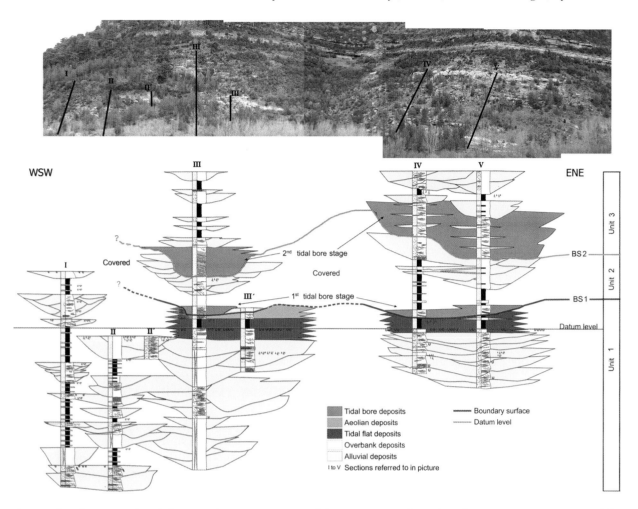

Fig. 11. Correlation panel with facies architecture. Position of the sections is indicated in the upper panoramic view of the outcrop (black lines). Fine red line marks the datum level. Thick red and green lines indicate two sharp erosive surfaces referred to in the text as boundary surfaces BS1 and BS2, that divide the entire succession in three units. Unit 1 shows a clear predominance of overbank and alluvial deposits that grade vertically into tidal flat deposits and subsequently aeolian dunes on top. Units 2 and 3 overly the laterally continuous erosive surfaces BS1 and BS2 and are characterised by the presence of isolated valley fills with tidal bore facies (TB); intermittent alluvial and tidal flat facies are also present but less than in Unit 1.

Facies A2 consists of cm-scale very fine well-sorted massive sandstone lenses interbedded within depressions in ephemeral alluvial (EA) and overbank (OB) deposits (see below). Their occurrence does not show a regular pattern. Usually erosive surfaces incise into facies A2.

A3 consists of discrete lags of quartzite pebbles, reworked wind-faceted pebbles (ventifacts) and polished pebbles, laterally persistent over tens of metres. These lags are typically one pebble thick (Fig. 8C), pebble-size is up to 6 cm and they do not show a preferred orientation.

Facies A2 and A3 usually appear as subordinate facies interbedded within the alluvial-ephemeral deposits described below.

Interpretation

Facies A1, A2 and A3 reveal the effects of aeolian activity in periods of no or minor alluvial discharge.

Facies A1 is interpreted as aeolian dunes (cf. Porter, 1987; Scherer, 2000; Mountney, 2006). The internal bounding surfaces, subparallel to the cross-strata (Fig. 8A and B), are reactivation surfaces that formed in response to the temporary degradation of the dune lee slope before the re-establishment of an active slipface (Mountney & Jagger, 2004). Erosive interdune surfaces were filled by bedforms migrating over each other, thus truncating cross strata (Fig. 8A and B; cf. Mountney, 2006).

V-shaped microscars seen in SEM pictures (Fig. 8D) are interpreted as percussion cracks on grain surfaces due to aeolian activity (cf. Rodríguez-López *et al.*, 2006). The high content of kaolinite cement in facies A1 has also been mentioned in other Albian aeolian deposits from the Iberian Ranges (Rodríguez-López *et al.*, 2009) and is interpreted as the product of telodiagenetic alteration of feldspars in aeolian dune sands.

Facies A2 is interpreted as windblown aeolian sand trapped in depressions during periods with no alluvial discharge but with a groundwater table high enough to enable their preservation. As they have usually been partly eroded by later alluvial processes, they may be classified as aeolian pods (cf. Porter, 1987).

The pebble lags of facies A3 (Fig. 8C) resulted from deflation (cf. Clemmensen & Abrahamsen, 1983; Mountney, 2006) of previous ephemeral alluvial deposits (EA1) and sheet flow deposits (EA2) (cf. Dávila & Astini, 2003; Mountney & Russel, 2004). Similar one-pebble thick deflation lags in Albian sandstones of the Iberian Basin were described by Rodríguez-López *et al.* (2010); having also been described from alluvial fan surfaces and wadi terraces in U.A.E. and Oman (Al-Farraj & Harvey, 2000)

Facies A2 and A3 usually appear as subordinate facies interbedded within the above alluvial-ephemeral deposits. Together, the three aeolian facies reflect aeolian deposition, deflation and wind reworking of previously deposited ephemeral deposits due to the alternating activity of wind and ephemeral runoff.

Ephemeral Alluvial Deposits with tidal influence (EA)

Description

The two facies in this category, EA1 and EA2 (Table 1 and Fig. 9), occur in all studied sections. They fill erosive depressions (channels) (Fig. 9A and B). Locally, aeolian facies A2 and A3 are interbedded.

Facies EA1 (Fig. 9) is represented by fine white, arkosic sandstone to microconglomerate with dispersed pebbles, abundant reworked faceted and polished pebbles up to 9 cm in diameter and fresh feldspars with a diameter up to 2 cm (Fig. 9D). Locally, reworked plant remains and mud clasts occur, as well as *Arenicolites*. This facies contains planar and trough cross-bedding, 20 cm to 50 cm-high, forming stacked dunes with foresets which generally consist of coarse sand (Fig. 9B, C and F). Also, 40 cm-high foresets with alternating coarse and medium sand occur (Fig. 9G and H).

Palaeocurrent directions are to the SE (N120 to N180), to the NE (N20 to N60) and to the SW (N240) and NW (N340 to N355). NW and SE directions dominate and alternate throughout the succession. Based on, amongst others, the palaeocurrent directions, four EA1 subfacies are distinguished:

Subfacies EA1a; with palaeocurrents exclusively towards the SE, containing dispersed pebbles.

Subfacies EA1b; with palaeocurrents exclusively towards the SE and foresets showing rhythmic alternations in grain-size (Fig. 9G and H).

Subfacies EA1c; with both NW and SE palaeocurrent directions.

Subfacies EA1d; with palaeocurrents exclusively towards the NW.

Each subfacies has sharp and flat erosive basal surfaces. There is no transitional grading from one to the other.

Facies EA2 (Fig. 9E) consists of decimetre-scale, massive, sandy-matrix-supported conglomeratic sandstones. Occasionally, clast-supported sandy conglomerates occur. Up to 8 cm faceted and polished pebbles are abundant. The generally massive facies locally shows planar cross bedding with palaeocurrent directions towards the SE and NW. This laterally persistent flat-based tabular facies shows erosive bases.

Vertical and lateral transitions between these two facies (EA1 and EA2) are sudden and sharp. Bounding surfaces are laterally continuous flat erosive bases. Both facies (and the related subfacies) stack unsystematically; there is no evidence for gradual transitions between them.

Facies EA1 and EA2 form lenticular sandy bodies (4 to 65 metres wide and up to 4 metres thick) (Fig. 9B) that amalgamate vertically. Aeolian facies A2 and A3 are locally interbedded as subordinate facies.

The bases of these sand channel bodies are erosive (Figs 9A, B, C, F and 11). Sandbodies amalgamate along the studied sections (Fig 11). The lower parts of the channel fills clearly show a predominance of ebb palaeocurrents whilst the upper parts show a flood prevalence.

Interpretation

EA1 is interpreted as ephemeral channel fills. The apparently chaotic stacking of facies, the absence of gradual transitions between the four

EA1 subfacies, as expected in a regular fluvial system, as well as the presence of arid features such as well-preserved feldspars, polished and faceted pebbles all reflect an arid ephemeral braid plain system. Interbedded aeolian facies with deflation lags indicate long dry stages (cf. Glennie, 1970; Rodríguez-López *et al.*, 2010). In such setting, short periods of high water discharge alternated with long periods of arid conditions and subaerial exposure (cf. Collinson, 1986).

Facies EA developed in a very arid system. The presence of feldspar grains up to 9 cm and deflation lags interbedded within this facies reflect the arid-climate conditions in which the ephemeral fluvial system developed. The presence of abundant *Arenicolites* ichnofacies is indicative of an intertidal environment where organisms cope with stressful conditions. Throughout geological history different animal species have developed habitat with U-shaped burrows in intertidal environments (cf. Ekdale *et al.*, 1984; Woolfe, 1990). This is in agreement with the locally preserved aeolian facies (A2 and A3), indicating a shallow groundwater level.

Variations in preserved palaeocurrent directions (NW and SE) in facies EA1 are interpreted to be the result of tidal activity, in agreement with the general palaeogeography.

EA1a Alluvial-dominated subfacies: This subfacies shows palaeocurrents exclusively to the SE, i.e. towards the sea and thus is interpreted as an alluvial-dominated system. Alluvial discharge and reworking was strong enough to prevent the preservation of possible tidal influences and/or the deposits were formed in periods with decreased tidal activity and/or a lower sea-level.

EA1b Tide-modulated, alluvial-dominated subfacies: This subfacies consists of dunes with palaeocurrent directions exclusively towards the SE. However, characteristic rhythmic alternations of coarse and medium-grained sand in successive foresets (Fig. 9G and H) in each dune indicate that alluvial discharge was regularly decelerated by the rising tide and accelerated during the falling tide (cf. van den Berg *et al.*, 2007; Martinius & van den Berg, 2011).

EA1c Tide-influenced, alluvial-dominated subfacies: This subfacies is characterised by both NW (flood) and SE (ebb/fluvial) palaeocurrent directions. Depending on the energy of the tide and/or the energy of the alluvial discharge, one or both were predominant. The change in dominant current direction from ebb to flood in the vertical

direction is ascribed to the asymmetry of the tide in tide-influenced channels. Due to the tidal asymmetry, maximum flood currents tend to occur in the second part of the flood period when intertidal flats are largely covered by water, while maximum ebb current strength occurs in the second part of the ebb period when tidal flats have been drained for a significant part and the ebb current is concentrated in the channels. Thus, lower parts of channel fills tend to be ebb dominated, whereas in higher parts the flood tends to be more dominant (cf. fig. 7.9 in de Boer, 1998).

EA1d Tide-dominated subfacies: Dune cross-bedding with palaeocurrent directions exclusively to the NW, that is inland, was formed by flood-dominated currents that reworked clastic material previously carried to the coast by the alluvial system and/or were transported alongshore; the presence in this subfacies of ventifacts, polished pebbles, large feldspars and admixtures of aeolian sand reflect the arid sediment source.

Thus, EA1 deposits record the interaction of ephemeral alluvial discharge and tidal processes in an arid braid plain context, with intermittent aeolian activity (facies A2 and A3).

Facies EA2 is interpreted to be the result of sheet flow events. This mostly massive facies shows erosive bases as the result of reactivation of alluvial discharge after dry stages. Strong ephemeral currents redistributed earlier unconsolidated ephemeral deposits within the channels (cf. Wakelin-King & Webb, 2007). Reworked ventifacts and polished pebbles, as well as high amounts (up to 40%) of unaltered feldspars (Fig. 9D) indicate a permanently arid, igneous and/or metamorphic source area and no or hardly any chemical weathering along the transport route.

Overbank/Ephemeral floodplain Deposits (OB)

Description

Intensely mottled reddish, massive sandy clays (facies OB1) and moderately mottled green and grey massive sandy clays (OB2) form the main components of this facies association (Table 1; Fig. 10A and B).

Gradual m-scale vertical transitions from green-grey clays (OB2) to red clays (OB1) are the main feature (Fig. 10A). Usually, facies OB1 shows different intensities of deep reddish mottling and root traces (Fig. 10B). Mottling intensity gradually increases upwards within the two facies.

Decimetre-scale massive lenticular sandstone beds (OB3) are occasionally interbedded in this facies alternation (Fig. 10B). Reddish mottling also frequently appears within ephemeral alluvial deposits (EA) and occasionally it is very concentrated at the top of sandbodies (Fig. 10C).

Interpretation

This facies association represents floodplain deposits. Fine-grained OB1 and OB2 sediments were deposited by settling in an overbank setting (cf. Spalletti & Colombo Piñol, 2005; Rodriguez-López *et al.*, 2006, 2012). Interbedded, lenticular, dm-scale sandstone bodies (OB3) are interpreted as crevasse splays (cf. Spalletti & Colombo Piñol, 2005) formed during alluvial discharge.

Root traces and (intense) reddish mottling and pedogenesis indicate that the floodplain was moist from time to time (cf. Teller, 1998) allowing colonisation by plants (cf. Langford & Chan, 1989). Redoximorphic features in OB2 indicate periodic water infiltration and saturation together with reducing conditions (Driese *et al.*, 1995). The gradual transitions from green-grey (OB2) to red clays (OB1) evidence a gradual change from reducing to oxidizing conditions, probably because of relative deepening of the groundwater level due to sediment accumulation.

FACIES ARCHITECTURE

A correlation panel of five logged sections at the Uña outcrop is shown in Fig. 11. A laterally continuous flat surface, characterised by intense U-shaped (*Arenicola*-type) bioturbation, has been used as correlation *datum* level.

The correlation panel shows the spatial distribution of the facies associations in the logged sections. All facies associations are present in all studied sections but differences in trends are observed. Above the *datum* level two irregular, concave-up, sharp erosion surfaces truncate the underlying strata.

The first erosive surface cuts down to 2 metres deep. Immediately below this, erosive truncations of alluvial facies (in sections I to III), aeolian facies (in section III') and tidal flat facies (in sections IV and V) are well exposed. Above this surface, tidal bore facies, related to the first tidal bore stage (in sections III, IV and V) onlap this erosive surface; locally, (section III') alluvial facies also overlies this erosive surface.

The second erosive surface incises up to 9 metres deep. Alluvial facies below this surface display sharp erosive truncations (in sections III to V). Tidal bore facies of the second tidal bore stage onlaps this erosive surface.

These two erosive surfaces are Boundary Surfaces (BS1 and BS2) that divide the entire succession in three units (Fig. 11): a lower unit (Unit 1), a middle unit (Unit 2) and an upper unit (Unit 3). The two tidal bore sequences identified (Fig. 11) occur above BS1 and BS2, infilling depressions generated during previous erosive periods when sea-level was lower.

Unit 1

Unit 1 is the lower unit. It unconformably overlies the Lower Cretaceous (Barremian) La Huérguina Limestone Fm (Meléndez *et al.*, 1994, fig. 2). The upper boundary of Unit 1 is the erosive bounding surface BS1. The *datum* level is located within Unit 1, near its top (Fig. 11).

The lower part of Unit 1 is characterised by ephemeral alluvial, overbank and tidal flat deposits, as well as minor aeolian activity. Ephemeral alluvial deposits with transported ventifacts grade laterally into overbank deposits. Channel incisions in the floodplain deposits are mainly filled with EA1 and EA2 alluvial and A2 and A3 aeolian deposits. Floodplain deposits show mature palaeosols and colour mottling (roots) occasionally penetrates into the ephemeral alluvial deposits below. Alluvial ephemeral bodies are erosive. They are stacked and amalgamate vertically over a distance of up to 30 m, they show frequent internal erosion surfaces and pinch out laterally into overbank deposits. *Arenicola*-type bioturbation is present in most tide-influenced alluvial facies.

The upper part of Unit 1 is characterised by up to 6 m-thick tidal flat deposits (facies TF1) with clear flood dominance. These deposits are laterally continuous and occur throughout the outcrop.

In most of the sections, tidal flat deposits are covered by aeolian deposits (facies A1). In addition, in section V interbedded tidal and aeolian deposits appear. The greatest preserved thickness of aeolian deposits, in section III', is 0.5 metres (Fig. 11). As the upper limit of Unit 1 is erosive (BS1), aeolian deposits are not preserved throughout the outcrop. In fact, in sections IV and V the erosive upper limit of Unit 1 (BS1) is represented by tidal facies.

Unit 2

Unit 2 overlies Unit 1 with a disconformity (BS1) and is topped by the erosive bounding surface BS2. Unit 2 starts with facies TB belonging to the first tidal bore stage (Fig. 11); this facies fills an erosive incision of up to 2 m.

Above the tidal bore deposits, Unit 2 consists of ephemeral alluvial and overbank deposits, with minor aeolian deposits (facies A1 and A2). Tidal flat deposits are absent. Alluvial ephemeral deposits grade laterally into overbank deposits, as is the case in Unit 1. Channelized sandbodies are narrower and less incised than in Unit 1, so they occur isolated within the overbank deposits. Ventifacts and polished pebbles are also present in Unit 2 but their abundance is fewer and their mean size less than in Unit 1. Alluvial ephemeral deposits in Unit 2 do not show tidal features but the intermittent occurrence of *Arenicola*-type bioturbation is indicative of tide-induced water-level fluctuations.

Unit 3

Unit 3 overlies Unit 2 with a disconformity (BS2). The upper limit of Unit 3 is a conformable contact with the Chera Marls Fm (Fig. 2). As for Unit 2, the tidal bore facies TB, corresponding to the second tidal bore stage, appears at the base of Unit 3, overlying BS2. The BS2 incision is much deeper than BS1. In section III, the TB facies association fills an erosive incision of up to 9 m-deep and 64 m-wide. In sections IV and V, the TB facies association alternates with minor facies EA elements.

Above the tidal bore deposits, Unit 3 consists of ephemeral alluvial and overbank deposits. Tidal flat deposits are absent. Alluvial ephemeral deposits display erosive surfaces at their base, eroding tidal bore facies. They are characterised by channelized bodies and grade laterally into overbank deposits, as is the case in Units 1 and 2. As in Unit 2, channelized sandbodies occur isolated within the overbank deposits and some ventifacts and polished pebbles are present Also, ephemeral alluvial deposits in Unit 3 do not show tidal features but *Arenicola*-type bioturbation was found.

PALAEOGRAPHICAL EVOLUTION

The spatial relationships of the five facies associations and the bi-directionality of palaeocurrent directions are the result of the interaction of ephemeral alluvial discharge, aeolian processes and tidal activity in a coastal to arid alluvial braid plain system. Variations and trends within the three units (Fig. 11) reflect varying palaeogeographical conditions through time (Fig. 12), as well as changes in tectonic, sea-level and/or climate controls.

Unit 1

Unit 1 represents a time interval characterised by high ephemeral alluvial discharge. Some tidal reworking and aeolian activity characterise the lower part and tidal flat deposits the upper part of this unit.

An active source area under semiarid conditions, as reflected by unweathered feldspars, ventifacts and polished pebbles (facies EA1 and EA2), gave rise to ephemeral sheet flow deposits in a broad braid plain environment, after incidental rainfall. In dry periods and despite the arid conditions, palaeosols could develop because of the nearness of the sea (Fig. 1) and a shallow groundwater level that followed the sea-level rise. Aeolian activity deflated previously deposited ephemeral alluvial sediments giving rise to deflation lags (facies A3); windblown sands accumulated in depressions (facies A1) and also were trapped within the palaeosols (facies A2). Several examples from the fossil record (cf. Porter, 1987; Langford & Chan, 1989; Spalletti & Colombo Piñol, 2005; Rodriguez-Lopez, *et al.* 2010) as well as from modern systems (cf. Langford, 1989; Bullard & McTainish, 2003; Al-Farraj & Harvey, 2000; 2004) corroborate this interpretation.

Besides alluvial-aeolian interaction, these deposits also show a clear interaction between alluvial discharge and variable tidal influence. The lowermost channel fills are alluvial dominated and seaward palaeocurrent directions dominate. Some tidal influence is inferred from the palaeocurrent patterns and the presence of *Arenicola*-type bioturbation. Towards the top, the tidal influence recorded in facies EA increases, from tidally modulated to tidally dominated. The fluvial-tidal transition zone shows a gradual change from purely fluvial to purely tidal dominance (cf. Cuevas-Gozalo & de Boer, 1991; Boyd *et al.*, 2006; Dalrymple & Choi, 2007). The length of the transition zone may depend on slope gradient (Martinius & Gowland, 2011). Also, a gradual decrease in alluvial ephemeral discharge in combination with a gradual relative sea-level rise may have allowed

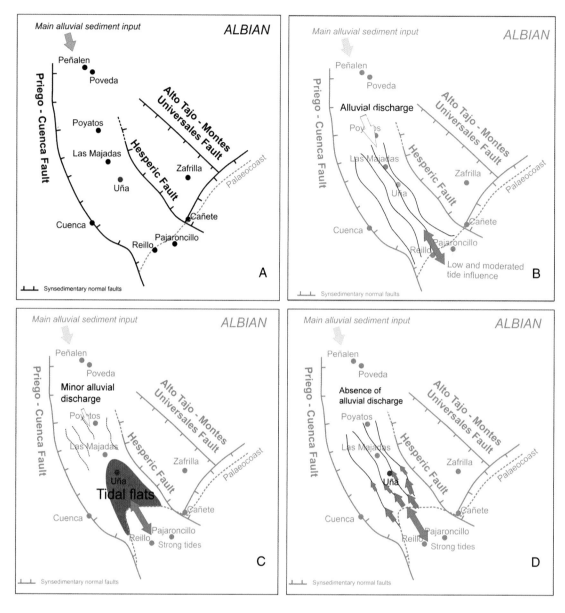

Fig. 12. Palaeogeography of the Serranía de Cuenca sub-basin during Albian. (A) Tectonic context of the area; palaeogeographical position of the source area in the NW and the Tethys in the SE. B, C and D show successive palaeogeographical maps: (B) corresponds to the lower part of Unit 1 (see correlation panel in Fig. 11) with alluvial discharge and floodplains in a fluvial-tidal transition zone; (C) refers to the upper part of Unit 1 with decreasing alluvial discharge and tides reaching the area with the development of tidal flats; (D) represents Units 2 and 3 where deep incisions (valleys) were created and subsequently filled with tidal bore deposits.

the purely tidal zone to migrate upstream along the river and to rework previously deposited sediments.

In the Iberian Basin, a continuous sea-level rise has been documented from the Albian to Turonian resulting in a marine transgression leading to development of extensive carbonate platforms during Upper Cretaceous (Alonso *et al.*, 1993; García & Mas, 2004; Martín-Chivelet *et al.*, 2002).

The Albian sedimentary record shows a desert system from the Iberian Massif (to the NW) to the Tethys (to the SE) (Rodríguez-López *et al.*, 2008, 2010, 2012). These authors documented the Tethys transgression over the coeval erg system under continuous arid conditions.

The interaction of alluvial, aeolian and tidal processes indicates that the lower part of Unit 1 was deposited in an arid-climate ephemeral alluvial system affected by tidal influences. Similar examples have been documented in the northern UAE (Al-Farraj & Harvey, 2004), where distal alluvial fans drain into a tide-influenced marine domain.

The gradual vertical transition in Unit 1, from alluvial to tidal dominance, gave way to the development of tidal flats. Signs of ephemeral alluvial discharge gradually disappear; well-preserved tidal flats with tabular beds, laterally extensive and thick (6 m) sandy deposits with bimodal palaeocurrents and *Arenicola*-type bioturbation developed and accumulated during relative sea-level rise. The relative sea-level rise and subsequent marine inundation of the previous alluvial plain led to tidal reworking of the sandy and gravely sediments and aeolian activity led to the accumulation of aeolian dunes.

The vertical evolution from a tidally reworked arid ephemeral alluvial system to tidal flats and aeolian dunes, as the result of a progressive diminishment of alluvial discharge in a transgressive setting, fits well with the relative sea-level rise in the Iberian Basin during the Albian (cf. Vilas *et al.*, 1983; Alonso *et al.*, l993; Segura *et al.*, 2004; García & Mas, 2004).

Units 2 and 3

From a palaeogeographical point of view, Unit 2 and 3 display a similar evolution. The most remarkable characteristics of these two units are; (i) the presence of deep incision surfaces (BS1 and BS2) at the base of both units (Fig. 11) and (ii) the conspicuous presence of strongly flood-tide induced sedimentation, interpreted as tidal bore stages, that carried sediment upstream through the deeper parts of (small) valleys previously incised during base-level fall.

The accumulation of facies TB (up to 9 m) in the incised valleys overlying boundary surfaces BS1 and BS2 is interpreted to be the result of tidal bore activity during a gradual sea-level rise. In both units a reactivation of ephemeral alluvial discharge is seen after the fill of the incised valleys. The alluvial bodies in Units 2 and 3 are narrower and less incised than those in the Unit 1. A lower amount of ventifacts and polished pebbles (but not of feldspars) also indicates weaker alluvial supply. These alluvial ephemeral sandbodies are isolated, not amalgamated, and are interbedded within and grade laterally into overbank deposits. The absence of tidal signatures in these alluvial

deposits indicates deposition in a proximal position beyond the tidal limit. Alternatively, tidal activity may have decreased. Thus, relative sea-level rise is only reflected by the tidal bore deposits in incised valleys.

Intermittently, when alluvial discharge was low or absent, palaeosols developed. Aeolian activity deflated previous ephemeral alluvial sediments giving rise to interbedded deflation lags. Windblown sands accumulated in depressions and were also trapped within palaeosols, as in Unit 1.

These conditions were maintained up to the top of the Utrillas Sandstone Fm, which is covered by marine marls and carbonates (Chera Fm, Fig. 2) representing the regional transgression of the Tethys, well-documented in the Iberian Basin, that culminated in the maximum global sea-level during the Turonian.

DISCUSSION

Deep incisions in Units 2 and 3, filled with tidal bore deposits (facies TB), differ from other alluvial channel incisions in Units 1, 2 and 3. The near-absence of ephemeral alluvial deposits within these incisions suggests no alluvial discharge or sediment bypass, possibly in relation to further aridification. The incisions, up to 9 m-deep, into sediments that were deposited at or close to sea-level, require a drop of sea-level. In the very arid climate surface runoff has probably been insufficient for eroding the incised valleys. Instead, sapping (cf. Higgins, 1982), with groundwater entering from aside, may well have caused the valley slope to be undermined and eroded.

The two deep incisions (BS1 and BS2) are filled with tidal bore deposits (TB) and no indications of other marine or fluvial deposits were found. This suggests that continental discharge was absent and that strong flood tides, such as tidal bores, brought in sediment from the seaside. This is corroborated by the minor amounts of feldspar and the absence of coarse ventifacts and polished pebbles in these TB deposits.

For the generation of tidal bores, certain conditions, such as tidal amplification in the adjacent sea, a funnel-shaped basin and/or a converging shallow channel with a gently sloping base, are favourable (Chanson, 2005, 2011; Lynch, 1982). The Serranía de Cuenca sub-basin (Figs 1 and 12) indeed had a NW-SE elongate form during the Albian due to the activity of NW-SE synsedimentary extension faults (Capote *et al.*, 2002). During deposition of the Utrillas Fm, in particular during deposition of Units 2 and 3, the sub-basin had a funnel-shaped configuration, controlled by the synsedimentary faults that confined a relatively narrow and long embayment (Fig. 12), connected to the Tethys towards the SE (Meléndez, 1983; Fig. 1). Although palaeogeographic information is insufficient to confirm the conditions for tidal amplification, the above characteristics are certainly not in conflict with the conditions needed for the generation of tidal bores (cf. Lynch, 1982; Bartsh-Winkler & Lynch, 1988; Kjerfve & Ferreira, 1993; Chanson, 2001, 2011).

In the case of a slow and gradual sea-level rise, quite probably because of the absence of (major) ice caps during the Albian, the valley depth may have been continuously shallow, allowing the continued activity of tidal bores. The rising sea-level continuously provided some extra accommodation space in which the tidal bore deposits brought in from the sea could be preserved, while the valley floor thus remained shallow.

In this way, the 9 metres of facies TB at the base of Unit 3 may have accumulated during an about 9 metre sea-level rise, which may have taken many thousands of years, considering the slow sea-level changes during the Cretaceous (cf. Schlager, 1981). Of course, only a minor part of the tidal bore deposits definitely became part of the sedimentary record as much of it will have been intermittently reworked and eroded. Obviously it is not possible to reconstruct whether tidal bores have been active continuously, i.e. during each flood tide, or if they only occurred around spring tides.

Tide-influenced sediments have been widely reported in Albian shallow marine sediments from this sub-basin (Meléndez, 1983) and in Albian deposits around the Iberian Basin (Martín-Chivelet *et al.*, 2002; García & Mas, 2004). In addition, Rodríguez-López *et al.* (2006, 2012) reported on semi-diurnal tides (but no tidal range was specified) in Albian shallow marine deposits some 120 km towards the NE, also connected to the Tethys. This allows inferring that tides in the study area have also been semi-diurnal; semi-diurnal tides generally have a much greater amplitude than diurnal tides (cf. Pugh, 1987).

Tidal bores are a very delicate phenomenon; any change in environmental conditions may lead to their disappearance (cf. Bartsh-Winkler & Lynch, 1988; Chanson, 2005; Martinius & Gowland,

2011). In Unit 3, reactivation of the alluvial discharge represented by facies EA above the tidal bore sediments may have hampered further tidal bore activity.

CONCLUSIONS

Albian deposits in the Uña outcrop represent a coastal complex formed under arid conditions in which ephemeral alluvial discharge interacted with tidal and aeolian processes. The relative influence of fluvial and tidal processes varied during deposition of the succession.

The correlation panel shows three units, separated by boundary surfaces BS1 and BS2: Unit 1 consists of alluvial, aeolian and tidal facies. In Units 2 and 3, tidal bore sediments filled incised valleys during a gradual increase of relative sea-level. During the tidal bore stages alluvial activity was absent or left no traces in the sedimentary record.

The formation of the incised valleys is interpreted to be the result of a relative sea-level fall.

Palaeogeography during deposition of Units 2 and 3 was characterised by a NW-SE-oriented embayment connected to the Tethys. The funnel-shaped geometry temporarily enhanced the amplitude of the incoming tides.

Tidal bore activity was favoured by diminished alluvial discharge and probably also by tidal amplification due to tidal resonance in a suitable basin configuration.

ACKNOWLEDGEMENTS

Kyangsik Choi, Allard Martinius and Editor Bernadette Tessier are thanked for their helpful comments and suggestions. This is a contribution to scientific project CGL2011-23717 (Ministerio de Economía y Competitividad, MINECO, Gobierno de España).

REFERENCES

Al-Farraj, **A.** and **Harvey**, **A.M.** (2000) Desert pavement characteristics on wadi terrace and alluvial fan surfaces: Wadi Al-Bih, U.A.E. and Oman. *Geomorphology*, **35**, 279–297.

Al-Farraj, **A.** and **Harvey**, **A.M.** (2004) Late Quaternary interactions between aeolian and fluvial processes: a case study in the northern UAE. *J. Arid Environments* **56**, 235–248.

Alonso, A., Floquet, M., Mas, J.R. and **Meléndez, A.** (1993) Late Cretaceous Carbonate Platforms: Origin and Evolution. Iberian Range. Spain. In: *Cretaceous Carbonate Platforms* (Eds **J.A.T. Simó, R.W. Scott** and **J.P. Masse**) *AAPG Spec. Publ.*, **56**, 297–316.

Bartsch-Winkler, S., Emmanuel, R.P. and **Winkler, G.R.** (1985) Reconnaissance Hydrology and Suspended Sediment Analysis, Turnagain Arm Estuary, Upper Cook Inlet. *US Geol. Surv. Circular*, **967**, 48–52.

Bartsch-Winkler, S. and **Lynch, D.K.** (1988) Catalog of Worldwide Tidal Bore Occurrences and Characteristics. *US Geol. Surv. Circular*, **1022**, 17 pp.

Boyd, R., Dalrymple, R.W. and **Zaitlin, B.A.** (2006) Estuarine and incised-valleys facies models. In: *Facies models revisited* (Eds **H.W. Posamentier** and **R.G. Walker**) *SEPM Spec. Pub.*, **84**, 171–235.

Bullard, J.E. and **McTanish, G.H.** (2003) Aeolian fluvial interactions in dryland environments: examples, concepts and Australia case study. *Progr. Phys. Geogr*, **27**, 471–501.

Capote, R., Muñoz, J.A., Simón, J.L., Liesa, L.C. and **Arlegui, L.E.** (2002) Alpine Tectonics I: The alpine system north of the Betic Cordillera. In: *The Geology of Spain* (Ed. **W. Gibbson, T. Moreno**). *Geol. Soc.* London. 367–400.

Chanson, H. (2011) Current knowledge in tidal bores and their environmental, ecological and cultural impacts. *Environ. Fluid Mech.*, **11**, 77–98.

Chanson, H. (2001) Flow field in a tidal bore: a physical model. In: *Proc. 29th IAHR Congress* (Ed. **G. Li**), pp. 365–373. Theme E, Tsinghua University Press, Beijing.

Chanson, H. (2005) Physical modelling of the flow field in an undular tidal bore. *J. Hydraul. Res.*, **43**, 234–244.

Chanson, H. and **Docherty, N.J.** (2012) Turbulent velocity measurements in open channel bores. *Eur. J. Mechanics B/Fluids*, **32**, 52–58.

Chanson, H., Reungoat, D. Simon, B. and **Lubin, P.** (2011) High-frequency turbulence and suspended sediment concentration measurements in the Garonne River tidal bore. *Estuar. Coast. Shelf Sci.*, **95**, 298–306.

Chen, J., Liu, C., Zhang, C. and **Walker, H.J.** (1990) Geomorphological development and sedimentation in Qiantang Estuary and Hangzhou Bay. *J. Coastal Res.*, **6**, 559–572.

Choi, K.S. (2010). Rhythmic climbing-ripple cross-lamination in inclined heterolithic stratification (IHS) of a macrotidal estuarine channel, Gomso Bay, west coast of Korea. *J. Sed. Res.* **80**, 550–561.

Clemmensen, L.B. and **Abrahamsen, K.** (1983) Aeolian stratification and facies association in desert sediments, Arran Basin, Permian (Scotland). *Sedimentology*, **30**, 311–339.

Collinson, J.D. (1986) Alluvial Sediments. In: *Sedimentary Environments and Facies* (Ed. **H.G. Reading**), pp. 20–62. Blackwell Scientific Publications.

Cuevas-Gozalo, M. and **de Boer, P.L.** (1991) Tide-influenced fluvial deposits; examples from Eocene of the southern Pyrenees. In: *Guidebook to the 4th International Conference on Fluvial Sedimentology* (Eds **M. Marzo** and **C. Puigdefàbregas**), *Publicacions del Servei Geològic de Catalunya*, Barcelona.

Dalrymple, R.W. and **Choi, K.** (2007) Morphologic and facies trends through the fluvial-marine transition in tide-dominated depositional systems: A schematic framework for environmental and sequence-stratigraphic interpretation. *Earth-Sci. Rev.*, **81**, 135–174.

Dávila, F. M. and **Astini, R.A.** (2003) Las eolianitas de la Sierra de Gamatina (Argentina): interacción paleoclimatectónica en el antepaís fragmentado andino central durante el Mioceno Medio. *Rev Geol Chile*, **30**, 187–204.

de **Boer, P.L.** (1998) Intertidal sediments: composition and structure. In: *Intertidal deposits: river mouths, Tidal flats and coastal lagoons* (Ed. **D. Eisma**), *CRC Marine Sciences Series*, pp. 345–361.

de **Boer, P.L., Oost, A.P.** and **Visser, M.J.** (1989) The diurnal inequality of the tide as a parameter for recognizing tidal influences. *J. Sed. Petrol.*, **59**, 912–921.

De Mowbray, T. and **Visser, M.J.** (1984) Reactivation surfaces in subtidal channel deposits, Oosterschelde, Southwest Netherlands. *J. Sed. Petrol.*, **54**, 811–824.

Donnelly, C. and **Chanson, H.** (2005) Environmental impact of undular tidal bores in tropical rivers. *Environ. Fluid Mech.*, **5**, 481–494.

Driese, S.G., Simpson, E.L. and **Eriksson, D.A.** (1995) Paleosols in alluvial and lacustrine deposits, 1.8GA Lochness Formation, Mount-Isa, Australia – Podogenic processes and implications for paleoclimate. *J. Sed. Res. Section A; Sedimentary Petrology and Processes*, **65**, 675–689.

Ekdale, A.A., Bromley, R.G. and **Pemberton, S.G.** (1984) Ichnology; trace fossils in sedimentology and stratigraphy. *SEPM Short Course* **15**, 317 pp.

Fan, D., Tu, J., Shang, S. and **Cai, G.** (2014) Characteristics of tidal bore deposits and facies associations in the Qiantang estuary, *China. Mar. Geol.*, **348**, 1–14.

Fan, D.D., Cai G F, Shang S, Wu, Y.J., Zhang, Y.W. and **Gao, L.** (2012) Sedimentation processes and sedimentary characteristics of tidal bores along the north bank of the Qiantang Estuary. *Chin. Sci. Bull*, **57**, 1578–1589.

Fan, Z. (2011) Is the small-scale turbulence an exclusive breaking product of oceanic internal waves? *Acta Oceanologica Sinica*, **30**, 1–11.

Furgerot, L., Mouazé, D., Tessier, B. and **Perez, L.** (2013) Suspended sediment concentration in relation to the passage of a tidal bore (Sée river estuary, Mont Saint Michel bay, NW France). *Proceedings of the 7th International Conference on Coastal Dynamics*, (Arcachon, France), 671–682.

García, A. and **Mas, J.R.** (coord), **García, A.** and **Mas, J.R., Segura, M., Carenas, B., García-Hidalgo, J.F., Gil, J., Alonso, A., Aurell, M., Bádenas, B., Benito, M.I., Meléndez, A.** and **Salas, R.** (2004) Segunda fase de post-rifting: Cretácico Superior. In: *Geología de España* (Ed. **J.A. Vera**). SGE-IGME, Madrid, 510–522.

Glennie, K.W. (1970) Desert sedimentary environments. *Dev. Sedimentol.*, **14**, 222 pp.

Greb, S.F. and **Archer, A.W.** (2007) Soft-sediment deformation produced by tides in a meizoseismic area, Turnagain Arm, Alaska. *Geology*, **35**, 435–438.

Gutiérrez-Marco, J.C. (2004) El Basamento Prealpino. In: *Geología de España* (Ed. **J.A. Vera**). SEG-IGME, Madrid, 470–484.

Higgins, C.G. (1982) Drainage systems developed by sapping on Earth and Mars. *Geology*, **10**, 147–152.

Khezri, N. and **Chanson, H.** (2012) Inception of bed load motion beneath a bore. *Geomorphology*, **153–154**, 39–47.

Kjerfve, B. and **Ferreira, H.O.** (1993) Tidal bores: first ever measurements. *Cienc. Cult.*, **45**, 135–138.

Langford, R.P. (1989) Fluvial-aeolian interactions: Part I, modern systems. *Sedimentology*, **36**, 1023–1035.

Langford, R.P. and **Chan, M.A.** (1989) Fluvial-aeolian interactions: Part II, ancient systems. *Sedimentology*, **36**, p. 1037–1051.

Lanier, W.P. and **Tessier, B.** (1998) Climbing-ripple bedding in the fluvio-estuarine transition: A common feature associated with tidal dynamics (modern and ancient analogues). In: *Tidalites: Processes and Products* (Eds C.R. Alexander, R.A. Davis and V.J. Henry). *SEPM Spec. Publ.*, **61**, 109–117.

Livingstone, I., Bullard, J.E., Wiggs, G.F.S. and **Thomas, D.S.G.** (1999) Grain-size variations on dunes in the southwest Kalahari, southern Africa. *J. Sed. Res.*, **69**, 546–552.

Lynch, D.K. (1982) Tidal bores. *Sci. Am.*, **247**, 134–143.

Martín-Chivelet, J. Berástegui, X., Rosales, I., Vilas, L., Vera, J.A., Caus, E., Gräfe, K., Mas, R., Puig, C., Segura, M., Robles, S., Floquet, M., Quesada, S., Ruiz,-Ortiz, P.A., Fregenal-Martinez, M.A., Salas, R., Arias, C., García, A., Martín-Algarra, A., Meléndez, M.N., Chacón, B., Molina, J.M., Sanz, J.L., Castro, J.M., García-Hernández, M., Carenas, B., García-Hidalgo, J., Gil, J. and Ortega, F. (2002) Cretaceous. In: *The Geology of Spain* (Eds W. Gibbson, T. Moreno), pp. 255–292. *Geol. Soc. London.*

Martinius, A.W. and **Gowland, S.** (2011) Tide-influenced fluvial bedforms and tidal bore deposits (Late Jurassic Lourinhã̃ Formation, Lusitanian Basin, Western Portugal). *Sedimentology*, **58**, 285–324.

Martinius, A.W. and **Van den Berg, J.H.** (2011) Atlas of sedimentary structures in estuarine and tidally-influenced river deposits of the Rhine-Meuse-Scheldt system – their application to the interpretation of analogous outcrop and subsurface depositional systems. *EAGE Publications*, Houten, 298 pp.

Meléndez, N. (1983) El Cretácico de la Región de Cañete - Rincón de Ademuz (Provincias de Cuenca y Valencia) [Tesis Doctoral]. *Seminarios de Estratigrafía, Serie Monografías* 9. Universidad Complutense, Madrid.

Meléndez, N., Meléndez, A and **Gómez-Fernández, J.C.** (1994) La Serranía de Cuenca Basin, Lower Cretaceous, Iberian Ranges, central Spain. In: *Global Geological Record of Lake Basins, Vol.* **1** (Eds E. Gierlowski-Kordesch, K. Kelts). Cambridge University Press, Cambridge, pp. 215–219.

Mountney, N.P. (2006) Eolian facies models. In: *Facies models revisited* (Eds R.G. Walker and H. Posamentier). *SEPM Mem.*, **84**, 19–83.

Mountney, N.P. and **Jagger, A.** (2004) Stratigraphic evolution of an Aeolian erg margin system: the Permian Cedar Mesa Sandstone, SE Utah, USA. *Sedimentology*, **53**, 789–823.

Mountney, N.P. and **Russell, A.** (2004) Sedimentology of cold-climate Aeolian sandsheet deposits in the Askja region of northeast Iceland. *Sed. Geol.*, **166**, 223–244.

Porter, M.L. (1987) Sedimentology of an ancient erg margin: The Lower Jurassic Aztec Sandstone, southern Nevada and southern California. *Sedimentology*, **34**, 661–680.

Pugh, D.T. (1987) *Tides, Surges and Sea-Level: a Handbook for Engineers and Scientists.* Wiley, New York, 472 pp.

Rodríguez-López, J.P., de Boer, P.L., Meléndez, N., Soria, A.R. and **Pardo, G.** (2006) Windblown desert sands in coeval shallow marine deposits a key for the recognition of coastal ergs; mid-Cretaceous Iberian Basin, Spain. *Terra Nova*, **18**, 314–320.

Rodríguez-López, J.P., Meléndez, N., de Boer, P.L. and **Soria, A.R.** (2008) Aeolian sand sea development along the Mid-Cretaceous Western Tethyan Margin (Spain): erg sedimentology and palaeoclimate implications. *Sedimentology*, **55**, 1253–1292.

Rodríguez-López, J.P., Meléndez, N., de Boer, P.L. and **Soria, A.R.** (2012) Controls on marine-erg margin cycle variability: aeolian-marine interaction in the mid-Cretaceous Iberian Desert System, Spain. *Sedimentology*, **59**, 466–501.

Rodríguez-López, J.P., Meléndez, N., de Boer, P.L. and **Soria, A.R.** (2010) The action of wind and water in a mid-Cretaceous subtropical erg-margin system close to the Variscan Iberian Massif, Spain. *Sedimentology*, **57**, 1315–1356.

Rodríguez-López, J.P., Meléndez, N., Soria, A.R. and **de Boer, P.L.** (2009) Reinterpretación estratigráfica y sedimentológica de las formaciones Escucha y Utrillas de la Cordillera Ibérica. *Rev. Soc. Geol. España*, **22**, 163–219.

Salas, R. and **Casas, A.** (1993) Mesozoic extensional tectonics, stratigraphy and crustal evolution Turing the Alpine cycle of the Eastern Iberian basin. *Tectonophysics*, **228**, 33–55.

Salas, R., Guimerá, J., Mas, R., Martín-Closas, C., Meléndez, A. and **Alonso, A.** (2001) Evolution of the Mesozoic Central Iberian Rift System and its cainozoic inversion (Iberian Chain). In: *Peri-Tethys Memoir 6: Peri-Tethyan Rift/Wrench Basins and Passive Margins* (Eds P.A. Ziegler, W. Cavazza, A.H.F. Robertson and S. Crasquin) *Mem. Mus. Nat. Hist.*, **186**, 145–185.

Scherer, C.M.S. (2000) Eolian dunes or the Botucatu Formation (Cretaceous) in Southernmost Brazil: morphology and origin. *Sed. Geol.*, **137**, 63–84.

Schlager, W. (1981) The paradox of drowned reefs and carbonate platforms. *Geol. Soc. Am. Bull.*, **92**, 197–211.

Segura, M., García-Hidalgo, J.F., Carenas, B., Gil, J. and **García, A.** (2004) Evolución paleogeográfica de la Cuenca Ibérica en el Cretácico Superior). *Geogaceta*, **36**, 103–106.

Soria, A.R., Meléndez, A., Aurell, M., Liesa, C.L., Meléndez, M.N. and **Gómez-Fernández, J.C.** (2000) The early Cretaceous of the Iberian Basin (northeastern Spain). In: *Lake basins through space and time* (Ed: E.H. Gierlowski-Kordesch and K.R. Kelts). *AAPG Studies in Geology*, **46**, 257–262.

Spalletti, L.A. and **Colombo Piñol, F.** (2005) From alluvial fan to playa: An Upper Jurassic ephemeral fluvial system, Neuquén Basin, *Argentina. Gondwana Res.*, **8**, 363–383.

Sztanó, O. and **de Boer, P.L.** (1995) Amplification of tidal motions in the Early Miocene North Hungarian bay. *Sedimentology*, **42**, 665–682.

Teller, J.T. (1998) Freshwater lakes in arid regions. In: *Palaeoecology of Africa and the Surrounding Islands* (Ed. K. Heine), pp. 241–253. A.A. Balkema, The Netherlands.

Tessier, B. and Terwindt, J.H.J. (1994) An example of soft-sediment deformations in an intertidal environment – the effect of a tidal bore. *CR Acad. Sci. Paris*, **319**, 217–233.

Trabucho-Alexandre, J., van Gilst, R.I., Rodríguez-López, J.P. and de Boer, P L. (2011) The sedimentary expression of oceanic anoxic event 1b in the North Atlantic. *Sedimentology*, **58**, 1217–1246.

Van den Berg, J.H., Boersma, J.R. and van Gelder, A. (2007) Diagnostic sedimentary structures of the fluvial-tidal transition zone – Evidence from deposits of the Rhine and Meuse. *Neth. J. Geosci.*, **86**, 287–306.

Vilas, L., Alonso, A., Arias, C., García, A., Mas, J.R., Rincón, R. and Meléndez, N. (1983) The Cretaceous of the Southwestern Iberian Ranges, (Spain). *Zitteliana*, **10**, 245–254.

Wakelin-King, G.A. and Webb, J.A. (2007) Upper-flow-regime mud floodplains, lower flow-regime sand channels: Sediment transport and deposition in a drylands mud-aggregate river. *J. Sed. Res.*, **77**, 702–712.

Wolanski, E., Moore, K., Spagnol, S., D'Adamo, N. and Pattiaratchi, C. (2001) Rapid, Human-Induced Siltation of the Macro-Tidal Ord River Estuary, Western Australia. *Estuarine, Coastal and Shelf Science*, **53**, 717–732.

Woolfe, K.J. (1990) Trace fossils as paleoenvironmental indicators in the Taylor Group (Devonian) of Antarctica. *Palaeogeogr. Palaeoclimatol. Palaeoecol.*, **80**, 301–310.

Wunderlich, F. (1969) Studien zur Sedimentbewegung. 1. Transportformen und Schichtbildung im Gebiet der Jade. *Senckenbergiana maritime*, **1**, 107–146.

Yokokawa, M., Kishi, M., Masuda, F. and Yamanaka, M. (1995) Climbing ripples recording the change of tidal current condition in the middle Pleistocene Shimosa Group. Japan. In: *Tidal signatures in modern and ancient sediments* (Eds B.W. Flemming and A. Bartholamä,). *Int. Assoc. Sedimentol. Spec. Publ.*, **24**, 301–311.

The Graafwater Formation, Lower Table Mountain Group, Ordovician, South Africa: Re-interpretation from a tide-dominated and wave-dominated depositional system to an alluvial fan/braidplain complex incorporating a number of tidal marine incursions

BURGHARD W. FLEMMING

Senckenberg Institute, Suedstrand 40, 26382 Wilhelmshaven, Germany
(E-mail: bflemming@senckenberg.de)

ABSTRACT

Due to sporadic occurrences of marine trace fossils, the early Ordovician (Tremadocian) Graafwater Formation, South Africa, had been interpreted to represent an intertidal to shallow-marine deposit in a mesotidal barrier-island setting; the latter environment having been claimed to be particularly prominent in the deposits of the Cape Peninsula. This interpretation is challenged in this paper on the basis of observations which are incompatible with the above model. Although the existence of several marine intercalations is acknowledged, the bulk of the deposits are interpreted to represent an alluvial fan environment, the succession on the Cape Peninsula representing the distal parts in proximity to the sea. Counter-indicative features to a tidal setting, in particular, include the occurrence of stacked lens-shaped sand bars facing downstream away from source, very rare wave ripples, deep sand-filled mud cracks, the complete lack of tidal rhythmites, very large standing wave and antidune deposits, the overall away-from-source fining trend and palaeocurrent directions, the large width of inferred intertidal zones (>50 km) and the thickness (~440 m) and depositional time scale (~10 Ma) of the succession as a whole. As a consequence the Graafwater Formation is reinterpreted to represent primarily an alluvial fan/braidplain/coastal plain environment which incorporates a number of transgressive marine incursions which have tentatively been linked with the global eustatic sea-level fluctuations of the early Ordovician.

Keywords: South Africa, Lower Table Mountain Group, Graafwater Formation, alluvial fan, braidplain deposits, marine transgression, trace fossils.

INTRODUCTION

In the Ordovician Cape Basin, South Africa (Fig. 1), the Lower Table Mountain Group commences with a ca. 800 m-thick conglomeratic coarse-grained sandstone (Piekenierskloof Formation), followed by 440 m of crossbedded quartzarenites interbedded with siltstones and mudstones locally containing marine trace fossils (Graafwater Formation) and 1800 m of medium-grained to coarse-grained, super-mature cross-bedded quartzarenites (Peninsula Formation). The latter formation is capped by a 120 m-thick

sequence comprising sandstones, conglomerates and a diamictite of glacial origin (Pakhuis Formation) (Table 1). Within the Graafwater and Peninsula Formations, horizons displaying marine trace fossils have been documented (Rust, 1967, 1973, 1977, 1981; Tankard & Hobday, 1977; Hobday & Tankard, 1978). As a result of these and other sedimentary structures, the succession as a whole had originally been interpreted to represent, from north to south, a prograding alluvial fan/braidplain/fan-delta to shallow-marine deposit along a mesotidal barrier-island coast (Rust, 1967, 1973, 1977, 1981; Visser, 1974;

Contributions to Modern and Ancient Tidal Sedimentology: Proceedings of the Tidalites 2012 Conference,
First Edition. Edited by Bernadette Tessier and Jean-Yves Reynaud.

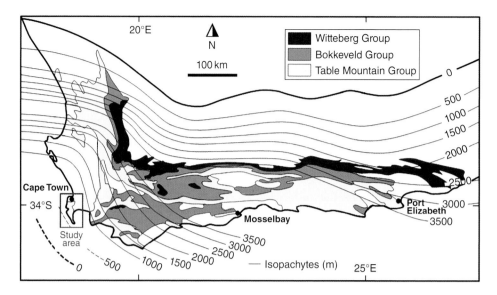

Fig. 1. Geological map of the Cape Supergroup (South Africa) with isopachytes of the Cape Basin superimposed (based on Rust, 1967; Tankard *et al.*, 1982). The study area (Cape Peninsula) is outlined on lower left.

Table 1. Stratigraphic relationships of the Ordovician Lower Table Mountain Group sedimentary succession as exposed on the Cape Peninsula (South Africa). Formations, thicknesses, lithology and approximate ages are based on Thamm & Johnson (2006).

Formation	Thickness	Lithology	Geological Period	Age
Pakhuis Formation	~10 m	diamictite (tillite)	Late Ordovician	~461–439 Ma
Peninsula Formation	~740 m	medium-coarse quartzarenite	Early–Middle Ordovician	~478–461 Ma
Graafwater Formation	~70 m	Quartzarenite-siltstone-mudstone	Early Ordovician	~488–478 Ma

--non-comformity--
Malmesbury Formation/Cape Granite Suite (late Proterozoic–Cambrian)

Tankard & Hobday, 1977; Hobday & Tankard, 1978; Tankard *et al.*, 1982). The 70 m-thick succession of the Graafwater Formation exposed on the Cape Peninsula, which is located near the south-western margin of the Cape Basin, has been interpreted to represent an intertidal to shallow subtidal deposit (Tankard & Hobday, 1977; Hobday & Tankard, 1978; Tankard *et al.*, 1982).

Doubts about this interpretation were first raised by Arthur O. Fuller in his 1985 Presidential Address to the Geological Society of South Africa, dealing with the conceptual modelling of pre-Devonian fluvial systems (Fuller, 1985). While accepting the evidence for sporadic marine influence within the Lower Table Mountain Group, he lists a number of depositional features, among them the occurrence of numerous pebbly units and the remarkably consistent palaeocurrent directions across the depositional strike, which he considered to be incompatible with a generalized tidal model. As a consequence, he suggested that a reinterpretation of the overall depositional setting in terms of an alluvial fan model with occasional marine transgressions was necessary. The counter-indicative evidence questioning the deposition in an essentially marine environment was subsequently reaffirmed with additional supporting evidence by Turner (1986, 1990), Flemming (1988), Broquet (1990) and Fuller & Broquet (1990). Later overview publications (e.g. Hartnady & Rogers, 1990; Hiller, 1992; Compton, 2004; Thamm & Johnson, 2006) cite the opposing views but remain neutral with respect to any preferred palaeoenvironmental interpretation. Tankard *et al.* (2009) acknowledge that the Peninsula Formation was largely of fluvial origin but ascribe the underlying Graafwater Formation to deposition in a mixed tide-dominated and wave-dominated back-barrier environment. They also discuss the overall tectonic setting of the Cape Basin, which they interpret as an extensional basin resulting from subduction-driven crustal thinning along the south-western margin of the Gondwana continent.

Very recently, an alternative depositional model for the Peninsula Formation has been proposed by Turner *et al.* (2011). It comprises a sequence of upward-fining, predominantly fluvial regressive-transgressive facies associations which are correlated with the global sea-level curve driven by orbitally-induced glacio-eustacy. In their paper, Turner *et al.* (2011) do not comment on the depositional nature of the Graafwater Formation, although one might expect that eustacy driven base-level changes should also apply in the latter case. A point of considerable concern is the fact that the chronostratigraphic positions allocated to the individual formations of the Cape Basin vary considerably between Thamm & Johnson (2006), Tankard *et al.* (2009) and Turner *et al.* (2011); the latter authors emphasizing that this uncertainty remains a major problem. In the present paper, the stratigraphy of Turner *et al.* (2011) is adopted because it offers a rational interpretation based on the well documented global eustatic sea-level trends for that period.

With respect to the Graafwater Formation, a closer look at the depositional criteria reveals that, at least as far as the exposures on the Cape Peninsula are concerned, the marine interpretation is based on mostly circumstantial evidence, very few of the primary sedimentary structures described in the literature being uniquely diagnostic of a tidal environment, although sequences displaying trace fossils are most probably of marine origin. In this context it should be noted that locally observed trace fossils have in very few cases been traced over large distances to establish their stratigraphic distribution (e.g. Rust, 1977; Fuller & Broquet, 1990). The purpose of the present paper, therefore, is to individually assess the features previously considered to support a tide-dominated and wave-dominated depositional model for the Graafwater Formation (Lower Table Mountain Group) and to weigh these up against highly diagnostic counter-indicative evidence arguing either individually or in unity against this model, at least for the greater part of the succession. In addition, an attempt is made to interpret the facies succession in terms of the global eustatic sea-level fluctuations characterizing early Ordovician times.

PHYSICAL SETTING OF THE STUDY AREA

Cape Town and the Cape Peninsula are located in the south-western corner of South Africa near the southern limit of the so-called Cape Basin which accommodates several thousand metres of Palaeozoic (pre-Carboniferous) sediments. According to Rust (1973) and Visser (1974), the lower part of the basin, which comprises the Table Mountain Group sedimentary succession, probably reached maximum thicknesses exceeding 4000 m (Fig. 1). Only part of the succession is represented on the Cape Peninsula itself (Table 1), the coarse-grained lowermost unit (Piekenierskloof Formation), for example, having never reached this region (Vos & Tankard, 1981). Instead, the succession begins with deposits of the Graafwater Formation which directly overlie a bevelled land surface that truncates the shales and intruded granites of the late Proterozoic Malmesbury Formation and Precambrian-Cambrian Cape Granite Suite (Fig. 2). Tankard *et al.* (1982) surmise that the Graafwater Formation could represent the more distal parts of the Piekenierskloof Formation. The contact between the granite and the basal deposits of the Graafwater Formation on the Cape Peninsula was until recently well exposed over a distance of more than 2 km along the scenic Chapmans Peak Drive which, for most part, follows the contact zone (Fig. 3).

The geology of the Cape Peninsula in plan view is illustrated in Fig. 4. Although the Graafwater Formation reaches a thickness of 70 m on the Cape Peninsula, it is mostly difficult to identify on true-to-scale geological maps because the exposures are generally very steep. To simplify matters, the black band marking the Graafwater Formation in Fig. 4 has been slightly exaggerated in width. Also shown are selected palaeocurrent roses and the locations of photographs presented in a number of figures documenting a variety of sedimentary structures arguing against the tidal model.

ASSESSMENT OF THE DEPOSITIONAL EVIDENCE

Evidence cited in support of a tide-dominated and wave-dominated depositional model

The evidence cited by Rust (1967, 1973, 1977, 1981), Visser (1974), Tankard & Hobday (1977), Hobday & Tankard (1978) and Tankard *et al.* (1982) in support of a tide-dominated and wave-dominated depositional setting for the sediments comprising the Graafwater Formation can be divided into two groups. A first group, comprising the occasional occurrence of a variety of trace fossils and sandstone beds displaying herringbone

Fig. 2. Aerial photograph of the Cape Peninsula (viewed towards the south-west) with the approximate boundaries and names of individual geological formations indicated in yellow.

Fig. 3. View of the scenic Chapmans Peak coastal road along which the lowermost Graafwater Formation directly overlying the Cape Granite is particularly well exposed over more than 2 kilometres (for location see label for Figs 7 to 9 on Fig. 4).

cross-stratification, can be readily accepted as being diagnostic of a tidal environment. A second group comprises certain lithological associations and a variety of primary sedimentary structures which have also been interpreted as representing tidal and shallow marine conditions, although none of these are unequivocally diagnostic of such a setting. Indeed, palaeocurrent directions determined from the dip directions of ripple and dune crossbedding overwhelmingly point

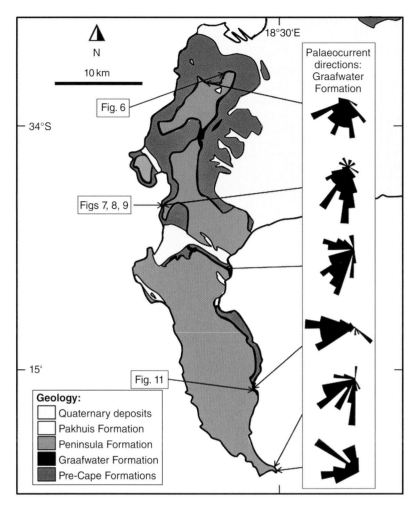

Fig. 4. Geological map of the Cape Peninsula (based on Tankard & Hobday, 1977). Also indicated are the locations of outcrop features illustrated in Figs 6 to 11 and of palaeocurrent measurements from the Graafwater Formation (based on Tankard & Hobday, 1977, and Hobday & Tankard, 1978). Note the rare occurrence of directly opposing (bipolar) flow directions that could be indicative of tidal current reversals. Instead, the diverging current pattern is typical of alluvial braidplain deposits.

towards the south, i.e. away from the terrestrial source located in the north.

The most common trace fossils found in the Graafwater Formation are of the *Skolithos* type, whereas *Scolicia, Cruziana, Platysolenites, Helminthoida* and occasional arthropod trails are less abundant. As far as so-called herringbone cross-stratification is concerned, it is necessary to distinguish between true herringbone structures and bipolar crossbedding indicating opposing flow directions (Flemming, 2012). Only the former, which are produced in the course of successive tides (either ebb-flood or flood-ebb), can be regarded as being unequivocally diagnostic of tidal current reversals, especially if several sets are stacked one above the other (cf. Conybeare & Crook, 1968; Ricci Lucchi, 1995; Stow, 2005; Flemming, 2012). Bipolar crossbeds, by contrast, can have formed at time intervals ranging from days to many years and are hence not strictly diagnostic of any particular environment, although

consistently alternating bipolar dip directions of crossbedded sets would argue strongly in favour of tidal current reversals. However, such sequences do not occur in the deposits of the Graafwater Formation. In this context it should be mentioned that ripple (or dune) trough-crossbedding, when cut along particular sections, may at first sight resemble herringbone cross-stratification. The only convincing example of herringbone cross-stratification illustrated in the cited literature is provided by Rust (1977). Although not expressly stated, it was probably found in stratigraphic proximity to marine trace fossils.

Of the second, non-diagnostic, group of features observed in the Graafwater Formation, the frequent occurrence of heterolithic sequences is particularly emphasized. However, the typical heterolithic facies elements (flaser, lenticular and wavy bedding sequences), besides also being a common depositional feature of tide-dominated environments (Reineck & Wunderlich, 1968; cf. also Flemming,

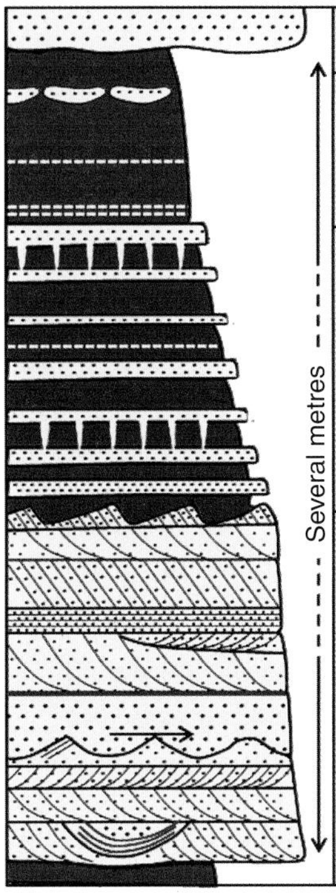

	Lithology	Sedimentary structures	Previous interpretation (Tankard & Hobday, 1977)	Present interpretation (this paper)
	Red-brown mudstone with thin quartz arenites; maroon-brown mudstone	Homogenous mud with ball & pillow structures, occasional concretions and wavy sandy-silt or silty-sand layers	Supratidal flat	Playa-lake deposits on distal alluvial fan or low-gradient coastal plain
			High intertidal mudflat	
	Alternating beds of mudstone and quartz arenite	Heterolithic facies consisting of thin, ripple cross-laminated sand beds and flaser, lenticular and wavy bedding; intercalated are homogenous mud beds occasionally displaying widely spaced and deep desiccation cracks	Mud-tide flat	Shallow playa-lake margin on distal alluvial fan or low-gradient coastal plain
	Quartz arenite beds, often separated by thin mud drapes	Cross-bedded sandstones of variable thickness, with rare apparently bipolar cross-lamination, interrupted by upper plane bed, standing wave and antidune bedforms; evidence for channelized flow	Low-tide terrace	Medial to distal alluvial fan
			Shallow subtidal	

Fig. 5. Synthetic diagram illustrating a typical facies sequence observed in the Graafwater Formation together with the original interpretation of Tankard & Hobday (1977) and the alternative interpretation proposed in this paper.

2003), have been found to occur frequently in fluvial and deltaic environments (Coleman & Gagliano, 1965; Conibeare & Crook, 1968), in non-tidal marginal seas (Werner, 1968) and even in the deep sea (Mutti, 1977; Ricci Lucchi, 1995).

In summary, a tide-dominated and wave-dominated interpretation of the entire depositional sequence comprising the Graafwater Formation could arguably be justified if unequivocal evidence to the contrary were lacking. However, as will be shown below, particular features which, either individually or in unity, are incompatible with such an interpretation have evidently been overlooked or ignored. This latter group of sedimentary structures, which are here argued to more consistently support a distal alluvial fan setting, is illustrated in the synthetic facies model of Fig. 5, in which the most characteristic bedding types, their stratigraphic arrangement in the Graafwater Formation and both the former and the new interpretation have been contrasted.

Evidence favouring a distal alluvial fan model

This section restricts itself to evidence which strongly and, in part, unequivocally argues against a tide-dominated and wave-dominated coastal setting for the depositional sequence as a whole. Among counter-indicative features are, in particular, (a) stacked lens-shaped sand bars facing downstream away from source, (b) the overall rare occurrence of wave ripples, (c) widely-spaced and unusually deep, sand-filled mud cracks, (d) the lack of tidal rhythmites (spring-neap cycles), (e) consistent away-from-source palaeocurrent directions, (f) the occurrence of very large standing waves and antidunes, (g) the overall away-from-source fining trend, (h) the spatial scale of inferred intertidal zones and, last but not least, (i) the thickness and depositional time scale of the succession as a whole.

(a) Stacked lens-shaped sandbodies: In many places the Graafwater Formation is characterized by laterally overlapping, vertically stacked lens-shaped sandbodies which have dimensions reaching over 10 m across, more than 20 m streamwise and up to 50 cm in thickness (Fig. 6). The surfaces of the cross-bedded bars are usually covered by large current ripples or small dunes with palaeocurrent directions consistently facing away from source, i.e. more or less southwards. The up to 50 cm-thick stacked bars are sometimes separated

Fig. 6. Stacked lens-shaped sand bars exposed in deposits of the Graafwater Formation along Tafelberg Road (for location see label of Fig. 6 on Fig. 4). (A) East-west section showing the stacked nature of the bars in cross-section normal to the flow direction. (B) North-south section through a bar located in close proximity and oriented parallel to the flow direction. Note the large current ripples/small dunes at the top of the bar indicating southward flow.

by thin mud drapes, although isolated bars overlying decimetre-thick mud beds also occur, such individuals being themselves frequently draped by thick, homogenous mud layers that fill the ripple and dune troughs. Such sandbodies are atypical of intertidal (and especially back-barrier) environments (Flemming, 2012). In the event, bars of this type may occur on tidal ebb deltas (as swash bars) or, less common, on intertidal sand flats well away from mud flats with inclined bedding dipping in the landward direction and, driven by wave-overwash processes, be devoid of superimposed larger-scale current-generated bedforms (Son *et al.*, 2011). The orientation of the observed bars and their association with underlying, intercalated and overlying mud beds, on the other hand, is entirely

compatible with a distal alluvial fan setting, the bars probably representing local delta-like sand splays deposited on top of mud beds forming the bottom of shallow playa-type water bodies.

(b) Wave-generated ripples: Although wave-generated ripple marks have been documented from the Graafwater Formation (e.g. Tankard & Hobday, 1977), their occurrence is surprisingly rare, especially in the deposits on the Cape Peninsula. A rare example from the Graafwater Formation on the Cape Peninsula is shown in Fig. 7. The troughs of the larger ripples display small ladderback ripples, which is indicative of very shallow water and shorter waves approaching at right angles to those that generated the larger ones. Such ripple systems are ubiquitous on intertidal sand flats (Chang *et al.*, 2006; Flemming, 2012) and, with the exception of channel margins, are often the dominant internal sedimentary structure of back-barrier sand flats (Davis & Flemming, 1995). Although wave ripples can also occur in freshly deposited muds (Chang & Flemming, 2013), none have been recorded in the muddy sequences of the Graafwater Formation. Their rare occurrence therefore speaks against an intertidal setting. A shallow upper shoreface setting, on the other hand, can also be excluded because the deposits of such environments are commonly characterized by upper plane bed, hummocky and swaley cross-stratification produced by storm-wave action (e.g. Chang & Flemming, 2006; Antia *et al.*, 1994; Son *et al.*, 2012a, 2012b). No such sedimentary structures have ever been recorded in deposits of the Graafwater Formation. In fact, already Du Toit (1926) had noted the dearth of wave ripples in these deposits. The rare occurrence of wave ripples would, on the other hand, not be incompatible with a distal alluvial fan setting, where they would be restricted to the shallow margins of playa-type water bodies.

(c) Desiccation cracks: A number of mud beds in the Graafwater Formation display variously sized and spaced, sand-filled desiccation cracks (Fig. 8A and B). Several features of these mudcracks argue strongly against their formation in an intertidal environment. This has also been noted by Turner (1986) and Turner *et al.* (2011). Particularly important counter-indicative features are the combination of the extraordinary width and depth of

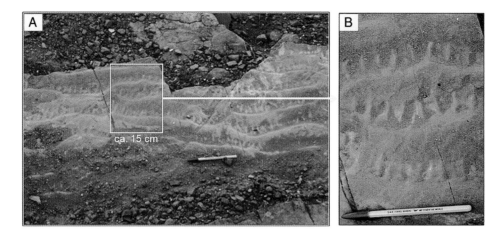

Fig. 7. (A) A rare example of wave-generated ripples preserved on a bedding plane of the Graafwater Formation exposed along the Chapmans Peak coastal road (for location see label for Figs 7 to 9 on Fig. 4). Note the much smaller ladderback ripples in the troughs of the larger ripples. (B) Enlarged section of panel (A). Such ripple marks are formed by short-period wind waves in shallow water and are not indicative of any particular environment.

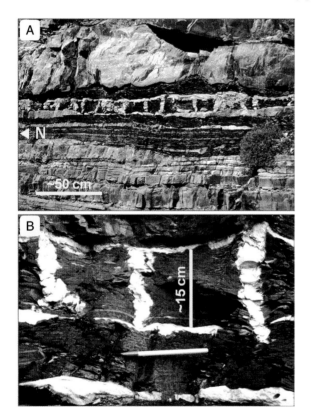

Fig. 8. (A) Sand-filled mud cracks are frequently observed in mud beds of the Graafwater Formation (for location see label for Fig. 8 on Fig. 4). (B) Example of deep mudcracks showing evidence of compaction. The homogenous nature of the mud bed and the lack of evidence for multiple exposure/submergence cycles, as would be expected in a tidal setting, suggests a singular depositional event followed by prolonged subaerial exposure. The cracks were subsequently infilled with sand, probably in the course of a single fluvial discharge event.

the mudcracks, their overall smooth sidewalls, the fact that they are filled with unstructured sand and the homogenous (non-laminated) nature of the mud beds in which they occur (Fig. 8B). In order to place this into perspective, it has to be recalled that intertidal mudcracks can only develop shoreward of the neap high-tide level, i.e. in the uppermost parts of intertidal mud flats which fall dry for some days between successive spring tides. This spatial and temporal limitation severely constrains intertidal mudcrack development. Firstly, the desiccation period is restricted to a few days. As a consequence, the mud rarely dries out completely and is usually still moist when again inundated by the rising high-water level approaching the next spring tide. Mudcrack development therefore proceeds in multiple steps from one neap tide to the next, interim spring tides adding new thin mud layers on each occasion or filling existing mudcracks with eroded mud chips and flakes if accompanied by wave action. Secondly, intertidal mudcracks are rarely filled with sand because sand flats are usually located some distance away. On the other hand, very similar sand-filled mudcracks interpreted as having formed in a tidal environment have been reported from the late Precambrian Batsfjord Formation in northern Norway (Siedlecka, 1978). However, the citations in that paper suggest a strong influence by the interpretations offered for the Graafwater Formation by Tankard & Hobday (1977) and,

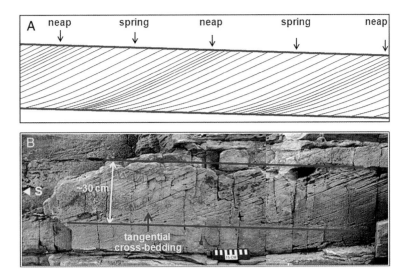

Fig. 9. The crossbedded sands of the Graafwater Formation are devoid of tidal rhythmites. (A) Idealized rhythmic thickness variations that would typically be observed in tide-generated crossbedded sequences. (B) A 2 m-long crossbedded sequence of the Graafwater Formation (for location see label for Fig. 9 on Fig. 4). Note the lack of rhythmic thickness variations in the bundle pattern.

as a consequence, similar doubts about the depositional environment can be raised. As the supratidal area landward of pre-Devonian tidal flats was not vegetated, one could conceive influx of sand by aeolian or fluvial action from the landward side, which would explain the southward (i.e. seaward) transport direction of the sand sheets revealed in the ripple or dune crossbedding. However, as already pointed out by Turner (1986), the nature of the mudcracks are better explained by singular events in shallow water (cf. also Donovan & Archer, 1975) where they formed after parts of the mud sheet became subaerially exposed due to a drop in the playa-lake water-level. The (probably still moist) mudcracked surface was subsequently covered by a thin sheet of fluvially supplied sand which infilled the cracks before the whole sequence again became draped in thick mud after a subsequent rise in the lake level. Significantly, and as illustrated in Fig. 8, mudcracked horizons are few and far between; neither the mud sheets below nor above individual mudcracked beds showing any signs of desiccation, irrespective of their thickness. Such a situation is difficult to visualize in an upper intertidal environment but is entirely compatible with a distal alluvial fan setting, besides being also in agreement with the arguments explaining the rare occurrence of wave ripples.

(d) Absence of tidal rhythmites: At the time of the earlier publications scrutinized here (1967 to 1978), tidal rhythmites covering several spring-neap-spring cycles had not yet been discovered, the credit for this going to Visser (1980), although Reineck (1967) had already referred to tidal bedding in the form of alternating sand-mud couplets. After the initial recognition of spring-neap cycles in the tidal bundle sequences of cross-bedded sands exposed in excavation pits of the Rhine Delta (The Netherlands), such tidal rhythmites were discovered on scales ranging from large current ripples to dunes in numerous ancient rocks dating back to the Precambrian. Good examples are illustrated and discussed in Allen & Homewood (1984), Williams (1989) and Archer (1998). A schematic illustration of what tidal rhythmites would look like in cross-bedded sands is shown in Fig. 9A. In spite of numerous attempts by the author to identify such rhythmites in any of the many excellent outcrops all over the Cape Peninsula, none were found that could confidently be assigned a tidal origin. An outcrop example is given in Fig. 9B which, besides lacking rhythmic thickness variations in the bundle sequence, does also not show any evidence of thin mud drapes on the lower slip faces indicative of regular tidal slack-water phases. The complete lack of tidal rhythmites in both

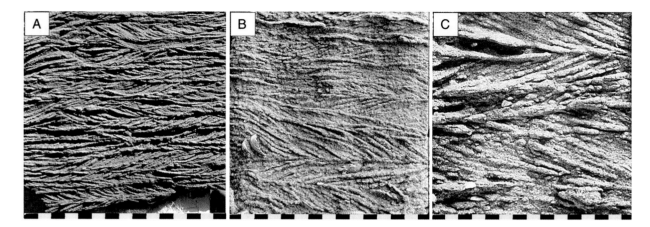

Fig. 10. Examples of true herringbone cross-stratificaton at different scales recorded in intertidal deposits of the German Wadden Sea (Jade Bay) (cf. Reineck 1963; Flemming, 2012). Source: Senckenberg am Meer boxcore archive.

cross-stratified sand beds and vertically stacked (heterolithic) sand-mud sequences therefore precludes a tidal origin for most of the Graafwater Formation.

(e) Palaeocurrent patterns: A predominantly alluvial origin of the Graafwater Formation is also reflected in the palaeocurrent directions which, as stated by Fuller (1985), show a consistent trend across the depositional strike. A particularly diagnostic type of tide-generated, bipolar sedimentary structure is herringbone cross-stratification. As already pointed out by Allen (1970), cross-bedding sets produced by successive tides are far more common than generally thought. Good fossil examples can be found in Conybeare & Crook (1968), Ricci Lucchi (1995) and Stow (2005), three modern examples from the Wadden Sea being illustrated in Fig. 10 (cf. Reineck, 1963; Flemming, 2012). The latter range in size from small (a) to large current ripples (B) and small dunes (c). In fact, as illustrated by a selection of palaeocurrent roses for the Graafwater Formation (Fig. 4), not a single truly bipolar current rose has been produced that would support deposition in a tidal environment. The attitude and variability of current directions, on the other hand, is entirely consistent with the constant directional switching of braided streams, which may occasionally also include channel sections where the flow is directed landward.

(f) Large standing waves and antidunes: Upper flow regime bedforms, i.e. plane beds, standing waves and antidunes, form at very high flow velocities. In the present context, i.e. when excluding wave-generated varieties,

they can be observed in both shallow fluvial (Kennedy, 1963) and shallow tidal flows (Flemming, 2012). In fact, Kennedy (1969) and Verbanck (2008) consider antidunes and standing waves as the final alluvial bedform when stream power is very high. Large upper regime bedforms have also been reported from deposits interpreted as representing an ancient wave-dominated open-shore tidal flat environment (Basilici *et al.*, 2012). However, the described facies associations do not fit those of the Graafwater example. In back-barrier tidal flows, upper flow regime bedforms rarely exceed a few decimetres in spacing and a few centimetres in height, their preservation potential in addition being very low. A tidal origin of the extraordinary large antidune (2 m wavelength, >10 cm height) and standing wave (1.8 m wavelength, >10 cm height) structures found on the southern Cape Peninsula (Fig. 11) is therefore highly unlikely, their formation in a high-discharge fluvial environment being much more plausible.

(g) Proximal to distal fining: The general away-from-source (southward) fining trend characterizing the Lower Table Mountain Group rocks, including those of the Graafwater Formation, had already been noted by Du Toit (1926) and was re-emphasized by Fuller (1985). This consistent trend is opposite to that typical of intertidal deposits which, irrespective of the tidal range, always fine towards the high-water line, in most cases therefore landwards. Indeed, this is a foremost diagnostic feature of tidal environments (Chang & Flemming, 2006; Flemming, 2012). The observed trend, on the other hand, is entirely compatible with an alluvial fan/braidplain setting, which in this case covers a source-to-sink distance of over 300 km (Figs 1 and 12).

(h) Spatial scale of inferred intertidal zones: An important feature of the Lower Table Mountain Group sedimentary succession, and which seems to have been overlooked by all proponents advocating a tide-dominated marine setting, is the sheer scale of the depositional system, the northern and (inferred) southern margins of the Cape Basin lying more than 400 km apart. An intertidal zone within this

Fig. 11. Large standing wave and antidune structures exposed in sandstone deposits of the Graafwater Formation (for location see label for Fig. 11 on Fig. 4). The scale of the features precludes a tidal current origin and, instead, favours formation under upper regime fluvial discharge conditions.

system, whether bounded by barrier islands or not, would be constrained in width by the hydraulics of tidal wave propagation and dissipation. This is illustrated by a simple calculation. In the Late Proterozoic (ca. 650 Ma BP) the lunar month was approx. 6% shorter than today (Williams, 1989, 1991). By the Middle Ordovician (ca. 450 Ma BP) this will have reduced to ca. 4%. As a consequence, a single semidiurnal flood or ebb tidal cycle was correspondingly shorter, amounting to about 6 h 6 min instead of the 6 h 21 min of today. Keeping in mind that both intertidal flooding and draining have to be completed within this period, and assuming a time-averaged flow velocity over a tidal cycle of 1 m s⁻¹ (which is on the high side), then the distance travelled by the water between the low-water and high-water line would amount to 21.96 km, measured normal to the coast. In reality, this distance would have been somewhat shorter due to the meandering nature of tidal channels. This crude estimate is supported by the results of recent numerical modelling exercises which postulate maximum equilibrium widths of mud flats to not exceed 5 to 6 km, irrespective of tidal range (Pritchard *et al.*, 2002; Friedrichs, 2011). Intertidal fringes/basins along open coasts comprising sand flats, mixed flats and mud flats can thus not have been wider than about 20 km. This renders a barrier-island model, as proposed by Tankard & Hobday (1977) and Hobday &

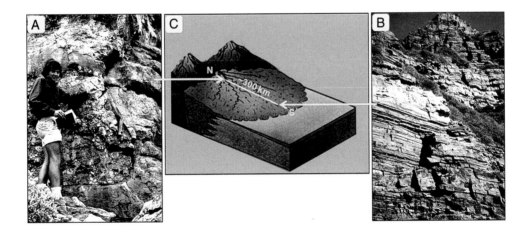

Fig. 12. The Lower Table Mountain Group sedimentary succession is characterized by progressive away-from-source fining over a distance of more than 300 kilometres. The proximal parts (A) consist of coarse-grained channel fills, whereas the distal parts (B) are composed of sandy-silty-muddy sequences. This facies architecture is typical of alluvial fan/braidplain deposits (C), the occurrence of occasional marine ingressions notwithstanding (block model adapted from McPherson *et al.*, 1987).

Tankard (1978) or a *ca.* 150km-wide 'tidal plain', as suggested by Rust (1967, 1973), unrealistic. Even a 20km-wide tidal basin or intertidal zone would not have straddled the Cape Peninsula which has a north-south extension of more than 50km. A back-stepping and upward-stepping marine transgression crossing the Cape Peninsula would thus, by necessity, have to have conserved terrestrial deposits below the ravinement surface. The above criticism, incidentally, also applies to the depositional model of Turner *et al.* (2011) for the Peninsula Formation, in that the horizontal scaling of their intertidal zone is excessively distorted, being disproportionately wide in comparison to the other depositional facies.

(i) Overall thickness and depositional time scales: The large surface area and width of the Cape Basin addressed above is associated with an equally impressive vertical sediment column. In the case of the Graafwater Formation this amounts to overall 430m (70m on the Cape Peninsula) over a time period of about 10Myr. Besides variable sediment supply and subsidence rates over time, orbitally-induced, glacio-eustatic sea-level fluctuations, as proposed by Turner *et al.* (2011) for the Peninsula Formation, could also have been a major factor controlling transgressions and regressions in the Graafwater Formation. This would have resulted in depositional cycles alternating between sea-level lowstands (terrestrial sedimentation) and highstands (marine sedimentation). Indeed, eustatically controlled alternations between terrestrial and marine depositional cycles would provide an excellent depositional model for the Graafwater Formation.

DISCUSSION AND CONCLUSIONS

Considering the counter-indicative evidence as a whole, and acknowledging the existence of marine ingressions revealed by trace fossils in association with bipolar cross-bedding, many facies successions and sedimentary structures recorded in the Graafwater Formation overwhelmingly favour an alluvial fan setting for large parts of the formation. As a consequence, the exclusive tidal interpretation of Rust (1967, 1973, 1977, 1981) and wave-dominated barrier-island depositional model proposed by Hobday & Tankard (1978) and Tankard *et al.* (1982) cannot be upheld, the facies model illustrated in Fig. 5 being reinterpreted as an alluvial fan/braidplain/coastal plain model in which the shoreline was mostly located some distance south of the Cape Peninsula. The sea-level lowstand situation reflected in the facies model of Fig. 5 is schematically illustrated in the palaeogeographic map of Fig. 13A.

During such lowstand intervals, the region around the Cape Peninsula was located on a coastal plain forming the distal parts of a large alluvial fan fringed by shallow playa-type water

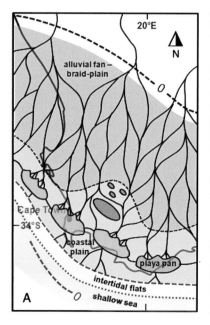 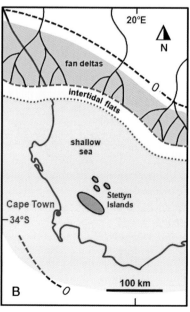

Fig. 13. Palaeogeography of the Graafwater period. (A) Braidplain deposition on a large alluvial fan/coastal plain complex. Note the position of the shoreline together with an intertidal fringe well south of the Cape Peninsula. (B) Palaeogeography during maximum marine transgression (modified after Rust, 1967, 1973).

bodies. When reaching the lake shore, fluvial feeder channels spread out and deposited their bedloads in the form of lens-shaped sandbodies representing deltaic splays, several such events producing the laterally overlapping stacked sequences illustrated in Fig. 6. From these, thinner sand sheets extended some distance into the shallow playa-lake, their surfaces subsequently developing wave-generated ripple marks in response to variable winds blowing from different directions (Fig. 7). These deposits became draped in mud as the water-level increased and the lake expanded. Sand sheets extending still farther into the lake occasionally collapsed into the soft mud bottoms, possibly triggered by earthquakes, to form ball and pillow structures. During low discharge periods, the lake level dropped and thereby exposed mud beds along the lake margin. Prolonged desiccation produced deep mudcracks that frequently penetrated down to the underlying sand sheets (Fig. 8). In the course of subsequent discharge events, the mud-cracked surfaces were locally covered by thin sand sheets which infilled the deep mudcracks. As the location of lake shores varied over time, mud-cracked surfaces appear randomly distributed in space and time. Lake levels, and hence water depths, were constrained by overflow towards the nearby sea. During such events the channelized flow reached upper regime velocities producing plane bed, large standing wave and antidune deposits (Fig. 11). The uppermost lowstand sequences were eventually reworked in the course of marine transgressions which penetrated progressively farther into the hinterland, a corresponding highstand scenario being illustrated in the palaeogeographic map of Fig. 13B. The lack of evidence for depositional sequences associated with barrier islands suggests that the tidal regime in the back-arc basin was probably macrotidal (<3.5 m), as also conceived by Turner *et al.* (2011) for the Peninsula Formation.

This overall picture finds support in the cyclical nature of the eustatic sea-level curve for the period in question. As illustrated in Fig. 14, the sea-level curve for the Tremadocian (based on Haq & Schutter, 2008) shows four transgression/regression cycles over a time period of about 10 Myr, the final one leading over into the Arenig which marks the onset of the Peninsula Formation and which was accompanied by an approximately six-fold increase in the average rate of deposition relative to that of the Graafwater Formation (0.7 m 10^5 yr^{-1}). As a consequence, the coastline was pushed far to

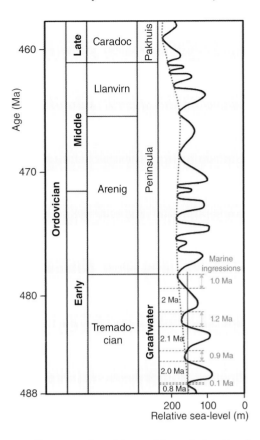

Fig. 14. Chronostratigraphic position of the Graafwater Formation (after Turner *et al.*, 2011) in relation to the global sea-level curve (after Haq & Schutter, 2008). The thin vertical black line indicates a possible position of the Cape Peninsula relative to the sea-level curve. Note the progressively deeper marine ingressions into the coastal hinterland and their approximate durations (blue Ma values) separated by prolonged terrestrial phases (black Ma values) characterizing the Graafwater Formation on the Cape Peninsula.

the south (Turner *et al.*, 2011). As the deposits of the lowermost Graafwater Formation are largely devoid of marine trace fossils, the coast must initially have been located some distance to the south of the study area (Fig. 13A).

A plausible location of the Cape Peninsula relative to the sea-level curve is indicated by the thin vertical black line in Fig. 14. Also indicated are the resulting periods dominated by either terrestrial (black values) or marine deposition (blue values). Terrestrial conditions would thus have prevailed for a total of 6.9 Myr or ~70% of the time and marine conditions for 3.2 Myr or ~30% of the time. As the exact position of the study area relative to the sea-level curve is not known, the values must be understood as rough estimates. According to this scenario, the first transgression barely reached the study area. This explains the dearth of animal traces and bioturbation characterizing the basal sequence.

The palaeogeography of the Graafwater Formation can thus be visualized as a massive braided alluvial fan/braidplain/coastal plain complex, the western boundary of which is not at all clear. The southward trending western shoreline reconstructed by Rust (1967, 1973), which suggests the existence of a large embayment, is purely speculative and actually inconsistent with the known westward continuation of the Cape Basin into what is today Argentina (Fig. 1), where it connected with the Ordovician back-arc to foreland Puna Basin of NW Argentina and Chile (Bahlburg, 1990; Zimmermann & Bahlburg, 2003). A more plausible interpretation would therefore be to assume that alluvial fan complexes also lined the basin westwards. However, a lithological equivalent of the Graafwater Formation does not appear to exist in Argentina (Tankard *et al.*, 2009). This may be explained by the fact that the basin became narrower towards the west and that, as a consequence, only proximal to medial alluvial fan sands were deposited, the finer fractions being carried away sideways to adjacent parts of the basin. By the same token, the marine transgressions would have shifted the shoreline progressively farther to the north on each occasion. These highstand situations correspond to the palaeogeography visualized by Rust (1967, 1973), except for a more realistic width of the 'tidal plain' (intertidal zone) fronting a more extensive shallow shelf environment as illustrated in Fig. 13B.

ACKNOWLEDGEMENTS

The author wishes to thank J.-F. Ghienne, a second anonymous referee, the guest editors and A. K. Martin of Repsol for constructive comments, which served to improve the manuscript.

REFERENCES

Allen, J.R.L. (1970) *Physical Processes of Sedimentation.* George Allen and Unwin, London, 248 pp.

Allen, P.A. and Homewood, P. (1984) Evolution and mechanics of a Miocene tidal sandwave. *Sedimentology*, **31**, 63–81.

Antia, E., Flemming, B. and Wefer, G. (1994) Transgressive facies sequence of a high energy, wave-tide-storm-influenced shoreface: a case study of the East Frisian barrier islands (southern North Sea). *Facies*, **30**, 15–24.

Archer, A.W. (1998) Hierarchy of controls on cyclic rhythmite deposition, Carboniferous basins of eastern and mid-continental USA. *Tidalites: Processes and Products* (Eds C.R. Alexander, R.A. Davis and V.J. Henry). *SEPM Spec. Publ.*, **61**, pp. 59–68.

Bahlburg, H. (1990) The Ordovician basin in the Puna of NW Argentina and Chile: geodynamic evolution from back-arc to foreland basin. *Geotekton. Forsch.*, **75**, 1–107.

Basilici, G., Vieira de Luca, P.H. and Oliveira, E.P. (2012) A depositional model for a wave-dominated open-coast tidal flat, based on analyses of the Cambrian-Ordovician Lagarto and Palmares formations, north-eastern Brazil. *Sedimentology*, **59**, 1613–1639.

Broquet, C.A.M. (1990) Trace fossils and ichno-sedimentary facies from the Lower Palaeozoic Peninsula Formation. *Geocongress '90, Geol. Soc. South Afr.*, Abstracts, pp. 64–67.

Chang, T.S. and Flemming, B.W. (2013) Ripples in intertidal mud – a conceptual explanation. *Geo-Mar. Lett.*, **33**, 449–461.

Chang, T.S. and Flemming, B.W. (2006) Sedimentation on a wave-dominated, open-coast tidal flat, southwestern Korea: summer tidal flat – winter shoreface – discussion. *Sedimentology*, **53**, 687–691.

Chang, T.S., Flemming, B.W., Tilch, E., Bartholomä, A. and Wöstmann, R. (2006) Late Holocene stratigraphic evolution of a back-barrier tidal basin in the East Frisian Wadden Sea, southern North Sea: transgressive deposition and its preservation potential. *Facies*, **52**, 329–340.

Coleman, J.M. and Gagliano, S.M. (1965) Sedimentary structures: Mississippi river deltaic plain. *Primary Sedimentary Structures and their Hydrodynamic Interpretation* (Ed G.V. Middleton). *SEPM Spec. Publ.*, **12**, 133–148.

Compton, J.S. (2004) *The Rocks and Mountains of Cape Town.* Double Story Books, Cape Town, 112 pp.

Conibeare, C.E.B. and Crook, K.A.W. (1968) Manual of sedimentary structures. *Bureau of Mineral Resources, Geology and Geophysics (Commonwealth of Australia) Bull.*, **102**, 1–327.

Davis, R.A., Jr. and Flemming, B.W. (1995) Stratigraphy of a combined wave- and tide-dominated intertidal sand body: Martens Plate, East Frisian Wadden Sea, *Germany. Int. Assoc. Sedimentol. Spec. Publ.*, **24**, 121–132.

Donovan, R.N. and Archer, R. (1975) Some sedimentological consequences of a fall in the level of Haweswater, Cumbria. *Proceedings of the Yorkshire Geological Society*, **40**, 547–562.

Du Toit, A.L. (1926) *The Geology of South Africa.* Oliver & Boyd, Edinburgh, 444 pp.

Flemming, B.W. (1988) Evidence for a fluvial rather than tidal origin of the lower Table Mountain Group sedimentary succession (Ordovician Cape basin, South Africa). *Terra Cognita*, **8**, p. 30.

Flemming, B.W. (2003) Flaser. *Encyclopedia of Sediments and Sedimentary Rocks* (Ed G.V. Middleton). Kluwer, Dordrecht, pp. 282–283.

Flemming, B.W. (2012) Siliciclastic back-barrier tidal flats. *Principles of Tidal Sedimentology* (Eds R.A. Davis, Jr. and R.W. Dalrymple). Springer, New York, pp. 231–267.

Friedrichs, C.T. (2011) Tidal flat morphodynamics: A synthesis. *Encyclopedia of Estuaries and Coasts, Vol. 3, Estuarine and Coastal Geology and Geomorphology* (Ed B.W. Flemming and J.D. Hansom). Elsevier, Amsterdam, pp. 138–170.

Fuller, A.O. (1985) A contribution to the conceptual modelling of pre-Devonian fluvial systems. *Trans. Geol. Soc. South Afr.*, **88**, 189–194.

Fuller, A.O. and Broquet, C.A.M. (1990) Aspects of the Peninsula Formation – Table Mountain Group. *Geocongress '90, Geol. Soc. South Afr.*, Abstracts, pp. 169–172.

Haq, B.U. and Schutter, S.R. (2008) A chronology of Paleozoic sea-level changes. *Science*, **322**, 64–68.

Hartnady, C.J.H. and Rogers, J. (1990) The scenery and geology of the Cape Peninsula. *Geocongress '90, Geol. Soc. South Afr.*, Guidebook **M1**, pp. 1–67.

Hiller, N. (1992) The Ordovician System of South Africa: a review. *Global Perspectives on Ordovician Geology* (Eds B.D. Webbey and J.R. Laurie). Balkema, Rotterdam, pp. 473–485.

Hobday, D.K. and Tankard, A.J. (1978) Transgressive-barrier and shallow-shelf interpretation of the lower Paleozoic Peninsula Formation, South Africa. *Geol. Soc. Am. Bull.*, **89**, 1733–1744.

Kennedy, J.F. (1963) Mechanics of dunes and antidunes in erodible-bed channels. *J. Fluid Mech.*, **16**, 521–544.

Kennedy, J.F. (1969) The formation of sediment ripples, dunes and antidunes. *Ann. Rev. Fluid Mech.*, **1**, 147–168.

McPherson, J.G., Shanmugam, G. and Moiola, R.J. (1987) Fan-deltas and braid deltas: Varieties of coarse-grained deltas. *Geol. Soc. Am. Bull.*, **99**, 331–340.

Mutti, E. (1977) Distinctive thin-bedded turbidite facies and related depositional environments in the Eocene Hecho Group (south-central Pyrenees, Spain). *Sedimentology*, **24**, 107–131.

Pritchard, D., Hogg, A.J. and Roberts W. (2002) Morphological modelling of intertidal mudflats: the role of cross-shore tidal currents. *Cont. Shelf Res.*, **22**, 1887–1895.

Reineck, H.-E. (1967) Layered sediments of tidal flats. *Estuaries* (Ed G.H. Lauff). American Association for the Advancement of Science, Washington, DC, pp. 191–206.

Reineck, H.-E. (1963) Sedimentgefüge im Bereich der südlichen Nordsee. *Abh. Senckenb. Naturf. Ges.*, **505**, 1–138.

Reineck, H.-E. and Wunderlich, F. (1968) Classification and origin of flaser and lenticular bedding. *Sedimentology*, **11**, 99–104.

Ricci Lucchi, F. (1995) *Sedimentographica. Photographic Atlas of Sedimentary Structures*, 2nd edition. Columbia University Press, New York, 225 pp.

Rust, I.C. (1977) Evidence of shallow marine and tidal sedimentation in the Ordovician Graafwater Formation, Cape Province, South Africa. *Sed. Geol.*, **18**, 123–133.

Rust, I.C. (1981) Lower Palaeozoic rocks of southern Africa. *Lower Palaeozoic of the Middle East, Eastern and Southern Africa* (Ed C.H. Holland). Wiley, New York, pp. 165–187.

Rust, I.C. (1967) On the sedimentation of the Table Mountain Group in the western Cape Province. DSc dissertation, University of Stellenbosch, South Africa, 110 pp.

Rust, I.C. (1973) The evolution of the Paleozoic Cape Basin, southern margin of Africa. *The Ocean Basins and Margins, I. The South Atlantic* (Eds A.E.M. Nairn and F.G. Stehli). Plenum Press, New York, pp. 247–276.

Siedlecka, A. (1978) Late Precambrian tidal-flat deposits and algal stromatolites in the Batsfjord Formation, East Finmark, North Norway. *Sed. Geol.*, **21**, 277–310.

Son, C.S., Flemming, B.W. and Bartholomä, A. (2011) Evidence for sediment recirculation on an ebb-tidal delta of the East Frisian barrier-island system, southern North Sea. *Geo-Mar. Lett.*, **31**, 87–100.

Son, C.S., Flemming, B.W. and Bartholomä, A. (2012a) Long-term changes of surface sediments and morphology in relation to energy variations on shoreface-connected ridges off Spiekeroog Island, southern North Sea. *J. Sed. Res.*, **82**, 385–399.

Son, C.S., Flemming, B.W. and Chang T.S. (2012b) Sedimentary facies of shore-face connected sand ridges off the East Frisian barrier-island coast, southern North Sea: climatic controls and preservation potential. *Int. Assoc. Sedimentol. Spec. Publ.*, **44**, 143–158.

Stow, D.A.V. (2005) *Sedimentary Rocks in the Field. A Colour Guide*. Manson Publ., London, 320 pp.

Tankard, A.J. and Hobday, D.K. (1977) Tide-dominated back-barrier sedimentation, early Ordovician Cape Basin, Cape Peninsula, South Africa. *Sed. Geol.*, **18**, 135–159.

Tankard, A.J., Jackson, M.P.A., Eriksson, K.A., Hobday, D.K., Hunter, D.R. and Minter, W.E.L. (1982) *Crustal Evolution of Southern Africa – 3.8 Billion Years of Earth History*. Springer-Verlag, New York, 523 pp.

Tankard, A., Welsink, H., Aukes, P., Newton, R. and Stettler, E. (2009) Tectonic evolution of the Cape and Karoo basins of South Africa. *Mar. Petrol. Geol.*, **26**, 1379–1412.

Thamm, A.G. and Johnson, M.R. (2006) The Cape Supergroup. *The Geology of South Africa* (Eds M.R. Johnson, C.R. Anhaeusser and R.J. Thomas), Council for Geoscience and Geological Society of South Africa, Pretoria, pp. 443–460.

Turner, B.R. (1990) Continental sediments in South Africa. *J. Afr. Earth Sci.*, **10**, 139–149.

Turner, B.R. (1986) Environmental significance of desiccation cracks in the Early Ordovician Graafwater Formation, Cape Peninsula, South Africa. *Geocongress '86, Geol. Soc. South Afr.*, Abstracts, pp. 433–435.

Turner, B.R., Armstrong, H.A. and Holt, P. (2011) Visions of ice sheets in the early Ordovician greenhouse world: Evidence from the Peninsula Formation, Cape Peninsula, South Africa. *Sed. Geol.*, **236**, 226–238.

Verbanck, M.A. (2008) How fast can a river flow over alluvium? *J. Hydr. Res.*, **46 Supplement** 1, 61–71.

Visser, J.N.J. (1974) The Table Mountain Group: a study in the deposition of quartz arenites on a stable shelf. *Trans. Geol. Soc. South Afr.*, **77**, 229–237.

Visser, M.J. (1980) Neap-spring cycles reflected in Holocene subtidal large-scale bedform deposits: preliminary note. *Geology*, **8**, 543–546.

Vos, R.G. and Tankard, A.J. (1981) Braided fluvial sedimentation in the lower Palaeozoic Cape Basin, South Africa. *Sed. Geol.*, **29**, 171–193.

Werner, F. (1968) Gefügeanalyse feingeschichteter Schlicksedimente in der Eckernförder Bucht (westliche Ostsee). *Meyniana*, **18**, 79–105.

Williams, G.E. (1989) Late Precambrian tidal rhythmites in South Australia and the history of the Earth's rotation. *J. Geol. Soc. London*, **146**, 97–111.

Williams, G.E. (1991) Upper Proterozoic tidal rhythmites, South Australia: sedimentary features, deposition, and implications for the earth's paleorotation. *Clastic Tidal Sedimentology* (Eds D.G. Smith, G.E. Reinson, B.A. Zaitlin and R.A. Rahmani). *Can. Soc. Petrol. Geol. Mem.*, **16**, 161–178.

Zimmermann, U. and Bahlburg, H. (2003) Provenance analysis and tectonic setting of the Ordovician clastic deposits in the southern Puna Basin, NW Argentina. *Sedimentology*, **50**, 1079–1104.

Tidal *versus* continental sandy-muddy flat deposits: Evidence from the Oncala Group (Early Cretaceous, N Spain)

I. EMMA QUIJADA[†‡*], PABLO SUAREZ-GONZALEZ[†‡], M. ISABEL BENITO[†‡] and RAMÓN MAS[†‡]

[†]*Departamento de Geología, Universidad de Oviedo, C/Jesus Arias de Velasco, s/n, 33005, Oviedo, Spain*
[‡]*Instituto de Geociencias IGEO (CSIC, UCM), C/José Antonio Novais 12, 28040, Oviedo, Spain*
[*]*Corresponding author, E-mail address: emma@geol.uniovi.es*

ABSTRACT

Recognizing and interpreting features attributable to tidal influence in the sedimentary record may be difficult in certain settings, such as fluvial-tidal estuaries or deltas and protected inland tidal embayments, because marine fossils and classic tidal features, such as bi-polar current indicators or tidal bundles, may be absent. Deposits of the central area of the Oncala Group, from the Cameros Basin (Early Cretaceous, N Spain), represent one of these cases in which the lack of several classic tidal features makes the recognition of tidal influence difficult. These deposits were developed at the terminus of a fluvial system and pass laterally to coeval, shallow, carbonate-evaporitic laminated deposits, which do not contain any definitive sedimentary feature or fossil remains that clearly indicate a marine depositional setting. The central area of the Oncala Group is characterized by laterally very continuous deposits, which can be followed along tens or hundreds of metres, predominance of non-channelled facies over channelled bodies and ubiquitous subaerial exposure evidence, such as desiccation cracks and vertebrate footprints. Although these characteristics might indicate deposition in a continental sandy-muddy flat, additional sedimentary features lead to interpret these deposits as formed in broad, low-gradient, tidal flats, traversed by meandering channels. These features include: (1) presence of meander loop bodies displaying low angle, inclined heterolithic stratification; (2) repetitive alternation of mudstone and siltstone/sandstone laminae both in meander loop bodies and non-channelled facies; (3) rhythmic vertical variations in the type of bedding (lenticular-wavy-flaser-wavy-lenticular) and in the thicknesses of the couplets; (4) presence of desiccation cracks at the top of numerous, consecutive, mudstone laminae or couplets. As a result, the Oncala Group represents a good example of tide-influenced deposits that lack several typical tidal features, such as marine fossils or common bi-polar current indicators, and thus yields criteria for recognition of tide influence in similar ancient sediments.

Keywords: Tidal flats, sandy-muddy flats, Inclined Heterolithic Stratification (IHS), tidal rhythmites, Cameros Basin, Lower Cretaceous.

INTRODUCTION

Recognition of tidal signatures in the sedimentary record is a powerful tool for reconstructing the palaeogeography of a basin, as well as the extent of marine influence in ancient settings. However, recognition of tidal deposits in the geological record may be difficult in certain settings where definitive tidal features, such as tidal bundles or bi-polar current indicators, are not present and marine fossil remains are absent (Kvale & Archer, 1990; Kvale & Mastalerz, 1998). As a consequence, ancient deposits formed in tidal environments could be interpreted as terrestrial sediments. In fact, numerous successions originally interpreted as formed in continental environments, such as meandering and braided fluvial systems or inland saline lakes, have been later reinterpreted as

Contributions to Modern and Ancient Tidal Sedimentology: Proceedings of the Tidalites 2012 Conference, First Edition. Edited by Bernadette Tessier and Jean-Yves Reynaud.

formed in fluvial-tidal estuaries or deltas and protected inland tidal embayments (e.g. Shanley *et al.*, 1992; Gingras *et al.*, 2002; Chakraborty *et al.*, 2003; Hovikoski *et al.*, 2005, Rebata-H. *et al.*, 2006). Ancient deposits formed in certain tidal settings could also be interpreted as formed in continental sandy-muddy flats developed around saline lakes in arid regions, such as in Death Valley and Saline Valley, California (Hardie *et al.*, 1978), or Salina de Ambargasta, Argentina (Zanor *et al.*, 2012), because both depositional environments have several sedimentary features in common. Laterally very continuous layering, ubiquitous subaerial exposure evidence, absence of marine fossils, location at the terminus of alluvial/fluvial systems and alternation of periods of bedload transport and settling from suspension are all characteristics that could lead researchers to ascribe a continental origin to deposits from tidal settings. The palaeogeographic implications of these contrasting palaeoenvironmental reconstructions are enormous as, for example, a coastal transitional system could be erroneously interpreted as an endorheic continental system.

The siliciclastic deposits of the Early Cretaceous Oncala Group (Cameros Basin, Northern Spain) pose a sedimentological challenge in this sense, as numerous sedimentary features could lead to a continental interpretation when analysed individually. However, the analysis of the sedimentary features as a whole shows that only in a tidal setting could the necessary sedimentary processes for development of such features have taken place. Therefore, this example provides useful clues to discriminate between a tidal depositional environment and a purely continental environment in ancient successions.

GEOLOGICAL SETTING

The Oncala Group is part of the sedimentary infill of the Cameros Basin, northern Spain (Fig. 1), formed during the intraplate rifting that took place in Iberia from the Late Jurassic to the Lower Cretaceous (Mas *et al.*, 2002; Martín-Chivelet *et al.*, 2002). As a consequence of the opening of the North Atlantic Ocean, a series of extensional basins were developed in Iberia (Martín-Chivelet *et al.*, 2002; Mas *et al.*, 2004). The Iberian Mesozoic Rift System, which includes the Cameros Basin, extended over a large part of Iberia in a north-west to south-east direction and the Basque-Cantabrian

Basin was formed in the northern part of Iberia (Fig. 1A). Broad tidal flats (Fig. 1B) were developed over large areas of these rift basins during Berriasian times (see García de Cortázar & Pujalte, 1982; Pujalte, 1982; Mas *et al.*, 1984; Salas, 1989; Aurell *et al.*, 1994; Bádenas *et al.*, 2004) and marine areas (Fig. 1B) were located in the south-easternmost area of the Iberian Mesozoic Rift System and in the northernmost areas of the Basque-Cantabrian Basin (see Salas, 1989; Aurell *et al.*, 1994; Martín-Chivelet *et al.*, 2002). The Cameros Basin was located in an intermediate position between the Basque-Cantabrian Basin and the south-eastern basins of the Iberian Mesozoic Rift System (Fig. 1A and B). During the Cainozoic Alpine Orogeny, the Cameros Basin was inverted and Tertiary basins were developed around it (Fig. 1A and C), which makes difficult to analyse in detail the relationships of the deposits of the Cameros Basin with those coeval in the rest of the Iberian Peninsula.

The infill of the Cameros Basin corresponds to a large cycle or super-sequence divided into eight depositional sequences (Fig. 2) and it recorded high subsidence and accumulation rates, with vertical thickness of more than 6000 m of sediments from the Tithonian to the early Albian (Mas *et al.*, 2011). These depositional sequences consist mainly of alluvial, fluvial, lacustrine and coastal deposits (Mas *et al.*, 1993, 2002; Quijada *et al.*, 2013b; Suarez-Gonzalez *et al.*, 2013, 2014, 2015).

The Oncala Group corresponds to the third depositional sequence of the basin (Fig. 2) and was deposited during the Berriasian, as has been established on the basis of the analysis of charophyte and ostracod assemblages (Salomon, 1982; Schudack & Schudack, 2009). The Oncala Group is one of the thickest units in the Cameros Basin, comprising up to 2500 m of sediments in the depocentral areas. It contains both siliciclastic and carbonate-sulphate deposits (Figs 1 and 2), which show gradual lateral changes from proximal facies in the western area to distal facies in the eastern area (Gómez-Fernández & Meléndez, 1994; Meléndez & Gómez-Fernández, 2000; Quijada *et al.*, 2013a, 2013b). The most proximal facies of the Oncala Group are located in the westernmost area of the basin, close to the town of Montenegro (Fig. 1) and they mainly consist of sandy, meandering fluvial deposits (Gómez-Fernández & Meléndez, 1994; Meléndez & Gómez-Fernández, 2000). Towards the east, these fluvial deposits pass to laterally very continuous, non-channelled

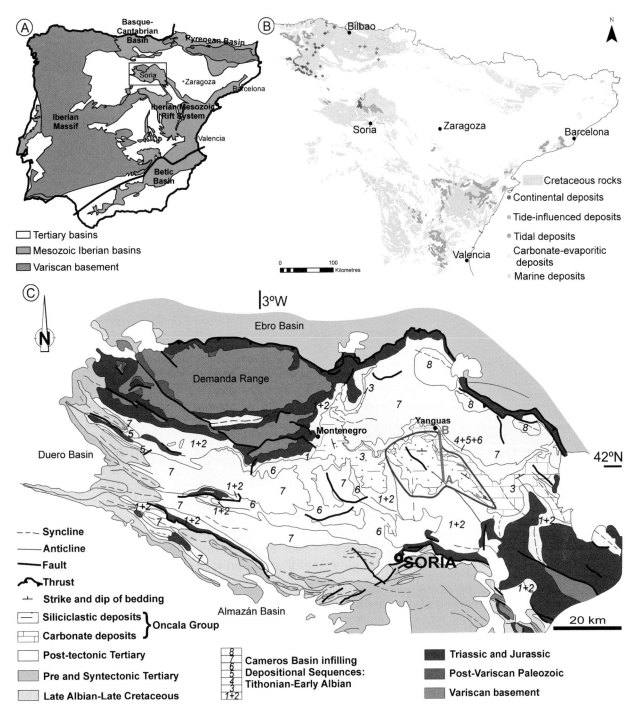

Fig. 1. (A) Synthetic map showing the location of the inverted Mesozoic basins of the Iberian Peninsula, the Variscan basement and the Duero, Ebro and Tajo Tertiary basins (modified from Quijada *et al.*, 2013b). The rectangle marks the location of the Cameros Basin and the area shown in Fig. 1C. (B) Synthetic map showing the main facies and location of Berriasian deposits in the Iberian Peninsula, based on data from outcropping areas and oil exploration wells (modified from Quijada *et al.*, 2013b). Note the large areas of the basins that contain Berriasian tidal deposits and the intermediate location of the Cameros Basin in relation to the marine areas. (C) Geological map of the Cameros Basin (modified from Mas *et al.*, 2002). The area highlighted in red colour marks the study area and the thick blue line indicates the location of the stratigraphic section shown in Fig. 3. Note the lateral changes of facies within the Oncala Group, which contains predominant siliciclastic deposits in the western area and in the lower and middle part of the succession, and carbonates in the eastern area. Carbonate deposits progressively extend towards the west in the middle and upper part of the Oncala Group.

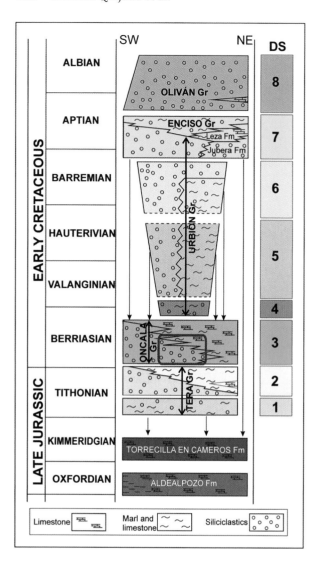

Fig. 2. Stratigraphic framework and depositional sequences (DS) filling the eastern sector of the Cameros Basin. The focus of this study, the siliciclastic deposits of the central area of the Oncala Group, is highlighted with a red rectangle. Modified from Mas *et al.* (2004). 'Gr' = group, 'Fm' = formation. The spacing between DS represents the stratigraphic gap between them and the broken line that limits some DS indicates that the age of the DS is uncertain.

mudstones and sandstones and less abundant sandy meander loop bodies. The most distal deposits are located in the eastern area of the basin and consist of laminated carbonate-sulphate deposits. The carbonate-sulphate deposits of the lower part of the unit were firstly interpreted as formed in shallow lakes and those of the upper part of the unit as formed in deep lakes (Gómez-Fernández & Meléndez, 1994; Meléndez & Gómez-Fernández, 2000). However, more recent studies have reinterpreted the carbonate-sulphate

deposits of both the lower and the upper part of the Oncala Group as formed in shallow, coastal water bodies and their peripheral mudflats (Quijada *et al.*, 2013a; 2013b).

Apart from the described lateral changes of facies, a retrograding vertical evolution is also evident in the Oncala Group (Quijada *et al.*, 2013b), as distal deposits overlie more proximal deposits and reach progressively more western areas (Figs 1, 2 and 3). As a consequence, siliciclastic deposits are more abundant in the lower and middle part of the unit and carbonate-sulphate deposits are more abundant upwards (Fig. 3), until they occupy most of the basin in the uppermost part of the unit (Figs 1 and 2).

This study is focused on the siliciclastic deposits developed in the central area of the basin in the lower and middle part of the unit (Figs 1, 2 and 3), which have been interpreted previously as deposited in continental sandy-muddy flats (Gómez-Fernández & Meléndez, 1994; Meléndez & Gómez-Fernández, 2000). They occupy an intermediate position between the most proximal fluvial deposits in the western area of the basin and the carbonate-sulphate deposits in the eastern area.

SANDY-MUDDY FLAT DEPOSITS OF THE ONCALA GROUP

The deposits of the central area of the Oncala Group are characterized by great lateral continuity, which can be followed tens or hundreds of metres (Fig. 4). These deposits are classified in three facies associations attending to their geometry, contacts with adjacent elements, internal surfaces and mineralogical composition (Figs 4 and 5). The three facies associations are: meander loop bodies, non-channelled heterolithic layers and tabular dolostone beds.

Meander loop bodies

Description

Meander loop bodies are tabular asymmetrical bodies with widths of up to 70 m and thicknesses of between 1 and 4 m (Fig. 4). Bases of the bodies are scoured flat surfaces and tops are also flat and pass gradually to overlying non-channelled heterolithic layers. In some transversal sections of the meander loop bodies, the erosion of underlying non-channelled heterolithic layers is observable at one end of the body and a channel plug is recognizable at the other end (Fig. 6A). The meander

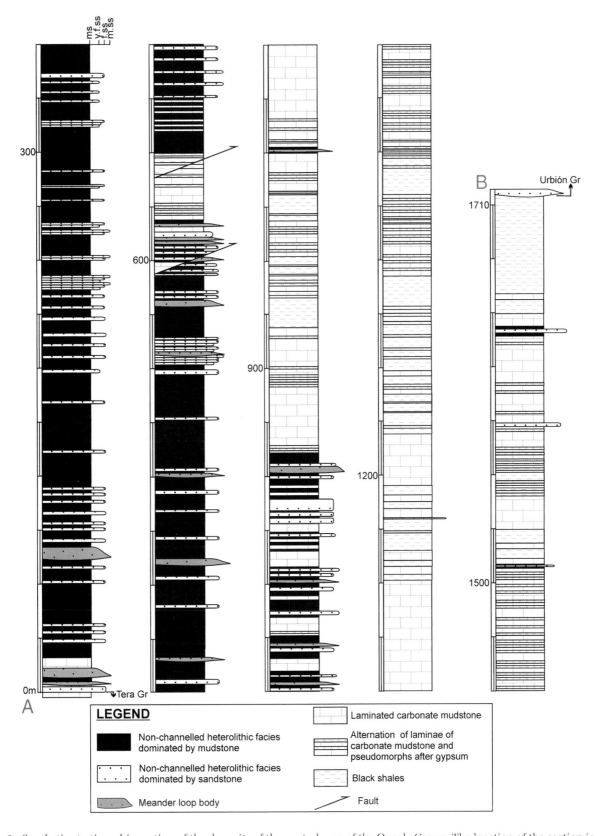

Fig. 3. Synthetic stratigraphic section of the deposits of the central area of the Oncala Group. The location of the section is indicated in Fig. 1. Note that siliciclastic facies are predominant in the lower and middle parts of the unit, carbonate and evaporites are more abundant in the middle part and they are predominant in the upper part. The studied deposits are the predominantly siliciclastic sediments of the lower and middle part of the Oncala Group.

Fig. 4. Field photograph showing the laterally continuous, siliciclastic deposits of the central area of the Oncala Group. Most of the sediments shown in the picture are non-channelled heterolithic layers, except the strata marked in blue and yellow, which are meander loop bodies and tabular dolostone layers, respectively. These deposits are stratigraphically located in the middle part of the Oncala Group (equivalent to ≈850 m of Fig. 3).

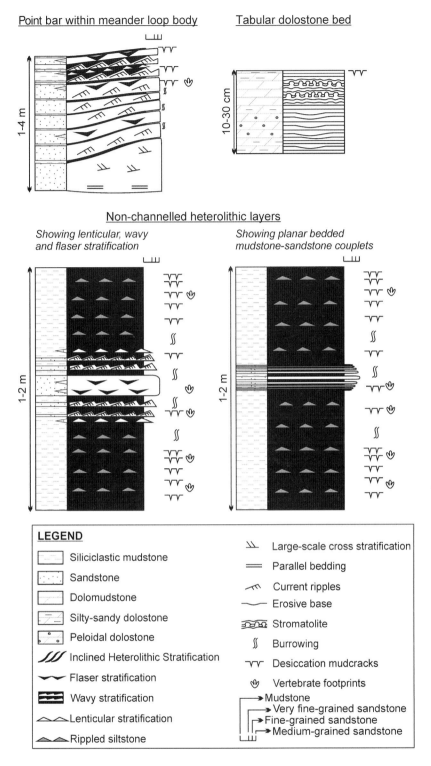

Fig. 5. Schematic representations of the three facies associations in the central area of the Oncala Group. Siliciclastic rocks are classified according to Wentworth granulometry scale (see Wentworth, 1922).

Fig. 6. (A) Field photograph of a meander loop body (highlighted with a broken white line) showing an asymmetric morphology and lateral accretion units dipping to the right part of the picture. Note the erosional surface at one end of the meander loop body (white arrow), which cuts the underlying non-channelled layers and the channel plug at the other end of the body (black arrow). The black rectangle marks the area shown in Fig. 8D. These deposits are located at 1020 m of the stratigraphic section (see Fig. 3). (B) Field photograph of a meander loop body made up of three point bar bodies separated by erosional surfaces (pink lines). Note the fining upwards trend of each point bar body, which is especially evident in point bar body 2. Left area of point bar body 2 is shown in Fig. 7A. These deposits are stratigraphically located in the middle part of the Oncala Group (equivalent to ≈590 m of Fig. 3).

Fig. 7. Field photographs of meander loop bodies. (A) Detail of the left area of the point bar body 2 shown in Fig. 6B and portion of the overlying point bar body 3. Note that point bar body 2 displays a low-angle erosional base (marked by the pink broken line) overlain by thick, medium-grained, sandstone units, which change upwards to Inclined Heterolithic Stratification (IHS). (B) Large-scale cross-stratification overlying the erosional base of a point bar body (yellow broken line). The red rectangle marks the area shown in Fig. 7C. (C) Small-scale cross-stratification in the bottomsets of the large-scale cross-stratified sandstone shown in Fig. 7B.

loop bodies are made up of one or more adjoined point bar bodies (*sensu* Díaz-Molina, 1993), which consist of sets of conformable lateral accretion units, separated by erosional surfaces (Fig. 6B).

Point bar bodies show a fining upwards trend, containing medium-grained sandstone in the base and fine-grained sandstone in the top (Figs 5 and 7A). The lower part of the point bar bodies is made up of medium-grained sandstone with parallel bedding or large-scale cross stratification (Fig. 7B). These sets of large-cross stratification commonly contain mud drapes and may be associated with ripples in the bottomsets (Fig. 7C), which commonly ascend high up the larger cross beds.

Fig. 8. Field photographs of meander loop bodies. (A) IHS within a channelled heterolithic body. The red arrows point to mudstone lateral accretion units. Note the gentle dip of the lateral accretion units. (B) Three point bar bodies (marked by red brackets) separated by erosional surfaces. Each point bar body is made up of conformable, sandstone and mudstone, lateral accretion units, displaying IHS. The white rectangle marks the area shown in Fig. 9A. Hammer for scale in central area of the photograph. (C) Sigmoidal, large-scale cross-stratification within a sandstone lateral accretion unit. (D) Large-scale cross stratification ascending up a lateral accretion surface, which dips to the right part of the photograph.

Above the basal sandstone unit, the point bar sequence is typically characterized by the presence of Inclined Heterolithic Stratification (IHS, *sensu* Thomas *et al.*, 1987), which consists of low angle (dips from 1 to 5° approximately), alternating sandstone and mudstone, lateral accretion units (Fig. 8A and B). Mudstone lateral accretion units are few centimetres to 20 cm-thick and they show thin lamination. Sandstone lateral accretion

units are 10 cm to 40 cm-thick and they display typically ripple and climbing ripple cross stratification. Large-scale, sigmoidal, cross stratification is also present in some sandstone units (Fig. 8C) and sets of large-scale cross stratification ascending up the lateral accretion surfaces are also evident in some sandstone units (Fig. 8D). Mud laminae commonly drape both small-scale and large-scale cross-stratification sets and foreset

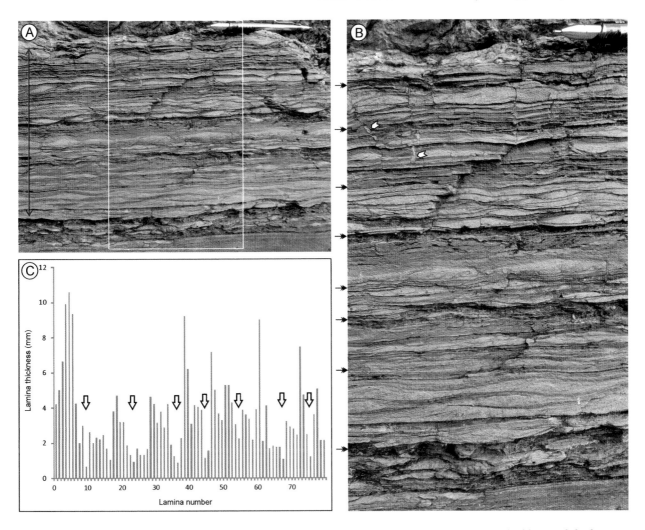

Fig. 9. (A) Flaser, wavy and lenticular bedding within IHS. Note the rhythmic variations in type of bedding and thicknesses of sandstone-mudstone couplets. The ripple cross stratification indicates palaeocurrents directions towards the left. The red arrow marks the stratigraphic interval in which measurements of the thicknesses of the couplets were conducted and the white rectangle points to the area shown in Fig. 9B. (B) Detail of sandstone-mudstone couplets shown in Fig. 9A. The black arrows point to the thinner couplets (compare with Fig. 9C). The white arrows point to (sub-) vertical burrows disrupting the bedding. (C) Plot of thicknesses of the couplets in Fig. 9A and B. Note the rhythmic variations, interpreted as probable tidal cyclicities. Arrows indicate probable neap-stage couplets. This graphic shows one arrow fewer than Fig. 9B, which corresponds to the lowermost arrow in that figure, because that interval could not be measured due to the poor condition of the outcrop.

laminae within the sandstone lateral accretion units.

The lateral accretion units fine upwards and in the intermediate to upper part they display a thin alternation of rippled sandstone and mudstone laminae developing flaser, wavy and lenticular bedding (Fig. 9A and B). Most of these small-scale structures are unidirectional, showing palaeocurrents to the east or south-east, which are approximately perpendicular to the lateral accretion surfaces. However, few bidirectional structures are also present. Lenticular bedding grades commonly to flaser and back to lenticular stratification, developing

cyclic changes in the type of bedding (Fig. 9A and B). Thicknesses of the sandstone-mudstone couplets have been measured in a 30 cm-thick IHS set showing exceptional outcropping conditions (Fig. 9). The couplets range from 0.7 to 10.6 mm in thickness and they exhibit rhythmic thickening and thinning (Fig. 9C). The thickest couplets coincide generally with the flaser bedding and the thinnest couplets correspond to the lenticular bedding. Considering that a complete cycle starts at a thinnest couplet and ends in the following thinnest couplet (Fig. 9C), the number of couplets per cycle ranges from 8 to 14.

Sparse bioturbation, which includes (sub-) vertical and horizontal burrows, is present in the meander loop bodies. (Sub-) vertical burrows are simple cylindrical structures (less than 2 cm long and less than 2 mm in diameter), contain structureless fills that contrast with the host sediment and their walls are unlined (Fig. 9B). Horizontal burrows are less than 1.5 cm in diameter, contain structureless fills differing from the host sediment and display unlined walls. Desiccation mudcracks and vertebrate footprints are observed at the top of the lateral accretion units, especially in the intermediate and upper part of the units. Fossil content is scarce and it includes fragments of bones, ostracods and rare charophytes. Ostracods and charophytes are poorly preserved due to dissolution and silicification.

Interpretation

The meander loop bodies are interpreted as the result of lateral migration in shallow (less than 5 m-deep) meandering channels due to the presence of lateral accretion units; the tabular morphology of the bodies and the channel plugs evident in some transversal sections (cf. Moody-Stuart, 1966; Puigdefabregas, 1973; Puigdefabregas & Van Vliet, 1978; Díaz-Molina, 1979; Nichols, 2009). The presence of sets of large-scale cross stratification ascending up the lateral accretion surfaces is also characteristic of meandering channels due to the helicoidal flow that occurs in meanders (see Allen, 1968 and references therein; Puigdefabregas, 1973; Puigdefabregas & Van Vliet, 1978; Díaz-Molina, 1979; Díaz-Molina & Bustillo, 1985). It is interpreted that relatively long duration of the meander belts in the Oncala Group caused development of wide meander loop bodies by juxtaposition of point bar bodies. The erosional surfaces that separate different point bars are interpreted as erosional reactivation surfaces (*sensu* Díaz-Molina, 1993) caused by fluctuations in the channel displacement direction. Nevertheless, it is not ruled out that seasonal higher channel discharges may have caused some erosional surfaces, as occurs in some modern, tidal point bars during winter rainfall (de Mowbray, 1983).

The low angles that characterize the lateral accretion units in the meander loop bodies suggest gently dipping point bar surfaces. The fining upwards infilling sequence and the upwards change from higher energy to lower energy stratification types in the point bar bodies indicates a relative decrease in the water current velocity from the lowermost part of the channel to the upper part. The presence of IHS indicates marked fluctuations of flow velocities, which allowed alternation of bedload sediment transport during high flow velocity and settling out of mud under conditions of much-reduced velocity (Thomas *et al.*, 1987). Moreover, the presence of lenticular, wavy and flaser bedding suggests also alternation of current action and slack water (Reineck & Wunderlich, 1968). The rhythmic variations of type of bedding and sandstone-mudstone couplet thicknesses suggest cyclic fluctuations of water current velocity. The presence of desiccation mudcracks and vertebrate footprints in the intermediate and upper part of the lateral accretion units indicate that at least the upper portion of the point bars was intermittently subaerially exposed.

Non-channelled heterolithic layers

Description

Non-channelled heterolithic layers are the most abundant facies in the studied area, representing around the 80% of the sediments (Fig. 4). They show laterally very continuous layering, which can be followed tens or hundreds of metres along the total length of an outcrop (Fig. 4). Non-channelled heterolithic layers are made up of interlaminated siliciclastic mudstone, siltstone and very fine-grained to medium-grained sandstone (Figs 5 and 10). Mudstone and siltstone predominate in most of the non-channelled layers (Fig. 10A) and they display lenticular and wavy bedding (Fig. 10B). Mudstone laminae are sub-millimetre or millimetre in thickness and siltstone layers are less than 2 cm-thick and display current and wave ripples (Fig. 10C). Less abundant non-channelled heterolithic layers show a larger proportion of sandstone and they display less than 2 cm-thick sandstone-mudstone couplets (Fig. 10D to G). Some couplets show planar bedding, which consist of a massive sandstone lamina capped by a mudstone lamina (Fig. 11A and B) and some others display flaser, wavy and lenticular bedding, which are made up of current ripple cross-stratified sandstone overlain by mudstone (Figs 10D, 10E, 10F, 11A, 11C and 11D). The thicknesses of the couplets and the type of bedding show commonly cyclic vertical variations (Fig. 10F and G), which result in thickening and thinning sequences and an evolution from lenticular to flaser bedding and back to lenticular stratification, as described in the meander loop bodies.

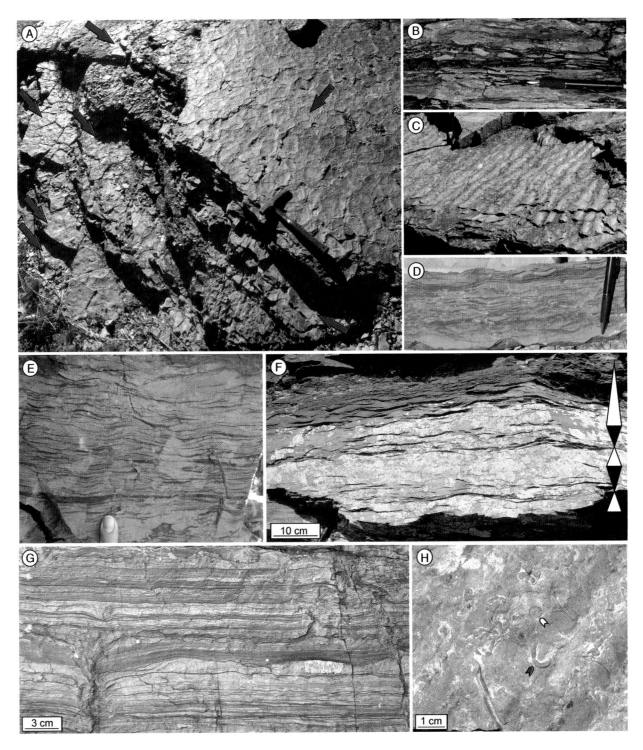

Fig. 10. Field photographs of the non-channelled heterolithic layers. (A) Desiccation mudcracks at the top of numerous successive mudstone laminae (red arrows). Hammer for scale in the lower right area of the picture. (B) Alternation of mudstone and siltstone laminae displaying lenticular stratification. (C) Consecutive, thin, rippled siltstone layers. Hammer for scale in the upper left area of the picture. (D) Alternation of sandstone and mudstone laminae displaying wavy bedding. (E) Alternation of sandstone and mudstone laminae displaying flaser bedding. (F) Cyclic variations in type of bedding (flaser, wavy or lenticular). White and black triangles indicate fining and coarsening trends, respectively. (G) Cyclic changes in the thicknesses of the mudstone-sandstone couplets. (H) Bioturbated layer displaying vertical (white arrow) and horizontal (red arrow) burrows with structureless fills that contrast with the host sediment and unlined walls (top view).

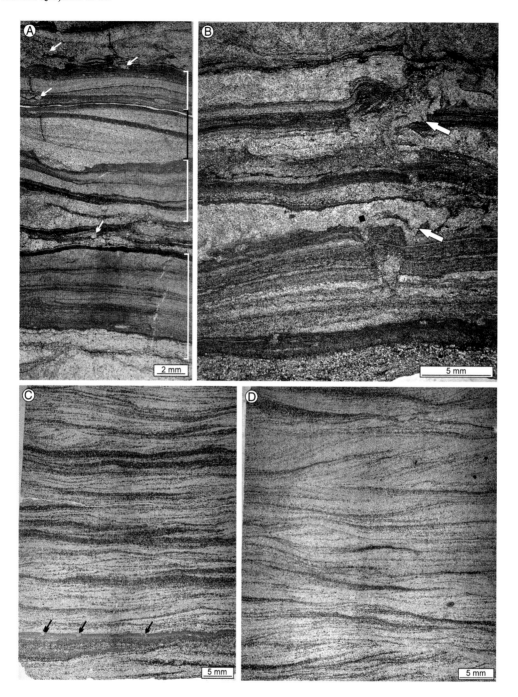

Fig. 11. Thin section photomicrographs of alternation of sandstone (light) and mudstone (dark) laminae. (A) Thin section photomicrograph showing two types of sandstone-mudstone couplets. White brackets mark couplets made up of a planar-bedded, sandstone lamina capped by a mudstone lamina and black bracket marks couplets consisting of cross-stratified sandstone overlain by a mudstone lamina. Arrows point to probable burrows. Some couplets are disturbed, probably by burrowing or/and compaction. (B) Planar-bedded, sandstone-mudstone couplets. Several couplets are disrupted in some areas (white arrows) probably due to burrowing. (C) Sandstone-mudstone couplets showing wavy bedding. Black arrows point to probable burrows. (D) Sandstone-mudstone couplets showing flaser bedding.

The non-channelled heterolithic layers are also characterized by very abundant desiccation mudcracks. Mudcracks are especially abundant at the top of the finest-grained laminae, in which they occur at the top of almost each lamina

(Fig. 10A). In these finest-grained laminae, mudcracks show concave-up curling or no curling, they develop polygons less than 10 cm in size and they are less than 2 cm in depth (Figs 10A and 12A). Coarser grained sediments made up of

Fig. 12. Field photographs of the non-channelled heterolithic layers displaying desiccation mudcracks. (A) Mudcracked mudstone layer showing no curling. Numerous dinosaur footprints are present at the top of the layer. (B) Convex-up desiccation polygons developed at the top of a layer that consists of alternating sandstone and mudstone laminae. (C) Downward-curling desiccation cracks deforming a package of alternating sandstone and mudstone. (D) Dinosaur footprints at the top of mudstone layers, which display also desiccation cracks. Hammer for scale in the lower central area of the picture.

alternating laminae of mudstone and fine-grained sandstone, commonly showing a vertical evolution from lenticular to flaser bedding, display less common desiccation mudcracks. The mudcracks developed in these coarser-grained sediments display convex-up, larger polygons (up to 30 cm in diameter) and disturb up to 5 cm-thick deposits (Fig. 12B and C).

Vertebrate footprints are also ubiquitous at the top of the non-channelled layers (Fig. 12A and D); 169 tracksites of theropod, ornithopod, sauropod, pterosaur, bird, crocodile and turtle have been found up to this time (Hernández *et al.*, 2005–2006; Moratalla & Hernán, 2010; Castanera *et al.*, 2013, 2015; Pascual-Arribas & Hernández-Medrano, 2015). Uncommon bioturbation, which includes (sub-) vertical and horizontal burrows and rare trails, is also evident in these layers (Figs 10H, 11A, 11B and 11C). (Sub-) vertical and horizontal burrows in non-channelled layers display similar features to those within the channelled bodies, differing only

in the size of some (sub-) vertical burrows which are larger in size (up to 2 cm in diameter). Trails are simple, sinusoidal and 1.5 mm in diameter. Fossil content is scarce and it includes ostracods, fragments of bones, rare charophytes and rare bivalves, which are poorly preserved due to dissolution and silicification. Rare raindrop impressions are present at the top of mudstone laminae.

Interpretation

The laterally very continuous layering that characterizes the non-channelled deposits suggests deposition in broad, flat areas. In these areas, bedload transport and settling from suspension alternated causing interlamination of mudstone and cross-laminated siltstone and sandstone. The presence of couplets displaying different types of bedding indicates varying water current velocities; parallel-bedded couplets were probably formed under lower velocity flows, whereas ripple-cross-bedded couplets formed under higher velocity flows (cf. Tessier, 1993). The rhythmic variation from lenticular bedding to flaser and back to lenticular stratification and cyclic changes in the thicknesses of the sandstone-mudstone couplets suggest a cyclic control on water velocity currents. The large amount of deposited mudstone suggests high suspended-sediment concentrations (see Dalrymple & Choi, 2007). The presence of ubiquitous desiccation mudcracks and vertebrate footprints indicate that these flat areas were frequently subaerially exposed. The presence of mudcracks at the top of each mudstone lamina or siltstone-mudstone couplet in abundant stratigraphic intervals suggests repetitive alternation of deposition of a thin mud lamina or silt-mud couplet and subsequent exposure and desiccation. The development of different types of desiccation mudcracks seems to be related with the grain size and textural grading of the sediment. Laboratory experiments and observations on modern deposits show that shrinkage capacity decrease with increasing grain size of the sediment (Allen, 1986; Plummer & Gostin, 1981, their fig. 1). As a result, the direction of curl of the polygons depends on the vertical textural grading of the affected layer; concave-up curling results in fining upwards packages and convex-up curling results in coarsening-upwards packages (Allen, 1986). This is consistent with the presence of convex-up polygons in layers of the Oncala Group made up of alternating laminae of mudstone and sandstone that show more abundant sandstone at the top. However, other factors, such as composition or tenacity of the mud (Kindle, 1917; Plummer & Gostin, 1981), may have also been determinant for the development of different types of mudcracks.

Tabular dolostone beds

Description

Tabular dolostone beds are interbedded with the non-channelled heterolithic layers (Fig. 13A) and they are not associated with the meander loop bodies. Dolostone layers are generally 10 to 30 cm-thick, but a few layers reach 70 cm in thickness. They display flat bottom and top and are laterally very continuous (individual layers can be followed tens to hundreds of metres; Fig. 4). They display common desiccation mudcracks at the top of the layers and rare calcite nodules made up of aggregates of pseudomorphs after lenticular and

acicular crystals. Three types of dolostone microfabrics are present and they are commonly interlayered (Fig. 5): silty-sandy dolostone, peloidal dolostone and stromatolitic dolostone.

Silty-sandy dolostone consists of ostracod mudstone to wackestone with 10% to 30% of coarse silt-sized to fine sand-sized grains of quartz and mica. Although the original lamination caused by variations in siliciclastic content is generally observed, occasionally the lamination has been disrupted as a consequence of nodulization, mudcracking and burrowing.

Peloidal dolostone consists typically of laminae displaying slightly erosional base and a gradation in textures from grainstone to packstone and, finally, wackestone at the top. Peloids are micritic and have no internal fabric (Fig. 13B). They are rounded elongated and less than 450 μm long, with long diameters 2 to 3 times their short diameter. They are well sorted and very well-rounded. Some of them contain a large amount of framboidal pyrite. Peloidal dolostone contains also dispersed silt-sized grains of quartz and mica and ostracod fragments.

Stromatolitic dolostone (Fig. 13C and D) show pseudocolumnar and undulatory structures (*sensu* Preiss, 1976) and display filamentous, dense micrite, clotted-peloidal and peloidal microfabrics. Peloidal microfabric is made up of identical peloids to those in the peloidal dolostone (described above) and occurs mostly in laterally continuous, undulatory laminae and less commonly within the columns. Peloids are also found filling depressions and the space between the columns of the stromatolites.

Interpretation

Tabular dolostone beds were probably deposited in very flat broad ponds, which were developed in shallow, inundated depressions of the extensive flats of the Oncala Group. The shallow depositional depth of these ponds is suggested by the common desiccation mudcracks at the top of the dolostone layers. These areas were affected episodically by currents that led to accumulation of graded peloidal laminae and silty-sandy laminae with up to 30% of siliciclastic grains. The rounded elongated shape of the peloids, their length: width ratio, the good sorting and their common high content in framboidal pyrite are characteristic of faecal pellets (Flügel, 1982; Scholle & Ulmer-Scholle, 2003). These faecal pellets were probably formed in these ponded areas, indicating adequate nutrients and

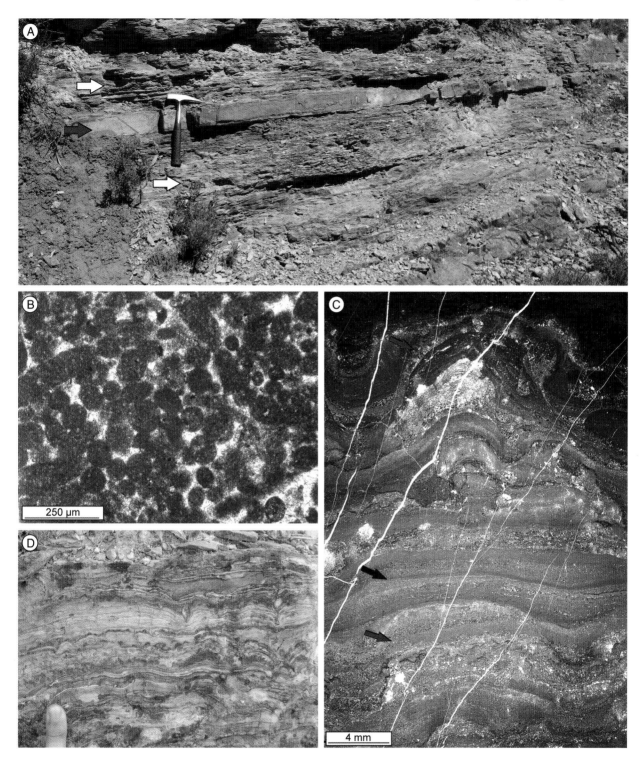

Fig. 13. (A) Field photograph of a tabular dolostone layer (red arrow) interbedded with non-channelled heterolithic layers (white arrows). (B) Thin section photomicrograph of peloids (probable faecal pellets) within the peloidal dolostones. Note the well-rounded shapes and the absence of internal fabric. (C) Thin section photomicrograph of stromatolitic dolostone displaying pseudocolumnar (red arrows) and undulatory structures (black arrow). (D) Field photograph of a tabular dolostone layer mainly formed by stromatolites.

oxygen conditions (Flügel, 1982), and afterwards were reworked by currents.

The stromatolitic facies indicate carbonate precipitation mediated by microbial activity (Riding, 2000; Dupraz *et al.*, 2009). The presence of continuous layers of pellets within the stromatolites suggests that trapping and binding was one of the accretional processes in the stromatolites of the Oncala Group and indicates that bedload transport occurred recurrently, also during the formation of the stromatolitic dolostones.

The calcite nodules are interpreted as pseudomorphs after displacive gypsum and anhydrite due to the lenticular and acicular habits of the original crystals (cf. Ciarapica *et al.*, 1985; Magee, 1991; Kendall, 1992; Ortí & Rosell, 1997). The precipitation of these minerals within the sediments indicates a source of sulphate ions and sporadic increase in salinity, which could be associated with intense evaporation.

DISCUSSION

Some of the most conspicuous features of the deposits of the central area of the Oncala Group are the laterally very continuous layering, which can be followed tens or hundreds of metres (Fig. 4), the predominance of non-channelled facies over meander loop bodies and the abundant desiccation mudcracks (Figs 10A and 12). These features suggest deposition in recurrently subaerially exposed, flat, broad areas, which were traversed by few channels. The lateral relationships of these deposits indicate that they were developed at the terminus of a meandering fluvial system located in the western area of the basin (Meléndez & Gómez-Fernández, 2000) and were laterally related with shallow, carbonate-sulphate water bodies to the East (Quijada *et al.*, 2013a, 2013b). The absence of marine fossil remains in these sediments could lead to interpret them as formed in a continental setting; and the fact that laterally related, coeval carbonates and evaporites do not contain definitive sedimentary features or fossils that clearly indicate marine influence seems to reinforce this hypothesis. The continental depositional setting that fits better with these characteristics, and therefore could be interpreted for the deposits of the central area of the Oncala Group, is a continental sandy-muddy flat developed around a saline lake. In fact, this is the interpretation given by previous authors for the studied deposits of the Oncala Group (Gómez-Fernández & Meléndez, 1994).

Continental sandy-muddy flats are roughly planar and horizontal regions with a sandy and muddy surface, in which unconfined flows are major sedimentary processes (cf. Hardie *et al.*, 1978; Hubert & Hyde, 1982; Gómez-Fernández & Meléndez, 1994; Fisher *et al.*, 2008; North & Davidson, 2012). As pointed out by North & Davidson (2012), these sandy-muddy flats can correspond to different landform elements, such as a) distal end of alluvial fans and subaerially exposed plain that fringes saline lakes ('sandflats' and 'dry mudflats' *sensu* Hardie *et al.*, 1978), b) large-scale, amalgamated sediment package at the terminus of a fluvial system ('terminal splay complex' *sensu* Fisher *et al.*, 2008), or c) ephemeral stream channels and floodplains (cf. Sneh, 1983; Deluca & Eriksson, 1989). Several sedimentary features developed in sandflats and dry mudflats *sensu* Hardie *et al.* (1978) and terminal splay complexes, such as deposition of mainly sandstone and mudstone layers with great lateral extent and sheet morphology, and abundant desiccation mudcracks (Hardie *et al.*, 1978; Robertson-Handford, 1982; Smoot & Lowenstein, 1991; Fisher *et al.*, 2008; Ainsworth *et al.*, 2012), resemble those observed in the deposits of the central area of the Oncala Group. Moreover, continuous laminae of mud capping the sand layers (Robertson-Handford, 1982; Smoot & Lowenstein, 1991; Fisher *et al.*, 2008; Ainsworth *et al.*, 2012) and flaser, wavy and lenticular bedding (Hardie *et al.*, 1978; Ainsworth *et al.*, 2012) have been described in modern sediments of sandflats, dry mudflats (*sensu* Hardie *et al.*, 1978) and terminal splay complexes; and they are interpreted as caused by changes in wind direction and velocity and discharge variations in the feeder river system (Ainsworth *et al.*, 2012). However, cyclic variations in type of bedding (lenticular-wavy-flaser-wavy-lenticular) and in thickness of sandstone-mudstone couplets as those evident in the Oncala Group (Figs 9, 10F and 10G), are not described in these continental environments, due to the episodic nature of water velocity changes in these areas. Furthermore, sandflats and dry mudflats *sensu* Hardie *et al.* (1978) and terminal splays are not traversed by meandering channels (Hardie *et al.*, 1978; Smoot & Lowenstein, 1991; Fisher *et al.*, 2008), as it is observed in the Oncala Group due to the association of meander loop bodies with non-channelled heterolithic layers.

Deposits from ephemeral stream channels and floodplains show also some similarities to the deposits of the Oncala Group, in that they consist of channel sediments interbedded with interchannel sheet-like sandstones and mudstones, which display cross-lamination, horizontal lamination, interlamination of mudstone, siltstone and sandstone and desiccation mudcracks (Robertson-Handford, 1982; Deluca & Eriksson, 1989). However, the alternation of sand and mud laminae in ephemeral fluvial systems shows no organization or rhythmicity, contrarily to what is observed in the Oncala Group. Furthermore, ephemeral stream channels are not meandering (cf. Sneh, 1983; Deluca & Eriksson, 1989) and, thus, their infilling is completely different from that of the meander loop bodies with lateral accretion units observed in the Oncala Group (Fig. 5).

Regarding all these considerations, sedimentation in a continental sandy-muddy flat cannot explain the sedimentary features of the deposits of the central area of the Oncala Group. As a consequence, a different palaeoenvironmental reconstruction based on the characteristics of the three facies associations present in the area (meander loop bodies, non-channelled heterolithic layers and tabular dolostone beds) is necessary.

The meander loop bodies of the Oncala Group, which display IHS, are interpreted as formed in meandering channels affected by periodic water current velocity fluctuations. Although IHS made up of interbedded sand and mud units may occur in various environments, including mixed load and high suspended load meandering rivers, it is particularly abundant in tide-influenced settings (Smith, 1987; Thomas *et al.*, 1987 and references therein; Allen, 1991; Dalrymple *et al.*, 2003; Choi *et al.*, 2004; Dalrymple & Choi, 2007). Development of IHS mostly in tidal settings "is explained by the favourable combination of the gradual accretion in bends of meandering channels, regular flow retardation by the tide and high-suspended mud concentration" (van den Berg *et al.*, 2007). Moreover, the IHS sets present in the Oncala Group display some sedimentary features that are much more common in tide-influenced meandering channels. Firstly, mudstone units within the IHS sets of the Oncala Group are continuous from the lower to the upper part of the lateral accretion units, which is rare in modern fluvial point bars, where deposition of mud drapes occurs primarily on their upper portions during falling flow stages

(Thomas *et al.*, 1987). Secondly, the presence of interlaminated sand and mud as well as flaser, wavy and lenticular bedding within the point-bar sequences, although not exclusive, is most common in tidal environments (Reineck & Wunderlich, 1968; Thomas *et al.*, 1987; Shanley *et al.*, 1992; Falcon-Lang, 1998; Beets *et al.*, 2003; Dalrymple *et al.*, 2003; Choi *et al.*, 2004). Very few examples of flaser bedding have been described from fluvial point bars and in these settings they occur "in a less systematic and regular manner than that of tidal origin" (Bhattacharya, 1997; Chakraborty, 2012). The observed cyclic patterns of variation in the type of bedding and thicknesses of sandstone-mudstone couplets (Fig. 9) are best explained in terms of neap-spring tidal fluctuations in tidal range, current velocities and sediment transport rates (cf. Thomas *et al.*, 1987; Allen, 1991; Nio & Yang, 1991; Tessier, 1993; Choi *et al.*, 2004). These tidal fluctuations would cause development of lenticular bedding and the thinnest couplets during neap tides and flaser bedding and the thickest couplets during spring tides (cf. Tessier, 1993). The presence of 8 to 14 couplets per cycle in the measured outcrop fits with neap-spring tidal rhythmicity. Although the theoretical number of dominant currents between two adjacent neap tides in a semidiurnal setting is approximately 28 (Visser, 1980; Kvale, 2012), modern and ancient examples of tidal deposits showing less than 14 couplets or laminae per cycle are common (Kvale & Archer, 1990; Lanier *et al.*, 1993; Tessier, 1993). A possible explanation is that in upper intertidal zones deposition of sand-mud couplets occurs only during half of the neap-spring cycle or less, because only stronger tides reach these areas (Tessier, 1993). Another possibility is that deposition of the couplets occurred in a diurnal tidal system, in which sands were transported and deposited only once a day by the dominant tidal current (Kvale & Archer, 1990). The rhythmic couplets would be similar in both cases and therefore, interpretations need to be based on additional data, as is discussed below.

Another feature that provides information about the depositional environment of the meander loop bodies is the presence of desiccation cracks and vertebrate footprints at the top of the lateral accretion units, especially in the upper part of the point bar bodies. This is also consistent with periodic variations of the water level due to tides that cause exposure of the point bars in upper intertidal areas (cf. Choi *et al.*, 2004). Moreover, gently dipping point bar surfaces (Figs 6B, 8A, 8B) are very common in tide-influenced meandering channels (Dalrymple *et al.*, 2003; Choi *et al.*, 2004; Corbett *et al.*, 2011). Although burrowing is not plentiful in the lateral accretion units, which allowed preservation of lamination, it is more abundant than in fluvial IHS sets and fits better with the interpretation of deposition in a tidal setting (Thomas *et al.*, 1987; Choi *et al.*, 2004; Dalrymple & Choi, 2007).

The observed, fining upwards, channel infilling sequence (Fig. 5), which indicates a relative decrease in the water current velocity from the lower to the upper part of the channels and the gradual transition between meander loop bodies and the overlying non-channelled hetero-lithic layers are typical characteristics of modern tidal channels (Choi *et al.*, 2004; Dalrymple & Choi, 2007), supporting also the tidal interpreta-tion for the channels of the Oncala Group. The coarser-grained sandstone present in the lower-most part of the point bar infilling sequence, which displays parallel bedding or large-scale cross stratification, was probably deposited within the subtidal zone. The IHS made up of sandstone and mudstone lateral accretion units in the intermediate part of the point bars was probably formed in the subtidal or intertidal zone and the finer-grained units, which display thin alternation of sandstone and mudstone laminae and common desiccation cracks and footprints at the top, were most probably deposited in the upper intertidal zone. In accordance with this interpretation, the height of the IHS fits roughly with the palaeotidal range of the area, which allows us to estimate that the tidal range was around 3 to 4 m. Furthermore, the measured tidal rhythmites would have been deposited in the upper intertidal zone which, according to mod-ern tidal settings (Tessier, 1993), was probably an area only reached by stronger tides. As a conse-quence, it is interpreted that sedimentation of only 8 to 14 tidal couplets between two adjacent neap tides was due to deposition within the upper intertidal zone in a semidiurnal setting and not due to sedimentation in a diurnal tidal system.

Although the influence of tidal cyclicities in the channels of the Oncala Group is evident, flu-vial processes had probably an important effect on sedimentation as well. Important sediment and freshwater discharges probably came from the rivers in the westernmost areas of the basin and favoured predominance of unidirectional

downstream (ebb) currents to the east to south-east in the tide-influenced channels of the central area of the Oncala Group. Moreover, the alterna-tion of 10 to 40 cm-thick sandstone and mudstone lateral accretion units within the meander loop bodies may be related with seasonal fluctuations in river discharge. As occurs in modern channels from the fluvial-tidal transition zone (e.g. van den Berg *et al.*, 2007; Sisulak & Dashtgard, 2012; Dashtgard *et al.*, 2012) and as it is interpreted for ancient tidal-fluvial channels (e.g. Smith, 1987; Hovikoski *et al.*, 2008; Musial *et al.*, 2012; Scasso *et al.*, 2012), predominantly sandy units could have deposited during freshet and dominantly muddy units could have sedimented during wan-ing freshet and base flow.

The meandering loop bodies of the Oncala Group are interbedded and cut through non-channelled heterolithic facies, which are made up of interlaminated siliciclastic mudstone, siltstone and very fine-grained to medium-grained sand-stone (Figs 10 and 11). The lateral continuity of these layers along tens or hundreds of metres sug-gests that they were deposited in broad, flat areas. The repetitive alternation between bedload trans-port and settling from suspension that character-izes these layers is typical in tidal settings related with fluctuations in water current velocity due to tides (Reineck & Wunderlich, 1968; Allen, 1991; Nio & Yang, 1991; Shanmugam *et al.*, 2000; Martinius & Van den Berg, 2011; Kvale, 2012). Moreover, rhythmic variations in the thicknesses of sand-mud couplets (Fig. 10G) and in the type of bedding (Fig. 10F) are observed in these layers. This type of rhythmicity is characteristic of tidal settings because lunar/solar cycles cause fluctua-tions in tidal range, current velocities and sediment transport rates (Kvale & Archer, 1990; Dalrymple *et al.*, 1991; Nio & Yang, 1991; Tessier, 1993; Tessier *et al.*, 1995; Kvale, 2012). Consequently, the presence of these rhythmic variations in the non-channelled layers of the Oncala Group is interpreted as an evidence of tidal influence. Lenticular bedding and the thinnest couplets within the rhythmites are interpreted as formed during neap tides and flaser bedding and the thickest couplets during spring tides. Another characteristic that suggests a tidal origin is the large amount of mudstone that was deposited, because it indicates high suspended-sediment concentrations, which are typical in tidal settings (Dalrymple & Choi, 2007). Moreover, the abundant desiccation mudcracks at the top of the layers

(Figs 10A and 12) indicate repetitive deposition of a thin lamina of sediment (less than few millimetres to 2 cm in thickness) and subsequent exposure and desiccation, which fits with sedimentation in intertidal zones that are recurrently flooded and exposed. Taking all these considerations into account, the non-channelled heterolithic layers of the Oncala Group are interpreted as deposited in broad, tidal, sandy-muddy flats.

The morphology of the tabular dolostone layers, which display flat bottom and top and are laterally continuous along tens to hundreds of metres, indicates that dolostones were deposited in broad, flat ponds. Considering the close association of the dolostone layers with the non-channelled heterolithic layers (Fig. 13A), it is interpreted that the carbonate ponds were developed within the tidal sandy-muddy flats. These ponds were shallow (less than few metres deep) so they desiccated episodically, as evidenced by the development of mudcracks at the top of the layers. The siliciclastic discharge into these areas was less abundant than in the areas were non-channelled heterolithic layers deposited, allowing precipitation of carbonates. Fluctuations in the water current velocity in these carbonate areas caused variations in the amount and grain size of the clastic particles within the silty-sandy dolostones, normal gradation in peloidal laminae and recurrent trapping of pelletal sand by stromatolites. Although these water current velocity variations may occur in a variety of settings, they are also consistent with a tidal setting (e.g. Shinn, 1983; Lasemi *et al.*, 2012). Moreover, modern stromatolites that trap and bind carbonate sand grains, as the ones in the Oncala Group frequently did, have been described almost exclusively in tidal environments, where tidal currents are an efficient mechanism for sediment supply onto the surface of the stromatolites (Logan, 1961; Dravis, 1983; Dill *et al.*, 1986; Suarez-Gonzalez *et al.*, this volume). Furthermore, the presence of abundant faecal pellets is also characteristic of shallow, low-energy, restricted marine environments (Wanless & Burton, 1981; Tucker & Wright, 1990); although pellets may also be abundant in saline lakes (Smoot & Lowenstein, 1991; Warren, 2006). In a tidal setting, the sulphate ions necessary for the precipitation of rare gypsum and anhydrite present in the dolostone layers could come from marine water input. According to these considerations and the interbedding of dolostone beds with non-channelled heterolithic layers, dolostone beds are interpreted as formed in inundated ponds developed within siliciclastic intertidal flats.

Given the preponderance of evidence, the interpretation of a tide-influenced sedimentary environment explains better the features of the deposits of the Oncala Group than any other palaeoenvironmental reconstruction. Moreover, the scarcity of fossil remains is not inconsistent with a tidal setting, as several examples of tidal environments with no marine fossils have been described from the ancient and modern geological record (e.g. Kvale & Archer, 1991; Wells & Goman, 1995; Kvale & Mastalerz, 1998; Chakraborty *et al.*, 2003; Choi *et al.*, 2004; Hovikoski *et al.*, 2005, 2007). High rates of sedimentation, high suspended-sediment concentrations, low or rapidly changing water salinities or presence of strongly acidic waters may cause scarcity or absence of carbonate fossil remains (Kvale & Archer, 1990; Kvale & Mastalerz, 1998; Dalrymple & Choi, 2007). Several of these conditions were probably achieved by the Oncala Group. High rates of sedimentation occurred likely during the deposition of the Oncala Group, as 2500 m of sediments were deposited in the depocentral areas during the Berriasian. High suspended-sediment concentrations are also interpreted for this unit considering the large amount of mud deposited. Moreover, the ostracod assemblage present in the Oncala Group indicates mixed fresh and brackish water conditions (Schudack & Schudack, 2009). Such fresh and brackish water conditions can be developed in tidal settings with important freshwater input, such as protected inland tidal embayments and fluvial-tidal transition estuaries or deltas (Wells & Goman, 1995; Dalrymple & Choi, 2007). Considering the lateral relationship of the tide-influenced setting developed in the central area of the Oncala Group with the fluvial system of the westernmost area of the basin, important amounts of freshwater probably discharged into the central area of the Oncala Group, which could have caused the fresh to brackish water conditions interpreted for the studied setting. Although the fossil record in the Oncala Group is restricted almost exclusively to ostracods and vertebrate bones, it is not ruled out that other fossils could have existed and that they dissolved during diagenesis because dissolution and silicification processes have been distinguished in the ostracods and the rare bivalves and charophytes present in the studied deposits.

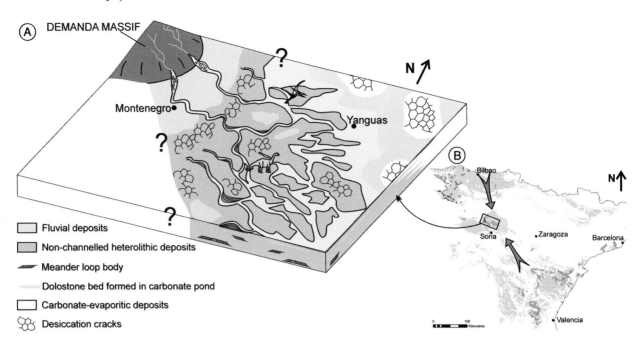

Fig. 14. (A) Palaeogeographic reconstruction of central and western areas of the Oncala Group. Broad, low-gradient, tidal flats, traversed by shallow, meandering channels were developed in the central area of the Oncala Group. The very shallow water depth of the basin and its flat topography caused development of very extensive intertidal flats. Scattered carbonate ponds formed in shallow depressions of the tidal flats. This tidal system passed westwards to a fluvial system and east-wards to carbonate-sulphate, shallow water bodies. See Fig. 1C for location of the towns indicated in this reconstruction. (B) Synthetic map showing the location and main facies of Berriasian deposits in the Iberian Peninsula (see Fig. 1B for more detail). The blue arrows mark the location of the possible seaways which connected the Cameros Basin with marine areas of the Basque-Cantabrian Basin and south-eastern basins of the Iberian Mesozoic Rift System. Modified from Quijada *et al.* (2013b).

Regarding all these considerations, the deposits of the central area of the Oncala Group provide a good example of a fine-grained sandy-muddy tide-influenced environment that lacks marine fossil remains. A sedimentary environment includ-ing broad, low-gradient, tidal flats, traversed by shallow, meandering channels and scattered shallow carbonate ponds is the interpretation that can better explain all the sedimentary facies developed in the studied deposits (Fig. 14A). Fluvial-tidal transition zones of modern inshore tidal systems, which are characterized by estua-rine point bars fringed by tidal mudflats, such as the Mont-Saint-Michel Bay, Gironde or Salmon River inner estuaries, may be good analogues for the studied deposits. Sediments of these modern fluvial-tidal areas typically contain alternations of sand and mud laminae, flaser, wavy and lenticu-lar bedding and tidal rhythmites and bioturbation is generally minimal (Allen, 1991; Tessier, 1993, 2002; Dalrymple *et al.*, 1991, 2012; Archer, 2013). Despite the similarities between these modern analogues and the sedimentary environment

proposed for the Oncala Group, a major difference is that the modern examples occur in very narrow palaeovalleys, which favour tidal amplification, whereas the tidal system developed in the Cameros Basin covered a wider, low-gradient area. In the case of the Oncala Group, tidal amplification was probably related with the palaeogeographic setting of the Cameros Basin at Berriasian times (Fig. 14B). The comparison of the Oncala Group with Berriasian deposits from other basins of the Iberian Peninsula suggests that the Cameros Basin was connected with the Basque-Cantabrian Basin and possibly with the south-eastern basins of the Iberian Mesozoic Rift System (Quijada *et al.*, 2013b), although the exact location of the con-nection areas cannot be determined because the uplifted Cameros Basin is nowadays surrounded by Tertiary basins (Fig. 1). According to this inter-pretation, the Cameros Basin at Berriasian times was located between two coastal areas, largely occupied by broad tidal flats, which passed later-ally to marine areas northwards in the Basque-Cantabrian Basin and towards the south-east in the Iberian Mesozoic Rift System (Fig. 14B). In such a context, the location of the Cameros Basin and its shallow water depth probably created the favourable conditions for tidal amplification, similarly to what has been described in other ancient, narrow, confined, rift basins (e.g. Mellere & Steel, 1996; Ravnås & Bondevik, 1997; Ravnås & Steel, 1998).

CONCLUSIONS

The deposits of the central area of the Oncala Group provide an example of laterally very continuous, fine-grained siliciclastic sediments with no marine fossils, deposited in a tidal setting, as opposed to continental sedimentary environments, such as sandy-muddy flats in saline lake basins. The criteria that in combination lead to this interpretation are: 1) Fine grain size of the sediments, ranging from clay to medium-sand, and large amount of mudstone. 2) Predominance of non-channelled facies over meander loop bodies. 3) Meander loop bodies displaying low angle, lateral accretion units and IHS. 4) Alternation of sandstone and mudstone laminae within non-channelled layers and meander loop bodies, forming lenticular, wavy and flaser bedding and planar bedded sandstone-mudstone couplets. 5) Rhythmic variations in the type of bedding (lenticular-wavy-flaser-wavy-lenticular) and in the thicknesses of the sandstone-mudstone couplets. 6) Abundant evidence of subaerial exposure, such as desiccation mudcracks and vertebrate footprints, and presence of desiccation cracks at the top of numerous, consecutive, mudstone laminae or couplets. These criteria may be useful for identifying other ancient tide-influenced deposits with no marine fossils or classical tidal features, which may have been more common in the rock record than previously thought.

ACKNOWLEDGMENTS

This work was funded by the Spanish DIGICYT projects CGL2008-01648/BTE, CGL2011-22709 and CGL2014-52670-P the 'Sedimentary Basin Analysis' Research Group of the Complutense University of Madrid-Madrid Community and a Spanish Department of Education FPU scholarship. The authors would like to thank the editor J.-Y. Reynaud and the reviewers G. Musial and E. Kvale for their careful and thoughtful reviews. We are also very grateful to B. Tessier for enriching and helpful discussion on tidal processes and sedimentary environments and to M. Díaz-Molina who largely improved this manuscript with her comments about processes and deposits of meandering channels. We would also like to thank A. Alonso and J. Bourgeois for scientific discussion in the field, S. Sacristán and S. Campos for help during field work and the staff of the Department of Stratigraphy of the Complutense University of Madrid for their technical support.

REFERENCES

Ainsworth, R.B., Hasiotis, S.T., Amos, K.J., Krapf, C.B.E., Payenber, T.H.D., Sandstrom, M.L., Vakarelov, B.K. and Lang, S.C. (2012) Tidal signatures in an intracratonic playa lake. *Geology*, **40**, 607–610.

Allen, G.P. (1991) Sedimentary processes and facies in the Gironde estuary: a recent model for macrotidal estuarine systems. In: *Clastic tidal sedimentology* (Eds D.G. Smith, G.E. Reinson, B.A. Zaitlin and R.A. Rahmani), *Can. Soc. Petrol. Geol. Mem.*, **16**, 29–39.

Allen, J.R.L. (1968) Current ripples: their relation to patterns of water and sediment motion. North-Holland Publishing Company, Amsterdam, 433 pp.

Allen, J.R.L. (1986) On the curl of desiccation polygons. *Sed. Geol.*, **46**, 23–31.

Archer, A.W. (2013) World's highest tides: Hypertidal coastal systems in North America, South America and Europe. *Sed. Geol.*, **284–285**, 1–25.

Aurell, M., Mas, R., Meléndez, A. and Salas, R. (1994) El tránsito Jurásico-Cretácico en la Cordillera Ibérica: relación tectónica-sedimentación y evolución paleogeográfica. *Cuad. Geol. Iber.*, **18**, 369–396.

Bádenas, B., Salas, R. and Aurell, M. (2004) Three orders of regional sea-level changes control facies and stacking patterns of shallow platform carbonates in the Maestrat Basin (Tithonian-Berriasian, NE Spain). *Geol. Rundsch.*, **93**, 144–162.

Beets, D.J., De Groot, T.A.M. and Davies, H.A. (2003) Holocene tidal back-barrier development at decelerating sea-level rise: a 5 millennia record, exposed in the western Netherlands. *Sed. Geol.*, **158**, 117–144.

Bhattacharya, A. (1997) On the origin of non-tidal flaser bedding in point bar deposits of the River Ajay, Bihar and West Bengal, NE India. *Sedimentology*, **44**, 973–975.

Castanera, D., Colmenar, J., Sauqué, V. and Canudo, J.I. (2015) Geometric morphometric analysis applied to theropod tracks from the Lower Cretaceous (Berriasian) of Spain. *Palaeontology*, **58**, 183–200.

Castanera, D., Pascual, C., Razzolini, N.L., Vila, B., Barco, J.L. and Canudo, J.I. (2013) Discriminating between Medium-Sized Tridactyl Trackmakers: Tracking Ornithopod Tracks in the Base of the Cretaceous (Berriasian, Spain). *PlosOne*, **8**, e81830.

Chakraborty, T. (2012) Reversing flow or fluctuating flow? A case study from the Atrai and Tista river of the sub-Himalayan alluvial plain. *Abstract,* 29th IAS Meeting of Sedimentology, Schladming, Austria.

Chakraborty, C., Ghosh, S.K. and **Chakraborty, T.** (2003) Depositional Record of Tidal-Flat Sedimentation in the Permian Coal Measures of Central India: Barakar Formation, Mohpani Coalfield, Satpura Gondwana Basin. *Gondwana Research,* **6,** 817–827.

Choi, K.S., Dalrymple, R.W., Chun, S.S. and **Kim, S.P.** (2004) Sedimentology of modern, Inclined Heterolithic Stratification (IHS) in the macrotidal Han River Delta, Korea. *J. Sed. Res.,* **74,** 677–689.

Ciarapica, G.L., Passeri, L. and **Schreiber, B.C.** (1985) Una proposta di classificazione delle evaporiti solfatiche. *Geol. Romana,* **24,** 219–232.

Corbett, M.J., Fielding, C.R. and **Birgenheier, L.P.** (2011) Stratigraphy of a Cretaceous coastal-plain fluvial succession: the Campanian Masuk Formation, Henry Mountains Syncline, Utah, U.S.A. *J. Sed. Res.,* **81,** 80–96.

Dalrymple, R.W., Baker, E.K., Harris, P.T. and **Hughes, M.** (2003) Sedimentology and stratigraphy of a tide-dominated, foreland-basin delta (Fly River, Papua New Guinea). In: *Tropical Deltas of Southeast Asia–Sedimentology, Stratigraphy and Petroleum Geology* (Eds **H. Sidi, D. Nummedal, P. Imbert, H. Darman** and **H.W. Posamentier**), *SEPM Spec. Publ.,* **76,** 147–173.

Dalrymple, R.W. and **Choi, K.** (2007) Morphologic and facies trends through the fluvial–marine transition in tide-dominated depositional systems: A schematic framework for environmental and sequence-stratigraphic interpretation. *Earth-Sci. Rev.,* **81,** 135–174.

Dalrymple, R.W., Mackay, D.A., Ichaso, A.A. and **Choi, K.S.** (2012) Processes, Morphodynamics and Facies of Tide-Dominated Estuaries. In: *Principles of Tidal Sedimentology* (Eds **R.A. Davis** and **R.W. Dalrymple**), pp. 79–107. Springer, New York.

Dalrymple, R.W., Makino, Y. and **Zaitlin, B.A.** (1991) Temporal and spatial patterns of rhythmite deposition on mud flats in the macrotidal Cobequid Bay-Salmon River estuary, Bay of Fundy, Canada. In: *Clastic tidal sedimentology* (Eds **D.G. Smith, G.E. Reinson, B.A. Zaitlin** and **R.A. Rahmani**), *Can. Soc. Petrol. Geol. Mem.,* **16,** 137–160.

Dashtgard, S.E., Venditti, J.G., Hill, P.R., Sisulak, C.F., Johnson, S.M. and **La Croix, A.D.** (2012) Sedimentation Across the Tidal-Fluvial Transition in the Lower Fraser River, Canada. *Sed. Rec.,* **10,** 4–9.

Deluca, J.L. and **Eriksson, K.A.** (1989) Controls on synchronous ephemeral- and perennial-river sedimentation in the middle sandstone member of the Triassic Chinle Formation, northeastern New Mexico, U.S.A. *Sed. Geol.,* **61,** 155–175.

de Mowbray, T. (1983). The genesis of lateral accretion deposits in recent intertidal mudflat channels, Solway Firth, Scotland. *Sedimentology,* **30,** 425–435.

Díaz-Molina, M. (1979) Características sedimentológicas de los paleocanales de la unidad detrítica superior al N. de Huete (Cuenca). *Estud. Geol.,* **35,** 241–251.

Díaz-Molina, M. (1993) Geometry and lateral accretion patterns in meander loops: examples from the Upper Oligocene-Lower Miocene, Loranca Basin, Spain. In: *Alluvial Sedimentation* (Eds **M. Marzo** and

C. Puigdefábregas), *Int. Assoc. Sedimentol. Spec. Publ.,* **17,** 115–131.

Díaz-Molina, M. and **Bustillo, A.** (1985) Wet fluvial fans of the Loranca Basin (Central Spain), channel models and distal bioturbated gypsum with chert. In: *6th European Regional Meeting of Sedimentology, Lérida, Excursion Guidebook* (Eds **M.D. Milá** and **J. Rosell**), pp. 147–185. Institut d'Estudis Ilerdencs, Lleida.

Dill, R.F., Shinn, E.A., Jones, A.T., Kelly, K. and **Steinen, R.P.** (1986) Giant subtidal stromatolites forming in normal salinity waters. *Nature,* **324,** 55–58.

Dravis, J.J. (1983) Hardened subtidal stromatolites, Bahamas. *Science,* **219,** 385–386.

Dupraz, C., Reid, R.P., Braissant, O., Decho, A.W., Norman, R.S., Visscher, P.T. (2009) Processes of carbonate precipitation in modern microbial mats. *Earth-Sci. Rev.,* **96,** 141–162.

Falcon-Lang, H. (1998) The impact of wildfire on an Early Carboniferous coastal environment, North Mayo, Ireland. *Palaeogeogr. Palaeoclimatol. Palaeoecol.,* **139,** 121–138.

Fisher, J.A., Krapf, C.B.E., Lang, S.C., Nichols, G.J. and **Payenberg, T.H.D.** (2008) Sedimentology and architecture of the Douglas Creek terminal splay, Lake Eyre, central Australia. *Sedimentology,* **55,** 1915–1930.

Flügel, E. (1982) *Microfacies Analysis of Limestone.* Springer-Verlag, Berlin, 633 pp.

García de Cortázar, A. and **Pujalte, V.** (1982) Litoestratigrafía y facies del Grupo Cabuérniga (Malm-Valanginiense inferior?) al S de Cantabria-NE de Palencia. *Cuad. Geol. Iber.,* **8,** 5–21.

Gingras, M., Räsänen, M.E. and **Ranzi, A.** (2002) The Significance of Bioturbated Inclined Heterolithic Stratification in the Southern Part of the Miocene Solimoes Formation, Rio Acre, Amazonia Brazil. *Palaios,* **17,** 591–601.

Gómez-Fernández, J.C. and **Meléndez, N.** (1994) Climatic control on Lower Cretaceous sedimentation in a playa-lake system of a tectonically active basin (Huérteles Alloformation, Eastern Cameros Basin, North-Central Spain). *J. Paleolimnology,* **11,** 91–107.

Hardie, L.A., Smoot, J.P. and **Eugster, H.P.** (1978) Saline lakes and their deposits: a sedimentological approach. In: *Modern and ancient lake sediments* (Eds **A. Matter** and **M.E. Tucker**), *Int. Assoc. Sedimentol. Spec. Publ.,* **2,** 7–41.

Hernández, N., Pascual, C., Latorre, P. and **Sanz, E.** (2005-2006) Contribución de los yacimientos de icnitas sorianos al registro general de Cameros. *Zubía,* **23-24,** 79–120.

Hovikoski, J., Gingras, M., Räsänen, M., Guerrero, J., Ranzi, A., Melo, J., Romero, L., Nuñez del Prado, H., Jaimes, F. and **Lopez, S.** (2007) The nature of Miocene Amazonian epicontinental embayment: High-frequency shifts of the low-gradient coastline. *Geol. Soc. Am. Bull.,* **119,** 1506–1520.

Hovikoski, J., Räsänen, M. Gingras, M., Ranzi, A. and **Melo, J.** (2008) Tidal and seasonal controls in the formation of Late Miocene inclined heterolithic stratification deposits, western Amazonian foreland basin. *Sedimentology,* **55,** 499–530.

Hovikoski, J., Räsänen, M.E., Gingras, M., Roddaz, M., Brusset, S., Hermoza, W., Romero Pittman, L. and

Lertola, K. (2005) Miocene semidiurnal tidal rhythmites in Madre de Dios, Peru. *Geology*, **33**, 177–180.

Hubert, J.F. and Hyde, M.G. (1982). Sheet-flow deposits of graded beds and mudstones on an alluvial sandflat-playa system: Upper Triassic Blomidon redbeds, St Mary's Bay, Nova Scotia. *Sedimentology*, **29**, 457–474.

Kendall, A.C. (1992) Evaporites. In: *Facies Models: Response to Sea Level Change* (Eds R.G. Walker and N.P. James), pp. 375-409. Geological Association of Canada, St. John's, Newfoundland.

Kindle, E.M. (1917) Some factors affecting the development of mud-cracks. *J. Geol.*, **25**, 135–144.

Kvale, E.P. (2012) Tidal Constituents of Modern and Ancient Tidal Rhythmites: Criteria for Recognition and Analyses. In: *Principles of Tidal Sedimentology* (Eds R.A. Davis, Jr. and R.W. Dalrymple), pp. 1–17. Springer, New York.

Kvale, E.P. and Archer, A.W. (1991) Characteristics of two, Pennsylvanian age, semidiurnal tidal deposits in the Illinois Basin, U.S.A. In: *Clastic Tidal Sedimentology* (Eds D.G. Smith, G.E. Reinson, B.A. Zaitling and R.A. Rahmani), *Can. Soc. Petrol. Geol. Mem.*, **16**, 179–188.

Kvale, E.P. and Archer, A.W. (1990) Tidal deposits associated with low-sulfur coals, Brazil Fm (Lower Pennsylvanian), Indiana. *J. Sed. Petrol.*, **60**, 563–574.

Kvale, E.P. and Mastalerz, M. (1998) Evidence of ancient freshwater tidal deposits. In: *Tidalites: Processes and Products* (Eds C.R. Alexander, R.A. Davis and V.J. Henry), *SEPM Spec. Publ.*, **61**, 95–107.

Lanier, W.P., Feldman, H.R. and Archer, A.W. (1993) Tidal sedimentation from a fluvial to estuarine transition Douglas Group, Missourian-Virgilian, *Kansas. J. Sed. Petrol.*, **63**, 860–873.

Lasemi, Y., Jahani, D., Amin-Rasouli, H. and Lasemi, Z. (2012) Ancient Carbonate Tidalites. In: *Principles of Tidal Sedimentology* (Eds R.A. Davis, Jr. and R.W. Dalrymple), pp. 567–607. Springer, New York.

Logan, B.W. (1961) Cryptozoon and associate stromatolites from the Recent, Shark Bay, western Australia. *J. Geol.*, **69**, 517–533.

Magee, J.W. (1991) Late Quaternary lacustrine, groundwater, aeolian and pedogenic gypsum in the Prungle Lakes, southeastern Australia. *Palaeogeogr. Palaeoclimatol. Palaeoecol.*, **84**, 3–42.

Martín-Chivelet, J., Berástegui, X., Rosales, I., Vilas, L., Vera, J.A., Caus, E., Gräfe, K.-U., Mas, R., Puig, C., Segura, M., Robles, S., Floquet, M., Quesada, S., Ruiz-Ortiz, P.A., Fregenal-Martínez, M.A., Salas, R., Arias, C., García, A., Martín-Algarra, A., Meléndez, M.N., Chacón, B., Molina, J.M., Sanz, J.L., Castro, J.M., García-Hernández, M., Carenas, B., García-Hidalgo, J., Gil, J. and Ortega, F. (2002) Cretaceous. In: *The Geology of Spain* (Eds W. Gibbons, T. Moreno), pp. 255–292. The Geological Society, London.

Martinius, A.W. and Van den Berg, J.H. (2011) *Atlas of sedimentary structures in estuarine and tidally-influenced river deposits of the Rhine-Meuse-Scheldt system.* EAGE Publications, Houten, The Netherlands, 298 pp.

Mas, J.R., Alonso, A. and Guimerà, J. (1993) Evolución tectonosedimentaria de una cuenca extensional intraplaca: La cuenca finijurásica-eocretácica de Los Cameros (La Rioja-Soria). *Rev. Soc. Geol. Esp.*, **6**, 129–144.

Mas, R., Alonso, A. and Meléndez, N. (1984) La Formación Villar del Arzobispo: un ejemplo de llanuras de marea siliciclásticas asociadas a plataformas carbonatadas. Jurásico terminal (NW. de Valencia y E. de Cuenca). *Publ. Geol. Univ. Aut. Barcelona*, **20**, 175–188.

Mas, R., Benito, M.I., Arribas, J., Alonso, A., Arribas, M.E., Lohmann, K.C., González-Acebrón, L., Hernán, J., Quijada, E., Suárez, P. and Omodeo, S. (2011) Evolution of an intra-plate rift basin: the Latest Jurassic-Early Cretacous Cameros Basin (Northwest Iberian Ranges, North Spain). In: *Post-Meeting Field trips 28th IAS Meeting, Zaragoza* (Eds C. Arenas, L. Pomar and F. Colombo), *Sociedad Geológica de España Geo-Guías*, **8**, 117–154.

Mas, R., Benito, M.I., Arribas, J., Serrano, A., Guimerà, A., Alonso, A. and Alonso-Azcarate, J. (2002) La Cuenca de Cameros: desde la extensión finijurásica-eocretácica a la inversión terciaria - implicaciones en la exploración de hidrocarburos. *Zubía*, **14**, 9–64.

Mas, R., García, A., Salas, R., Meléndez, A., Alonso, A., Aurell, M., Bádenas, B., Benito, M.I., Carenas, B., García-Hidalgo, J.F., Gil, J. and Segura, M. (2004) Segunda fase de rifting: Jurásico Superior-Cretácico Inferior. In: *Geología de España* (Ed. J.A. Vera), pp. 503–510. Sociedad Geológica de España - Instituto Geológico y Minero de España, Madrid.

Meléndez, N. and Gómez-Fernández, J.C. (2000) Continental deposits of the eastern Cameros Basin (northern Spain) during Tithonian-Berriasian time. In: *Lake basins through space and time* (Eds E. Gierlowski-Kordesch and K.R. Kelts), *AAPG Studies in Geology*, **46**, 263–273.

Mellere, D. and Steel, R.J. (1996) Tidal sedimentation in Inner Hebrides half grabens, Scotland: the Mid-Jurassic Bearreraig Sandstone Formation. In: *Geology of Siliciclastic Shelf Seas* (Eds M. De Batist and P. Jacobs), *The Geol. Soc.*, **117**, 49–79.

Moody-Stuart, M. (1966). High- and low-sinuosity stream deposits, with examples from the Devonian of Spitsbergen. *J. Sed. Petrol.*, **36**, 1102–1117.

Moratalla, J.J. and Hernán, J. (2010) Probable palaeogeographic influences of the Lower Cretaceous Iberian rifting phase in the Eastern Cameros Basin (Spain) on dinosaur trackway orientations. *Palaeogeogr. Palaeoclimatol. Palaeoecol.*, **295**, 116–130.

Musial, G., Reynaud, J.-Y., Gingras, M.K., Féniès, H., Labourdette, R. and Parize, O. (2012) Subsurface and outcrop characterization of large tidally influenced point bars of the Cretaceous McMurray Formation (Alberta, Canada). *Sed. Geol.*, **279**, 156–172.

Nichols, G. (2009) Sedimentology and Stratigraphy. Wiley-Blackwell, Oxford, 419 pp.

Nio, S.D. and Yang, C.S. (1991) Diagnostic attributes of clastic tidal deposits: a review. In: *Clastic Tidal Sedimentology* (Eds D.G. Smith, G.E. Reinson, B.A. Zaitlin and R.A. Rahmani), *Can. Soc. Petrol. Geol. Mem.*, **16**, 3–28.

North, C.P. and Davidson, S.K. (2012) Unconfined alluvial flow processes: Recognition and interpretation of their deposits and the significance for palaeogeographic reconstruction. *Earth-Sci. Rev.*, **111**, 199–223.

Ortí, F. and Rosell, L. (1997) Sulfatos evaporíticos de interés petrológico. In: *Atlas de asociaciones minerales en lámina delgada* (Ed. **J.C. Melgarejo**), pp. 210–235. Fundació Folch, Barcelona.

Pascual-Arribas, **C.** and **Hernández-Medrano**, **N.** (2015) Las huellas de tortuga del Grupo Oncala (Berriasiense, Cuenca de Cameros, España). *Estud. Geol.*, **71**, e030.

Plummer, **P.S.** and **Gostin**, **V.A.** (1981) Shrinkage cracks: desiccation or synaeresis?: *J. Sed. Petrol.*, **51**, 1147–1156.

Preiss, **W.V.** (1976) Basic field and laboratory methods for the study of stromatolites. In: *Stromatolites* (Ed. **M.A. Walter**), *Dev. Sedimentol.*, **20**, 5–13.

Puigdefabregas, **C.** (1973) Miocene point-bar deposits in the Ebro Basin, Northern Spain. *Sedimentology*, **20**, 133–144.

Puigdefabregas, **C.** and **Van Vliet**, **A.** (1978) Meandering stream deposits from the Tertiary of the southern Pyrenees. In: *Fluvial Sedimentology* (Ed. **A.D. Miall**), pp. 469–485. *Can. Soc. Petrol. Geol.*, Calgary.

Pujalte, **V.** (1982) Tránsito Jurásico-Cretácico, Berriasiense, Valanginiense, Hauteriviense y Barremiense. In: *El Cretácico de España* (Ed. **A. García**), pp. 51–63. Universidad Complutense de Madrid, Madrid.

Quijada, **I.E.**, **Suarez-Gonzalez**, **P.**, **Benito**, **M.I.** and **Mas**, **R.** (2013a) Depositional depth of laminated carbonate deposits: insights from the Lower Cretaceous Valdeprado Formation (Cameros Basin, N Spain). *J. Sed. Res.*, **83**, 241–257.

Quijada, **I.E.**, **Suarez-Gonzalez**, **P.**, **Benito**, **M.I.** and **Mas**, **R.** (2013b) New insights on stratigraphy and sedimentology of the Oncala Group (eastern Cameros Basin): implications for the paleogeographic reconstruction of Iberia at Berriasian times. *J. Iberian Geol.*, **39**, 313–334.

Ravnås, **R.** and **Bondevik**, **K.** (1997) Architecture and controls on Bathonian–Kimmeridgian shallow-marine synrift wedges of the Oseberg–Brage area, northern North Sea. *Basin Res.*, **9**, 197–226.

Ravnås, **R.** and **Steel**, **R.J.** (1998) Architecture of Marine Rift-Basin Successions. *AAPG Bull.*, **82**, 110–146.

Rebata-H., **L.A.**, **Gingras**, **M.K.**, **Räsänen**, **M.E.** and **Barbieri**, **M.** (2006) Tidal-channel deposits on a delta plain from the Upper Miocene Nauta Formation, Marañón Foreland Sub-basin, Peru. *Sedimentology*, **53**, 971–1013.

Reineck, **H.E.** and **Wunderlich**, **F.** (1968) Classification and origin of flaser and lenticular bedding. *Sedimentology*, **11**, 99–104.

Riding, **R.** (2000) Microbial carbonates: the geological record of calcified bacterial-algal mats and biofilms. *Sedimentology*, **47** (Suppl. 1), 179–214.

Robertson-Handford, **C.** (1982) Sedimentology and evaporite genesis in a Holocene continental-sabkha playa basin-Bristol Dry Lake, California. *Sedimentology*, **29**, 239–253.

Salas, **R.** (1989) Evolución estratigráfica secuencial y tipos de plataformas de carbonatos del interval Oxfordiense-Berriasiense en las cordilleras ibérica oriental y costero catalana meridional. *Cuad. Geol. Iber.*, **13**, 121–157.

Salomon, **J.** (1982) El Cretácico inferior de Cameros-Castilla. In: *El Cretácico de España* (Ed. **A. García**), pp. 345–387.Universidad Complutense de Madrid, Madrid.

Scasso, **R.A.**, **Dozo**, **M.T.**, **Cuitiño**, **J.I.** and **Bouza**, **P.** (2012) Meandering tidal-fluvial channels and lag concentration of terrestrial vertebrates in the fluvial-tidal transition of an ancient estuary in Patagonia. *Latin Am. J. Sedimentol. Basin Anal.*, **19**, 27–45.

Scholle, **P.A.** and **Ulmer-Scholle**, **D.S.** (2003) *A Color Guide to the Petrography of Carbonate Rocks: Grains, textures, porosity, diagenesis.* AAPG, Tulsa, Oklahoma, 474 pp.

Schudack, **U.** and **Schudack**, **M.** (2009) Ostracod biostratigraphy in the Lower Cretaceous of the Iberian chain (eastern Spain). *J. Iberian Geol.*, **35**, 141–168.

Shanley, **K.W.**, **McCabe**, **P.J.** and **Hettinger**, **R.D.** (1992) Tidal influence in Cretaceous fluvial strata from Utah, USA: a key to sequence stratigraphic interpretation. *Sedimentology*, **39**, 905–930.

Shanmugam, **G.**, **Poffenberger**, **M.** and **Toro Álava**, **J.** (2000) Tide-Dominated Estuarine Facies in the Hollin and Napo ("T" and "U") Formations (Cretaceous), Sacha Field, Oriente Basin, Ecuador. *AAPG Bull.*, **84**, 652–682.

Shinn, **E.A.** (1983) Tidal Flat. In: *Carbonate depositional environments* (Eds **P.A. Scholle**, **D.G. Bebout** and **C.H. Moore**), *AAPG Mem.*, **33**, 171–210.

Sisulak, **C.F.** and **Dashtgard**, **S.E.** (2012) Seasonal controls on the development and character of inclined heterolithic stratification in a tide-influenced, fluvially dominated channel: Fraser River, *Canada. J. Sed. Res.*, **82**, 244–257.

Smith, **D.G.** (1987) Meandering river point bar lithofacies models: modern and ancient examples compared. In: *Recent developments in fluvial sedimentology* (Eds **F.G. Ethridge**, **R.M. Flores** and **M.D. Harvey**), *SEPM Spec. Publ.*, **39**, 83–91.

Smoot, **J.P.** and **Lowenstein**, **T.K.** (1991) Depositional environments of non-marine evaporites. In: *Evaporites, Petroleum and Mineral Resources* (Ed. **J.L. Melvin**), *Dev. Sedimentol.*, **50**, 189–347.

Sneh, **A.** (1983). Desert stream sequences in the Sinai Peninsula. *J. Sed. Petrol.*, **53**, 1271–1279.

Suarez-Gonzalez, **P.**, **Quijada**, **I.E.**, **Benito**, **M.I.** and **Mas**, **R.** (2013) Eustatic *versus* tectonic control in an intraplate rift basin (Leza Fm, Cameros Basin). Chronostratigraphic and paleogeographic implications for the Aptian of Iberia. *J. Iberian Geol.*, **39**, 285–312.

Suárez-Gonzalez, **P.**, **Quijada**, **I.E.**, **Benito**, **M.I.**, **Mas**, **R.**, **Merinero**, **R.** and **Riding**, **R.** (2014) Origin and significance of lamination in Lower Cretaceous stromatolites and proposal for a quantitative approach. *Sed. Geol.*, **300**, 11-27.

Suarez-Gonzalez, **P.**, **Quijada**, **I.E.**, **Benito**, **M.I.** and **Mas**, **R.** (2015) Sedimentology of ancient coastal wetlands: insights from a Cretaceous multifaceted depositional system. *J. Sed. Petrol.*, **85**, 95–117.

Suarez-Gonzalez, **P.**, **Quijada**, **I.E.**, **Benito**, **M.I.** and **Mas**, **R.** (2016) Do stromatolites need tides to trap ooids? Insights from oolitic stromatolites of a Cretaceous system of coastal-wetlands. In: Contributions to Modern and Ancient Tidal Sedimentology (Eds B. Tessier and J.Y. Reynaud), *Int. Assoc. Sedimentol. Spec. Publ.*, **47**, 155–184.

Tessier, **B.** (2002) The depositional facies of the inner estuary: the tidal rhythmites in the *tangues* at Gué de l'Èpine. In: *The Bay of Mont-Saint-Michel and the Rance Estuary: Recent development and evolution of* *depositional environments*. (Eds **C. Bonnot-Courtois**, **B. Caline**, **A. L'Hommer** and **M. Le Vot**), *Bull. Centre Rech. Elf Explor. Prod.*, **26**, 82–89.

Tessier, B. (1993) Upper intertidal rhythmites in the Mont-Saint-Michel Bay (NW France): Perspectives for paleo-reconstruction. *Mar. Geol.*, **110**, 355–367.

Tessier, B., Archer, A.W., Lanier, W.P. and Feldman, H.R. (1995) Comparison of ancient tidal rhythmites (Carboniferous of Kansas and Indiana, USA) with modern analogues (The Bay of Mont-Saint-Michel, France). In: *Tidal Signatures in Modern and Ancient Sediments* (Eds **B.W. Flemming** and **A. Bartholomä**), *Int. Assoc. Sedimentol. Spec. Publ.*, **24**, 259–271.

Thomas, R.G., Smith, D.G., Wood, J.M., Visser, M.J., Calverley-Rang, E.A. and Koster, E.H. (1987) Inclined heterolithic stratification – Terminology, description, interpretation and significance. *Sed. Geol.*, **53**, 123–179.

Tucker, M.W. and Wright, V.P. (1990) *Carbonate Sedimentology*. Blackwell Scientific Publications, Oxford, 482 pp.

van den Berg, J.H., Boersma, J.R. and van Gelder (2007) Diagnostic sedimentary structures of the fluvial-tidal transition zone – Evidence from deposits of the Rhine and Meuse. *Geol. Mijnbouw*, **86**, 287–306.

Visser, M.J. (1980) Neap-spring cycles reflected in Holocene subtidal large-scale bedform deposits: A preliminary note. *Geology*, **8**, 543–546.

Wanless, H.R. and Burton, E.A. (1981) Hydrodynamics of carbonate fecal pellets. *J. Sed. Petrol.*, **51**, 27–36.

Warren, J.K. (2006) *Evaporites: Sediments, Resources and Hydrocarbons*. Springer, Berlin, 1036 pp.

Wells, L.E. and Goman, M. (1995) Late Holocene Environmental Variability in the Upper San Francisco Estuary as Reconstructed from Tidal Marsh Sediments. In: *Proceedings of the Eleventh Annual Pacific Climate (PACLIM) Workshop* (Eds **C.M. Isaacs** and **V.L. Tharp**), *Technical Report of the Interagency Ecological Program for the Sacramento-San Joaquín Estuary*, **40**, 185–198.

Wentworth, C.K. (1922) A scale of grade and class terms for clastic sediments. *J. Geol.*, **30**, 377–392.

Zanor, G.A., Piovano, E.L., Ariztegui, D. and Vallet-Coulomb, C. (2012) A modern subtropical playa complex: Salina de Ambargasta, central Argentina. *J. S. Am. Earth Sci.*, **35**, 10–26.

Do stromatolites need tides to trap ooids? Insights from a Cretaceous system of coastal-wetlands

PABLO SUAREZ-GONZALEZ[*†], I. EMMA QUIJADA[*†],
M. ISABEL BENITO[*†] and RAMÓN MAS[*†]

*Departamento de Estratigrafía, Universidad Complutense de Madrid, C/José Antonio Novais 12, 28040, Madrid, Spain
†Instituto de Geociencias IGEO (CSIC, UCM), C/José Antonio Novais 12, 28040, Oviedo, Spain
E-mail address: pablosuarez@geo.ucm.es

ABSTRACT

Stromatolites associated with ooids are often described in the literature, both in marine and continental environments. However, a lateral relationship between them does not necessarily entail that ooids are trapped within the stromatolites. For example, present-day stromatolites that trap ooids (agglutinated oolitic stromatolites) are only found in tidal environments of the Bahamas and Shark Bay, whereas non-agglutinated stromatolites laterally related with oolitic facies are common in different modern and fossil environments. The Leza Fm carbonates (Cameros Basin, N Spain) were formed in a system of coastal-wetlands during Barremian-Aptian times (Early Cretaceous) and they offer an opportunity to elucidate the role of tides in ooid-trapping processes because they contain examples of both agglutinated and non-agglutinated stromatolites associated with oolitic facies. Agglutinated stromatolites are found in tide-influenced oolitic facies from the eastern Leza Fm and they show very scarce calcified microbial filaments. Non-agglutinated stromatolites are found in freshwater-dominated oolitic facies from the western Leza Fm and these stromatolites contain calcified filamentous microfabrics (i.e. skeletal stromatolites) without significant ooids trapped in them. The textural and sedimentological differences between both stromatolites suggest that water chemistry and hydrodynamics were different during their formation. The carbonate saturation state of the water might have been low enough to prevent intense microbial calcification of the tide-influenced stromatolites, developing soft microbial mats; moreover, the cyclic hydrodynamic changes of tides allowed the periodic supply of grains to be trapped by the soft mats. In contrast, the higher carbonate saturation of meteoric waters, which passed through and dissolved the Jurassic carbonate substrate of the basin, probably led to the stronger mat calcification of skeletal stromatolites from the western Leza Fm, without tidal influence. This together with the lower hydrodynamic changes of the environment prevented ooids from being trapped within these stromatolites. The Leza Fm example is therefore a step forward for understanding the processes involved in the development of stromatolites in tidal oolitic depositional environments. Moreover, its study together with a review of the literature suggests that conditions for ooid-trapping by stromatolites may be preferentially achieved in tidal environments.

Keywords: tidal carbonates, coastal-wetlands, oolitic stromatolites, calcified cyanobacteria, trapping and binding, stromatolite accretion.

Contributions to Modern and Ancient Tidal Sedimentology: Proceedings of the Tidalites 2012 Conference,
First Edition. Edited by Bernadette Tessier and Jean-Yves Reynaud.
© 2016 International Association of Sedimentologists. Published 2016 by John Wiley & Sons, Ltd.

INTRODUCTION

Stromatolites are laminated organosedimentary benthic structures formed by the interaction of microbial metabolism, mineral precipitation and external sediment (Riding, 1999; Tewari & Seckbach, 2011). However, the role of external sediment in stromatolite accretion is a controversial issue (Logan *et al.*, 1962; Monty, 1977; Awramik & Riding, 1988; Fairchild, 1991; Ehrlich, 1998; Altermann, 2008; Browne, 2011).

The Leza Formation, a Lower Cretaceous carbonate unit from the Cameros Basin (Northern Spain) interpreted to be deposited in a system of coastal-wetlands (Suarez-Gonzalez *et al.*, 2013; 2015), contains two different stromatolite types, both occurring in oolitic facies. One type of stromatolite occurs in tide-influenced facies and shows abundant ooids trapped in its laminae. The other type occurs in facies with no signs of tidal influence and its laminae are dominated by calcified filaments without significant ooids trapped in them.

Stromatolites and oolites are facies that commonly occur together both in marine and continental environments of the sedimentary record (e.g. Kalkowsky, 1908; Peryt, 1975; Horodyski, 1976; Surdam & Wray, 1976; Buck, 1980; Grotzinger, 1989; Riding *et al.*, 1991b; Smith & Mason, 1991; Camoin *et al.*, 1997; Chow & George, 2004; Matyszkiewicz *et al.*, 2006; Bourillot, 2009; Arenas & Pomar, 2010; Rodríguez-Martínez *et al.*, 2012; Mercedes-Martín *et al.*, 2013; Seard *et al.*, 2013; Woods, 2013), as well as in present-day continental (e.g. Halley, 1976) and marine environments (e.g. Black, 1933; Logan, 1961; Dravis, 1983; Reid & Browne, 1991; Feldmann & McKenzie, 1998; Reid *et al.*, 2003); although, in most of these examples ooids are not incorporated into the stromatolite laminae. Only a few shallow marine stromatolites, showing tidal influence, are predominantly formed by trapping and binding of the surrounding ooids (Logan, 1961; Dravis, 1983; Riding *et al.*, 1991b; Matyszkiewicz *et al.*, 2006; Bourillot, 2009; Arenas & Pomar, 2010; Mercedes-Martín *et al.*, 2013; Woods, 2013).

This study of the Leza Fm stromatolites and their depositional environments, together with a detailed bibliographical analysis, aims to elucidate the influence that tides might have on the processes that control ooid-trapping by stromatolites.

GEOLOGICAL SETTING

The Cameros Basin, located in northern Spain, is the north-westernmost basin of the Mesozoic Iberian Rift System (Fig. 1A). It was formed over a Triassic-Jurassic substrate from Late Jurassic to Early Cretaceous times (Fig. 2) and was inverted during the Cenozoic Alpine Orogeny (Mas *et al.*, 1993; Guimerà *et al.*, 1995). The sedimentary infill of the basin has a vertical thickness of up to 6500 m and is composed of continental and transitional sediments with episodes of clear marine influence (Guiraud & Seguret, 1985; Mas *et al.*, 1993; Quijada *et al.*, 2010; 2013a; 2013b; this volume; Suarez-Gonzalez *et al.*, 2010; 2013; 2015).

The sedimentary record of the Cameros Basin has been divided in eight depositional sequences (Fig. 2). The Leza Fm crops out along the northern margin of the basin and it belongs to the seventh depositional sequence (DS7), late Barremian to early Aptian in age (Figs 1B and 2). The sedimentary record of DS7 (Fig. 1B) is composed of siliciclastic fluvial facies of the Urbión Group in the SW, which change gradually to the NE to mixed siliciclastic-carbonate fluvio-lacustrine facies of the Enciso Group (Mas *et al.*, 2002; 2011; Figs 1B and 2). In the northernmost area of the basin a series of faults on the Jurassic limestone substrate controlled the sedimentation during DS7, creating small depressions where two different units were deposited: the siliciclastic Jubera Fm and the carbonate Leza Fm (Alonso & Mas, 1993; Suarez-Gonzalez *et al.*, 2011; 2013) (Figs 1B, 2, 3 and 4). The Jubera Fm is interpreted as alluvial-fan deposits related to the erosion of the faulted Jurassic substrate (Alonso & Mas, 1993; Mas *et al.*, 2002, 2011). The Leza Fm overlies and changes laterally to the Jubera Fm (Figs 1B and 2) and it is interpreted as deposited in a system of coastal-wetlands with influence of both fresh water and sea water during its sedimentation, as shown by the presence of continental fossils (charophytes, terrestrial vertebrates) and marine fossils (dasyclad algae, miliolid foraminifera) (Suarez-Gonzalez *et al.*, 2010; 2013; 2015). The Leza Fm changes laterally to the SW to the fluvio-lacustrine deposits of the Enciso Gr (Figs 1B and 2; Suarez-Gonzalez *et al.*, 2013). The fossil content, the sedimentary evolution and the stratigraphic architecture of the Leza Fm, as well as correlations with other units, have allowed dating it as early Aptian (Suarez-Gonzalez, *et al.*, 2013). The early Aptian was a period of rising global sea-level

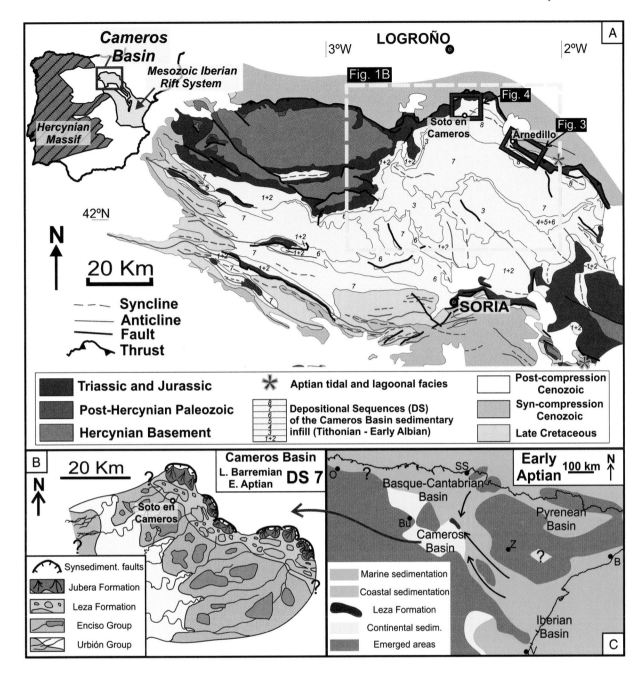

Fig. 1. (A) Geological map of the Cameros Basin, showing its location on the Iberian Peninsula and in the Mesozoic Iberian Rift System (upper left). Yellow dashed square shows approximate location of Fig. 1B. Red squares show approximate situation of the geological maps of Figs 3 and 4. Modified after Mas *et al.*, 2002; (B) Palaeogeographic sketch of the E Cameros Basin during its Depositional Sequence (DS) 7 (Late Barremian – Early Aptian), modified after Mas *et al.*, 2011. See Fig. 2 and text for explanation; (C) Palaeogeographic reconstruction of NE Iberia for the middle-upper part of the Early Aptian. Black arrows mark probable transgressive seaways for this period. O: Oviedo, SS: San Sebastián, Bu: Burgos, Z: Zaragoza, B: Barcelona, V: Valencia. Modified after Suarez-Gonzalez *et al.*, 2013.

(Mutterlose, 1998; Huang *et al.*, 2010). This trend is also apparent in the sedimentary basins adjacent to the Cameros Basin (Fig. 1C), the Basque-Cantabrian Basin to the NW (e.g. García-Mondéjar, 1990; García Garmilla, 1990; Rosales, 1999) and the Iberian Basin to the SE (e.g. Salas *et al.*, 2001; Peropadre *et al.*, 2007; Bover-Arnal *et al.*, 2010; Navarrete *et al.*, 2013), in which marine and transitional sedimentary environments predominated during Barremian-Aptian times, including

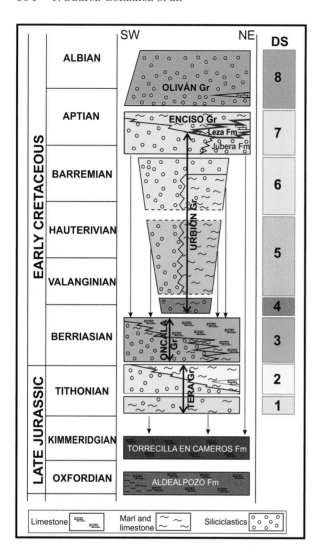

Fig. 2. Chronostratigraphic chart of the Cameros Basin, showing part of the Upper Jurassic substrate of the basin and the Tithonian-Albian sedimentary record, divided in eight depositional sequences (DS). The Leza Fm (DS7) is outlined in red. Modified after Mas *et al.*, 2011.

abundant tidal deposits. The early Aptian transgression allowed marine influence to reach the northern Cameros Basin during this period (Fig. 1C; Suarez-Gonzalez *et al.*, 2013).

The Leza Fm is, therefore, a tectonically and eustatically controlled unit mainly formed by limestones and dolomites but containing some siliciclastic intervals (Fig. 5). The tectonic control produced significant changes in thickness of the unit (Figs 3, 4 and 5; Suarez-Gonzalez *et al.*, 2013), which ranges from less than 20 m to more than 270 m. Various facies and abundant microbial carbonates are found in the Leza Fm (Suarez-Gonzalez *et al.*, 2013; 2015). Interesting differences in facies

are observed between the eastern and western outcrops of the Leza Fm (Figs 5 and 6), especially regarding the stromatolite-bearing facies, which will be described in this work.

METHODOLOGY

This work is based on the detailed sedimentological analysis of the Leza Fm and the revision of literature concerning carbonate sedimentology, modern and ancient tidal environments and stromatolite processes and fabrics. For the Leza Fm, geological mapping has been performed (Figs 3 and 4) using field observations, aerial photographs and satellite images. Twelve complete stratigraphic sections of the Leza Fm were measured along the eastern and western sectors of the unit (Figs 3, 4 and 5). Nine sections contain stromatolite levels, six from the eastern sector and three from the western sector (Fig. 5). The most representative stratigraphic sections of the facies observed in each sector are shown in detail in Fig. 6. All stratigraphic sections were measured at the decimetre scale and observations were taken at the centimetre and even millimetre scale. 752 rock samples were collected throughout all the stratigraphic sections of the Leza Fm, including 42 samples of the stromatolitic levels: 29 from the eastern sector and 13 from the western sector. A polished and uncovered thin section (30 μm-thick) was prepared for each sample, in order to conduct petrographic analysis. Thin sections were partially stained with Alizarin Red S and potassium ferrcyanide (Dickson, 1966), for accurate distinction between calcite and dolomite. Nomenclature of depositional texture of carbonates (both at macroscopic and microscopic scale) follows the classification of carbonate rocks of Dunham (1962).

SEDIMENTOLOGY OF THE LEZA FORMATION

Sedimentological analysis of the Leza Fm reveals many different facies that have been classified into five facies associations (Figs 5 and 6; Table 1; Suarez-Gonzalez *et al.*, 2015). Here, a short description of the main features of each facies association is given, as well as their interpretation in terms of depositional systems, with emphasis on the stromatolite-bearing facies.

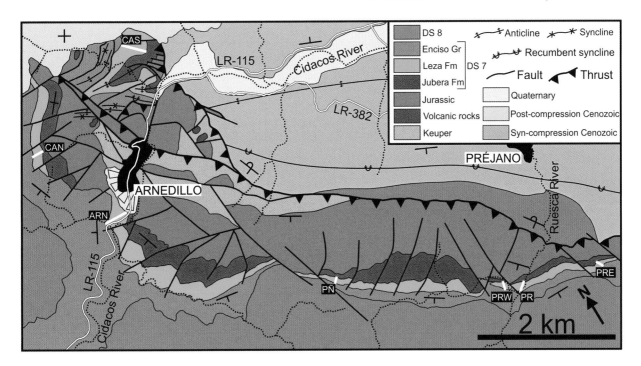

Fig. 3. Geological map of the eastern outcrops of the Leza Fm, which are located around the towns of Arnedillo and Préjano (La Rioja, Spain). See area context in Fig. 1. White lines marked in the Leza Fm outcrops represent measured stratigraphic sections of Fig. 5. CAS: Castellar. CAN: Canteras. ARN: Arnedillo. PÑ: Peñalmonte. PRW: Préjano West. PR: Préjano. PRE: Préjano East. DS: Depositional sequence of the Cameros Basin (see Fig. 2).

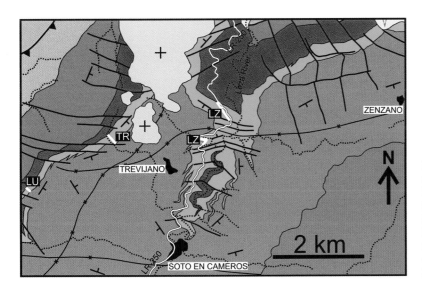

Fig. 4. Geological map of the westernmost outcrops of the Leza Fm, which are located around the towns of Trevijano, Soto en Cameros and Zenzano (La Rioja, Spain). See area context in Fig. 1 and Fig. 3 for legend. White lines marked in the Leza Fm outcrops represent measured stratigraphic sections of Fig. 5. LU: Luezas. TR: Trevijano. LZ: Leza River.

Clastic facies association

Description: The clastic facies association is observed in the lower and middle part of the measured sections of the Leza Fm and it is more abundant in the western sector of the unit than in the eastern sector (Fig. 5; Table 1). It includes conglomerates, sandstones and marls. Conglomerates are poorly sorted and are composed of medium to very coarse pebbles (Jurassic limestone lithoclasts, quartzite pebbles, carbonate intraclasts and oncoids) within a sandy and/or micritic matrix. Sandstones are coarse-grained to fine-grained and they typically show irregular bases, fining-upward trends and cross-bedding. They also contain abundant lithoclasts of Jurassic limestone. Fossils are rare: scattered fragments of charophytes, ostracods and gastropods.

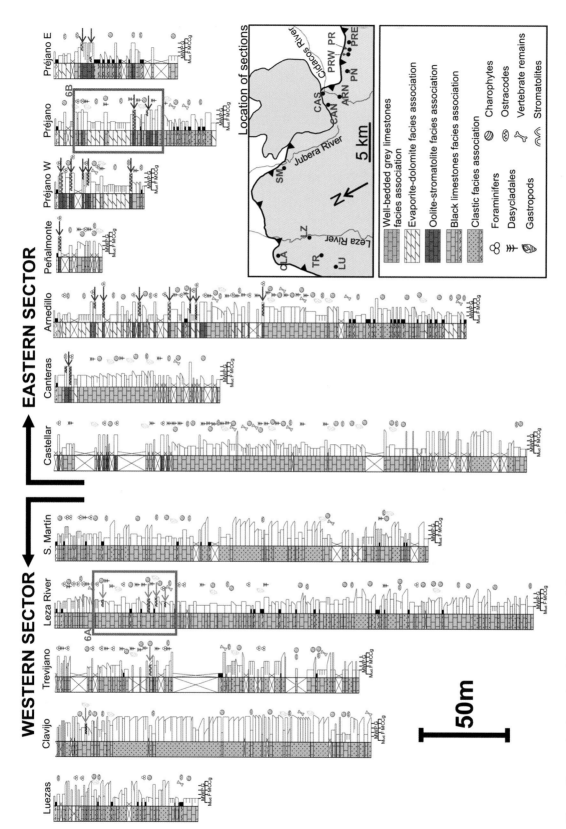

Fig. 5. Summarized stratigraphic sections of the Leza Fm showing only the most palaeoecologically relevant fossils. Small map of the N Cameros Basin (green) on the right side shows location of the sections, marked with the same abbreviations as in Figs 3 and 4 (plus CLA: Clavijo. SM: San Martín). Note the two sectors of the Leza Fm on the map, separated by an extensive outcrop of Cenozoic conglomerates (yellow). Blue arrows on the logs mark the location of skeletal stromatolite levels and red arrows mark the location of agglutinated stromatolite levels. Orange rectangles correspond to detailed sections of Fig. 6.

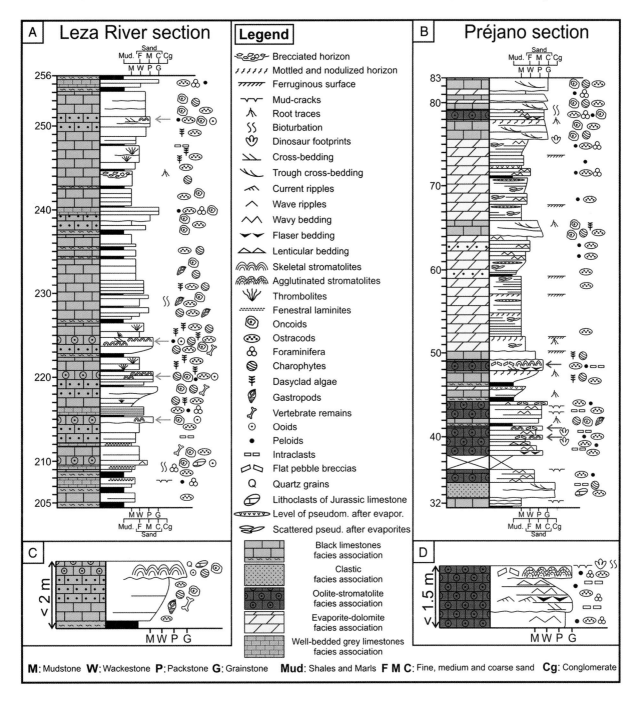

Fig. 6. Detailed logs of representative stratigraphic sections from both sectors (W and E) of the Leza Fm. Both logs have the same scale in metres: (A) Portion of the Leza River section that contains skeletal stromatolites (blue arrows); (B) Portion of the Préjano section that contains agglutinated stromatolites (red arrows); (C) Idealized sequence showing the main features of the stromatolite-bearing facies of the western sector of the Leza Fm; (D) Idealized sequence showing the main features of the stromatolite-bearing facies of the eastern sector of the Leza Fm.

Interpretation: This facies association is interpreted as deposited in an alluvial system whose main source area was the marine Jurassic limestones and sandstones that form the substrate of the Cameros Basin (Alonso & Mas, 1993; Suarez-Gonzalez *et al.*, 2010; 2013).

Black limestones facies association

Description: The black limestones facies association includes the most abundant and characteristic facies association of the Leza Fm (Fig. 5; Table 1). It is typically arranged in thickening-upwards sequences, 1 to 4 m-thick (Fig. 7), formed

Table 1. Summary of the main features of the five facies associations defined in the Leza Fm. Includes geographic and stratigraphic distribution and sedimentological interpretation.

Facies Association	Distribution	Lithologies, components and sedimentary structures	Fossil content	Interpretation
Clastic facies association	Eastern and Western sectors. Lower and middle part of sections.	Poorly sorted conglomerates, cross-bedded sandstones and marls. Jurassic lithoclasts, quartz, intraclasts and oncoids.	Oncoids. Rare fragments of charophytes, ostracods and gastropods.	Alluvial system with source area in the Jurassic substrate of the basin.
Black limestones facies association	Eastern and Western sectors. Throughout the sections.	Bioclastic limestones and marls. Common oncoids. Locally grainy with quartz grains, Jurassic lithoclasts, intraclasts, ooids and peloids (west). Mud-cracks, ferruginous, nodular, mottled and brecciated horizons.	Oncoids. Skeletal stromatolites (W). Thrombolites (W). Ostracods, charophytes, gastropods, vertebrate remains. Dasycladales (upper part of sections). Bioturbation, root traces, vertebrate footprints.	Coastal-wetlands formed by shallow water-bodies separated by vegetated areas. Influence of fresh water and sea water.
Oolite-Stromatolite facies association	Eastern sector. Middle and upper part of eastern sections.	Cross-bedded oolitic grainstones with carbonate mudstone creating flaser, wavy and lenticular bedding. Stromatolites and flat-pebble breccias. Ooids (mostly 'superficial ooids'), peloids, intraclasts and bioclasts.	Agglutinated stromatolites. Ostracods miliolid foraminifers. Bioturbation, vertebrate footprints.	Tide-influenced, shallow water-bodies, dominated by sea water, at the seaward-most area of coastal-wetlands.
Evaporite-Dolomite facies association	Eastern sector. Middle and upper part of eastern sections.	Thinly-bedded to laminated dolomites with carbonate and quartz pseudomorphs after sulphates. Locally silty and peloidal with wavy lamination. Mud-cracks, ferruginous and nodular horizons.	Scattered ostracods and miliolid foraminifers.	Relatively restricted, shallow water-bodies dominated by sea water, with high salinity. Probable tidal influence.
Well-bedded grey limestones facies association	Western sector. Middle and upper part of western sections.	Thinly-bedded limestones. Peloids, bioclasts, fenestral porosity. Mud-cracks.	Ostracods and miliolid foraminifers. Rare charophytes and dasycladales. Vertebrate footprints.	Shallow, commonly desiccated, water-bodies dominated by sea water.

by black bioclastic limestones and marls. Limestone beds are 0.1 to 2 m-thick and they generally show mudstone-wackestone textures at the lower part of the sequences and wackestone-packstone, locally grainstone, textures at the upper part. The top surface of the sequences is marked by features such as mud cracks, bioturbation, root traces, nodular and mottled horizons, brecciated horizons, ferruginous surfaces and vertebrate footprints.

This facies association shows certain variability throughout the unit. Root traces, nodular, mottled and brecciated horizons are more common in the lower half of the unit than in the upper half. Black limestones of the western sector are typically

sandy, containing abundant quartz grains, lithoclasts of Jurassic limestone and carbonate intraclasts (Fig. 6A). In fact, sandy-oolitic grainstone levels are found at the top of some sequences from the upper part of western sections (Fig. 6A), where they are associated with stromatolites (termed "skeletal stromatolites", *sensu* Riding, 1977, as described below). On the other hand, black limestones from the eastern sector limestones are only sandy at the lower half of the stratigraphic sections and they are not associated with stromatolites.

The fossil content of the black limestone facies association is ostracods, charophytes, gastropods and vertebrate remains. Dasycladales are found in sequences from the upper half of the unit and they

Fig. 7. Thickening-upwards sequences of the black limestone facies association from the Leza River section.

are especially abundant in outcrops of the eastern sector (Fig. 5). Charophytes and dascycladales can occur together but, typically, sequences show predominance of one over the other (Suarez-Gonzalez *et al.*, 2012; 2015). The black limestones facies association contains abundant microbial carbonates. Oncoid-rich facies occur throughout all sections and are especially common in the western sector. They typically display irregular bases, packstone-grainstone textures and cross-bedding; and they contain oncoids, bioclasts (ostracods, charophytes and gastropods), quartz grains, intraclasts and ooids. Skeletal stromatolites occur within this facies association only in the upper part of the western sections, typically in charophyte-rich sequences (Fig. 6A).

Stromatolite-bearing sequences are relatively thin (<2 m) and are topped by sandy-oolitic grainstone levels up to 50 cm-thick (Figs 6A and C, 8A, B and C). Stromatolites are found laterally related to the grainstone levels (Fig. 8A and B) and this interrelation is observed both at macroscopic scale (Fig. 8A and B) and at microscopic scale (Fig. 8D, E, F and G). The sandy-oolitic grainstones

are poorly-sorted and do not generally show internal sedimentary structures (Fig. 8A and C), but locally horizontal bedding and small-scale cross-bedding have been observed. They are composed of quartz grains, ooids, lithoclasts of Jurassic limestone, oncoids, bioclasts (ostracods, charophytes and gastropods) and micritic intraclasts (Fig. 8C). Mean particle size is ~1 mm, except for the oncoids, which are typically much larger, up to 3 cm in diameter. Ooids from this sandy-oolitic grainstone levels show nuclei of quartz grains, micritic particles (peloids and intraclasts) and less commonly bioclasts and fragments of Jurassic limestone. Cortices of the ooids are well developed, with typically more than 5 cortex laminae of radial-fibrous crystalline microfabric and micritic microfabric (Fig. 8C).

Interpretation: This facies association is interpreted as shallowing-upwards sequences deposited in water-bodies that could have influence of fresh water and/or sea water since these sequences can contain either freshwater fossils, marine fossils or both (Suarez-Gonzalez *et al.*, 2012; 2013; 2015). Areas between water-bodies were probably covered by vegetation, since the top of the sequences shows characteristic features of edaphic alteration (e.g. Platt & Wright, 1992; Freytet & Verrechia, 2002; Alonso-Zarza & Wright, 2010 and references therein). This system of water-bodies with variable salinities separated by broad vegetated areas can be classified as a coastal wetland (e.g. Wolanski *et al.*, 2009; Suarez-Gonzalez *et al.*, 2015).

The stromatolite-bearing sequences of this facies association only occur in the western sections (Fig. 6C) and do not typically include abundant marine fossils but do include charophytes and, therefore, were probably deposited in water-bodies dominated by fresh water. The sandy-oolitic grainstone levels that occur at the top of these sequences, associated with skeletal stromatolites (Fig. 8A, B, C, and D), are interpreted as deposited at the marginal areas of the water-bodies, where ooids were developed. The abundance of quartz grains and lithoclasts of Jurassic limestone suggests that these areas had a strong alluvial input, related to the erosion of the Jurassic substrate of the basin. The fact that sandy-oolitic grainstones occur between columns of the skeletal stromatolites (Fig. 8B) and interfingering with stromatolite laminae (Fig. 8E, F and G) shows that stromatolite accretion and sedimentation of the sandy-oolitic grainstones were coetaneous.

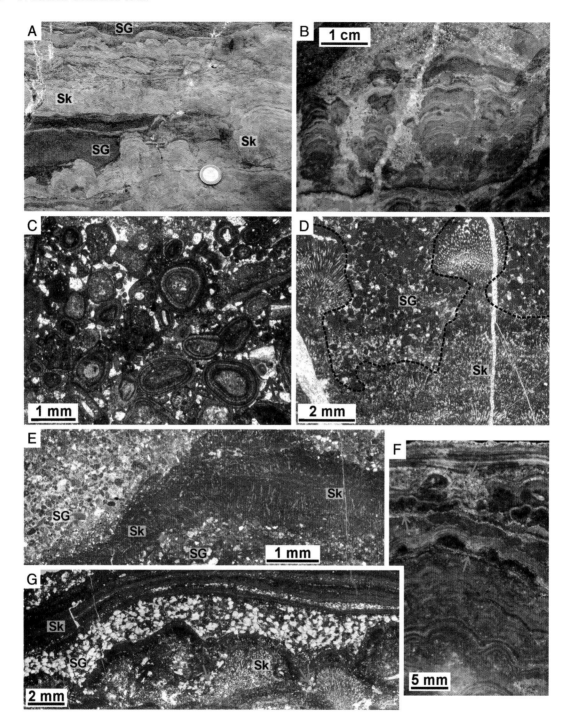

Fig. 8. Sandy-oolitic grainstone levels of the western sector black limestone facies association: (A) Domal and laterally continuous skeletal stromatolites (Sk) associated with sandy-oolitic grainstones (SG). Coin is 2.3 cm in diameter; (B) Small skeletal stromatolites with columnar shape within a sandy-oolitic grainstone level. Note how grains are deposited in between the stromatolite columns; (C) Photomicrograph of a sandy-oolitic grainstone mainly composed of ooids, micritic grains (peloids and intraclasts), quartz grains and lithoclasts (fragments of Jurassic limestone, red arrow); (D) Photomicrograph of a skeletal stromatolite (Sk) topped by a sandy-oolitic grainstone level (SG). Dashed line marks the contact between both. Note how grains are not included in the stromatolite microfabrics; (E) Thin skeletal stromatolite intercalation (Sk) between two sandy-oolitic grainstone levels (SG); (F) Polished hand specimen of a skeletal stromatolite. Blue arrows point to thin laminae of sandy-oolitic grainstone that locally interrupt stromatolite accretion; (G) Photomicrograph of a thin laminae of sandy-oolitic grainstone (SG) interrupting accretion of a skeletal stromatolite (Sk) and filling the irregularities of the stromatolite surface. Blue arrows point to grains between individual filament fans.

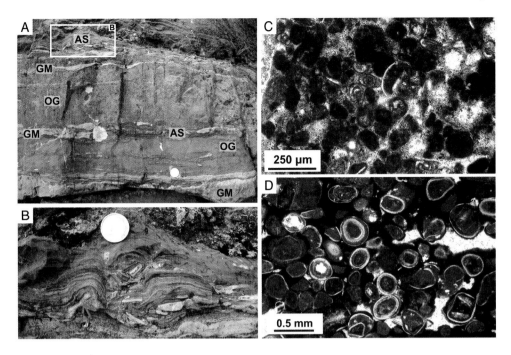

Fig. 9. Oolite-stromatolite facies association from the eastern sector of the Leza Fm: (A) Field photograph. AS = agglutinated stromatolite. GM = grey mudstones. OG = Oolitic grainstones. Coin is 2.3 cm wide; (B) Detail of A showing agglutinated stromatolites associated with flat-pebble breccia (orange arrows). Coin is 2.3 cm wide; (C) Photomicrograph of the oolite-stromatolite facies association showing its fossil content, which includes ostracods and benthic miliolid foraminifera; (D) Superficial ooids (*sensu* Illing, 1954) of the Leza Fm oolite-stromatolite facies association.

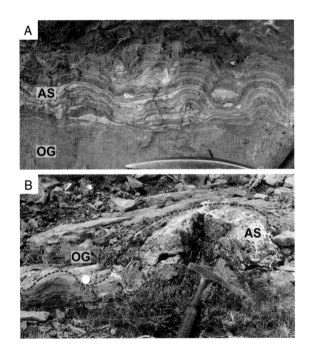

Fig. 10. Field photographs showing the relationship of the agglutinated stromatolites (AS) of the Leza Fm with the oolitic grainstones (OG) of the oolite-stromatolite facies association. Scale in (A) is the top of the rock-hammer in (B). Red line in (B) outlines stromatolite shape.

Oolite-stromatolite facies association

Description: This facies association is only observed in the eastern outcrops of the Leza Fm and it occurs at the middle and upper parts of the stratigraphic sections (Fig. 5; Table 1). It is typically arranged in sequences up to 1.5 m-thick that are mostly formed by oolitic grainstones (Figs 6D and 9). Stromatolites (termed "agglutinated stromatolites", *sensu* Riding 1991a, as described below) occur at the top of the sequences, changing laterally and vertically to the oolitic grainstones (Figs 9A and 10). Flat-pebble breccias, composed of micritic intraclasts and fragments of stromatolites, are also common at the top of the sequences, laterally associated to the stromatolites (Fig. 9B). Some of the pebbles are observed within the stromatolite beds (Fig. 9B). The top surface of the sequences of the oolite-stromatolite facies association is marked by vertical bioturbation, mudcracks and/or dinosaur footprints.

The oolitic grainstones are composed of ooids, peloids, micritic intraclasts and bioclasts (ostracods and benthic miliolid foraminifers, Fig. 9C); dasycladales and charophytes occur rarely, mostly as small fragments forming ooid nuclei. Mean

172   *P. Suarez-Gonzalez* et al.

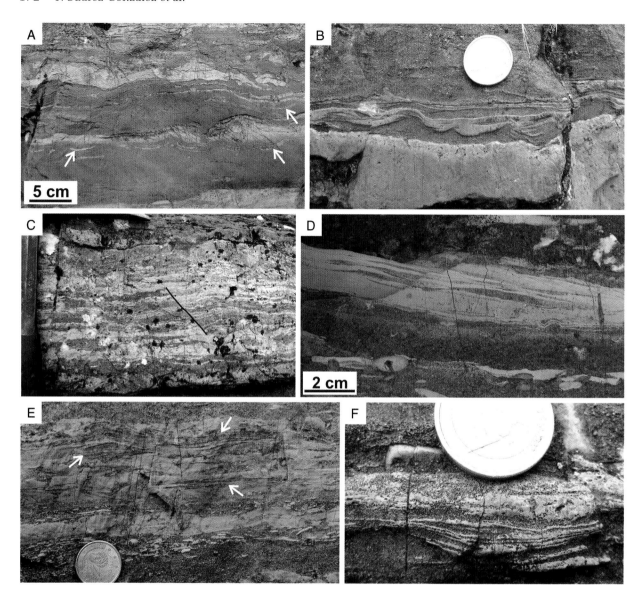

Fig. 11. Field photographs showing sedimentary structures of the oolite-stromatolite facies association. Oolitic grainstones are darker coloured and mudstones are lighter coloured: (A) Cross-bedding with flasers (white arrows) (*sensu* Reineck & Wunderlich, 1968); (B) Wavy flaser bedding (*sensu* Reineck & Wunderlich, 1968). Coin is 2.3 cm wide; (C) Wavy bedding. Small fracture (drawn in black) alters layer continuity; (D) Lenticular bedding; (E) Lenticular bedding (white arrows). Coin is 2.4 cm wide; (F) Thin mudstone layers draping foreset laminae of a grainstone ripple. Coin is 2.3 cm wide.

particle size is ~500 µm. The ooids have nuclei typically made of peloids and intraclasts and also quartz grains and bioclast fragments (Fig. 9D). Cortices of the ooids are generally poorly developed ("superficial ooids" *sensu* Illing, 1954), with a low number of cortex laminae of radial-fibrous crystalline microfabric and less abundant micritic microfabric (Fig. 9D). Ostracods and foraminifers are abundant but show very low diversity (Fig. 9C). The oolitic grainstones show wave and current ripples and they contain inter-

calations of grey mudstone, which generate the characteristic sedimentary structures of this facies association (Figs 6D, 9A and 11). Flaser bedding occurs in rippled grainstones where ripple trains are partially covered by mudstone flasers. Two different types of flaser bedding are observed: a) Cross-bedding with flasers (*sensu* Reineck & Wunderlich, 1968), when concave mudstone flasers are scattered in the rippled grainstone (Fig. 11A); and b) Wavy flaser bedding (*sensu* Reineck & Wunderlich, 1968), when flasers are more abundant and linked to each other, being concave and thicker as they fill ripple troughs and convex and thinner as they overlie ripple crests (Fig. 11B). Wavy bedding occurs in beds formed by alternation of thin (up to 2 cm-thick) irregularly-shaped layers of rippled grainstone and layers of mudstone (Fig. 11C). Lenticular bedding occurs in mudstone levels, which contain small lenses (up to 6 cm long) of rippled grainstone, which are either disconnected or partially linked (Fig. 11D and E). This relationship between grainstone and mudstone can even be observed at millimetre-scale in cases where many foreset laminae of rippled grainstone are draped by very thin micritic laminae (Fig. 11F). These sedimentary structures typically show a fining-upwards gradation: the base is dominated by grainstone levels (up to 40 cm-thick) with flaser bedding, grading upwards to wavy bedding and to mudstone-rich levels (up to 25 cm-thick) with lenticular bedding (Figs 6D, 9A and 11A).

Interpretation: The presence of flaser, wavy and lenticular bedding in this facies association implies periodical changes in water agitation during deposition. Moreover, vertical changes within single sequences point to progressive decrease in the water agitation as flaser bedding predominates in the lower part of the sequences and lenticular bedding predominates in the upper part. Breccias composed of mudstone pebbles at the top of the sequences (Figs 6D, 9A and 11A), suggests desiccation and subsequent erosion of the already indurated top mud layer by a return to agitated conditions (Demicco & Hardie, 1994). These continuous and cyclic changes in water agitation, as well as the sedimentary structures and sequences they produce, are typically found in tidal environments (Reineck & Singh, 1980; Demicco & Hardie, 1994), where the rippled grainy layers are deposited during agitated periods of flood or ebb tides and the muddy layers are deposited during calm periods of slack waters (Reineck & Wunderlich, 1968). Other interpretations (McCave, 1970; Hawley, 1981) suggest that this heterolithic stratification of grainy and muddy layers is formed by processes of longer period than diurnal tides, such as storm cycles or neap-spring tides cycles (see Allen, 1984 and Demicco & Hardie, 1994 for discussion). However, the fact that in the Leza Fm the grainstone-mudstone alternation is found at different scales (Figs 9A and 11) might mean that all these processes with different cyclicity (diurnal tides,

monthly tidal cycles, storm cycles...) are recorded in the oolite-stromatolite facies association.

These sedimentary structures and sequences of the oolite-stromatolite facies association are found in many present-day and fossil examples of carbonate tidal environments (Hagan & Logan, 1977; Hardie & Ginsburg, 1977; Laporte, 1977; Zamarreño, 1977; Demicco, 1983; 1985; King & Chafetz, 1993; Arenas & Pomar, 2010; Lasemi *et al.*, 2012). In addition, the fossil content of this facies association (ostracods and miliolid foraminifera, with very rare charophytes) points to a clear influence of marine waters and the high abundance and very low diversity of these microfossils is a palaeontological indicator of stressful environments (Brenchley & Harper, 1998). This biotic stress is commonly caused by anomalous salinity and/or rapid changes in salinity. The fact that this facies association lacks significant freshwater biota but also lacks fossils indicative of water with normal marine salinity (e.g. echinoderms) suggests that mixture of both fresh water and sea water may have occurred in the sedimentary environment, providing the stressful salinity. Similarly, Hardie & Garrett (1977) noted that low diversity of marine biota in the Bahamian coastal-wetlands and tidal-flats was due to changes in salinity produced by mixture of marine water and meteoric water during the rainy season. In addition, the agglutinated stromatolites of this facies association of the Leza Fm locally contain pseudomorphs after sulphates (Suarez-Gonzalez *et al.*, 2014), which suggests that salinity of the sedimentary environment of these facies was anomalous not only by mixture with fresh water, but also by evaporation leading to hypersalinity, at least sporadically.

Due to these palaeontological and sedimentological evidences, we interpret that the oolite-stromatolite facies association of the Leza Fm was deposited in a coastal area with tidal influence, under shallow to very shallow water conditions. Although the fossil content of this facies association suggests some influence of fresh water, this is not incompatible with the presence of tidal currents. Tide-influenced water-bodies are common in the seaward area of present-day carbonate systems of coastal-wetlands, such as the tidal-flats and marshes of the Bahamas (Black, 1933; Monty, 1972; Ginsburg *et al.*, 1977; Hardie & Garrett, 1977) and the Florida Everglades (Gebelein, 1977; Platt & Wright, 1992). In addition, modern coastal settings with

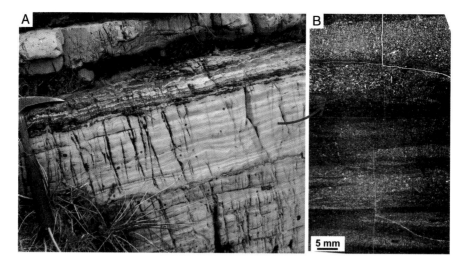

Fig. 12. (A) Field photograph of the evaporite-dolomite facies association from the Préjano section showing wavy bedding. Orange arrows point to levels of pseudomorphs after sulphates; (B) Thin section of the wavy-bedded silty-peloidal facies of the evaporite-dolomite facies association from the Préjano section.

strong discharge of meteoric water commonly contain areas with brackish water and even fresh water that undergo periodic changes in water agitation due to tidal currents, such as the Bahamas (Hardie & Garrett, 1977), many estuaries and deltas (Dalrymple & Choi, 2007; Dashtgard *et al.*, 2012) and Atlantic tidal fresh-water wetlands (Whigham *et al.*, 2009; and references therein).

Evaporite-dolomite facies association

Description: The evaporite-dolomite facies association is only observed in the eastern outcrops of the Leza Fm and it mainly occurs at the upper half of the stratigraphic sections (Fig. 5; Table 1). It is formed by thinly-bedded to laminated grey dolomites that display dense micritic or peloidal textures, locally including silt-sized quartz grains. Micritic and peloidal or silty laminae commonly alternate creating wavy lamination (Figs 6B and 12). These dolomites contain very abundant euhedral pseudomorphs, which are currently composed of quartz, calcite and dolomite, but preserve the original lenticular and tabular morphology of evaporitic sulphates. Pseudomorphs are variable in size and they are either scattered in the dolomites or grouped in centimetre-scale laterally-continuous layers. They grow both displacing and replacing the dolomitic matrix. Fossils are scarce in this facies association; only scattered ostracods and miliolid foraminifers are found. Ferruginous surfaces, nodular horizons

and mud-cracks are observed at the top of some beds.

Interpretation: This facies association is interpreted as deposited in relatively restricted coastal water-bodies. The abundance of pseudomorphs after sulphates and the scarcity and nature of the fossil content suggest that salinity of these water-bodies was typically high (Suarez-Gonzalez *et al.*, 2013; 2015). The common presence of wavy lamination suggests continuous alternation of agitated and non-agitated moments in the water-bodies, as is typical of tide-influenced environments (Reineck & Singh, 1980; Demicco & Hardie, 1994).

Well-bedded grey limestones facies association

Description: The well-bedded grey limestones facies association has only been observed in western outcrops of the Leza Fm, at the middle and upper parts of the stratigraphic sections (Figs 5 and 6A). It is formed by 10 to 30 cm-thick beds of grey limestones with common mud-cracks and/or vertebrate footprints at their top surfaces. The texture of limestones is either mudstone-wackestone of ostracods and miliolid foraminifers; rippled packstone-grainstone of peloids, ostracods and miliolid foraminifers; or fenestral laminites with micritic, clotted-peloidal and agglutinated microfabrics. Charophytes and dasycladales occur very rarely in this facies association.

Interpretation: This facies association is interpreted as deposited in very shallow, commonly

Table 2. Summary of the main features and associated facies of both stromatolite types observed in the Leza Fm.

Stromatolites		Macrostructure	Classification	Microfabrics	Interpretation of accretion processes	Associated facies	Environmental interpretation
Eastern sector		Stratiform and laterally-linked domes (up to 40 cm high).	**Agglutinated stromatolites** (*sensu* Riding, 1991a). Coarse-grained stromatolites (*sensu* Awramik & Riding, 1988)	Oolitic microfabric / Clotted-peloidal microfabric / Thin micritic crusts	Trapping and binding of surrounding grains. Calcification of microbial mats that trapped few grains. Microbial alteration during hiatuses in accretion.	*Oolite-stromatolite facies association.* Oolitic grainstones with flaser, wavy and lenticular beddings. Peloids, intraclasts and bioclasts (ostracods and forams). Flat-pebble breccias.	Tide-influenced shallow water-bodies at the seaward-most area of coastal-wetlands.
Western sector		Stratiform, laterally-linked domes (up to 30 cm high), and columnar (columns up to 5 cm-thick and 10 cm high).	**Skeletal stromatolites** (*sensu* Riding, 1977). Porostromate stromatolites (*sensu* Monty, 1981) Filamentous-calcimicrobial stromatolites (*sensu* Turner *et al.*, 2000)	Filament fans microfabric / Micrite with filaments microfabric	Very early photosynthetically-induced calcification of *Rivularia*-like cyanobacteria. Calcification of cyanobacteria and of microbial mats.	*Black limestones facies association.* Poorly sorted sandy-oolitic grainstone levels with quartz, ooids, oncoids. Jurassic lithoclasts and bioclasts (ostracods, charophytes, and gastropods).	Marginal areas of freshwater-dominated shallow water-bodies of coastal-wetlands. Strong alluvial influence.

176   *P. Suarez-Gonzalez* et al.

desiccated, water-bodies dominated by sea water (Suarez-Gonzalez *et al.*, 2013; 2015).

DESCRIPTION OF STROMATOLITES AND INTERPRETATION OF ACCRETION PROCESSES

The main features and interpretations of the studied stromatolites are summarized in Table 2.

Stromatolites of the eastern sector: Agglutinated stromatolites

Description

Macrostructure: The stromatolites observed in the oolite-stromatolite facies association from the eastern sector of the Leza Fm show morphologies that range from stratiform (Fig. 10A) to laterally-linked domes, which are up to 70 cm across and 40 cm high (Fig. 10B). Flanks of the domes have generally strong dipping angles reaching even 90° (Figs 9B and 10A). Stromatolites are laterally continuous up to 100 m. They show distinct macroscopic lamination (Figs 9B and 10A) formed by alternation of darker and lighter laminae with

thicknesses ranging from 0.5 to 4 mm (Suarez-Gonzalez *et al.*, 2014).

Microstructure: Stromatolites from the eastern sector of the Leza Fm are mainly characterized by the abundance of medium-grained to coarse-grained carbonate particles (ooids, peloids and bioclasts) observed inside them under the microscope (Fig. 13A). Therefore, they can be classified as "agglutinated stromatolites" (*sensu* Riding 1991a) and also as "coarse-grained stromatolites" (*sensu* Awramik & Riding, 1988). Laminae observed within these stromatolites show three different microfabrics: oolitic microfabric (Fig. 13B), clotted-peloidal microfabric (Fig. 13C) and thin micritic crusts (Fig. 13B). Rare pseudomorphs after sulphates are observed in some stromatolite samples (Suarez-Gonzalez *et al.*, 2014). The oolitic microfabric is the most abundant (Fig. 13A) and is composed of the same grains as the surrounding facies (Figs 9C, D and 13B) but typically finer, with a mean particle size of ~350 μm (Fig. 13B). Intergranular space is filled by micrite, clotted-peloidal micrite and/or sparite cement (Fig. 13B). Fossilized filaments are rarely found and, when observed, they are poorly preserved (Suarez-Gonzalez *et al.*, in press). The oolitic microfabric is observed throughout the

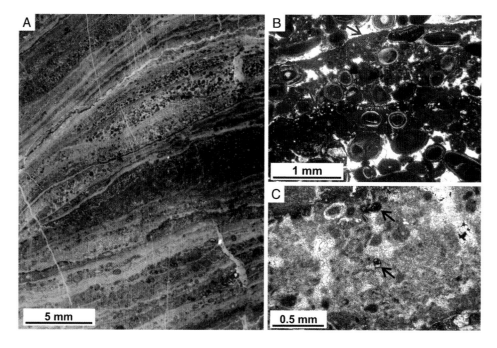

Fig. 13. Photomicrographs of the Leza Fm agglutinated stromatolites; (A) General view of a thin section. Note the predominance of carbonate grains in the composition of the stromatolite laminae; (B) Detail of laminae of oolitic microfabric. Red arrows point to the top surfaces of thin micritic crusts; (C) Detail of a lamina of clotted-peloidal microfabric with scattered grains. Black arrows point to foraminifers.

stromatolites, including steeply dipping laminae on the flanks of the domes (Figs 9B, 10A and 13A). Clotted-peloidal microfabric consists of irregular clusters of micrite (micrite clots) and peloids (20 to 80 μm across), embedded in calcite cement (Fig. 13C). Grains (ooids, intraclasts, ostracods and foraminifers) are found scattered in laminae of this microfabric (Fig. 13C). The clotted-peloidal microfabric is also observed throughout the stromatolites. Thin micritic crusts are dark laminae with a mean thickness of ~120 μm, which occur at the top of some stromatolite laminae (Fig. 13B). These crusts show micritic and/or clotted-peloidal textures, sharp upper surfaces and diffuse and irregular lower surfaces. The upper surface of the thin micritic crusts is commonly associated with micritized and truncated grains (Fig. 9B; Suarez-Gonzalez *et al.*, 2014).

Interpretation: Grains of the oolitic microfabric of these agglutinated stromatolites are the same as those of the surrounding oolitic grainstones, but typically finer. In addition, grainy laminae dip as much as 90° (Figs 9B, 10A and 13A). These facts point to an external origin of the grains, which are interpreted as trapped and bound by the mucilaginous microbial mats of the stromatolites. Therefore, we infer that laminae of this microfabric developed when carbonate sand (preferentially the finer particles) was remobilized by water currents and grains were trapped within the microbial mat, as observed in microfabrics of present-day agglutinated oolitic stromatolites of Bahamas (Black, 1933; Dravis, 1983; Dill *et al.*, 1986) and Shark Bay (Logan, 1961; Monty, 1976). Very rare and poorly preserved filaments are found in the oolitic microfabric of the Leza Fm agglutinated stromatolites, which might be relics of mat microbes, as is also observed in present-day examples (Planavsky *et al.*, 2009).

Laminae of clotted-peloidal microfabric show none or few carbonate grains scattered in the clotted-peloidal matrix (Fig. 13C). Clotted-peloidal micrite is typically explained, both in present-day and fossil microbial carbonates, as formed by the anaerobic calcification of microbial mats mainly favoured by the metabolism of heterotrophic bacteria (Chafetz & Buczynski, 1992; Reitner, 1993; Dupraz *et al.*, 2004; Riding & Tomás, 2006; Spadafora *et al.*, 2010). Thus, we interpret this microfabric as formed when sediment supply onto the stromatolites was low and its microbial mats accreted but trapped very few grains, being subsequently calcified.

The thin micritic crusts of the Leza Fm agglutinated stromatolites (Fig. 13B) are very similar to micritic crusts found in present-day agglutinated stromatolites of Shark Bay (Monty, 1976; Reid *et al.*, 2003) and the Bahamas (Reid & Browne, 1991; Reid *et al.*, 1995; Macintyre *et al.*, 1996; Visscher *et al.*, 1998; Reid *et al.*, 2000; Dupraz *et al.*, 2009; 2011). These authors interpret the crusts as lithified horizons formed by microbially induced surface alteration (mainly endolithic micritization) and surface and subsurface carbonate precipitation during hiatuses in stromatolite accretion.

In conclusion, agglutinated stromatolites of the eastern sector of the Leza Fm were mainly formed by trapping and binding of ooids and other carbonate particles available in their surrounding environment. Nevertheless, during periods of lower grain supply stromatolites also accreted generating grain-poor laminae that were calcified, developing clotted-peloidal microfabrics. Hiatuses on stromatolite accretion favoured microbial alteration of the stromatolite surface, which produced thin micritic crusts (Suarez-Gonzalez, *et al.*, 2014).

Stromatolites of the western sector: Skeletal stromatolites

Description

Macrostructure: The stromatolites associated with sandy-oolitic grainstone levels from western outcrops of the black limestone facies association are either developed on top of a surface of sandy-oolitic grainstone (Fig. 8A and B) or over an oncoid-rich layer, as stromatolitic overgrowths in continuity with the oncoid laminae. Various stromatolite levels are typically superposed in the same bed (Fig. 8A). The largest and most continuous stromatolites (outcrop conditions allow only 5 m of observable lateral continuity) show a morphology of laterally linked domes up to 30 cm high (Fig. 8A), whereas smaller stromatolites show stratiform or columnar morphology, with small columns up to 5 cm-thick and 10 cm high (Fig. 8B). All of them show distinct macroscopic lamination (Fig. 8B and F) formed by alternation of darker and lighter laminae with thicknesses ranging from 0.3 to 8 mm.

Microstructure: Stromatolites of the western sector of the Leza Fm are characterized by their internal composition: strikingly preserved and

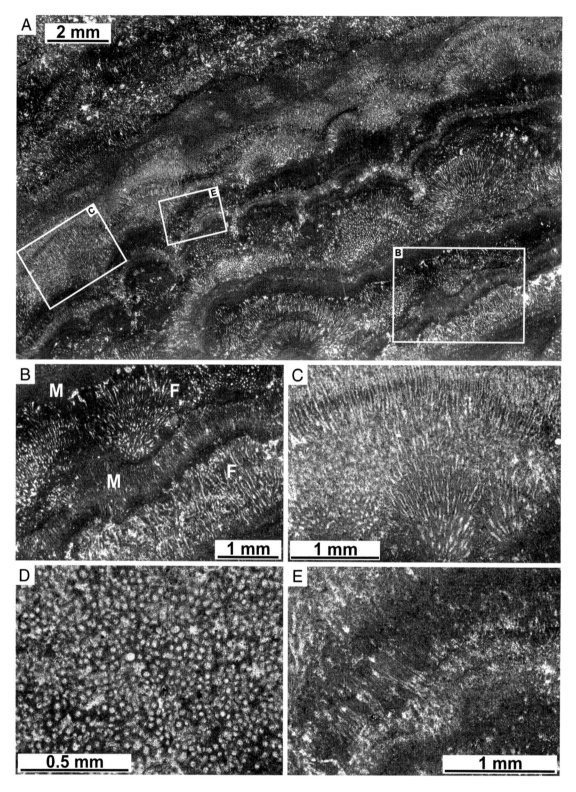

Fig. 14. Photomicrographs of the Leza Fm skeletal stromatolites: (A) General view of a thin section. Note the predominance of calcified filaments in the composition of the stromatolite laminae; (B) Detail of A showing alternation of laminae of the filament fans microfabric (F) and laminae of the micrite with filaments microfabric (M); (C) Detail of the filament fans microfabric; (D) Transversal section of the filament fans microfabric; (E) Detail of the micrite with filaments microfabric. Note thinner and poorer preserved filaments (red arrows).

abundant calcified filaments (Fig. 14A). This feature makes them classifiable as "skeletal stromatolites" (*sensu* Riding, 1977), "porostromate stromatolites" (*sensu* Monty, 1981), or "filamentous-calcimicrobial stromatolites" (*sensu* Turner *et al.*, 2000). When observed under the microscope, laminae of these skeletal stromatolites display two main microfabrics: filament fans microfabric and micrite with filaments microfabric (Fig. 14A and B). Filament fans microfabric is the most prominent microfabric of the skeletal stromatolites and is composed of erect calcified filaments that generally radiate from a small area, forming fan-like structures with an average radiating angle of 107° (Fig. 14A, B and C). Filaments are long (<2 mm), have circular cross sections (Fig. 8D) and are preserved as calcified tubes with a lumen 9 to 38 µm-thick, filled with calcite cement and surrounded by a thinner (<10 µm) dark micritic wall (Fig. 14C and D). Filaments seem to be entangled and locally show branching (Fig. 14D) with an average divergence angle of 29°. The micrite with filaments microfabric is formed by micrite traversed by erect filaments (Fig. 14B and E). Fenestrae (up to 300 µm wide) filled with sparite cement are common (Fig. 14A, B and E). Filaments of the micrite with filaments microfabric grow in palisade arrangement (Fig. 14B and E) and are typically preserved as moulds, 3 to 10 µm-thick, within the micrite. Locally, a very thin (<4 µm) dark micritic wall surrounds the molds. Branching is scarcely seen in filaments of this microfabric.

Despite the fact that these skeletal stromatolites occur within sandy-oolitic grainstone, and laterally changing to them, grains of this facies have not been found included as the main constituent in their microfabrics, as occurred in the previously described agglutinated stromatolites (Figs 8D, E and 14). Locally, two levels of sandy-oolitic grainstones are separated by a thin stromatolite level (Fig. 8E). In columnar forms, ooids and other grains are commonly observed in the space between the columns (Fig. 8B). In the filament fans microfabric, grains are locally found in the pores between individual filament fans (Fig. 8G). In domal forms, thin grainy laminae (up to 4 mm) are locally observed filling concave irregularities of the stromatolite surface (Fig. 8F and G). Stromatolite accretion by calcified filaments is resumed after these thin grainy laminae (Fig. 8F and G).

Interpretation

Filaments similar to those of the Leza Fm skeletal stromatolites have traditionally been interpreted as the calcified sheath of fossil cyanobacteria (Pollock, 1918; Klement & Toomey, 1967; Riding, 1975). Sheath calcification is a process influenced by the photosynthetic metabolism of the cyanobacteria but it is also strongly controlled by environmental conditions (Merz, 1992; Arp *et al.*, 2001; Riding, 2006; Dupraz *et al.*, 2009). Filaments of the filament fans microfabric of the skeletal stromatolites are very well preserved and show clearly the sheath morphology (Fig. 14B, C and D). Using the general morphological classification of fossil cyanobacteria of Riding (1991b), these structures could be classed inside the Hedstroemia Group, probably as *Cayeuxia*. These fossil structures are commonly related with the extant genus *Rivularia* (Schäfer & Stapf, 1978; Nickel, 1983; Leinfelder, 1985; Dragastan, 1985; Kuss, 1990). We interpret that cyanobacteria that formed this microfabric were calcified during very early stages of their development, probably when they were still alive in the surface of the stromatolite. The micrite with filaments microfabric shows less abundance and poorer preservation of cyanobacterial filaments (Fig. 14B and E). Still, some remnants of calcified sheaths can be observed but carbonate precipitation is clearly not restricted to the cyanobacterial sheaths (Fig. 14E). Calcified cyanobacteria similar to those of the micrite with filaments microfabric have been assigned to the extant genera *Scytonema*, *Phormidium*, *Calothrix* or *Dichothrix* (e.g. Monty, 1976; Schäfer & Stapf, 1978; Monty & Mas, 1981; Nickel, 1983; Leinfelder, 1985).

Differences in sheath preservation between the two microfabrics of these skeletal stromatolites might be explained by the timing of calcification: if sheath calcification does not occur at very early stages, cyanobacteria and their mucilaginous sheaths are prone to degradation by heterotrophic bacteria, which might contribute to postmortem calcification (Chafetz & Buczynski, 1992; Pratt, 2001; Planavsky *et al.*, 2009). However, the differences between the two microfabrics may also be produced by slight environmental changes or by changes in the biotic composition of the original mat.

In conclusion, we interpret that the skeletal stromatolites of the western sector of the Leza Fm accreted mainly by early calcification of cyanobacterial filaments of the microbial mats. The thin sandy-oolitic intercalations that locally occur between laminae of domal skeletal stromatolites (Fig. 8F and G) might have formed when grainy sediment was remobilized by currents (e.g. storms, waves) and partly covered the stromatolite surface. Nevertheless, stromatolites typically continued their filament accretion after these events (Fig. 8G).

DISCUSSION

Depositional system of the Leza Fm

Given the interpretations of sedimentary environments for the main facies associations of the Leza Fm (Table 1) and the interpretation of the particular sedimentary environments of the Leza Fm stromatolites (Table 2), the sedimentary system of the Leza Fm can be defined as a system of coastal-wetlands (Suarez-Gonzalez et al., 2013; 2015). Such systems are typically formed by shallow water-bodies separated by broad areas with variable vegetation, depending on the climate or the geographic setting (Wolanski et al., 2009). Coastal-wetlands include many subenvironments with variable inputs of fresh water and sea water, producing areas of differing salinity within the same system (Black, 1933; Monty & Hardie, 1976; Gebelein, 1977; Ginsburg et al., 1977; Hardie & Garrett, 1977; Platt & Wright, 1992; Waterkeyn et al., 2008). This explains the variety of facies of the Leza Fm and the fact that freshwater facies and marine-influenced facies typically alternate vertically in each stratigraphic section and also change laterally from one section to the next (Figs 5 and 6; Suarez-Gonzalez et al., 2015). However, some general trends are observed:

1) Freshwater facies are more abundant in the western outcrops of the unit, where the clastic and the black limestone facies associations predominate (Table 1; Figs 5 and 6) and the common presence of edaphic features in the latter suggests the development of broad marsh areas between water-bodies.
2) On the other hand, facies with stronger marine influence, as shown by higher abundance of marine fossils and by sedimentary structures of tidal origin, are more abundant in the eastern outcrops of the Leza Fm (Figs 5, 6, 9, 11 and 12). Tidal influence is common in the seaward areas of modern coastal-wetlands, laterally associated with areas that lack tidal influence; and even with fresh water areas (Black, 1933; Monty, 1972; Monty & Hardie, 1976; Ginsburg et al., 1977; Hardie & Garrett, 1977; Gebelein, 1977; Platt & Wright, 1992; Reed, 2002; Maloof & Grotzinger, 2012). This is probable for the case of the Leza Fm, since many of the coeval units of adjacent basins were deposited in tidal environments (e.g. García-Mondéjar, 1990; García Garmilla, 1990; Rosales, 1999; Salas et al., 2001; Peropadre et al., 2007; Bover-Arnal et al., 2010). In fact, tidal facies occur in the two marine Aptian outcrops closest to the Leza Fm (Fig. 1A): the Pellejera-Grávalos outcrop, to the NE (Arribas et al., 2009), and the Ciria outcrop, to the SW (Alonso & Mas 1988).
3) Furthermore, the Leza Fm shows a general vertical trend of increasing marine influence towards the top of the unit, which points to an eustatic control, in addition to tectonics, due to the early Aptian global rising sea-level (Mutterlose, 1998; Huang et al., 2010; Suarez-Gonzalez et al., 2013). However, palaeoenvironmental differences between eastern and western sectors are still observed at the upper part of the unit, such as the differences between both stromatolite-bearing facies described above (Fig. 6; Table 2): those of the eastern outcrops show tidal influence, those of the western outcrops lack tidal influence and their features suggest an environment dominated by meteoric water. Interestingly, the stromatolites also present great differences in their accretion process, especially in their ability to trap grains. Therefore, the Leza Fm presents a unique opportunity to discuss the influence of tides in stromatolite development.

Influence of tides on the development of Leza Fm stromatolites

Analyses of modern agglutinated oolitic stromatolites led Riding (2011) to recognize that their facility to trap ooids and other carbonate particles is provided by a) constant grain availability in the environment and b) the presence of thick, soft and sticky microbial mats on the surface of the stromatolites. These two factors can be used to interpret the differences between the stromatolites of the Leza Fm.

Grain availability

Both stromatolite types are associated with grain-stone facies, so grains were theoretically available for the stromatolites to trap them. As interpreted above, agglutinated stromatolites were developed in a shallow-water tide-influenced environment, which implies that grains were continuously and periodically mobilized by tides, in addition to waves and episodic storms. On the other hand, skeletal stromatolites were developed at the margins of water-bodies with strong alluvial input, where grains were probably only mobilized by waves and storms. In present-day environments, the only examples of agglutinated oolitic stromatolites we are aware of occur in intertidal and subtidal areas of Shark Bay (Australia) and the Bahamas, where tidal currents are regarded as an efficient mechanism for sediment supply onto the surface of the stromatolites (Logan, 1961; Dravis, 1983; Dill *et al.*, 1986; Riding *et al.*, 1991a). Some studies also cite waves and storms as contributors to the grain supply (Black, 1933; Dravis, 1983; Riding *et al.*, 1991a; Feldmann & McKenzie, 1998). In these present-day agglutinated examples it has been noted that sediment trapped within the stromatolites is consistently finer than the surrounding loose sediment (Logan, 1961; Dravis, 1983; Riding *et al.*, 1991a; Reid *et al.*, 1995), as it happens in agglutinated stromatolites of the Leza Fm. In consequence, we interpret that the continuous and cyclic hydrodynamic changes due to tidal currents were important for the supply of grains (preferentially the finer particles) onto the Leza Fm agglutinated stromatolites. However, the role of waves and storms in the mobilization and supply of sediment to the Leza Fm agglutinated stromatolites should also be taken into account.

Calcification of microbial mats

The softness and stickiness of microbial mats from the surface of present-day agglutinated oolitic stromatolites, which allows them to trap grains easily, is explained as due to the fact that they are "largely uncalcified" (Riding, 2011) when they trap the grains. Since carbonate precipitation in microbial mats takes place inside the extracellular polymeric substances (EPS) secreted by cyanobacteria and other mat microbes (Arp *et al.*, 1998; Braissant *et al.*, 2003; Decho *et al.*, 2005; Dupraz *et al.*, 2009), absence of intense very early carbonate precipitation in the surface mats entails that large quantities of mucous EPS are available for trapping the grains. The lithification required for the preservation of present-day agglutinated stromatolites is mainly reached after the trapping of the grains and it is produced by carbonate precipitation induced by the anaerobic metabolism of heterotrophic bacteria that degrade EPS (e.g. Visscher *et al.*, 1998, 2000; Reid *et al.*, 2000). By analogy with the present-day examples, we interpret that the Leza Fm agglutinated stromatolites had poorly calcified microbial mats with abundant EPS that were able to trap the ooids and other carbonate particles available in the environment. This explains the scarcity of calcified filaments in the Leza Fm agglutinated stromatolites (Fig. 13).

In contrast, microfabrics of the Leza Fm skeletal stromatolites show a predominance of heavily calcified cyanobacterial filaments (Fig. 14), suggesting that the main role on stromatolite lithification was played by their phototrophic metabolism. Since cyanobacterial calcification takes place within or upon their mucous sheath (Golubic, 1973; Riding, 1977; Pentecost, 1978; Pentecost & Riding, 1986), very early calcification of the mat cyanobacteria would cause an important decrease in the trapping and binding ability of the stromatolites because less extracellular mucous substances would be available due to the precipitation of micrite within them. This would explain why strongly calcified skeletal stromatolites (e.g. Black, 1933; Hudson, 1970; Bertrand-Sarfati, 1976; Riding, 1977; Pentecost & Riding, 1986; Rasmussen *et al.*, 1993; Arp, 1955; Freytet, 2000) include very scarce trapped grains in their microfabrics, as occurs in the Leza Fm examples. Some of the published examples (Hudson, 1970; Bertrand-Sarfati, 1976; Rasmussen *et al.*, 1993; Freytet, 2000) describe sediment commonly filling the space between stromatolite columns, between filamentous fan-like colonies, or between branching calcified filaments, as is also locally observed in the Leza Fm skeletal stromatolites (Fig. 8G). This filling process takes place after filament growth and thus it should not be considered an accretionary process of the stromatolite.

Therefore, early calcification of the surface microbial mats seems to be an important control on stromatolite trapping and binding ability. Cyanobacteria calcification is favoured by their photosynthetic metabolism, but it is a process largely controlled by environmental factors, mainly $CaCO_3$ saturation state ($\Omega CaCO_3$), Ca_{2+}

concentration, pH buffering and dissolved inorganic carbon (DIC) content (Golubic, 1973; Riding, 1982; Kempe & Kazmierczak, 1990; Merz, 1992; Arp *et al.*, 2001; Riding & Liang, 2005; Riding, 2006; Aloisi, 2008; Dupraz *et al.*, 2009; Planavsky *et al.*, 2009). Therefore, differences in the palaeoenvironments of the Leza Fm stromatolites might be used to explain the differences in early calcification of their microbial mats. However, since microbial mats are complex biological structures formed by the interaction of many different microbes, (Freytet & Plet, 1996; Riding, 2000; Dupraz *et al.*, 2009; Foster *et al.*, 2009; Reitner, 2011), the possibility of intrinsic biotic controls on these differences between both stromatolite types of the Leza Fm should also be considered. Unfortunately, the only clear palaeontological indicators of the Leza Fm mat microbes are the calcified cyanobacterial filaments of the skeletal stromatolites, which show only two different morphotypes. Therefore, a biotic control on the differences between stromatolites and on their ability to trap grains is probable but difficult to test. Nevertheless, calcification of cyanobacteria shows very little specificity, which means that the same species can be calcified or not depending on the hydrochemistry of the environment it is in (see further discussion in Golubic, 1973; Monty, 1973), but even if both stromatolite types of the Leza Fm were probably formed by different microbial communities, microbial mats of both were surely rich in mucous and sticky EPS substances, prone to trap grains, as are most modern mats (e.g. Decho *et al.*, 2005; Dupraz *et al.*, 2009; Reitner, 2011). Given the relevance of environmental conditions on the lithification of cyanobacteria and EPS (e.g. Dupraz *et al.*, 2009), it is interesting to discuss the differences in sedimentary environment that may underlie the differences in stromatolite microfabric of the Leza Fm stromatolites.

Amongst all the environmental factors influencing cyanobacterial calcification, Ω_{CaCO_3} of the waters is considered as of major importance (Kempe & Kazmierczak, 1990; Merz, 1992; Riding, 2000; Arp *et al.*, 2001; Riding & Liang, 2005; Aloisi, 2008). Supersaturated waters are needed to favour effective calcification of cyanobacterial sheaths (Merz, 2000; Arp *et al.*, 2001). Therefore, it is plausible that waters where skeletal stromatolites of the Leza Fm developed were more saturated in $CaCO_3$ than those where agglutinated stromatolites grew. Skeletal stromatolites occurred in water-bodies mainly supplied with meteoric water, which was probably supersaturated in $CaCO_3$ since it had passed through the marine Jurassic limestones of the basin substrate, as shown by the common Jurassic limestone lithoclasts of the stromatolite-bearing sandy-oolitic grainstones of the black limestones facies association. On the other hand, agglutinated stromatolites occurred in water-bodies of the same system but in the eastern sector, under stronger marine influence as shown by the fossil content and the tidal sedimentary structures. Works dealing with geochemical models of Phanerozoic sea water evolution (Arp *et al.*, 2001; Riding & Liang, 2005) have discussed the Ω_{CaCO_3} of Cretaceous seas, concluding that there was a significant drawdown in Ω_{CaCO_3} of shallow platform waters starting in this period. This decrease in saturation has been proposed as the main factor influencing the progressive scarcity of marine calcified cyanobacteria since Cretaceous times (Monty, 1973; Gebelein, 1976; Arp *et al.*, 2001; Riding & Liang, 2005; Planavsky *et al.*, 2009) and it suggests that marine waters entering the Leza Fm coastal-wetlands could have had lower Ω_{CaCO_3} than meteoric waters of the system. Furthermore, agglutinated stromatolites of the Leza Fm were formed in water-bodies of the eastern sector with a probable mixture of water sources, producing anomalous salinities and salinity changes. Mixture of waters with different Ω_{CaCO_3} is a common cause of decrease in saturation in coastal settings (Runnel, 1969; Smart *et al.*, 1988). This decrease in Ω_{CaCO_3} would explain the absence of early cyanobacterial calcification in the Leza Fm agglutinated stromatolites.

Differences in Ω_{CaCO_3} between both stromatolitic environments of the Leza Fm could also explain the differences between the ooids that occur in both environments, since ooid development is also controlled by Ω_{CaCO_3} (Opdyke & Wilkinson, 1990): ooids of the freshwater facies generally have well-developed and thick cortices (Fig. 8C), suggesting origin in more saturated waters, whereas ooids of the tide-influenced facies generally have poorly-developed cortices (Figs 9D and 13B), suggesting origin in less saturated waters.

Summarizing, early lithification of microbial mats differed significantly in both stromatolite types of the Leza Fm, explaining their contrasting microfabrics and their ability or inability to trap ooids. These differences were probably controlled by environmental factors, suggesting that tidal environments of the Leza Fm favoured the development of partially-uncalcified, soft and sticky surface microbial mats that allowed the agglutinated stromatolites of the Leza Fm to trap surrounding ooids. Nevertheless, biotic differences between both stromatolite types should not be discarded as a probable control on their ability to trap grains and on the development of microfabrics.

Is trapping of ooids limited to tidal environments?

We have interpreted that tides might have been an important control on the development of the Leza Fm agglutinated stromatolites because a) they provide constant and regular remobilization of the oolitic sediment, producing a source of grains to be trapped by the surface mats of the stromatolites; and b) water of the tide-influenced water-bodies probably had a lower $CaCO_3$ saturation state than meteoric water, impeding cyanobacterial early calcification and thus keeping surface mats of the stromatolites soft and sticky, able to trap ooids.

Interestingly, Black (1933) made similar interpretations while studying modern coastal-wetlands in the Bahamas. He studied abundant agglutinated stromatolites in brackish-water lagoons influenced by the tides but he also found skeletal stromatolites, formed by colonies of "radiating filaments, without much interstitial sediment", on freshwater ponds (these have also been studied by Monty & Hardie, 1976). He interpreted the different microfabrics of these stromatolites and their lack of trapped particles as due to "firstly, the absence of sediment (...) and secondly, the different properties of the water, which is rain-water with a certain amount of freshly dissolved calcium carbonate in solution" (Black, 1933, p. 170). Is then trapping of ooids an accretion process limited to stromatolites in tidal environments?

The first discoveries of modern agglutinated oolitic stromatolites were from intertidal environments of the Bahamas (Black, 1933) and Shark Bay, Australia (Logan, 1961). Further research in these localities (Monty, 1976; Dravis, 1983; Dill *et al.*, 1986; Awramik & Riding, 1988; Reid & Browne, 1991; Riding *et al.*, 1991a; Reid *et al.*, 1995; Macintyre *et al.*, 1996; Feldmann & McKenzie, 1998; Reid *et al.*, 2000; Reid *et al.*, 2003; Jahnert & Collins, 2012) confirmed that agglutinated stromatolites occur in intertidal and subtidal environments and that their main accretion process is trapping and binding of ooids and

other carbonate particles. Awramik & Riding (1988) noted that these modern examples were poor analogues of all fossil stromatolites, since most of these are not formed almost exclusively by microfabrics of trapped ooids. In fact, very few fossil analogues of the modern agglutinated oolitic stromatolites have been recognized until now. Riding *et al.* (1991b) studied fossil agglutinated oolitic stromatolites in the Miocene of SE Spain and they interpreted the stromatolites as formed under normal marine salinity in a shallow, wave-swept platform dominated by oolitic shoals. Bourillot (2009) and Bourillot *et al.* (2010) studied the same Miocene stromatolites and interpreted that they were deposited in environments ranging from subtidal oolitic shoals to intertidal mudflats and supratidal sabkhas. Arenas & Pomar (2010) studied deposits of the same age in Mallorca (E Spain) containing agglutinated oolitic microbialites that occur in cross-bedded oolitic facies with wavy and flaser beddings interpreted as formed in intertidal and subtidal environments. Rodríguez-Martínez *et al.* (2012) described Upper Cretaceous stromatolites from N Spain that are partially formed by trapping and binding of carbonate particles and quartz grains, developed in shallow subtidal deposits. Lower Cretaceous tide-influenced deposits studied by Quijada *et al.* (this volume) also contain stromatolites partially formed by trapping of carbonate particles. Matyszkiewicz *et al.* (2006, 2012) described Upper Jurassic agglutinated oolitic stromatolites in Poland, formed in a sedimentary environment compared to intertidal examples of the same age. Middle Triassic agglutinated oolitic-peloidal stromatolites have been described by Mercedes-Martín *et al.* (2014) from moderate to high energy intertidal to subtidal deposits including herringbone and wavy lamination, in NE Spain. Woods (2013) described Early Triassic ooid-rich intertidal to shallow subtidal deposits from SW USA, which contain agglutinated microbialites that include peloids, oncoids, ooids, intraclasts and skeletal grains. To our knowledge, no clear examples of agglutinated oolitic stromatolites have been described older than Early Triassic. Nevertheless, Chow & George (2004) described Upper Devonian "tepee-shaped microbial mounds" in Australia, formed in high-energy shallow-subtidal conditions, containing wavy-laminated fabrics interpreted to have formed by trapping and binding of carbonate particles by microbial mats, compared to some of the younger agglutinated stromatolites listed before.

All the present-day and fossil examples of agglutinated oolitic stromatolites discussed here were formed in agitated shallow marine environments and only one example is not interpreted to be formed in tidal-related environments (Riding *et al.*, 1991b), although Bourillot (2009) assigned it a tidal origin. This fact raises questions regarding the relationships between tides and the trapping of ooids by stromatolites. The coexistence of stromatolites and oolitic facies dates back to the Archean (e.g. Buck, 1980) and the Proterozoic (e.g. Horodyski, 1976; Grotzinger, 1989). The popular Lower Triassic outcrops of central Germany, where the terms stromatolite (Kalkowsky, 1908) and oolite (Brückmann, 1721) were first defined, contain examples of stromatolites passing laterally to oolitic facies interpreted to have formed in lacustrine environments. In these examples, ooids and other grains typically occur filling the space between stromatolite columns and also as irregular grainy levels intercalated with laminar stromatolites (Paul & Peryt, 2000; Paul *et al.*, 2011). These authors mention that grains can be found inside the stromatolite fabrics but they clearly state that "most carbonate was precipitated *in situ* and the detrital phase never controlled the stromatolitic lamination" (Paul & Peryt, 2000). Therefore, they should not be considered agglutinated stromatolites. Upper Jurassic lacustrine oolitic deposits contain stromatolites in the Morrison Fm, SW USA (Neuhauser *et al.*, 1987) but the stromatolites are predominantly filamentous and do not seem to include significant trapped sediment. Lacustrine deposits from the Upper Cretaceous of Bolivia also show lateral relationship between stromatolites and ooids, but ooids are not commonly included in the stromatolite laminae (Camoin *et al.*, 1997). On the Eocene Green River Fm (USA) stromatolites occur in lacustrine oolitic facies (Surdam & Wray, 1976; Seard *et al.*, 2013) but their microfabrics do not seem to be predominantly formed by trapping of ooids (Cole & Picard, 1978; Seard *et al.*, 2013). Pleistocene stromatolites and oncoids occur in oolitic facies of Etosha Pan (Namibia) but their accretion is dominated by carbonate precipitation, although some ostracods can be found in them (Smith & Mason, 1991; Brook *et al.*, 2011). Nowadays, stromatolites grow surrounded by rippled ooid sand in the Great Salt Lake of Utah (USA) but the stromatolites are not formed by trapping and binding the neighbouring ooids (Halley, 1976; Chidsey *et al.*, 2015).

Therefore, only a few geologically young examples of ooid-related stromatolites clearly show abundant agglutinated microfabrics, formed by trapping and binding of ooids, and these examples are formed in tidal environments. However, not all oolitic tidal environments show development of agglutinated oolitic stromatolites. This entails that some other factors, both intrinsic and extrinsic to the stromatolites themselves (Browne, 2011), should be taken into account to explain the scarcity of these stromatolites. For example, Awramik & Riding (1988) considered that the origin and diversification of diatoms in microbial mats was important for the development of agglutinated oolitic stromatolites because these eukaryotes enhance their trapping ability. The presence of grazing metazoans has also been used to explain the scarcity of present-day marine stromatolites, implying that these are restricted to stressful environments where metazoans are not abundant. In Shark Bay and the Bahamas this environmental stress has been explained as caused either by abnormal or changing salinity (Logan, 1961; Garrett, 1970; Hardie & Garrett, 1977) or by tidal currents that constantly move loose oolitic substrate (Garrett, 1970; Dravis, 1983). Salinity and currents have been used to explain the development of some fossil agglutinated oolitic stromatolites (Riding *et al.*, 1991b; Bourillot, 2009) and they could also be applied to the Leza Fm examples. This concept of refuges from grazing metazoans supports the idea of tidal environments as most suitable for the development of agglutinated oolitic stromatolites, since these environments are especially prone to provide constant currents that mobilize the sediment substrate and salinities different than normal marine.

In conclusion, we propose that tidal environments are especially suitable for the development of stromatolites that mainly accrete by trapping and binding of ooids and other carbonate particles, i.e. agglutinated oolitic stromatolites. We do not intend to exclude the presence of trapped grains in stromatolites from other environments. Nevertheless, considering the distribution of present-day and fossil examples of agglutinated oolitic stromatolites and the environmental implications of this study of Lower Cretaceous stromatolites, tidal environments seem to be the best ones for these conditions to be easily achieved, at least since Mesozoic times.

CONCLUSIONS

The Leza Fm (Early Cretaceous, N Spain) was deposited in a system of coastal-wetlands under the influence of both marine and meteoric waters. Oolitic facies associated with stromatolites are found in many sections of the unit but differences are observed between the eastern and western sections. Tide-influenced oolitic facies occur in the eastern sector and are associated with aggluti-nated stromatolites, which are mainly formed by trapping and binding of ooids and other carbonate particles. Oolitic facies of the western sector show no signs of tidal influence and are associated with skeletal stromatolites, mainly formed by calcified cyanobacteria and do not typically include grains. Modern and ancient examples of agglutinated oolitic stromatolites are relatively rare and almost entirely restricted to tidal environments. Analysis of the Leza Fm stromatolites and examples from the literature suggests that the main conditions that favour the development of agglutinated oolitic stromatolites are a) continuous and cyclic sediment mobilization, providing availability of particles to be trapped; b) water-chemistry condi-tions (mainly relatively low $CaCO_3$ saturation state) necessary for development of soft surface mats with partially uncalcified mucous EPS; and c) factors limiting metazoan abundance, such as continuous substrate mobilization or anomalous and/or rapidly changing salinity. Any oolitic envi-ronment providing these conditions would be suitable for the development of stromatolites mainly formed by trapping and binding of ooids, but the fact that these conditions are easily achieved in tidal settings might explain why agglutinated oolitic stromatolites are almost entirely restricted to tidal environments both at the present-day and in the geological record.

ACKNOWLEDGEMENTS

This study was funded by the Spanish DIGICYT projects CGL2011-22709 and CGL2014-52670-P, by the Research Group 'Sedimentary Basin Analysis' UCM-CM 910429 of the Complutense University of Madrid and by a FPU scholarship of the Spanish Department of Education. We would like to thank Beatriz Moral, Gilberto Herrero and Juan Carlos Salamanca for preparing the thin sections, and Laura Donadeo for finding many of the references used. Benjamin Brigaud and an anonymous reviewer pro-vided comments that helped us improve the original manuscript. We also thank Jean-Yves Reynaud for editorial support.

REFERENCES

Allen, J.R.L. (1984) *Sedimentary structures: their character and physical basis.* (Unabridged one-volume edition). Elsevier, Amsterdam, 1256 pp.

Allwood, A.C., Walter, M.R., Kamber, B.S., Marshall, C.P. and **Burch, I.W.** (2006) Stromatolite reef from the Early Archaean era of Australia. *Nature*, **441**, 714–718.

Aloisi, G. (2008) The calcium carbonate saturation state in cyanobacterial mats throughout Earth's history. *Geochim. Cosmochim. Acta*, **72**, 6037–6060.

Alonso, A. and **Mas, J.R.** (1993) Control tectónico e influ-encia del eustatismo en la sedimentación del Cretácico inferior de la cuenca de Los Cameros. *Cuad. Geol. Ibérica*, **17**, 285–310.

Alonso, A. and **Mas, J.R.** (1988) La transgresión Aptiense al sur del Moncayo (límite de las provincias de Soria y Zaragoza). *Comunicaciones del II Congreso Geológico de España*, Vol. **1**, pp. 11–14, Universidad de Granada, Granada.

Alonso-Zarza, A.M. and **Wright, V.P.** (2010) Palustrine Carbonates. In: *Carbonates in continental settings. Facies, environments and processes* (Eds A.M. Alonso-Zarza and L.H. Tanner), *Dev. Sedimentol.*, **61**, pp. 103–132, Elsevier, Amsterdam.

Altermann, W. (2008) Accretion, trapping and binding of sediment in Archean stromatolites – Morphological expression of the antiquity of life. *Space Sci. Rev.*, **135**, 55–79.

Arenas, C. and **Pomar, L.** (2010) Microbial deposits in upper Miocene carbonates, Mallorca, Spain. *Palaeogeogr. Palaeoclimatol. Palaeoecol.*, **297**, 465–485.

Arp, G. (1995) Lacustrine bioherms, spring mounds, and marginal carbonates of the Ries-Impact-Crater (Miocene, Southern Germany). *Facies*, **33**, 35–90.

Arp, G., Hofmann, J. and **Reitner, J.** (1998) Microbial fabric formation in spring mounds ("Microbialites") of alka-line salt lakes in the Badain Jaran Sand Sea, PR China. *Palaios*, **13**, 581–592.

Arp, G., Reimer, A. and **Reitner, J.** (2001) Photosynthesis-induced biofilm calcification and calcium concentra-tions in Phanerozoic oceans. *Science*, **292**, 1701–1704.

Arribas, M.E., Mas, R., Arribas, J., Benito, M.I. and **Le Pera, E.** (2009) Marine influence at the last rifting stages of a continental basin. The northernmost Cameros Basin record (Early Cretaceous, North Spain). In: *Abstracts, 27th IAS Meeting of Sedimentology, Alghero, Italy* (Eds V. Pascucci and S. Andreucci), p. 434.

Awramik, S.M. and **Riding, R.** (1988) Role of algal eukary-otes in subtidal columnar stromatolite formation. *Proc. Natl Acad. Sci. USA*, **85**, 1327–1329.

Bertrand-Sarfati, J. (1976) An attempt to classify Late Precambrian stromatolite microstructures. In: *Stromatolites* (Ed. M.R. Walter), *Dev. Sedimentol.*, **20**, pp. 251–259, Elsevier, Amsterdam.

Black, M. (1933) The algal sedimentation of Andros Island Bahamas. *Phil. Trans. Roy. Soc. London*, Series B: Biological Sciences, **222**, 789–803.

Bourillot, R. (2009) Evolution des plates-formes carbonatées pendant la crise de salinité messinienne: de la déformation des evaporites aux communautés microbialithiques (Sud-Est de l'Espagne). PhD thesis of the University of Bourgogne, 384 pp. (Available online at: http://www.asf.epoc.u-bordeaux1.fr/theses/theses_Prix_Gubler_2011.html).

Bourillot, R.; **Vennin**, E., **Rouchy**, J.M., **Durlet**, C., **Rommevaux**, V., **Kolodka**, C. and **Knap**, F. (2010) Structure and evolution of a Messinian mixed carbonate-siliciclastic platform: the role of evaporites (Sorbas Basin, South-east Spain). *Sedimentology*, **57**, 477–512.

Bover-Arnal, T., **Moreno-Bedmar**, J.A., **Salas**, R., **Skelton**, P.W., **Bitzer**, K. and **Gili**, E. (2010) Sedimentary evolution of an Aptian syn-rift carbonate system (Maestrat Basin, E Spain): effects of accommodation and environmental change. *Geol. Acta*, **8**, 249–280.

Braissant, O., **Cailleau**, G., **Dupraz**, C. and **Verrecchia**, E.P. (2003) Bacterially induced mineralization of calcium carbonate in terrestrial environments: the role of exopolysaccharides and amino acids. *J. Sed. Res.*, **73**, 485–490.

Brenchley, P.J. and **Harper**, D.A.T. (1998) *Palaeoecology: Ecosystems, environments and evolution.* Chapman & Hall, London, 402 pp.

Brook, G.A., **Railsback**, L.B. and **Marais**, E. (2011) Reassessment of carbonate ages by dating both carbonate and organic material from an Etosha Pan (Namibia) stromatolite: Evidence of humid phases during the last 20 ka. *Quatern. Int.*, **229**, 24–37.

Browne, K.M. (2011) Modern marine stromatolitic structures: The sediment dilemma. In: *Stromatolites: Interaction of microbes with sediments* (Eds **V.C. Tewari** and **J. Seckbach**), pp. 291–312, Springer, New York.

Brückmann, F.E. (1721) *Specimen Physicum exhibens Historam naturalem Oolithi.* Salomonis Schnorri, Helmestadii, **28**, pp.

Buck, S.G. (1980) Stromatolite and ooid deposits within the fluvial and lacustrine sediments of the Precambrian Ventersdorp Supergroup of South Africa. *Precambrian Res.*, **12**, 311–330.

Camoin, G., **Casanova**, J., **Rouchy**, J.M., **Blanc-Valleron**, M.M. and **Deconinck**, J.F. (1997) Environmental controls on perennial and ephemeral carbonate lakes: the central palaeo-Andean Basin of Bolivia during Late Cretaceous to early Tertiary times. *Sed. Geol.*, **113**, 1–26.

Chafetz, H.S. and **Buczynski**, C. (1992) Bacterially induced lithification of microbial mats. *Palaios*, **7**, 277–293.

Chidsey, T.C., **Vanden Berg**, M.D. and **Eby**, D.E. (2015) Petrography and characterization of microbial carbonates and associated facies from modern Great Salt Lake and Uinta Basin's Eocene Green River Formation in Utah, USA. In: *Microbial Carbonates in Space and Time* (Eds **D.W.J. Bosence**, **K.A. Gibbons**, **D.P. Le Heron**, **W.A. Morgan**, **T. Pritchard** and **B.A. Vining**), Geological Society, London, Special Publications, **418**, pp. 261–286.

Chow, N. and **George**, A.D. (2004) Tepee-shaped agglutinated microbialites: an example from a Famennian carbonate platform on the Lennard Shelf, northern Canning Basin, Western Australia. *Sedimentology*, **51**, 253–265.

Cole, R.D. and **Picard**, M.D. (1978) Comparative mineralogy of nearshore and offshore lacustrine lithofacies, Parachute Creek Member of the Green River Formation, Piceance Creek Basin, Colorado and eastern Uinta Basin, *Utah. Geol. Soc. Am. Bull.*, **89**, 1441–1454.

Dalrymple R.W. and **Choi**, K. (2007) Morhpologic and facies trends through the fluvial-marine transition in tide-dominated depositional systems: A schematic framework for environmental and sequences-stratigraphic interpretation. *Earth-Sci. Rev.*, **81**, 135–174.

Dashtgard, S.E., **Venditti**, J.G., **Hill**, P.R., **Sisulak**, C.F., **Johnson**, S.M. and **La Croix**, A.D. (2012) Sedimentation across the tidal-fluvial transition in the Lower Fraser River, Canada. *The Sed. Record*, **10**, 4–9.

Decho, A.W., **Visscher**, P.T. and **Reid**, R.P. (2005) Production and cycling of natural microbial expolymers (EPS) within a marine stromatolite. *Palaeogeogr. Palaeoclimatol. Palaeoecol.*, **219**, 71–86.

Demicco, R.V. (1985) Platform and off-platform carbonates of the Upper Cambrian of western Maryland, U.S.A. *Sedimentology*, **32**, 1–22.

Demicco, R.V. (1983) Wavy and lenticular-bedded carbonate ribbon rocks of the Upper Cambrian Conococheague Limestone, Central Appalachians. *J. Sed. Res.*, **53**, 1121–1132.

Demicco, R.V. and **Hardie**, L.A. (1994) Sedimentary structures and early diagenetic features of shallow marine carbonate deposits. *SEPM Atlas Series*, **1**, 265 pp.

Dickson, J.A.D. (1966) Carbonate identification and genesis as revealed by staining. *J. Sed. Petrol.*, **36**, 491–505.

Dill, R.F., **Shinn**, E.A., **Jones**, A.T., **Kelly**, K. and **Steinen**, R.P. (1986) Giant subtidal stromatolites forming in normal salinity waters. *Nature*, **324**, 55–58.

Dragastan, O. (1985) Review of Tethyan Mesozoic algae of Romania. In: *Paleoalgology: contemporary research and applications* (Eds D.F. Toomey and M.H. Nitecki), pp. 101–161, Springer-Verlag, Berlin.

Dravis, J.J. (1983) Hardened subtidal stromatolites, Bahamas. *Science*, **219**, 385–386.

Dunham, R.J. (1962) Classification of carbonate rocks according to depositional texture. In: *Classification of carbonate rocks* (Ed. W.E. Ham). *AAPG Mem.*, **1**, 108–121.

Dupraz, C., **Reid**, R.P., **Braissant**, O., **Decho**, A.W., **Norman**, R.S. and **Visscher**, P.T. (2009) Processes of carbonate precipitation in modern microbial mats. *Earth-Sci. Rev.*, **96**, 141–162.

Dupraz, C., **Reid**, R.P. and **Visscher**, P.T. (2011) Modern microbialites. In: *Encyclopedia of Geobiology* (Eds J. Reitner and V. Thiel), pp. 617–635, Springer, Berlin.

Dupraz, C., **Visscher**, P.T., **Baumgartner**, L.K. and **Reid**, R.P. (2004) Microbe-mineral interactions: early carbonate precipitation in a hypersaline lake (Eleuthera Island, Bahamas). *Sedimentology*, **51**, 745–765.

Ehrlich, H.L. (1998) Geomicrobiology: its significance for geology. *Earth-Sci. Rev.*, **45**, 45–60.

Fairchild, I.J. (1991) Origins of carbonate in Neoproterozoic stromatolites and the identification of modern analogues. *Precambrian Res.*, **53**, 281–299.

Feldmann, M. and **McKenzie**, J.A. (1998) Stromatolite-thrombolite associations in a modern environment, Lee Stocking Island, Bahamas. *Palaios*, **13**, 201–212.

Foster, J.S., **Green**, S.J., **Ahrendt**, S.R., **Golubic**, S., **Reid**, R.P., **Hetherington**, K.L. and **Bebout**, L. (2009) Molecular and morphological characterization of cyanobacterial diversity in the stromatolites of Highborne Cay, Bahamas. *The ISME Journal*, **3**, 573–587.

Freytet P. (2000) Distribution and palaeoecology of non marine algae and stromatolites: II, The Limagne of Allier Oligo-Miocene lake (central France). *Ann. Paléontol.*, **86**, 3–57.

Freytet, P. and Plet, A. (1996) Modern freshwater microbial carbonates: the Phormidium stromatolites (tufa-travertine) of Southeastern Burgundy (Paris Basin, France). *Facies*, **34**, 219–238.

Freytet, P. and Verrecchia, E.P. (1998) Freshwater organisms that build stromatolites: a synopsis of biocrystallization by prokaryotic and eukaryotic algae. *Sedimentology*, **45**, 535–563.

Freytet, P. and Verrechia, E.P. (2002): Lacustrine and palustrine carbonate petrography: an overview. *J. Paleolimn.*, **27**, 221–237.

García-Mondéjar, J. (1990) The Aptian-Albian carbonate episode of the Basque-Cantabrian Basin (northern Spain): general characteristics, controls and evolution. *Int. Assoc. Sedimentol. Spec. Publ.*, **9**, 257–290.

García Garmilla, F. (1990) Evolución paleogeográfica d elos sistemas deposicionales Wealdenses y del Aptiense inferior en el sector central de la región Vasco-Cantábrica. *Est. Mus. Cienc. Nat. de Álava*, **5**, 5–26.

Garrett, P. (1970) Phanerozoic stromatolites: noncompetitive ecologic restriction by grazing and burrowing animals. *Science*, **169**, 171–173.

Gebelein, C.D. (1976) The effects of the physical, chemical and biological evolution of the Earth. In: *Stromatolites* (Ed. M.R. Walter), *Dev. Sedimentol.*, **20**, pp. 499–515, Elsevier, Amsterdam.

Gebelein, C.D. (1977) *Dynamics of recent carbonate sedimentation and ecology. Cape Sable, Florida.* Brill, Leiden, 120 pp.

Gierlowski-Kordesch, E.H. (2010) Lacustrine carbonates. In: *Carbonates in continental settings. Facies, environments and processes* (Eds A.M. Alonso-Zarza and L.H. Tanner), *Dev. Sedimentol.*, **61**, pp. 1–101, Elsevier, Amsterdam.

Ginsburg, R.N. and Hardie, L.A. (1977) Tidal and storm deposits, northwestern Andros Island, Bahamas. In: *Tidal deposits: a casebook of recent examples and fossil counterparts* (Ed. R.N. Ginsburg), pp. 201–208. Springer-Verlag, Berlin.

Golubic, S. (1973) The relationship between blue-green algae and carbonate deposits. In: *The biology of blue-green algae* (Eds N.G. Carr and B.A. Whitton), pp. 434–472. Blackwell, Oxford.

Grotzinger, J.P. (1989) Facies and evolution of Precambrian carbonate depositional systems: emergence of the modern platform archetype. In: *Controls on carbonate platform and basin development* (Ed. P.D. Crevello), *SEPM Spec. Publ.*, **44**, 79–106.

Guimerà, J., Alonso, A. and Mas, J.R. (1995) Inversion of an extensional-ramp basin by a newly formed thrust: the Cameros basin (N. Spain). In: *Basin Inversion* (Eds J.G. Buchanan and P.G. Buchanan), *Geol. Soc. London Spec. Publ.*, **88**, 433–453.

Hagan, G.M. and Logan, B.W. (1977) Prograding tidal-flat sequences: Hutchison Embayment, Shark Bay, Western Australia. In: *Tidal deposits: a casebook of recent*

examples and fossil counterparts (Ed. R.N. Ginsburg), pp. 215–222. Springer-Verlag, Berlin.

Halley, R.B. (1976) Textural variation within Great Salt Lake algal mounds. In: *Stromatolites* (Ed. M.R. Walter), *Dev. Sedimentol.*, **20**, pp. 435–445, Elsevier, Amsterdam.

Hardie, L.A. and Garrett, P. (1977) General environmental setting. In: *Sedimentation on the modern carbonate tidal flats of Northwest Andros Island, Bahamas* (Ed. L.A. Hardie), pp.12–49. The John Hopkins University Press, Baltimore.

Hardie, L.A. and Ginsburg, R.N. (1977) Layering: the origin and environmental significance of lamination and thin bedding. In: *Sedimentation on the modern carbonate tidal flats of Northwest Andros Island, Bahamas* (Ed. L.A. Hardie), pp. 50–123. The John Hopkins University Press, Baltimore.

Hawley, N. (1981) Flume experiments on the origin of flaser bedding. *Sedimentology*, **28**, 699–712.

Horodyski, R.J. (1976) Stromatolites of the upper siyeh Limestone (Middle Proterozoic), Belt Supergroup, Glacier National Park, *Montana. Precambrian Res.*, **3**, 517–536.

Huang, C., Hinnov, L., Fischer, A.G., Grippo, A. and Herbert, T. (2010) Astronomical tuning of the Aptian Stage from Italian reference sections. *Geology*, **38**, 899–902.

Hudson, J.D. (1970) Algal limestones with pesudomorphs after gypsum from the Middle Jurassic of Scotland. *Lethaia*, **3**, 11–40.

Illing, L.V. (1954) Bahaman calcareous sands. *AAPG Bull.*, **38**, 1–52.

Jahnert, R.J. and Collins, L.B. (2012) Characteristics, distribution and morphogenesis of subtidal microbial systems in Shark Bay, Australia. *Marine Geology* **303–306**, 115–136.

Kalkowsky, E. (1908) Oölith und Stromatolith im norddeutschen Buntsandstein. *Z. Deut. Geol. Ges.*, **60**, 68–125.

Kempe, S. and Kazmierczak, J. (1990) Calcium carbonate supersaturation and the formation of in situ calcified stromatolites. In: *Facets of modem biogeochemistry* (Eds V. Ittekkot, S. Kempe, W. Michaelis and A. Spitzy), pp. 255–278. Springer, Berlin.

King, D.T. and Chafetz, H.S. (1993) Tidal-flat to shallow-shelf deposits in the Cap Mountain Limestone Member of the Riley Formation, upper Cambrian of central Texas. *J. Sed. Petrol.*, **53**, 261–273.

Klement, K.W. and Toomey, D.F. (1967) Role of the blue-green alga Girvanella in skeletal grain destruction and lime-mud formation in the Lower Ordovician of west Texas. *J. Sed. Petrol.*, **37**, 1045–1051.

Kuss, J. (1990) Middle Jurassic calcareous algae from the circum-arabian area. *Facies*, **22**, 59–86.

Laporte, L.F. (1977) Carbonate tidal-flat deposits of the Early Devonian Manlius Formation of New York State. In: *Tidal deposits: a casebook of recent examples and fossil counterparts* (Ed. R.N. Ginsburg), pp. 243–250. Springer-Verlag, Berlin.

Lasemi, Y., Jahani, D., Amin-Rasouli, H. and Lasemi, Z. (2012) Ancient carbonate tidalites. In: *Principles of Tidal Sedimentology* (Eds R.A. Davis, Jr. and R.W. Dalrymple), pp. 567–607. Springer, Dordrecht.

Leinfelder, R.R. (1985) Cyanophyte calcification morphotypes and depositional environments (Alenquer oncolite, Upper Kimmeridgian?, Portugal). *Facies*, **12**, 253–274.

Logan, B.W. (1961) Cryptozoon and associate stromatolites from the Recent, Shark Bay, western Australia. *J. Geol.*, **69**, 517–533.

Logan, B.W., Rezak, R. and Ginsburg R.N. (1964) Classification and environmental significance of algal stromatolites. *J. Geol.*, **72**, 68–83.

Macintyre, I.G., Reid, R.P. and Steneck, R.S. (1996) Growth history of stromatolites in a Holocene fringing reef, Stocking Island, Bahamas. *J. Sed. Res.*, **66**, 231–242.

Maloof, A.C. and Grotzinger, J.P. (2012) The Holocene shallowing-upward parasequence of north-west Andros Island, Bahamas. *Sedimentology*, **59**, 1375–1407.

Mas, J.R., Alonso, A. and Guimerà, J. (1993) Evolución tectonosedimentaria de una cuenca extensional intraplaca: La cuenca finijurásica-eocretácica de Los Cameros (La Rioja-Soria). *Rev. Soc. Geol. España*, **6**, 129–144.

Mas, R., Benito, M.I., Arribas, J., Alonso, A., Arribas, M.E., Lohmann, K.C., González-Acebrón, L., Hernán, J., Quijada, I.E., Suarez-Gonzalez, P. and Omodeo, S. (2011) Evolution of an intra-plate rift basin: the Latest Jurassic-Early Cretaceous Cameros Basin (Northwest Iberian Ranges, North Spain). In: Post-Meeting Field trips Guidebook, 28th IAS Meeting, Zaragoza (Eds C. Arenas, L. Pomar and F. Colombo). *Sociedad Geológica de España, Geoguías*, **8**, 117–155.

Mas, R., Benito, M.I., Arribas, J., Serrano, A., Guimerà, J., Alonso, A. and Alonso-Azcárate, J. (2002) La Cuenca de Cameros: desde la extensión finijurásica-eocretácica a la inversión terciaria – Implicaciones en la exploración de hidrocarburos. *Zubía*, **14**, 9–64.

Mas, R., García, A., Salas, R., Meléndez, A., Alonso, A., Aurell, M., Bádenas, B., Benito, M.I., Carenas, B., García-hidalgo, J.F., Gil, J. and Segura, M. (2004) Segunda fase de rifting: Jurásico Superior-Cretácico Inferior. In: *Geología de España* (Ed. J.A. Vera), pp. 503–510. Sociedad Geológica de España - Instituto Geológico y Minero de España, Madrid.

Matyszkiewicz, J., Kochman, A. and Duś, A. (2012) Influence of local sedimentary conditions on development of microbialites in the Oxfordian carbonate buildups from the southern part of the Kraków-Czestochowa Upland (South Poland). *Sed. Geol.*, **263–264**, 109–132.

Matyszkiewicz, J., Krajewski, M. and Kedzierski, J. (2006) Origin and evolution of an Upper Jurassic complex of carbonate buildups from Zegarowe Rocks (Kraków-Wielun Upland, Poland). *Facies*, **52**, 249–263.

McCave, I.N. (1970) Deposition of fine-grained suspended sediment from tidal currents. *J. Geophys. Res.*, **75**, 4151–4159.

Mercedes-Martín, R., Arenas, C. and Salas, R. (2014) Diversity and factors controlling widespread occurrence of syn-rift Ladinian microbialites in the western Tethys (Triassic Catalan Basin, NE Spain). *Sedimentary Geology*, **313**, 68–90.

Merz M.U.E. (1992) The biology of carbonate precipitation by cyanobacteria. *Facies*, **26**, 81–102.

Merz-Preiß, M. (2000) Calcification in cyanobacteria. In: *Microbial sediments* (Eds R.E. Riding and S.M. Awramik), pp. 50–56, Springer-Verlag, Berlin.

Monty, C. (1977) Evolving concepts on the nature and the ecological significance of stromatolites. In: *Fossil algae: recent results and developments* (Ed. E. Flügel), pp. 15–35, Springer-Verlag, Berlin.

Monty, C. (1981) Spongiostromate vs. Porostromate stromatolites and oncolites. In: *Phanerozoic stromatolites: case histories* (Ed. C. Monty), pp. 1–4, Springer-Verlag, Berlin.

Monty, C. and Mas, J.R. (1981) Lower Cretaceous (Wealdian) blue-green algal deposits of the province of Valencia, Eastern Spain. In: *Phanerozoic stromatolites: case histories* (Ed. C. Monty), pp. 85–120, Springer-Verlag, Berlin.

Monty, C.L.V. (1972) Recent algal stromatolitic deposits andros Island, Bahamas. Preliminary Report. *Geol. Rundsch.*, **61**, 742–783.

Monty, C.L.V. (1973) Precambrian background and Phanerozoic history of stromatolitic communities, an overview. *Ann. Soc. Géol. Belg.*, **96**, 585–624.

Monty, C.L.V. (1976) The origin and development of cryptalgal fabrics. In: *Stromatolites* (Ed. M.R. Walter), *Dev. Sedimentol.*, **20**, pp. 193–249, Elsevier, Amsterdam.

Monty, C.L.V. and Hardie, L.A. (1976) The geological significance of the freshwater blue-green algal calcareous marsh. In: *Stromatolites* (Ed. M.R. Walter), *Dev. Sedimentol.*, **20**, pp. 447–477, Elsevier, Amsterdam.

Mutterlose, J. (1998) The Barremian-Aptian turnover of biota in northwestern Europe: evidence from belemnites. *Palaeogeogr. Palaeoclimatol. Palaeoecol.*, **144**, 161–173.

Navarrete, R., Rodríguez-López, J.P., Liesa, C.L., Soria, A.R. and Veloso, F.L. (2013) Changing physiography of rift basins as a control on the evolution of mixed siliciclastic-carbonate back-barrier systems (Barremian Iberian Basin, Spain). *Sed. Geol.*, **289**, 40–61.

Neuhauser, K.R., Lucas, S.G., de Albuquerque, J.S., Louden, R.J., Hayden, S.N., Kietzke, K.K., Oakes, W. and Des Marais, D. (1987) Stromatolites of the Morrison Formation (Upper Jurassic), Union County, New Mexico: a preliminary report. *New Mexico Geological Society Guidebook*, **38**, 153–159.

Nickel, E. (1983) Environmental significance of freshwater oncoids, Eocene Guarga Formation, Southen Pyrenees, Spain. In: *Coated Grains* (Ed. T.M. Peryt), pp. 308–329. Springer-Verlag, Berlin.

Opdyke, B.N. and Wilkinson, B.H. (1990) Paleolatitude distribution of Phanerozoic marine ooids and cements. *Palaeogeogr. Palaeoclimatol. Palaeoecol.*, **78**, 135–148.

Paul, J. and Peryt, T.M. (2000) Kalkowsky's stromatolites revisited (Lower Triassic Buntsandstein, Harz Mountains, Germany). *Palaeogeogr. Palaeoclimatol. Palaeoecol.*, **161**, 435–458.

Paul, J., Peryt, T.M. and Burne, R.V. (2011) Kalkowsky's stromatolites and oolites (Lower Buntsandstein, Northern Germany). In: *Advances in stromatolite geobiology* (Eds J. Reitner, N.V. Quéric and G. Arp), *Lect. Notes Earth Sci.*, **131**, 13–28.

Pentecost, A. (1978) Blue-green algae and freshwater carbonate deposits. *Proc. Roy. Soc. London B*, **200**, 43–61.

Pentecost, A. and Riding, R. (1986) Calcification in cyanobacteria. In: *Biomineralization in lower plants and animals* (Eds B.S.C. Leadbeater and R. Riding), pp. 73–90, Claredon Press, Oxford.

Peropadre, C., Meléndez, N. and Liesa, C.L. (2007) Heterogeneous subsidence and paleogeographic elements in an extensional setting revealed through the correlation of a storm deposit unit (Aptian, E Spain). *J. Iberian Geol.*, **33**, 79–91.

Peryt, T.M. (1975) Significance of stromatolites for the environmental interpretation of the Buntsandstein (Lower Triassic) rocks. *Geol. Rundsch.*, **64**, 143–158.

Planavsky, N., Reid, R.P., Lyons, T.W., Myshrall, K.L. and **Visscher, P.T.** (2009) Formation and diagenesis of modern marine calcified cyanobacteria. *Geobiology*, **7**, 566–576.

Platt, N.H. and **Wright, V.P.** (1991) Lacustrine carbonates: facies models, facies distributions and hydrocarbon aspects. In: *Lacustrine facies analysis* (Eds P. Anadón, L. Cabrera and K. Kelts), *Int. Assoc. Sedimentol. Spec. Publ.*, **13**, 57–74.

Platt, N.H. and **Wright, V.P.** (1992) Palustrine carbonates and the Florida Everglades: towards and exposure index for the fresh-water environment? *J. Sed. Petrol.*, **62**, 1058–1071.

Pollock, J.B. (1918) Blue-green algae as agents in the deposition of marl in a Michigan lake. *Michigan Ac. Sci. Report*, **20**, 247–259.

Pratt, B.R. (2001) Calcification of cyanobacterial filaments: Girvanella and the origin of lower Paleozoic lime mud. *Geology*, **29**, 763–766.

Quijada, I.E., Suarez-Gonzalez, P., Benito, M.I. and **Mas, J.R.** (2013a) Depositional depth of laminated carbonate deposits: insights from the Lower Cretaceous Valdeprado Formation (Cameros Basin, Spain). *J. Sed. Res.*, **83**, 241–257.

Quijada, I.E., Suarez-Gonzalez, P., Benito, M.I. and **Mas, J.R.** (2013b) New insights on stratigraphy and sedimentology of the Oncala Group (eastern Cameros Basin): implications for the paleogeographic reconstruction of NE Iberia at Berriasian times. *J. Iberian Geol.*, **39**, 115–135. 10.5209/rev_JIGE.2013.v39.n2.42503

Quijada, I.E., Suarez-Gonzalez, P., Benito, M.I. and **Mas, J.R.** (2011) Tidal influence in Berriasian deposits from the Cameros Basin (N Spain). Paleogeographic implications. In: Abstracts, 28th IAS Meeting of Sedimentology, Zaragoza, Spain (Eds B. Bádenas, M. Aurell and A.M. Alonso-Zarza), p. 178.

Quijada, I.E., Suarez-Gonzalez, P., Benito, M.I., Mas, J.R. and **Alonso, A.** (2010) Un ejemplo de llanura fluvio-deltaica influenciada por las mareas: el yacimiento de icnitas de Serrantes (Grupo Oncala, Berriasiense, Cuenca de Cameros, N. de España). *Geogaceta*, **49**, 15–18.

Quijada, I.E., Suarez-Gonzalez, P., Benito, M.I. and **Mas, J.R.** (this volume) Tidal versus continental sandy-muddy flat deposits: evidence from the Oncala Group (Early Cretaceous, N Spain). In: *Contributions to Modern and Ancient Tidal Sedimentology* (Eds **B. Tessier** and **J.Y. Reynaud**), *Int. Assoc. Sedimentol. Spec. Publ.*, **48**.

Rasmussen, K.A., Macintyre, I.G. and **Prufert, L.** (1993) Modern stromatolite reefs fringing a brackish coastline, Chetumal Bay, Belize. *Geology*, **21**, 199–202.

Reed, D.J. (2002) Understanding Tidal Marsh Sedimentation in the Sacramento-San Joaquin Delta, California. *J. Coastal Res.*, SI **36**, 605–611.

Reid, R.P. and **Browne, K.M.** (1991) Intertidal stromatolites in a fringing Holocene reef complex, Bahamas. *Geology*, **19**, 15–18.

Reid, R.P., James, N.P., Macintyre, I.A., Dupraz, C. and **Burne, R.V.** (2003) Shark Bay stromatolites: microfabrics and reinterpretation of origins. *Facies*, **49**, 299–324.

Reid, R.P., Macintyre, I.G., Browne, K.M., Steneck, R.S. and **Miller, T.** (1995) Modern marine stromatolites in the Exuma Cays, Bahamas: uncommonly common. *Facies*, **33**, 1–18.

Reid, R.P., Visscher, P.T., Decho, A.W., Stolz, J.F., Bebout, B.M., Dupraz, C., Macintyre, I.G., Paerl, H.W., Pinckney, J.L., Prufert-Bebout, L., Steppe, T.F. and **DesMarais, D.J.** (2000) The role of microbes in accretion, lamination and early lithification of modern marine stromatolites. *Nature*, **406**, 989–992.

Reineck, H.E. and **Singh, I.B.** (1980) *Depositional sedimentary environments.* 2nd edn, Springer-Verlag, Berlin, 549 pp.

Reineck, H.E. and **Wunderlich, F.** (1968) Classification and origin of flaser and lenticular bedding. *Sedimentology*, **11**, 99–104.

Reitner, J. (2011) Biofilms. In: *Encyclopedia of Geobiology* (Eds **J. Reitner** and **V. Thiel**), pp. 134–135, Springer, Berlin.

Reitner, J. (1993) Modern cryptic microbialite/Metazoan facies from Lizard Island (Great Barrier Reef, Australia). Formation and concepts. *Facies*, **29**, 3–40.

Riding, R. (1991a) Calcified cyanobacteria. In: *Calcareous algae and stromatolites* (Ed. **R. Riding**), pp. 55–87, Springer-Verlag, Berlin.

Riding, R. (1991b) Classification of microbial carbonates. In: *Calcareous algae and stromatolites* (Ed. **R. Riding**), pp. 21–52, Springer-Verlag, Berlin.

Riding, R. (2006) Cyanobacterial calcification, carbon dioxide concentrating mechanisms and Proterozoic-Cambrian changes in atmospheric composition. *Geobiology*, **4**, 299–316.

Riding, R. (1982) Cyanophyte calcification and changes in ocean chemistry. *Nature*, **299**, 814–815.

Riding, R. (2000) Microbial carbonates: the geological record of calcified bacterial-algal mats and biofilms. *Sedimentology*, **47** (Suppl. 1), 179–214.

Riding, R. (2011) Microbialites, stromatolites and thrombolites. In: *Encyclopedia of Geobiology* (Eds **J. Reitner** and **V. Thiel**), pp. 635–654, Springer, Berlin.

Riding, R. (1977) Skeletal stromatolites. In: *Fossil algae: recent results and developments* (Ed. **E. Flügel**), pp. 57–60, Springer-Verlag, Berlin.

Riding, R. (1999) The term stromatolite: towards an essential definition. *Lethaia*, **32**, 321–330.

Riding, R., Awramik, S.M., Winsborough, B.M., Griffin, K.M. and **Dill, R.F.** (1991a) Bahamian giant stromatolites: microbial composition of surface mats. *Geol. Mag.*, **128**, 227–234.

Riding, R., Braga, J.C. and **Martín, J.M.** (1991b) Oolite stromatoiltes and thrombolites, Miocene, Spain: analogues of Recent giant Bahamian examples. *Sed. Geol.*, **71**, 121–127.

Riding, R. and **Liang, L.** (2005) Seawater chemistry control of marine limestone accumulation over the past 550 million years. *Rev. Esp. Micropaleontol.*, **37**, 1–11.

Riding, R. and **Tomás, S.** (2006) Stromatolite reef crusts, Early Cretaceous, Spain: bacterial origin of *in situ*-precipitated peloid microspar? *Sedimentology*, **53**, 23–34.

Rodríguez-Martínez, M., Sánchez, F., Walliser, E.O. and **Reitner, J.** (2012) An Upper Turonian fine-grained shallow marine stromatolite bed from the Muñecas Formation, Northern Iberian Ranges, Spain. *Sed. Geol.*, **263–264**, 96–108.

Rosales, I. (1999) Controls on carbonate-platform evolution on active fault blocks: The Lower Cretaceous Castro Urdiales platform (Aptian-Albian, Northern Spain). *J. Sed. Res.*, **69**, 447–465.

Runnel, D.D. (1969) Diagenesis, chemical sediments and the mixing of natural waters. *J. Sed. Petrol.*, **39**, 1188–1201.

Salas, R., Guimerà, J., Mas, R., Martín-Closas, C., Meléndez, A. and Alonso, A. (2001) Evolution of the Mesozoic Central Iberian Rift System and its Cainozoic inversion (Iberian chain). In: *Peri-Tethys Memoir 6: Peri-Tethyan Rift/Wrench Basins and Passive Margins* (Eds P.A. Ziegler, W. Cavazza, A.H.F. Robertson and S. Crasquin-Soleau), *Mém. Mus. Natn. Hist. Nat.*, **186**, 145–185.

Schäfer, A. and Stapf, K.R. (1978) Permian Saar-Nahe Basin and Recent Lake Constance (Germany): two environments of lacustrine algal carbonates. In: *Modern and Ancient Lake Sedimentes* (Eds A. Matter and M.E. Tucker), *Int. Assoc. Sedimentol. Spec. Publ.*, **2**, 83–107.

Seard, C., Camoin, G., Rouchy, J.M. and Virgone, A. (2013) Composition, structure and evolution of a lacustrine carbonate margin dominated by microbialites: Case study from the Green River formation (Eocene; Wyoming, USA). *Palaeogeogr. Palaeoclimatol. Palaeoecol.*, **381–382**, 128–144.

Smart, P.L., Dawans, J.M. and Whitaker, F. (1988) Carbonate dissolution in a modern mixing zone. *Nature*, **335**, 811–813.

Smith, A.M. and Mason, T.R. (1991) Pleistocene, multiple-growth, lacustrine oncoids from the Poacher's Point Formation, Etosha Pan, northern Namibia. *Sedimentology*, **38**, 591–599.

Spadafora, A. Perri, E., Mckenzie, J.A. and Vasconcelos C. (2010) Microbial biomineralization processes forming modern Ca:Mg carbonate stromatolites. *Sedimentology*, **57**, 27–40.

Suarez-Gonzalez, P., Martin-Closas, C., Quijada, I.E., Benito, M.I. and Mas, J.R. (2012) Calcarous algae (dasycladales and charophytes), essential for the sedimentological interpretation of ancient water-bodies systems. The Barremian-Aptian Leza Fm, Cameros Basin, N Spain. *Abstracts, 29th IAS Meeting of Sedimentology*, Schladming, Austria, p. 467.

Suarez-Gonzalez, P., Quijada, I.E., Benito, M.I. and Mas, J.R. (2013) Eustatic versus tectonic control in an intraplate rift basin (Leza Fm, Cameros Basin). Chronostratigraphic and paleogeographic implications for the Aptian of Iberia. *J. Iberian Geol.*, **39**, 89–114.

Suarez-Gonzalez, P., Quijada, I.E., Benito, M.I., Mas, R., Merinero Palomares, R. and Riding, R. (2014) Origin and significance of lamination in Early Cretaceous stromatoloites and proposal for a quantitative approach. *Sed. Geol*, **300**, 11–27.

Suarez-Gonzalez, P., Quijada, I.E., Benito, M.I. and Mas, J.R. (2015) Sedimentology of ancient coastal wetlands: Insights from a Cretaceous multifaceted depositional system. *J. Sed. Res.*, **85**, 95–117.

Suarez-Gonzalez, P., Quijada, I.E., Benito, M.I., Mas, J.R. and Omodeo-Salé, S. (2011) Textbook example of tectonically controlled carbonate sedimentation at the active margin of a rift basin: the Leza Fm (Early Cretaceous, Cameros Basin, Spain). In: *Abstracts, 28th IAS Meeting of Sedimentology, Zaragoza, Spain* (Eds B. Bádenas, M. Aurell and A.M. Alonso-Zarza), p. 455.

Suarez-Gonzalez, P., Quijada, I.E., Mas, J.R. and Benito, M.I. (2010) Nuevas aportaciones sobre la influencia marina y la edad de los carbonatos de la Fm Leza en el sector de Préjano (SE de La Rioja). Cretácico Inferior, Cuenca de Cameros. *Geogaceta*, **49**, 7–10.

Surdam, R.C. and Wray, J.L. (1976) Lacustrine stromatolites, Eocene Green River Formation, Wyoming. In: *Stromatolites* (Ed. M.R. Walter), *Dev. Sedimentol.*, **20**, pp. 535–541, Elsevier, Amsterdam.

Tewari, V.C. and Seckbach, J. (2011) Stromatolites: interaction of microbes with sediments. Springer, New York, 752 pp.

Turner, E.C., James, N.P. and Narbonne, G.M. (2000) Taphonomic control on microstructure in early Neoproterozoic reefal stromatolites and thrombolites. *Palaios*, **15**, 87–111.

Visscher, P.T., Reid, R.P. and Bebout, B.M. (2000) Microscale observations of sulfate reduction: Correlation of microbial activity with lithified micritic laminae in modern marine stromatolites. *Geology*, **28**, 919–922.

Visscher, P.T., Reid, R.P., Bebout, B.M., Hoeft, S.E., Macintyre, I.G. and Thompson, J.A. (1998) Formation of lithified micritic laminae in modern marine stromatolites (Bahamas): The role of sulfur cycling. *Am. Mineral.*, **83**, 1482–1493.

Waterkeyn, A., Grillas, P., Vanschoenwinkel, B. and Brendonck, L. (2008): Invertebrate community patterns in Mediterranean temporary wetlands along hydroperiod and salinity gradients. *Freshwater Biol.*, **53**, 1808–1822.

Whigham, D.F., Baldwin, A.H. and Barendregt, A. (2009) Tidal freshwater wetlands. In: *Coastal Wetlands. An integrated ecosystem approach* (Eds G.M.E. Perillo, E. Wolanski, D.R. Cahoon and M.M. Brinson), pp. 515–533. Elsevier, Amsterdam.

Wolanski, E., Brinson, M.M., Cahoon, D.R. and Perillo G.M.E. (2009) Coastal wetlands: A synthesis. In: *Coastal Wetlands. An integrated ecosystem approach* (Eds G.M.E. Perillo, E. Wolanski, D.R. Cahoon and M.M. Brinson), pp. 1–62. Elsevier, Amsterdam.

Woods, A.D. (2013) Microbial ooids and cortoids from the Lower Triassic (Spathian) Virgin Limestone, Nevada, USA: Evidence for an Early Triassic microbial bloom in shallow depositional environments. *Global Planet. Change*, **105**, 91–101.

Zamarreño, I. (1977) Peritidal origin of Cambrian carbonates in Northwest Spain. In: *Tidal deposits: a casebook of recent examples and fossil counterparts* (Ed. R.N. Ginsburg), pp. 289–298. Springer-Verlag, Berlin.

Angular and tangential toeset geometry in tidal cross-strata: An additional feature of current-modulated deposits

DOMENICO CHIARELLA*

* Pure E&P Norway AS, Grundingen 3, N-0250 Oslo (Norway)
e-mail: domenico.chiarella@pure-ep.no

ABSTRACT

Cyclicity is one of the main characteristics used to recognise tidal current-modulation in clastic deposits. Usually, cyclicity detected in cross-strata consists of sand-mud or siliciclastic-bioclastic couplets, thick-thin bundles or lateral and vertical thickness variation of rhythmites. The aim of this paper is to direct attention to the cyclicity showed by the toeset geometry, which has not been previously emphasized. Recognition of the toeset geometry can be useful in sand-rich cross-strata when other tidal sedimentary signals are faint or ambiguous. The lower Pleistocene cross-strata of the Catanzaro Strait contain a wide suite of sedimentary structures supporting tidal origin. In particular, sand-rich cross-strata of the Pianopoli Unit are prone to record cyclicity in toesets geometry. Especially along the forward migration path of the bedform, toesets can be angular or tangential in shape, related to the cyclic increase and decrease of flow velocities during one tidal cycle.

Keywords: Angular toeset, Tangential toeset, Tidal cross-strata, Neap-Spring cycle, Catanzaro Strait.

INTRODUCTION

Tides have captivated attention, probably for their cyclic change that occurs in an orderly fashion. From a geological point of view, considering the positive correspondence between tidal-signal and basin geometry, modern and ancient tidal deposits have been extensively documented (e.g. Tessier & Gigot, 1989; Dalrymple, 1992; Tessier, 1993; Anastas et al., 1997; Alexander et al., 1998; Pontén & Plink-Björklund, 2007; Chiarella & Longhitano, 2012; Chiarella et al., 2012b; Davis & Dalrymple, 2012; Longhitano et al., 2012c; Olariu et al., 2012a; Reynaud et al., 2013; Longhitano et al., 2014). Moreover, recognition of tidal deposits is not only an important element of palaeo-environment reconstruction but it is also an important target in terms of hydrocarbon-reservoir potential (Longhitano, 2013; Reynaud et al., 2013; Messina et al., 2014).

Diagnostic criteria to recognise tidal deposits have been studied extensively (e.g. De Raaf & Boesma, 1971; Allen, 1980; Boersma & Terwindt, 1981; Kreisa & Moiola, 1986; Nio & Yang, 1991; Flemming & Bartholomä, 1995; Longhitano et al., 2012c and references therein). Common tide-diagnostic features in sedimentary deposits include bidirectional cross-strata, reactivation surfaces, tidal rhythmites and tidal bundles (de Boer et al., 1989; Nio & Yang, 1991, Dalrymple, 1992; Longhitano et al., 2012a). In addition to these well-established criteria for recognizing tidal influence in sedimentary deposits, less attention has been paid to the geometry of cross-strata. In particular, the toeset geometry variations can be related to the cyclical variation in the current competence (Jopling, 1965; Sato et al., 2006), as well as tidal-rhythmites.

In this paper, the angular versus tangential toeset geometry of cross-strata is documented as a diagnostic criterion to recognise tide-generated sedimentary structures. This criterion can be used with the already existing recognition criteria for tidal deposits to support a tidal interpretation.

Cross-stratified 2D cross-strata of the Catanzaro succession were chosen for this study because

Contributions to Modern and Ancient Tidal Sedimentology: Proceedings of the Tidalites 2012 Conference, First Edition. Edited by Bernadette Tessier and Jean-Yves Reynaud.
© 2016 International Association of Sedimentologists. Published 2016 by John Wiley & Sons, Ltd.

192   D. Chiarella

they developed under tidal currents (Chiarella, 2011; Longhitano, 2011; Longhitano *et al.*, 2012b; Longhitano *et al.*, 2014). In addition, the 3D exposure of deposits permits researchers to fully investigate the geometry of cross-strata. This excludes the possibility that the different observed toeset geometries can be related to a particular outcrop exposure effect.

GEOLOGICAL SETTING

The dataset selected for this study is exposed in natural outcrops and quarries in a lower Pleistocene strait of the Calabrian Arc (Fig. 1). The

Calabrian Arc is a fragment of continental crust of the western Mediterranean that rifted off the southern margin of the European plate and drifted south-eastwards (Oligocene to Present) to collide with the African plate (Malinverno & Ryan, 1986; Doglioni, 1991; Gueguen *et al.*, 1997; Cello & Mazzoli, 1999; Critelli, 1999).

During the south-eastward migration of the Calabrian Arc terranes, the interaction between subducting oceanic crust, collisional processes and back-arc extension suggest the presence of lithospheric discontinuities bordering the Ionian slab, which result in the different behaviour of the foreland due to its different rheological properties (Finetti & Del Ben, 1986; Finetti, 2005 and

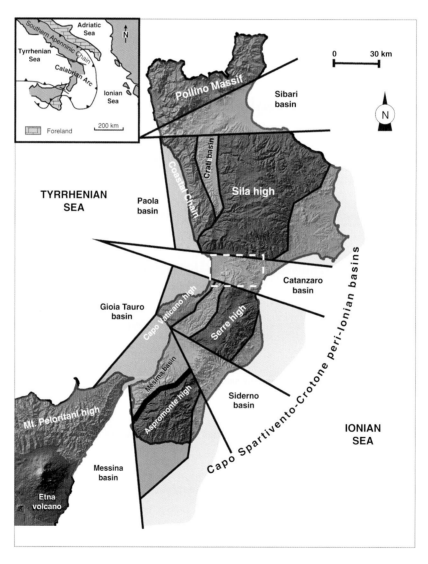

Fig. 1. Late Pliocene-Early Quaternary block-segmentation of the Calabrian Arc (modified after Ghisetti, 1979). The main sedimentary basins are in yellow. Data for the present study comes from the Catanzaro basin (dashed white square).

references therein). According to Del Ben *et al.* (2008) these discontinuities have produced regional strike-slip systems. These strike-slip systems, responsible for the current arcuate shape of the Calabrian Arc, dissect the chain in several blocks, allowing to the opening of the Catanzaro Basin and other similar basins (Fig. 1). The Catanzaro area, during the early Pleistocene (Calabrian) experienced a palaeogeographic configuration of a strait, where persistent tidal currents controlled the distribution of sediments (Chiarella, 2011; Longhitano, 2011; Longhitano *et al.*, 2012c). In particular, the Pianopoli Unit (Chiarella *et al.*, 2012a; Longhitano *et al.*, 2012b; Longhitano *et al.*, 2014) of the Catanzaro mixed deposits consists of sandy-rich 2D cross-strata accumulated in a current modulated setting. The present-day Catanzaro Basin has an area ~30 km in length and 9 to 13 km in width. This basin configuration is a result of active tectonics, with persistent reactivation of normal and left-lateral faults (Tansi *et al.*, 2007).

Lower Pleistocene Catanzaro deposits

The Lower Pleistocene (Calabrian) Catanzaro succession (up to 100 m-thick) in the Catanzaro Basin records accumulation of tidal deposits composed by siliciclastic sediments derived from a metamorphic basement in the Sila and Serre Massifs (Fig. 2A) and bioclastic sediments derived from an *in situ* carbonate factory (Chiarella, 2011; Longhitano *et al.*, 2014). The lower Pleistocene cross-stratified deposits occur above mudstone and marls (Fig. 2B) considered time-correlative with the Trubi Formation and representing the stratigraphic record of the transgressive event that ~5.3 Ma ago re-established open-marine conditions in the Mediterranean area after the Messinian salinity crisis (Krijgsman *et al.*, 1999; Bache *et al.*, 2012).

The cross-stratified deposits, in turn, are overlain by highly bioturbated fine-grained sandstone, siltstone and claystone with abundant articulated bivalves and ichnofauna (Chiarella, 2011).

The cross-stratified deposits contain a broad suite of tidally generated sedimentary structures (Chiarella, 2011; Longhitano *et al.*, 2014). The related bedforms consist of a series of stacked bidirectional large-scale tidal dunes (0.4 to 6 m-thick) of mixed siliciclastic-bioclastic composition that accumulated in a narrow basin (Chiarella, 2011).

According to the internal organization, the cross-stratified interval can be divided into two

laterally-stacked and vertically-stacked stratal units (Fig. 2B), the Vena di Maida and the Pianopoli Units (Chiarella *et al.*, 2012a; Longhitano *et al.*, 2014). The Vena di Maida Unit (up to 30 m-thick) is composed of mixed siliciclastic-bioclastic deposits organised into vertically-stacked 3D cross-strata. The ~50 m of Pianopoli Unit (Fig. 2C) which is the focus of this study is composed mainly of siliciclastic sandstones and is weakly fossiliferous. The Pianopoli Unit deposits occur as aggrading vertically-stacked bidirectional 2D cross-strata.

A tidal origin for these deposits has been extensively documented by several papers and it is indicated by bimodal palaeocurrent pattern, reactivation surfaces and tidal rhythmites composed by mixed siliciclastic-bioclastic couplets (Chiarella, 2011; Longhitano, 2011; Chiarella *et al.*, 2012a; Longhitano *et al.*, 2012b; Longhitano *et al.*, 2012c; Longhitano, 2013; Longhitano *et al.*, 2014).

ANGULAR-TANGENTIAL TOESET GEOMETRIES

Tide-generated sedimentary structures have been documented extensively (Allen, 1980; Visser, 1980; Boersma & Terwindt, 1981; Kreisa & Moiola, 1986; Nio & Yang, 1991; Longhitano *et al.*, 2012a). Among them, less attention has been paid to the different characters pertaining of the toeset area of cross-strata (Boerma & Terwindt, 1981; Kohsiek & Terwindt, 1981; Martinius & Gowland, 2011). Some authors recognised that toesets appear to be particularly sensitive to current-modulation (Kohsiek & Terwindt, 1981; Martinius & Gowland, 2011). For example, Kohsiek & Terwindt (1981) describe changes in the internal angle of cross-strata related to changes in the current-velocity during a neap-spring cycle. Recently, Martinius & Gowland (2011) document change in foreset geometries from concave to convex stratification, interpreted to be caused by successively decreasing and increasing flood retardation in tide-influenced fluvial bedforms.

Description

In the Cavaliere section (Fig. 2C) a spectacular 3D exposure of 2D cross-strata (Pianopoli Unit) outcrops laterally for ~33 m with a height of ~15 m. Sediments consist of moderately-sorted

194   *D. Chiarella*

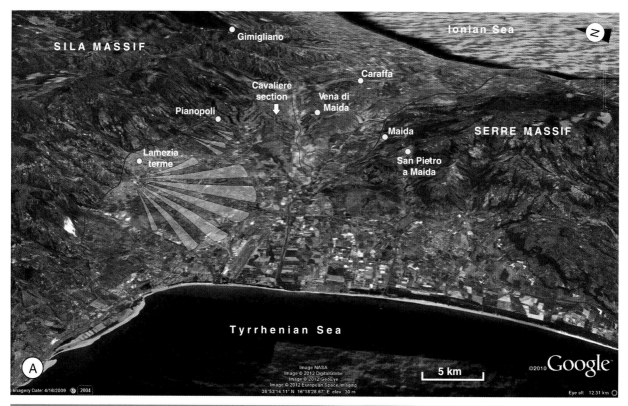

Fig. 2. A) 3D view of the Catanzaro Strait showing the areal distribution (yellow) of the lower Pleistocene cross-stratified deposits (Photos is from Google Earth™). B) Stratigraphic column of the Catanzaro basin. C) Photograph of tidal dunes (Pianopoli Unit) at the Cavalieri section. Person for scale is *ca* 1.8 m-tall.

to well-sorted medium-grained to coarse-grained sand with rare granules. A key feature of these 2D cross-strata is a marked difference in bedform morphology along the forward migration path of the bedform. In particular, systematic lateral variations of toeset geometries can be recognised in cross-strata (Fig. 3A). The bedform shows a cyclic alternation of angular toesets, followed by tangential toesets along the forward migration direction (Figs. 3B and 4). Angular toesets are defined here as consisting of straight foreset laminae with dip angles between 20 and 35 degrees, forming a sharp contact with the basal surface. Tangential toesets consist of foreset laminae with a dip angle between 5 and 20 degrees that merge gently into the basal surface. Generally, foresets characterized by toesets with tangential geometry are thicker than foresets with angular toesets geometry (Fig. 4A and B). Accordingly, foresets with tangential toesets migrate approximately twice the distance over the same time interval than foresets with angular toesets.

Interpretation

The repeated alternation of angular and tangential toesets described in this paper is comparable with bedform migration observations reported in Jopling (1963; 1964; 1965). Jopling (1963) suggests that bedform migration is influenced by variable velocity of currents and bed shear stress. For given sediment grain size and a constant height of the dune, an increasing velocity of flow (bed shear stress) generates a geometric transition from an angular to a tangential toeset shape. Accordingly, the resulting cyclic alternation between angular and tangential toeset geometry depends on (i) the proportion of the suspended *versus* bed load on the leeside of the dunes; (ii) eddies developed at the base of leeside of the dunes and (iii) associated strength of current (Kohsiek & Terwindt, 1981, Sato *et al.*, 2006). Thinner/angular and thicker/tangential foresets correspond to the neap-spring cycles (Fig. 4A and B) interpreted by Longhitano (2011) and Longhitano *et al.* (2012c) in these deposits.

Analogues

Internal architecture of cross-strata has been investigated in different tide-dominated and tide-influenced settings (Kreisa & Moiola, 1986; Tape *et al.*, 2003; Chiarella & Longhitano, 2012; Longhitano *et al.*, 2012c; Olariu *et al.*, 2012b). The recent study of the tide-influenced bedforms in the Lusitanian

Basin (Martinius & Gowland, 2010) shows a gradual transition in foreset geometry from convex to concave lamination along the forward migration path of the bedform (Fig. 5A). Martinius & Gowland (2010) recognise that (i) sections with convex lamination are characterized by angular toesets geometry and (ii) sections with concave lamination are characterized by low angle (tangential) toesets geometry. This cyclical alternation from concave to convex foreset geometry is interpreted as related to successive fluctuations in flow regime from lower to higher current velocities (Martinius & Gowland, 2010). Similar alternating neap-spring angular and tangential basal contacts of tidal cross-strata occur in other tidal formations but have not been highlighted (Fig. 5B, C and D). In these formations, the alternation of angular and tangential toesets overlaps the neap-spring cyclicity documented on the basis of other tidal signals. The widespread occurrence of cyclical alternation of angular and tangential toesets along the forward migration path of 2D tidal dunes supports the idea that alternation in toesets geometry is strictly related to current modulation during neap-spring cycles.

DISCUSSION

Most of the classic criteria for recognizing tidal deposits are based on observations related to the specific behaviours of heterolithic sediments (Nio & Yang, 1991; Longhitano, 2011 and references therein). However, tidal currents also occur in setting with abundant sandy sediments where the mud fraction is almost absent. Thus, many of the diagnostic criteria for tidal sedimentation are not directly applicable in sand-dominated deposits. For this reason, the recognition of tidal cyclicity from the observation of the toesets geometry along the forward migration path of the bedform could be very useful in such sand-dominated tidal systems with cross-strata lacking mud drapes, unless the available size range of sand grains is sufficiently wide as to allow grain size segregation in a given hydraulic regime of flow-power changes (Longhitano & Nemec, 2005).

Relationship between toeset geometries and others tidal signatures

Tidal signals are recognised in several sedimentary structures. In many cases, different tidal signatures overlap each other to build compound tidal sedimentary structures.

Fig. 3. A) Panoramic view of tidal cross-strata in the Catanzaro Strait (Cavalieri section). Arrow indicates the main palaeo-current direction. See cow on the bottom for scale. B) Close up view showing the down-current migration of foresets with tangential and angular toesets. Cross-strata are ~60 cm-thick.

The thickness variation recognised in tidal rhythmites, interpreted as related to neap-spring cycles, can also be detected in sand-dominated tidal deposits. Usually, thick foresets develop during the spring cycle and thin foresets during the neap cycle (Tape *et al.*, 2003; Olariu *et al.*, 2012b). Analogously, the rhythmical alternation of angular and tangential toesets geometry follows the thickness variation of foresets. In particular, foresets with angular toesets geometry are thinner than foresets with tangential toeset geometries.

Moreover, the segregation between the two heterolithic fractions recognised in mixed siliciclastic/bioclastic tidal dunes (Chiarella, 2011; Longhitano, 2011; Chiarella & Longhitano, 2012; Longhitano

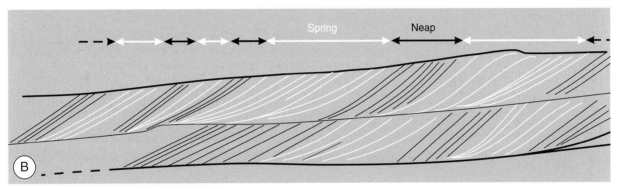

Fig. 4. A) Outcrop example of lateral variations in toeset shape in cross-strata (Cavalieri section). B) Line-drawing showing cyclic alternation of angular (black) and tangential (white) toesets. Alternation has been interpreted as related to neap-spring tidal periods.

et al., 2012c) is affected by the current modulation (i.e. different flow regime) controlling the development of angular *versus* tangential toeset geometries. The result is that during spring period (high flow regime) the segregation ratio is lower than during neap period (low flow regime). Following the classification proposed by Chiarella & Longhitano (2012), segregation is good in foresets with angular toesets geometry (low flow regime) and moderate in foreset characterized by tangential toesets geometry (high flow regime).

Current modulation and toeset geometry

Fluctuations in the strength of tidal currents are reflected in cyclic variability in the toeset geometry of cross-strata. Cyclic change between angular and tangential toeset geometry along the forward migration path of the bedform have been interpreted as related to current modulation during a tidal cycle. In particular, cyclic alternance of toeset geometry from angular to tangential and vice versa is interpreted as a response to neap/spring tides. The neap tide produces an interval characterized

by angular toeset geometry and the spring tide the tangential toeset geometry (Fig. 4A and B).

At low velocities (neap tide) of flow, the sediment particles move along the topset and settle on the foreset. The rate of sediment transport is minimal and transport takes place mainly by bedload movement. Sediment therefore accumulates on the upper foreset where it is distributed downslope by gravity (Fig. 6A). The foreset front is essentially a planar slip-face that abuts the basal surface with an angular contact (Jopling, 1965; Harms *et al.*, 1975). When the strength of the tidal current is at a maximum, a greater proportion of sediment is taken into suspension and carried beyond the front of the cross-strata. Some of these sediments accumulate in the toeset sector at the base of the foreset slope and the contact is tangential (Fig. 6B). Deposition of suspended sediment in the toe sector and the reworking of the basal slope by eddy action are the primary causes of tangential toesets formation (Jopling, 1965; Harms *et al.*, 1975; Kostaschuk *et al.*, 2008).

The cyclic alternation between angular and tangential toeset geometry (Fig. 4B) might thus

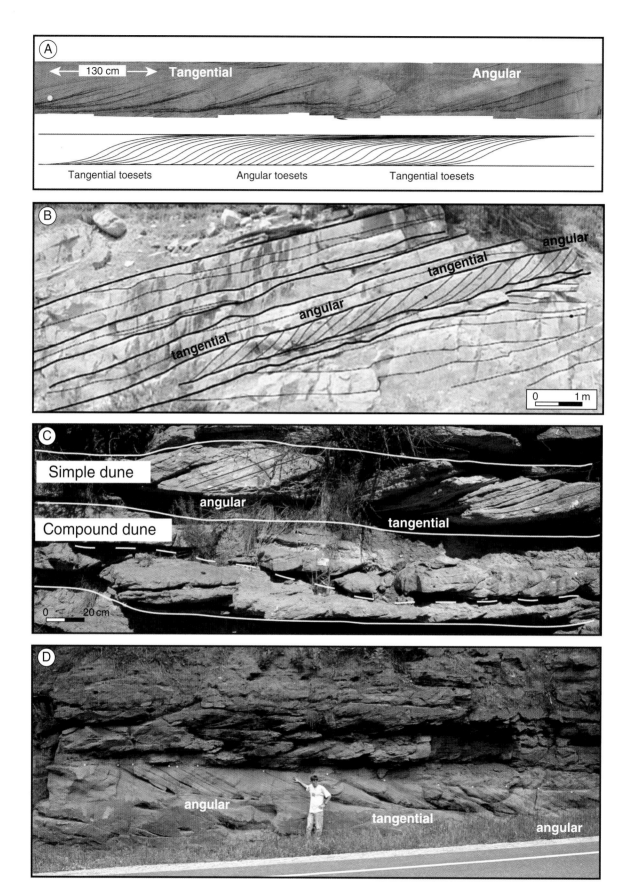

Fig. 5. Angular and tangential toeset geometries in ancient tidal deposits. A) Fluvial dunes of the Lusitanian basin (Upper Jurassic, Portugal) affected by tidal modulation (modified from Martinius & Gowland, 2011, reprinted by permission of John Wiley and Sons, whose permission is required for further use). B) Tide-dominated 2D cross-stratified sandstone of the Lower Cambrian Cog Group, Canada (modified from Desjardins *et al.*, 2012, reprinted by permission of John Wiley and Sons, whose permission is required for further use). C) 2D dunes pertaining to the Lower Eocene Esdolomada tidal bar, Spain (modified from Olariu *et al.* 2012b, reprinted by permission of John Wiley and Sons, whose permission is required for further use). D) Tabular (2D) cross-bed set of the Upper Cambrian Jordan Sandstone, Minnesota (modified from Tape *et al.*, 2003, reprinted by permission from SEPM whose permission is required for further use).

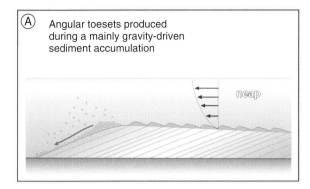

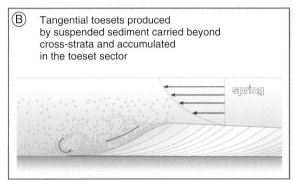

Fig. 6. Model for the deposition of cross-strata under tidally modulated current. A) During the neap phases sediments are mainly transported as bed-load on the topset of cross-strata and accumulated on the slip-face by gravity producing angular toeset geometries. B) During the spring phases sediment is kept in suspension on the front area of foresets permitting sediments to accumulate in the toeset sector. Moreover, in the toeset area eddies occur reworking sediments and generating tangential toeset geometries. Foreset profile is exaggerated to better highlight the tangentiality.

represent, respectively, deposition under the minimum and maximum hydraulic energy of a tidally modulated current.

Although much effort has been devoted to accounting for, or predicting, the sediment transport rate, no single formula has gained a universal acceptance and the normal procedures involve the use of reference graphs in addition to mathematical equations. However, several authors (Le Mehaute & Hans, 1990; Nio & Yang, 1991; Van Rijn, 2005) state that the sand transport rate is related to the 3rd or 4th power of the current velocity. This means that a relatively minor change in current velocity can bring about major change in sediment transport rate. Changes in dominant current velocity during a neap-spring cycle are therefore recorded in the toeset geometry variation of successive bundles. The short-term cyclic repetition of these conditions resulted in the alternation of angular and tangential toeset geometry within migrating cross-strata

(Fig. 4). The same type of cyclicity is observable in other types of tidal rhythmites recording neap-spring tidal cyclicity.

A similar style of toeset geometry (angular and tangential) can be caused by the change in the direction of a 3D dune progradation and/or outcrop exposure. However, instead of cyclical alternation between angular and tangential toesets along the forward migration path of the bedform, change in the direction of 3D dunes could develop a local variation from angular to tangential toesets geometry or *vice versa*. These toesets variation will tend to be non-cyclical and not in accordance with others tidal signal (e.g. foresets thickness variation and/or heterolithic segregation), as they depend on the local conditions and outcrop cut, rather than on the astronomical cycles or climatic seasonality alone. In the present study, the possibility to investigate the outcrops in a 3D view and the occurrence of other tidal signals avoids the possibility of misinterpretation.

CONCLUSIONS

Being able to disentangle neap-spring cycle in the rock record is most important in recognizing tidal generated deposits.

This study shows the occurrence of angular and tangential toeset geometry cyclic alternation along the same cross-strata in tide-dominated deposits of the Catanzaro Strait. This specific pattern has been interpreted to be controlled by tidal currents cyclicity. Foresets with angular toesets indicate low flow velocities (lower flow regime) with bedload transport and grain-flow dominating on the slip-face. The tangentially shaped toesets are formed when the current velocity increases, caused by settling of sediments from suspension as bottomsets and toesets. Then, cyclical changes in toeset geometry can be used and tested to recognise tidal environments and facies where other tidal sedimentary signals are faint or ambiguous.

ACKNOWLEDGEMENTS

The basic idea benefited from stimulating discussion about tidal deposits with Sergio Longhitano and Wojciech Nemec. Thanks also to Grazia, Elena and Giuseppe who stimulated me every day. The manuscript was critically reviewed by Piret Plink-Bjorklund, Cornel Olariu and the volume editor

Jean-Yves Reynaud, whose helpful comments are much appreciated by the author. Clinton A. Cowan, Patricio R. Desjardins, Allard A. Martinius and Mariana I. Olariu are acknowledged for providing pictures for Fig. 5. The author would also like to thanks Stephen Eaton for checking the quality of the English language.

REFERENCES

Alexander, C.R., Davis, R.A. and **Henry, V.J.** (1998) Tidalites: Processes and Products. *SEPM Spec. Publ.*, **61**, 171 pp.

Allen, J.R.L. (1980) Sandwaves: a model of origin and internal structure. *Sed. Geol.*, **26**, 281–328.

Anastas, A.S., Dalrymple, R.W., James, N.P. and **Nelson, C.S.** (1997) Cross-stratified calcarenites from New Zealand: subaqueous dunes in a cool-water, Oligo-Miocene seaway. *Sedimentology*, **44**, 5, 869–891.

Bache, F., Popescu, S-P., Rabineau, M., Gorini, C., Suc, J-P., Clauzon, G., Olivet, J-L., Rubino, J-L., Melinte-Dobrinescu, M.C., Estrada, F., Londeix, L., Armijo, R., Meyer, B., Jolivet, L., Jouannic, G., Leroux, E., Aslanian, D., Dos Reis, A.T., Mocochain, L., Dumurdžanov, N., Zagorchev, I., Lesić, V., Tomić, D., Çağatay, M.N., Brun, J-P., Sokoutis, D., Csato, I., Ucarkus., G. and **Çakır, Z.** (2012) A two-step process for the reflooding of the Mediterranean after the Messinian Salinity Crisis. *Basin. Res.*, **24**, 125–153.

Boersma, J.R. and **Terwindt, J.H.J.** (1981) Neap–spring tide sequences of intertidal shoal deposits in a mesotidal estuary. *Sedimentology*, **28**, 151–170.

Cello, G. and **Mazzoli, S.** (1999) Apennine tectonics in southern Italy: a review. *J. Geodyn.*, **27**, 191–211.

Chiarella, D. (2011) *Sedimentology of Pliocene–Pleistocene mixed (lithoclastic–bioclastic) deposits in southern Italy (Lucanian Apennine and Calabrian Arc): depositional processes and palaeogeographic frameworks.* Ph.D. Thesis, Univ. of Basilicata, Potenza, 216 pp.

Chiarella, D. and **Longhitano, S.G.** (2012) Distinguishing depositional environments in shallow water mixed, bio-siliciclastic deposits on the base of the degree of heterolithic segregation (Gelasian, southern Italy). *J. Sed. Res.*, **82**, 969–990.

Chiarella, D., Longhitano, S.G. and **Muto, F.** (2012a) Sedimentary features of the Lower Pleistocene mixed siliciclastic-bioclastic tidal deposits of the Catanzaro Strait (Calabrian Arc, south Italy). *Rend. Online Soc. Geol. It.*, **21**, 919–920.

Chiarella, D., Longhitano, S.G., Sabato, L. and **Tropeano M.** (2012b) Sedimentology and hydrodynamics of mixed (siliciclastic–bioclastic) shallow-marine deposits of Acerenza (Pliocene, Southern Apennines, Italy). *Boll. Soc. Geol. Ital.*, **131**, 136–151.

Critelli, S. (1999) The interplay of lithospheric flexure and thrust accommodation in forming stratigraphic sequences in the southern Apennines foreland basin system, Italy. *Rend. Accad. Nazion. Lincei*, **9**, 257–326.

Dalrymple, R.W. (1992) Tidal depositional system. In: *Facies Models* (Eds **R.G. Walker** and **N.P. James**). Geological Association of Canada, St John's, Newfoundland, Canada, 195–218.

Davis, R.A. and **Dalrymple, R.W.** (2012) *Principles of tidal sedimentology.* Springer, Berlin, 621 pp.

De Boer, P.L., Oost, A.P. and **Visser, M.J.** (1989) The diurnal inequality of the tide as a parameter for recognising tidal influences. *J. Sed. Petr.*, **59**, 912–921.

Del Ben, A. (2008) Strike-slip systems as the main tectonic features in the Plio-Quaternary kinematics of the Calabrian Arc. *Mar. Geophys. Res.*, **29**, 1–12.

De Raaf, J.F.M. and **Borsma, J.R.** (1971) Tidal deposits and their depositional structures. *Geol. Mijnbouw*, **50**, 479–504.

Desjardins, P.R., Buatois, L.U., Pratt, B.R. and **Gabriela Màngano, M.** (2012) Sedimentological–ichnological model for tide-dominated shelf sandbodies: Lower Cambrian Gog Group of western Canada. *Sedimentology*, **59**, 1452–1477.

Doglioni, C. (1991) A proposal of kinematic modelling for Wdipping subductions - Possible applications to the Tyrrhenian-Apennine system. *Terra Nova*, **3**, 423–434.

Finetti, I.R. (2005) *CROP Project: deep seismic exploration of the Central Mediterranean and Italy.* Atlases in Geoscience, Elsevier, v. **1**.

Finetti, I.R. and **Del Ben, A.** (1986) Geophysical study of the Tyrrhenian opening. *Boll. Geofis. Teor. Appl.*, **28**, 75–155.

Flemming, B.W. and **Bartholomä, A.** (1995) Tidal signature in modern and ancient sediments. *Int. Assoc. Sedimentol., Spec. Publ.*, **24**, 358 pp.

Ghisetti, F. (1979) Evoluzione neotettonica dei principali sistemi di faglie della Calabria centrale: *Boll. Soc. Geol. Ital.*, **98**, 387–430.

Gueguen, E., Doglioni, C. and **Fernandez, M.** (1997) Lithospheric boudinage in the Western Mediterranean back-arc basin. *Terra Nova*, **9**, 184–187.

Harms, J.C., Southard, J.B., Spearing, D.R. and **Walker, R.G.** (1975) *Depositional environments as interpreted from primary sedimentary structures and stratification sequences. SEPM*, Short Course **2**, Dallas, Texas, 161 pp.

Jopling, A.V. (1963) Hydraulic studies on the origin of bedding. *Sedimentology*, **2**, 115–121.

Jopling, A.V. (1964) Laboratory study of sorting processes related to flow separation. *J. Geophys. Res.*, **69**, 3403–3418.

Jopling, A.V. (1965) Hydraulic factors controlling the shape of the laminae in laboratory deltas. *J. Sed. Petr.*, **35**, p. 777–791.

Kohsiek L.H.M. and **Terwindt J.H.J.** (1981) Characteristics of foreset and topset bedding in megaripples related to hydrodynamic conditions on an intertidal shoal. In: *Holocene marine sedimentation in the North Sea Basin* (Eds S.D. Nio, R.T.E. Shüttenhelm and Tj.C.E. Van Weering), *Int. Assoc. Sedimentol. Spec. Publ.*, **5**, 27–37.

Kostaschuk, R., Shugar, D., Best, J.L., Parsons, D.R., Lane, S.N., Hardy, R.J. and **Orfeo, O.** (2008) Suspended sediment transport over a dune. In: *Marine and River Dune Dynamics*, 1–3 April 2008, Leeds.

Kreisa, R.D. and **Moiola, R.J.** (1986) Sigmoidal tidal bundles and other tide-generated sedimentary structures of the Curtis Formation, Utah. *Geol. Soc. Am. Bull.*, **97**, 381–387.

Krijgsman, **W.**, **Hilgen**, **F.J.**, **Raffi**, **J.**, **Sierro**, **F.J.** and **Wilson**, **D.S.** (1999) Chronology, causes and progression of the Messinian salinity crisis. *Nature*, **400**, 652–655.

Le Mehaute, **B.** and **Hanes**, **D.M.** (1990) *Ocean Engineering Science*. Harvard University Press, 1334 pp.

Longhitano, **S.G.** (2011) The record of tidal cycles in mixed silici-bioclastic deposits: examples from small Plio-Pleistocene peripheral basins of the microtidal central Mediterranean Sea. *Sedimentology*, **58**, 691–719.

Longhitano, **S.G.** (2013) A facies-based depositional model for ancient and modern, tectonically-confined tidal straits. *Terra Nova*, **25**, 446–452.

Longhitano, **S.G.** and **Nemec**, **W.** (2005) Statistical analysis of bed-thickness variation in a Tortonian succession of biocalcarenitic tidal dunes, Amantea Basin, Calabria, southern Italy. *Sed. Geol.*, **179**, 195–224.

Longhitano, **S.G.**, **Chiarella**, **D.** and **Muto**, **F.** (2014) Three-dimensional to two-dimensional cross-strata transition in the lower Pleistocene Catanzaro tidal strait transgressive succession (southern Italy). *Sedimentology*, DOI: 10.1111/sed.12138.

Longhitano, **S.G.**, **Mellere**, **D.**, **Steel**, **R.J.** and **Ainsworth**, **R.B.** (2012a) Tidal depositional systems in the rock record: A review and new insights, *Sed. Geol.*, **279**, 2–22.

Longhitano, **S.G.**, **Sabato**, **L.**, **Tropeano**, **M.** and **Gallicchio**, **S.** (2010) A mixed bioclastic/siliciclastic flood-tidal delta in a microtidal setting: depositional architectures and hierarchical internal organization (Pliocene, Southern Apennine, Italy). *J. Sed. Res.*, **80**, 36–53.

Longhitano, **S.G.**, **Zecchin**, **M.**, **Chiarella**, **D.**, **Prosser G.** and **Muto**, **F.** (2012b) Tectonics and 5 Sedimentation in Neogene-to-Quaternary Sedimentary Basins of Central Calabria (South 6 Italy). A 4-days-long Field Course for Petroleum Geologists and Geophysicists, 62 p.

Longhitano, **S.G.**, **Chiarella**, **D.**, **Di Stefano**, **A.**, **Messina**, **C.**, **Sabato**, **L.** and **Tropeano**, **M.** (2012c) Tidal signatures in Neogene to Quaternary mixed deposits of southern Italy straits and bays. *Sed. Geol.*, **279**, 74–96.

Malinverno, **A.** and **Ryan**, **W.B.F.** (1986) Extension in the Tyrrhenian Sea and shortening in the Apennines as result of arc migration driven by sinking of the lithosphere. *Tectonics*, **5**, 227–245.

Martinius, **A.W.** and **Gowland**, **S.** (2011) Tide-influenced fluvial bedforms and tidal bore deposits (Late Jurassic Lourinhã Formation, Luisitanian Basin, Western Portugal). *Sedimentology*, **58**, 285–324.

Messina, **C.**, **Nemec**, **W.**, **Martinius**, **A.** and **Elfenbein.**, **C.** (2014) The Garn Formation (Bajocian-Bathonian) in the Kristin Field, Halten Terrace: its origin, facies architecture and primary heterogeneity model. In: *From Depositional Systems to Sedimentary Successions on the Norwegian Continental Shelf* (Eds A.W. Martinius, J. Howell, T. Olsen, R. Ravnås, R.J. Steel and J. Wonham). *Int. Assoc. Sedimentol. Spec. Publ.*, **46**, 513–550.

Nio, **S.D.** and **Yang**, **C.S.** (1991) Diagnostic attributes of clastic tidal deposits: a review. In: *Clastic Tidal Sedimentology* (Eds D.G. Smith, G.E. Reinson, B.A. Zaitlin and R.A. Rahmani). *Can. Soc. Petrol. Geol. Mem.*, **16**, 3–28.

Olariu, **C.**, **Steel**, **R.J.**, **Dalrymple**, **R.W.** and **Gingras**, **M.K.** (2012a) Tidal dunes versus tidal bars: The sedimentological and architectural characteristics of compound dunes in a tidal seaway, the lower Baronia Sandstone (Lower Eocene), Ager Basin, Spain. *Sed. Geo.*, **279**, 134–155.

Olariu, **I.M.**, **Olariu**, **C.**, **Steel**, **R.J.**, **Dalrymple**, **R.W** and **Martinius**, **A.W.** (2012b) Anatomy of a laterally migrating tidal bar in front of a delta system: Esdolomada Member, Roda Formation, Tremp-Graus Basin, Spain. *Sedimentology*, **59**, 356–378.

Pontén, **A.** and **Plink-Björklund**, **P.** (2007) Depositional environments in an extensive tide influenced delta plain, Middle Devonian Gauja Formation, Devonian Baltic Basin. *Sedimentology*, **54**, 969–1006.

Reynaud, **J.V.**, **Ferrandini**, **M.**, **Ferrandini**, **J.**, **Santiago**, **M.**, **Thinon**, **I.**, **Andrè**, **P.**, **Barthet**, **Y.**, **Guennoc**, **P.** and **Tessier**, **B.** (2013) From non-tidal shelf to tide-dominated strait: The Miocene Bonifacio Basin, Southern Corsica. *Sedimentology*, **60**, 599–623.

Sato, **T.**, **Masuda**, **F.** and **Nishio**, **T.** (2006) Estimation of the depositional time by comparison of foreset lamina cross-stratification angles and ocean current fluctuations: The Pleistocene Ichijiku Formation, Kazusa Group. *Mar. Geol.*, **235**, 241–245.

Tansi, **C.**, **Muto**, **F.**, **Critelli**, **S.** and **Iovine**, **G.** (2007) Neogene-Quaternary strike-slip tectonics in the central Calabrian Arc (southern Italy). *J. Geodyn.*, **43**, 393–414.

Tape, **C.H.**, **Cowan**, **C.A.** and **Runkel**, **A.C.** (2003) Tidal-bundle sequences in the Jordan sandstone (Upper Cambrian), southeastern Minnesota, U.S.A.: evidence for tides along inboard shorelines of the Sauk epicontinental sea. *J. Sed. Res.*, **73**, 354–366.

Tessier, **B.** (1993) Upper intertidal rhythmites in the Mont-Saint-Michel Bay (NW France): Perspectives for paleoreconstruction. *Mar. Geol.*, **110**, 355–367.

Tessier, **B.** and **Gigot**, **P.** (1989) A vertical record of different tidal cyclicities: an example from the Miocene Marine Molasse of Digne (Haute Provence, France). *Sedimentology*, **36**, 767–776.

Van Rijn, **L.C.** (2005) *Principles of Sediment Transport in Rivers, Estuaries and Coastal Seas*. Aqua Publications, Blokzijl, The Netherlands, 580 pp.

Visser, **M.J.** (1980) Neap-spring cycles reflected in Holocene subtidal large-scale bedform deposits: a preliminary note. *Geology*, **8**, 543–546.

Hierarchy of tidal rhythmites from semidiurnal to solstitial cycles: Origin of inclined heterolithic stratifications (IHS) in tidal channels from the Dur At Talah Formation (upper Eocene, Sirte Basin, Libya) and a facies comparison with modern Mont-Saint-Michel Bay deposits (France)

JONATHAN PELLETIER*, ASHOUR ABOUESSA[†], MATHIEU SCHUSTER[†], PHILIPPE DURINGER[†] and JEAN-LOUP RUBINO*

*Total, Centre Scientifique et Technique Jean Feger, Avenue Larribau, 64000, Pau, France (jonathan.pelletier@total.com; jean-loup.rubino@total.com)
[†]Institut de Physique du Globe de Strasbourg (IPGS)-UMR 7516; Université de Strasbourg (UdS)/ École et Observatoire des Sciences de la Terre (EOST), Centre National de la Recherche Scientifique (CNRS), 1 rue Blessig, Strasbourg, 67084, France (ashour.abouessa@etu.unistra.fr; duringer@eost.u-strasbg.fr; mschuster@unistra.fr)

ABSTRACT

The Dur At Talah sequence outcrops in the Abu Tumayam Trough, in the southern part of the Sirt Basin (Libya). This formation consists of two units, the New Idam Unit at the base and the Sarir Unit at the top. This stratigraphic succession highlights a regressive trend attributed to the upper Eocene. Sedimentological investigations based on lithofacies and ichnofacies suggest that the depositional environments were mainly dominated by a tidal dynamic in the New Idam Unit. Several palaeoenvironments have been defined in this unit: The basal part is built up of an intertidal to supratidal flat system associated with oyster patches. The medium part is characterized by an estuarine channels belt. The upper part exposes typical facies of tidal delta plain and a prograding bar system (mouth-bars). The extremely good preservation of sedimentary structures and sequences allows investigation of the recording of tidal cycles at various scales of time, from the elementary tidal cycle to the solstitial cycle. A comparison with modern Mont-Saint-Michel Bay (France) deposits is proposed to validate facies and palaeoenvironment interpretations in the New Idam Unit. Tidal overprint in the New Idam Unit is expressed through tidal rhythmites composed of horizontal laminations, ripples (flaser, wavy or lenticular) as well as climbing ripples. The elementary recording is made of a mud-sand couplet corresponding to one tide event (slack and flood/ebb). The next scale of recording corresponds to the semi-lunar neap-spring cycle (fortnightly) up to, occasionally, the lunar cycle. Finally, a higher wavelength cycle is recognized corresponding to semi-annual cyclicity. All these cycles are identified in large-scale sedimentary bodies attributed to tidal channels and are best expressed within IHS. The superimposition of the different tidal cycles controls the heterolithic nature of the channel infill. In addition, the tidal rhythmites constitute an accurate chronometer to estimate the duration of tidal channel migration and infilling. This estimation is possible due to the very good preservation of the rhythmites and because of the scarcity of internal erosions within the channel infill.

Keywords: Tidal rhythmites, tidal channel, IHS, Eocene, Dur At Talah, Libya.

Contributions to Modern and Ancient Tidal Sedimentology: Proceedings of the Tidalites 2012 Conference, First Edition. Edited by Bernadette Tessier and Jean-Yves Reynaud.

INTRODUCTION

Dur At Talah outcrops (Fig. 1) show tidal sedimentary structures throughout the entire stratigraphic sequence, with the exception of the upper part. This type of sedimentary dynamic, rarely mentioned until now in this formation, has never been studied in detail for sedimentological features. The present article is devoted to the study and recognition of a hierarchy of tidal rhythmites. Tides have a great effect on the distribution of sediments in the coastal fringe, i.e. the marginal-marine part of the marine foreshore. The type of structures can be used to define the deposition dynamics and associated water depth (e.g. Van den Berg *et al.*, 2007). In some cases, sedimentary structures help to estimate the tidal range and to define the tidal regime. Tidal cycles are generated by astronomical rhythms; consequently, the tidal domain is one of the few environments that allows the quantification of deposition rates at a very short time-scale. These scales can range from the order of a few hours to a half day as well as a longer term (semi-annual cycles). At the Dur At Talah (DAT) tidal rhythmites are very well preserved and recorded into inclined heterolithic stratifications of tidal channels. Recognition of the hierarchical tidal cycles will allow appreciation of the time value in sediments. The description and interpretation of sedimentary structures observed in the DAT are argued partially by a comparative approach with sedimentary facies of the modern Bay of Mont-Saint-Michel (coast of the English Channel, France; Fig. 1). Such a modern analogue provides the actualistic basis for ancient rhythmite research (cf. Zaitlin, 1987; Dalrymple & Makino, 1989; de Boer *et al.*, 1989; Tessier *et al.*, 1989; Dalrymple *et al.*, 1991; Tessier, 1993; Tessier *et al.*, 1995; Kvale *et al.*, 1995; Archer & Johnson, 1997). These two sites have many similar facies and sedimentary structures. An actualistic approach, thus, is used as calibration of depositional environments as well as time recording through sedimentary structures. Modern facies are presented in parallel to ancient tidal structures of the DAT. It is worth noting that the comparison approach displays some limits. It is useful at the scale of sedimentary structures and short-time sedimentary successions. At the larger scale of depositional sequences, the DAT Formation can be compared better to systems in subtropical to tropical environments with mangroves.

Geological setting of the studied areas

Brief presentation of the modern Mont-Saint-Michel Bay, France

The Mont-Saint-Michel Bay (MSMB) is located in north-western France, in the funnel shaped embayment formed by the Cotentin peninsula (Normandy)

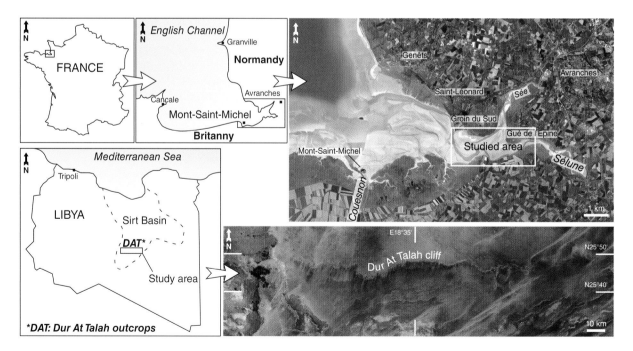

Fig. 1. Location maps of the study areas. The three upper maps locate the Bay of Mont-Saint-Michel (France) and the two lower maps locate the Dur At Talah outcrops (Libya).

at the east and the Brittany to the south and west (Fig. 1). In this region, the tidal regime is semidiurnal with almost no diurnal equality. Tidal range reaches up to 15 m (Larsonneur, 1989). The system is thus hypertidal according to the classification of Archer (2013). The study area is focused on the estuarine channelized belt located in the eastern part of the MSMB (Larsonneur, 1989; Bonnot-Courtois *et al.*, 2002). Three small perennial rivers feed the MSMB; the Couesnon, the Sée and the Sélune. The Sée and Sélune channels merge into a main channel running north to south, between Grouin du Sud and Roche Thorin (Fig. 1).

Tidal rhythmites are commonly observed in the estuarine zone, along tidal channels and creeks (Tessier, 1993; Tessier *et al.*, 1995; Lanier & Tessier, 1998). Pictures of tidalites illustrating this paper have been taken in the inner estuarine part, between Grouin du Sud (south of Saint-Léonard) and Gué de l'Epine (Fig. 1). In this area, sediment grain-size varies between medium to very fine sands with a non-negligible mud content accumulated during slack water periods. This area of the estuary is dominated by braided-bar system (Lanier & Tessier, 1998). Meandriform tributary tidal creeks form a dendritic network crossing salt-marsh around and along the distributary channel belt system. The best preserved tidal rhythmites are observed on meanders of tidal channels coming out onto the distributaries. As already mentioned by Lanier & Tessier (1998), the model of Dalrymple *et al.* (1992) can be extrapolated to both channels Sée and Sélune rivers. The study area corresponds to the contact between the 'tidal zone' and the 'fluvio-tidal transition zone', in which the two channels display a well-developed straight-meandering-straight morphology (Dalrymple *et al.*, 1992).

The Eocene Dur At Talah Formation (Sirte Basin, Libya)

The Dur At Talah (DAT) outcrop is an escarpment exposed in the Abu Tumayam Trough in the southern part of the Sirte Basin in central Libya (Fig. 1). The cliff (≈125 m high and ≈150 km in length) is oriented along an E-W transect facing southward. The complete succession is built up of marine to fluvial deposits crossing intermediate series such as tidal fringe deposits and deltaic sandstones (Fig. 2). This stratigraphic succession highlights a regressive trend attributed to the uppermost Eocene. Only a few field studies have been carried out previously in this area and these were mainly focused on its vertebrate fossil-content (see Jaeger

et al., 2010a; 2010b). Hence, the sedimentology and depositional environments remain poorly documented. The latest age dating given by Jaeger *et al.* (2010b) is Bartonian.

The lower part of the cliff is principally represented by the New Idam Unit (Abouessa *et al.*, 2012; Pelletier, 2012) and is defined as tide-dominated environments with evidence for estuarine/bay settings (Fig. 2). These sedimentary rocks can be interpreted as deposited in a tidal coastal fringe of a large palaeo-sirtic embayment. This long and large embayment probably behaved as a funnel amplifying the tidal range. In terms of facies, the New Idam Unit starts from the base by an alternation of metric-scale green mud layers and oyster shell banks, corresponding to a thick tidal mudflat. Progressively, sandy to muddy heterolithic channels are developed up to an amalgamated channel belt. These are attributed to tidal channels from inner to outer estuary and mainly run northwards. The meandering nature of these tidal channels is expressed through lateral accretions and the occurrence of point bars. Then, the series display a succession of metric fining-upward sequences from whitish sand layers to greenish mud intervals (Fig. 2). These small sequences are densely bioturbated by *Thalassinoides isp.* (Abouessa *et al.*, 2012; Pelletier, 2012) and, upward, show gradually increasing occurrences of root traces attributed to mangrove vegetation developed in tidal-influenced delta plain (Pelletier, 2012). Heterolithic thickening-upward intervals are vertically and laterally associated to these previous bioturbated mud-sand sequences and correspond to prograding deltaic bars. The following Sarir Unit is much sandier and displays an evolution from a tidal-influenced foreshore/shoreface with unequivocal megaripple mud couplet-bearing, to very coarse grain sandstone corresponding to fluvial channels at the top of the whole sequence (Fig. 2).

This paper is focused on tidal channels and their internal recordings observed in the lower part of the New Idam Unit (Fig. 2). Such migrating channels on a well-developed mudflat define a short succession comparable to that observed on modern systems such as the Mont-Saint-Michel Bay. The latter is thus used as a modern analogue to validate the Eocene intertidal facies interpretation. In terms of palaeogeography, the palaeo-gulf of Sirte was directly connected to the Tethys Ocean through an arm of the sea. This arm was a few hundred kilometres in length. The funnel-shaped morphology was particularly narrow and is supposed to be the origin of the high intensity

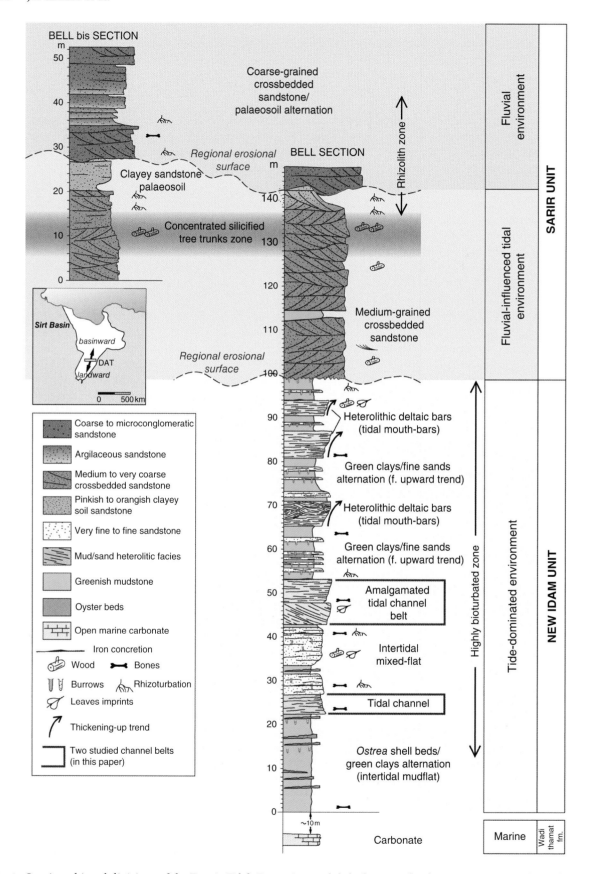

Fig. 2. Stratigraphic subdivisions of the Dur At Talah Formation with lithofacies and palaeoenvironments (from Abouessa *et al.*, 2012; Pelletier, 2012). Studied intervals are highlighted by red lines and correspond to tidal channel belt facies.

of tides at the DAT. The Sirte Basin was initiated by a fractural rifting subsidence mainly during the Cretaceous and Palaeocene ages (see Abouessa *et al.*, 2012). The basin limits are controlled by a succession of horsts and grabens, orientated NW-SE in the northern part of the basin and NE-SW in the southern part. The DAT escarpment belongs to the Abu Tumayam Trough (Abouessa *et al.*, 2012), which is located in the southernmost part of the Sirte Basin.

At the local scale of the DAT outcrop, the vertical vicinity of tidal sedimentary structures and root traces of mangrove swamps constrain the depositional environment in which the studied tidal rhythmites were deposited. The environment was an intertidal coastal domain under sub-tropical to tropical climatic conditions. Moreover, sedimentary facies demonstrate that wave action was minor or absent. This is consistent with the presence of mangroves and, possibly, of a very gently sloping foreshore (tidal flat) that damped wave action.

At the DAT, tidal signals are well preserved in smaller channels. These channels are present at the lower part of the New Idam Unit (Fig. 2) and exhibit neap-spring cycles as well as longer-term, annual cycles. Conversely, it seems that the overlying larger tidal channels show less obvious tidal cyclicities. This is particularly true in regards to solstitial semi-annual cycles. Inclined Heterolithic Stratification (IHS) was formed in the innermost estuary. In the larger channels IHS was formed in the outer zones of the estuary. This interpretation is consistent with the study of Hovikoski *et al.* (2008) and suggests that the annual cycles are particularly well preserved and visible in very proximal domains, i.e. inner estuary areas.

HIERARCHY OF TIDAL RHYTHMITES IN THE DUR AT TALAH FORMATION

Several scales of tidal cyclicity were recorded in the tidal channel fills of the New Idam Unit of the DAT Formation (Fig. 2). A channel filling sequence consists of a fining-upward trend of heterolithic deposits with an alternation of cm-scale sand layers separated by thin, mm-scale mud drapes. The base of the sequence displays lag composed of bones (crocodilian, fish...) and oyster-shell fragments. At the top, sedimentary facies are much muddier and commonly include root traces. This typical sequence does not exceed 5 m in thickness. Thus, the total thickness can be attributed to a lateral-accretion bar deposited within a meandering channel. Multi-scale tidal rhythmites are

preserved into these point bars, forming IHS (Thomas *et al.*, 1987, Choi *et al.*, 2004; Hovikoski *et al.*, 2008). Typically, the rhythmites (*sensu* Eriksson & Simpson, 2004) are best developed in the muddiest zones of the point bars. Such areas are sheltered from very rapid and erosive tidal currents. This is particularly obvious in the inner parts of hypertidal estuaries (Dalrymple *et al.*, 1991; Tessier, 1993; Choi *et al.*, 2004; Archer, 2013).

In addition, modern and ancient RCR (rhythmic climbing ripples, Kvale & Archer, 1991; Lanier & Tessier, 1998; Choi, 2010) commonly occur in the upper part of the fill sequences. The RCR was deposited at the top of the point-bar IHS. The occurrence of mud and RCR at the top of the depositional sequence is diagnostic of decreasing energy. This is confirmed by the presence of root marks (plant colonisation).

Daily cyclicity

The elementary cyclicity corresponds to a sand-mud couplet (Fig. 3). The sandy interval (rippled or laminated) is a response to tractive, flood or ebb currents. The mud drapes indicate a slack phase dominated by settling processes between reverse current flow (Figs 3, 4A and B). At this time, it is not possible to determine whether the tidal regime is diurnal or semi-diurnal. Most of the time between two successive tides one current is dominant and the following is subordinate, creating an asymmetry of ripple thicknesses. The dominant current forms a well-developed ripple whilst the subordinate current is weakly expressed through small ripple caps developed on the crest of the underlying ripples (Fig. 3C and D) similar to those described by Kvale & Archer (1990; 1991). In some cases, the subordinate current is too weak to generate a sandy ripple or the following dominant currents are erosive and delete underlying deposits, creating daily hiatuses.

Fortnightly cyclicity

The fortnightly-scale cyclicity corresponds to the neap-spring cycles (~14.7 days). This cyclicity is expressed by a progressive variation of the sand-mud ratio (Figs 4 and 5). Thickening of the sandy intervals, combined with a thinning or disappearance of mud laminae, corresponds to the passage of neap to spring tides. Conversely, the progressive thinning of sandy intervals combined with a thickening of mud-rich laminae corresponds to the passage from spring to neap tides. The number of sand-mud couplets deposited during a complete

208   *J. Pelletier* et al.

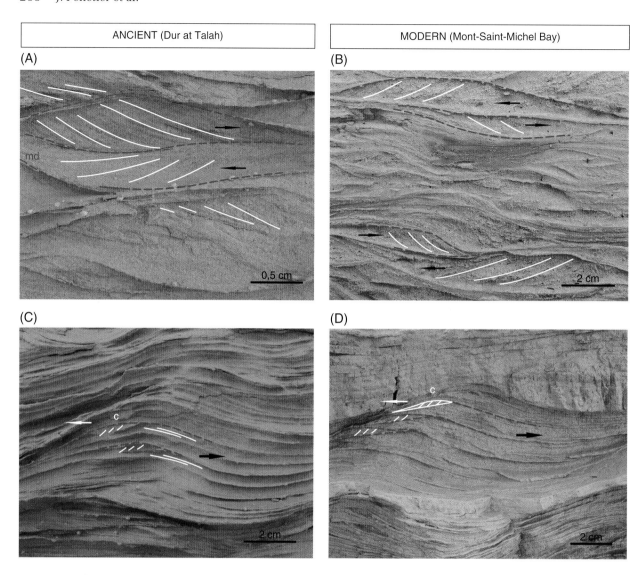

Fig. 3. Examples of the elementary tidal cycle: (A and B) Opposite tidal ripples formed under symmetrical currents separated by thin mud drapes; (C and D) Rhythmic climbing ripples structure showing a diagnostic asymmetry. Dominant tidal currents generate thick sand intervals while subordinate currents form small ripple caps (c) on the crest of the previously deposited ripple.

neap-spring cycle is the only diagnostic criterion for defining the tidal regime (Boersma, 1969; Visser, 1980; Nio *et al.*, 1983; Yang & Nio, 1985, Kvale & Archer, 1991). At DAT, up to 23 mud drapes were measured within a single neap-spring cycle (Fig. 5A). This indicates that more than one tide per day occurred and therefore the tidal regime was semi-diurnal. As already described in modern facies, as well as in rock records (e.g. Lanier & Tessier, 1998; Choi, 2010; 2011; Kvale & Archer, 1990; 1991), fortnightly tidal rhythmites are commonly preserved in climbing ripple stratifications (Fig. 5). Such fortnightly rhythmic climbing ripples (RCR) are commonly developed at DAT (Fig. 5A). These rippled bundles show thickening

and thinning as well as a gradual variation of the climbing angle. Many field-based measurements in the DAT deposits showed that the number of mud drapes (slack period of low or high tide) in the fortnightly cycles preserved in RCR regularly reaches 20, sometimes 25, demonstrating unequivocally that the tidal regime was semi-diurnal.

Monthly cyclicity

This cyclicity is expressed in IHS of the tidal channels of the DAT. It consists of an alternation of thick sandy intervals (large spring tides, perigean spring tides or apogean spring tides) and thinner sandy intervals (small spring tides, apogean spring tides

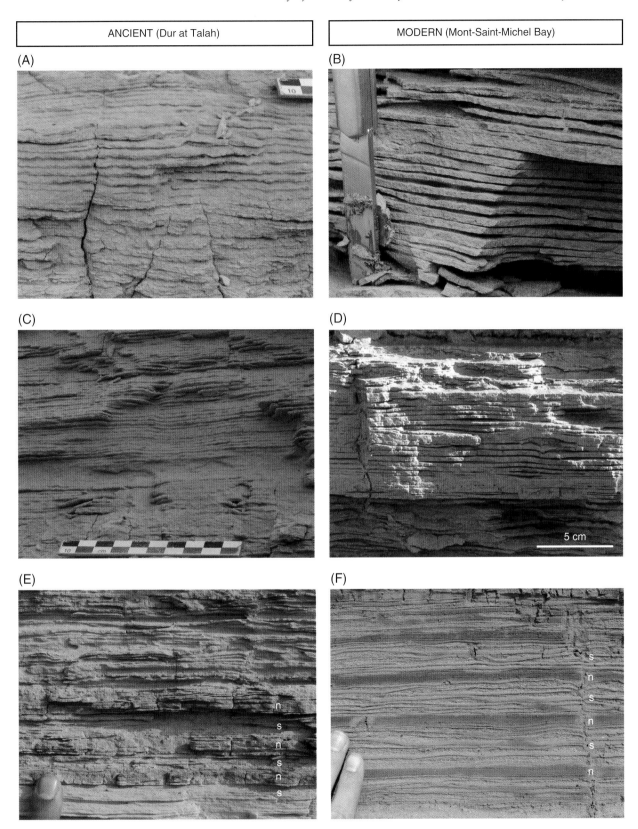

Fig. 4. Planar tidal bedding: (A and B) Ancient and modern planar bedding showing a millimetric-scale mud/sand alternation; (C and D) Ancient and modern planar bedding showing an evolution of the mud/sand ratio; (E and F) Ancient and modern planar bedding recording neap/spring (n, s) cyclicity.

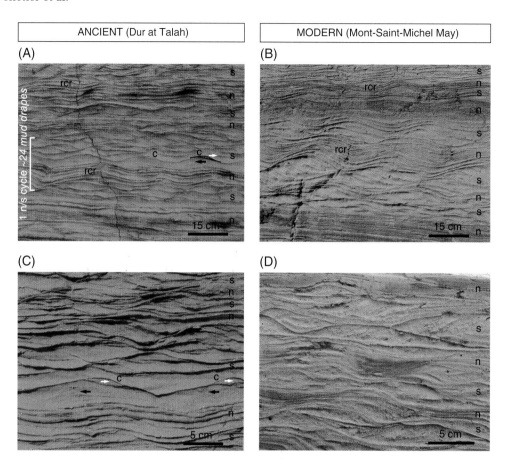

Fig. 5. Modern (Mont-Saint-Michel Bay) and ancient (Dur At Talah Formation) semi-lunar tidal rhythmites (A, B, C and D). Spring intervals are much sandier than neap periods. Ancient neap/spring cycles record about 24 tides. Dominant currents are represented by black arrow and subordinate currents are represented by white arrow. In A and B 'rcr' stands for rhythmic climbing ripples interval and 'c' is for ripple caps corresponding to the subordinate current.

or perigean spring tides), separated by muddier intervals with an invariable thickness, corresponding to the neap periods (Fig. 6). The wavelength of this cycle is about 29 days. These synodic lunar cycles have been observed only locally in the DAT outcrop. Synodic tidal rhythmites have been observed commonly in ancient rock records (e.g. Williams, 1989; Tessier & Gigot, 1989; Kvale *et al.*, 1989; Brown *et al.*,1990; Martino & Sanderson, 1993; Archer, 1996; Miller & Eriksson, 1997; Kvale *et al.*, 1999; Choi *et al.*, 2001; Choi *et al.*, 2004; Couëffé *et al.*, 2004. In MSM, this lunar cyclicity is rarely observed (Tessier, 1993).

Semi-annual cyclicity

The largest-scale cycles at DAT outcrop were noted from several repeating units containing a dozen fortnightly cycles (Fig. 7). These cycles correspond to 6-month periods. They are delimited by brownish-blackish muddy and oxidized horizons. Generally, the thickness between two muddy horizons varies from 40 to 55 cm (Fig. 7A and E). Laterally, a single interval can display a variety of sedimentary structures. These include laminated rhythmites or rippled rhythmites (Fig. 7), depending of the relative position into the channel. Towards the channel axis, rhythmites are expressed by rippled surfaces whereas, towards the channel bank, cycles are expressed by laminated surfaces. This cyclicity has been interpreted as an expression of a semi-annual periods related to the equinoxial and solstitial tides. The muddy horizons are assumed to reflect periods of very low energy. Moreover, a blackish rust-colour may have been produced by sub-aerial exposure significantly longer than common neap periods; and is interpreted as a signature of solstitial phases. These dark horizons emphasize and underline the IHS. This interpretation is supported by the continuous recording during several years. Locally, 10 cycles at least, i.e. 5 years, were measured

(A)

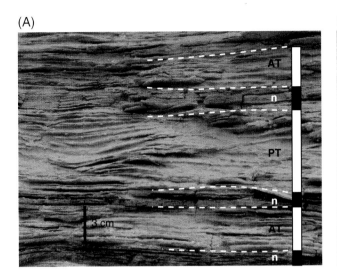

(B)

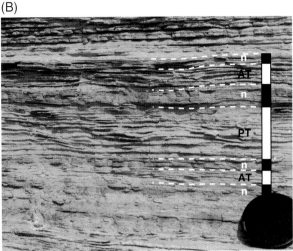

Fig. 6. Apogean/perigean lunar tidal rhythmites in the Dur At Talah Formation (A and B). Apogean (small) spring tides (AT) correspond to thinner spring tides intervals whilst perigean (large) spring tides (PT) are represented by thicker spring tides intervals. Neap tides intervals (n) are thinner, muddier and brownish.

(Fig. 7E). Within the intertidal domain, the duration of neap emersion is higher during equinoctial tides. However, during solstitial periods, tidal range being generally low, upper intertidal to supratidal areas can remain emerged for several days or even weeks for the highest topographies, especially in macrotidal settings (e.g. in the Bay of Mont-Saint-Michel). The longest periods of sub-aerial exposure thus occur during solstitial time in the upper part of the intertidal to supratidal zone. Consequently, the upper intertidal and inner estuary tidal flat could be the most favourable depositional environment for these semi-annual tidal rhythmites to be deposited and preserved (e.g. Pugh, 1987; Greb & Archer, 1998).

RATE OF SEDIMENTATION AND OF LATERAL MIGRATION OF TIDAL CHANNELS IN THE DUR AT TALAH FORMATION

Time preservation in sediments through the study of tidal rhythmites has been investigated by Couëffé *et al.* (2001, 2004). In the DAT, our study of tidal rhythmites preserved in channel IHS, allows an accurate estimation of the sedimentation rate and therefore of the average velocity of tidal channel lateral migration. Semi-annual cycle intervals are 40 to 60 cm-thick, corresponding to a mean sedimentation rate of about 3 mm per day. The IHS inclination, that features the channel lateral

accretion, ranges from 3° to 10° in average. These data allow determination of the rate of channel migration thanks to the relationship below (Fig. 8):

$\sin \alpha =$ thickness of one semi-annual cycle/distance of migration (x)
with $\alpha = 3°$ to 10° and the thickness = 4 to 60 cm.

From this relationship, it is found that the semi-annual rate of lateral migration ranges from 2.30 to 11.40 m, i.e. from about 0.4 to almost 2 m per month. Hence, for a 200 m-wide channel, as illustrated in Fig. 9, the estimation of the lateral accretion rate allows definition of the duration of 'life' or activity of the channel, i.e. the duration for its migration and filling. Extreme values give a duration ranging from 9 to 42 years but, using the most frequently observed and measured values, most common filling duration varies between 15 and 20 years. Tidal rhythmites can thus be used in the DAT as a relative chronometer of the tidal channel dynamic and infilling as well as to constrain palaeoenvironmental reconstructions.

DISCUSSION

The estimation of the rate of sedimentation and of tidal channel migration in the DAT was made in intervals containing tidal rhythmites as continuous as possible. However, the highest parts of HIS are characterized by incomplete or amalgamated tidal rhythmites. Neap tides usually do not reach the

Tidal rhythmites in ripple bedding
(vertical & lateral accretion)

Parallel laminated tidal rhythmites
(vertical accretion)

A

50 cm

B

D

45 cm

C

D

10 to 12 semi-lunar cycles (neap/spring cycles) between two argillaceous brown layers

Limits of semi-annual cycles (solstices)

Neap periods

6 month interval

E

40–50 cm

Fig. 7. Semi-annual cycles in the Dur At Talah Formation: (A) Tidal channel rippled sequence showing a cyclic succession of muddier, brownish intervals (red arrows); (B) Tidal channel sequence showing parallel laminated tidal rhythmites marked by well-developed muddy surfaces (red arrows); (C) Close view of A showing about 10 to 12 neap/spring cycles (a six-month period) included between two main muddy intervals; (D) Close view of B showing 10 to 12 neap/spring cycles (a six-month period) between two argillaceous brown layers. These well-developed muddy to argillaceous layers correspond to the long lasting aerial exposures that occur during solstitial time; (E) General view of a channel infill showing a dozen semi-annual cycles.

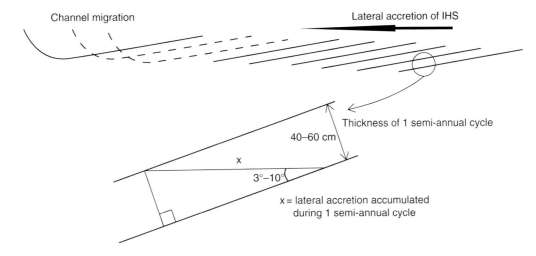

Fig. 8. Geometric elements for lateral migration rate calculation of the tidal channel.

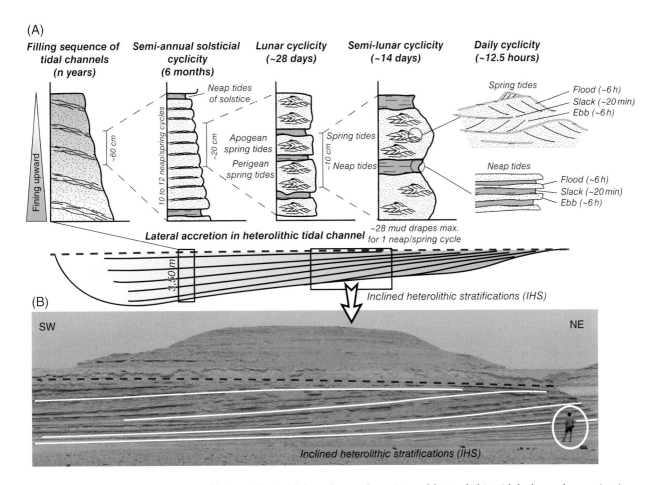

Fig. 9. Schematic representation of hierarchical tidal cycles at the origin of heterolithic tidal channel organization: (A) Semi-annual (solstitial) cycle to daily (semi-diurnal) cycle; (B) Heterolithic tidal channel.

highest parts of the tidal flat, explaining the incompleteness (Tessier, 1993; Archer, 1995) and tidal currents on top of the point bars are generally highly turbulent, inducing erosion and amalgamation. The lowest parts of the IHS are not favourable for the preservation of complete and continuous tidal rhythmites, as erosional processes are intense in the axis of the channel. Therefore, only the middle part of IHS potentially offers a continuous record preserving all tidal cyclicities.

Several authors have attempted to quantify the time and rate of migration for tidal channels, but almost exclusively in modern environments (e.g. Choi, 2011). Couëffé *et al.* (2001, 2004) quantified rates of deposition based on the recognition facies and tidal rhythmite interpretation in the Miocene molasse of Digne-lès-Bains (Alps, SE France). Due to the limited horizontal extent of the outcrop, the geometry of studied sedimentary bodies was poorly constrained, preventing quantification of lateral migration rate. At DAT, the large lateral extensions of the outcrop in addition to the exceptionally well preserved tidal rhythmites with different wavelengths of cyclicity allow for an accurate time quantification of tidal channel migration. The study of Choi (2011) on tidal rhythmites preserved in IHS of tidal channel in the modern Gomso Bay (South Korea) estimated a migration rate of about 2.40 m per month. Although slightly higher, this value is quite comparable to that measured into the DAT deposits. The duration of 'life' and infill of the channels in the DAT were estimated thanks to the semi-annual tidal rhythmites and IHS, to 15 to 20 years. This is a minimum duration since hiatuses may exist during the filling. However, these values of the order of a few decades are consistent with the duration found for tidal channels in the inner estuary of MSMB (Billeaud *et al.*, 2007). Thus, these results confirm that the DAT environment was a wide embayment with tidal (estuarine) channels migrating actively. The tidal range was probably not too high (low macrotidal to mesotidal), as in the inner estuary of the MSMB, in order to prevent erosional processes and overall to enhance the preservation potential of tidal rhythmites.

CONCLUSION

Based upon an analysis of tidal rhythmites, there is a clear delineation of the impact of tidal dynamics during the deposition of the Eocene DAT formation. Four main tidal cyclicities can be recognized: semi-diurnal, fortnightly, monthly and semi-annual. The hierarchy of these discrete cycles is best expressed within inclined heterolithic stratification (IHS). These bedforms occur in tidal channels and the features can be used to quantify the channel migration rates. In this study, migration rates ranged from about 0.5 to 2 m per month. Tidal rhythmites are the basic component of the heterolithic nature of the IHS deposits. Thus, the different tidal wavelengths control the distribution of mud in the studied tidal meandering channel system. Overall, the following conclusions can be drawn about the original tidal depositional environments: Firstly, the tidal regime was semi-diurnal. Secondly, several criteria (size range of sedimentary structures, thickness of tidal sequences) suggest that tidal range was probably mesotidal (between 2 and 4 m) in the area of deposition of the studied tidal rhythmites. Thirdly, in modern deposits, such as those observed in the Mont-Saint-Michel Bay, the preservation of tidal rhythmites and more precisely of tidal rhythmic climbing ripples (RCR) is a diagnostic feature of deposition in an inner estuary environment characterised by relatively low energy conditions. Consequently, the similarity with the modern deposits of the Mont-Saint-Michel allows for an accurate definition of the palaeoenvironments of deposition within the Dur At Talah tidal facies.

Tidal rhythmites are known and analysed as very reliable tools for time quantification in rock records. In this case, tidal rhythmites are contained in very well constrained bodies in terms of geometry, i.e. tidal channels. This allows calculation of the rate of migration and the duration of functioning of these channels, independently from external factors such as regional subsidence. In that way, this study can be considered as innovative in ancient deposits. However, it is worth noting that time quantification based on tidal rhythmite analysis is only possible within the inclined heterolithic stratification facies. At the scale of the whole Dur At Talah succession, about 100 m-thick, on the basis of subsidence rate and age information (Jaeger *et al.*, 2010a and 2010b), duration of the order of one million years can be estimated. This discrepancy evidences the disconnection between the controls of very long term basin infill and very short-term tidal channel infill.

ACKNOWLEDGEMENTS

Thanks go to Total to their financial support to J. Pelletier and A. Abouessa. Funds were also provided by the Centre National de la Recherche Scientifique (CNRS) and the University of Strasbourg (UDS/EOST). Great thanks to the CNRS - Caen University M2C Lab, for organising the Tidalites 2012 Conference. Finally, thanks to the reviewers (A. Archer, P. de Boer and B. Tessier) for their suggestions regarding improvement of the manuscript.

REFERENCES

Abouessa, A., Pelletier, J., Duringer P., Schuster, M., Schaeffer, P., Métais, E., Benammi, M., Salem, M., Hlal, O., Brunet, M., Jaeger, J.-J. and Rubino, J.-L. (2012) New insight into the sedimentology and stratigraphy of the Dur At Talah tidal-fluvial transition sequence (Eocene-Oligocene, Sirte Basin, Libya). *Journal of African Earth Sciences*, **65**, 72–90.

Archer, A.W. (2012) Comparison of hypertidal systems in Europe, South and North America. 8th International Conference on Tidal Environments, Tidalites 2012, Caen. Oral presentation.

Archer, A.W. (1995) Modelling of cyclic tidal rhythmites based on a range of diurnal to semidiurnal tidal-station data. *Mar. Geol.*, **123**, 1–10.

Archer, A.W. (1996) Reliability of lunar orbital periods extracted from ancient cyclic tidal rhythmites. *Earth Planet. Sci. Lett.*, **141**, 1–10.

Archer, A.W. and Johnson, T.W. (1997) Modelling of cyclic tidal rhythmites (Carboniferous of Indiana and Kansas, Precambrian of Utah, USA) as a basis for reconstruction of intertidal positioning and palaeotidal regimes. *Sedimentology*, **44**, 991–1010.

Billeaud, I., Tessier, B, Lesueur, P. and Caline, B. (2007) Preservation potential of highstand coastal sedimentary bodies in a macrotidal basin: example from the Bay of Mont-Saint-Michel, France. *Sed. Geol.*, **202**, 754–775.

Boersma, J.R. (1969) Internal structures of some tidal megaripple on a shoal in the Westerschelde estuary, the Netherlands. *Geol. Mijnbouw*, **48**, 409–414.

Bonnot-Courtois, C., Caline, C., L'Homer, A. and Le Vot, M. (2002) The bay of Mont-Saint-Michel and the Rance estuary. Recent development and evolution of depositional environments. *Bull. Centre Rech. Elf Explor. Prod.*, **26**, 256 pp.

Brown, M.A., Archer, A.W. and Kvale, E.P. (1990) Neap-spring tidal cyclicality in laminated carbonate channel-fill deposits and its implications: Salem limestone (Mississippian), South-Central Indiana, U.S.A. *J. Sed. Petrol.*, **60**, 152–159.

Choi, K. (2010) Rhythmic climbing-ripple cross-lamination in inclined heterolithic stratification (IHS) of a macrotidal estuarine channel, Gomso Bay, West Coast of Korea. *J. Sed. Res.*, **80**, 550–551.

Choi, K. (2011) Tidal rhythmites in a mixed-energy, macrotidal estuarine channel, Gomso Bay, west coast of Korea. *Mar. Geol.*, **280**, 105–115.

Choi, K.S., Dalrymple, R.W., Chun, S.S. and Kim, S.P. (2004) Sedimentology of modern inclined heterolithic stratifications (IHS), in the macrotidal Han River delta, Korea. *J. Sed. Res.*, **74**, 677–689.

Choi, K.S., Kim, B.O. and Park, Y.A. (2001) Late Pleistocene tidal rhythmites in the Kyunggi Bay, west coast of Korea: a comparison with simulated rhythmites based on modern tides and implications for intertidal positioning. *J. Sed. Res.*, **71**, 680–691.

Couëffé, R., Tessier, B., Beaudoin, B. and Gigot P. (2001) Le temps préservé sous forme de sédiments: résultats semi-quantitatifs obtenus dans la Molasse Marine miocène du bassin de Digne (Alpes-de-Haute-Provence, Sud-Est de la France). *CR Acad. Sci. Paris*, **332**, 5–11.

Couëffé, R., Tessier, B., Gigot, P. and Beaudoin, B. (2004) Tidal rhythmites as possible indicators of very rapid subsidence in a foreland basin: an example from the Miocene marine molasse formation of the Digne foreland basin, SE France. *J. Sed. Res.*, **74**, 746–759.

Dalrymple, R.W and Makino, Y. (1989) Description and genesis of tidal bedding in the Cobequid Bay Salmon river estuary, Bay of Fundy, Canada. In: *Sedimentary Facies of the active Plate Margin*. (Eds A. Taira and F. Masuda), Terra Publishing, Tokyo, pp. 151–177.

Dalrymple, R.W., Makino, Y. and Zaitlin, B.A. (1991) Temporal and spatial patterns of rhythmite deposition on mud flats in the macrotidal Cobequid Bay-Salmon River estuary, Bay of Fundy, Canada. In: *Clastic Tidal Sedimentology* (Eds D.G. Smith, G.E. Reinson, B.A. Zaitlin, R.A. Rahmani), *Can. Soc. Petrol. Geol. Mem.*, **16**, 137–160.

Dalrymple, R.W., Zaitlin, B.A. and Boyd, R. (1992) Estuarine facies models: conceptual basis and stratigraphic implications. *J. Sed. Petrol.*, **62**, 1130–1146.

Davies, J.L. (1964) A morphogenetic approach to world shorelines. *Geomorphology.*, **8**, 127–142.

De Boer, P.L., Oost, A.P. and Visser, M.J. (1989) The diurnal inequality of the tide as a parameter for recognising tidal influences. *J. Sed. Petrol.*, **59**, 912–921.

Eriksson, K.A. and Simpson, E.L. (2004) Precambrian tidalites: recognition and significance. In: *The Precambrian Earth: Tempos and Events* (Ed Elsevier) pp. 631–642.

Greb, S.F. and Archer, A.W. (1998) Annual sedimentation cycles in rhythmites of Carboniferous tidal channels. In: *Tidalites: Processes and Products* (Eds C.R. Alexander, R.A. Davis and V.J. Henry), *SEPM Spec. Publ.*, **61**, 75–83.

Hovikoski, J., Räsänen, M., Gingras, M., Ranzi, A. and Melo, J. (2008) Tidal and seasonal controls in the formation of Late Miocene inclined heterolithic stratification deposits, western Amazonian foreland basin. *Sedimentology*, **55**, 499–530.

Jaeger, J.J., Beard, K.C., Chaimanee, Y., Salem, M.J., Benammi, M., Hlal, O., Coster, P., Bilal, A.A., Duringer, Ph., Schuster, M., Valentin, X., Marandat, B., Marivaux, L., Métais, E., Hammuda, O. and Brunet, M. (2010a) Late middle Eocene epoch of Libya yields earliest known radiations of African anthropoids. *Nature*, **467**, 1095–1099.

Jaeger, J.J., Marivaux, L., Salem, M., Bilal, A.A., Benammi, M., Chaimanee, Y., Duringer, Ph., Marandat, B., Métais, E., Schuster, M., Valentin, X. and **Brunet, M.** (2010b) New rodent assemblages from the Eocene Dur At-Talah escarpment (Sahara of central Libya): systematic, biochronological and Palaeobiogeographical implications. *Zoological Journal of Linnean Society*, **160**, 195–213.

Kvale, E.P. and **Archer, A.W.** (1991) Characteristics of two Pennsylvanian-age, semidiurnal tidal deposits in the Illinois Basin, USA. In: *Clastic Tidal Sedimentology* (Eds D.G. Smith, G.E. Reinson and B.A. Zaitlin), *Can. Soc. Petrol. Geol.*, **16**, 179–188.

Kvale, E.P. and **Archer, A.W.** (1990) Tidal deposits associated with lowsulfur coals, Brazil Formation (lower Pennsylvanian), Indiana. *J. Sed. Petrol.*, **60**, 563–574.

Kvale, E.P., Archer, A.W. and **Johnson, H.R.** (1989) Daily, monthly and yearly tidal cycles within laminated siltstones of the Mansfield Formation (Pennsylvanian) of Indiana. *Geology*, **17**, 365–368.

Kvale, E.P., Cutright, J., Bilodeau, D., Archer, A.W., Johnson, H.R. and **Pickett, B.** (1995) Analysis of modern tides and implications for ancient tidalites. *Cont. Shelf Res.*, **15**, 1921–1943.

Kvale, E.P., Johnson, H.W., Sonett, C.P., Archer, A.W. and **Zawistoski, A.** (1999) Calculating lunar retreat rates using tidal rhythmites. *J. Sed. Res.*, **69**, 1154–1168.

Lanier, W.P. and **Tessier, B.** (1998) Climbing ripple bedding in fluvio-estuarine system; a common feature associated with tidal dynamics. Modern and ancient analogues. In: *Tidalites: Processes and Products* (Eds C. Alexander, R.A. Davis Jr. and V.J. Henry) *SEPM Spec. Publ.*, **61**, 109–117.

Larsonneur, C. (1989) La Baie du Mont-Saint-Michel. *Bull. Inst. Géol. Bassin Aquit.*, **46**, 5–74.

Martino, R.L. and **Sanderson, D.D.** (1993) Fourrier and autocorrelation analysis of estuarine tidal rhythmites, lower Breathitt Formation (Pennsylavanian), eastern Kentucky, U.S.A. *J. Sed. Petrol.*, **63**, 105–119.

Miller, D. and **Eriksson, K.A.** (1997) Late Mississippian Prodeltaic rhythmites in the Appalachian basin: a hierarchical record of tidal and climatic periodicities. *J. Sed. Res.*, **67**, 653–660.

Nio, S.D., Siegenthaler, C. and **Yang, C.S.** (1983) Megaripple cross-bedding as a tool for the reconstruction of the palaeo-hydraulics in a Holocene subtidal environment, SW Netherlands. Special issue in the honour of J.D. de Jong. *Geol. Mijnbouw*, **62**, 499–510.

Pelletier, J. (2012) Faciès, architecture et dynamique d'un système margino-littoral tidal: exemple de la Formation du Dur At Talah (Eocène supérieur, Bassin de Syrte, Libye). [Unpublished Ph.D. thesis], University of Strasbourg, 470 pp.

Pugh, D.T. (1987) Tides, Surges and Mean Sea Level. (Eds John Wiley and Sons) 472 pp., New York.

Tessier, B. (1993) Upper intertidal rhythmites in the Mont-Saint Michel Bay (NW France): perspectives for paleoreconstruction. *Mar. Geol.*, **110**, 355–367.

Tessier, B., Archer, A.W., Lanier, W.P. and **Feldman, H.R.** (1995) Comparison of modern analogues (The Bay of Mont-saint-Michel) with ancient tidal rhythmites (Carboniferous of Kansas and Indiana, U.S.A.), (Eds B.W. Flemming and A. Bartholomä), *Int. Assoc. Sedimentol. Spec. Publ.* **24**, 259–271.

Tessier, B. and **Gigot, P.** (1989) A vertical record of different tidal cyclicities: an example from the Miocene Marine Molasse of Digne (Haute Provence, France). *Sedimentology*, **36**, 767–776.

Tessier, B., Monfort, Y., Gigot, P. and **Larsonneur, C.** (1989) Enregistrement des cycles tidaux en accrétion verticale, adaptation d'un outil de traitement mathématique. Exemples en baie du Mont-Saint-Michel et dans la molasse marine miocène du bassin de Digne. *Bull. Soc. Géol.*, **5**, 1029–1041.

Thomas, R.G., Smith, D.G., Wood, J.M., Visser, J., Calverley-Range, E.A. and **Koster, E.H.** (1987) Inclined heterolithic stratification-terminology, description, interpretation and significance. *Sed. Geol.*, **53**, 123–179.

Van den Berg, J.H., Boersma, J.R. and **Van Gelder, A.** (2007) Diagnostic sedimentary structures of the fluvial-tidal transition zone – Evidence from deposits of the Rhine and Meuse. *Neth. J. Geosci., Geol. Mijnbouw*, **86**, 287–306.

Visser, M.J. (1980) Neap-spring cycles reflected in Holocene subtidal large-scale bedform deposits: a preliminary note. *Geology*, **8**, 543–546.

Williams, G.E. (1989) Late Precambrian tidal rhythmites in South Australia and the history of the Earth's rotation. *J. Geol. Soc. London*, **146**, 97–111.

Yang, C.S. and **Nio, S.D.** (1985) The estimation of palaeohydrodynamic processes from subtidal deposits using time series analysis methods. *Sedimentology*, **32**, 41–57.

Zaitlin B.A. (1987) Sedimentology of the Cobequid Bay-Salmon river estuary, Bay of Fundy, Canada. [Unpublished PhD thesis], Queen's University, Kingston, Ontario, 391 pp.

Cataclysmic burial of Pennsylvanian Period coal swamps in the Illinois Basin: Hypertidal sedimentation during Gondwanan glacial melt-water pulses

ALLEN W. ARCHER*[†], SCOTT ELRICK[‡], W. JOHN NELSON[‡] and WILLIAM A. DIMICHELE[§]

[†] *Department of Geology, Kansas State University, Manhattan, Kansas, 66506, USA*
[‡] *Illinois State Geological Survey, Champaign, Illinois, 61820, USA*
[§] *Department of Paleobiology, NMNH, Smithsonian Institution, Washington, D.C., 20560, USA*
* *E-mail: aarcher@k-state.edu*

ABSTRACT

Roof facies of widespread, Pennsylvanian Period, economically important coals in the Eastern Interior Coal Basin (Illinois Basin) commonly exhibit evidence of extensive and cataclysmic tidal sedimentation during rapid marine-flooding events. Upright Trees and smaller plants were covered and entombed by swiftly deposited, thin-bedded and laminated mud and sand. This facies has historically been interpreted as fluvial-style cross bedding. Detailed examination clearly shows that these facies consist of inclined heterolith stratifications (IHS). Entire forests of upright trunks have been documented in the roof strata of surface and underground coal mines. Detailed sedimentological analyses of mine-roof facies indicate a pervasive and significant tidal influence. In some cases, daily and semi-monthly tidal periods have been preserved within laminated facies ('tidal rhythmites'). Based upon modern analogues, the tidal facies are indicative of hypertidal conditions. The hyper-dynamic tidal regime resulted in rapid sediment accumulation, particularly along pre-existing drainages within the ancient coastal swamp. Stratigraphic successions indicate a recurring pattern of very rapid change from widespread coal-swamp conditions to tidally-influenced deposition. The repetition of these stratigraphic phenomena throughout the Pennsylvanian Period suggests significant external controls on sea-level, probably related to Gondwanan deglaciations and resultant large-scale meltwater pulses. Despite the rapid changes of sea-level rise, the hypertidal depositional dynamics resulted in conditions whereby inundated coastal forests were buried by tidally influenced sedimentation.

Keywords: Pennsylvanian, coal swamps, hypertidal sedimentation, Illinois Basin, Gondwana glaciations, melt-water pulses, Holocene.

INTRODUCTION

Objectives and study area

The objectives of this research include the documentation of the common occurrence of upright plants within shale-rich lithofacies that directly overly coals. From a sedimentological perspective, very rapid generation of accommodation space would apparently be required to preserve rapid plants. Lacking clearly defined modern analogues, the authors suggest that the deglaciations and meltwater surges during the Late Cenozoic Era (Pleistocene and Neogene Epochs) might contain meaningful 'modern' analogues.

Pennsylvanian Period coal forests, in terms of taxonomic composition, diversity and long-term dynamics, lack clearly-defined modern counterparts (Falcon-Lang & DiMichele, 2010). These palaeoforests were unique in the history of the Earth's terrestrial vegetation and were amongst the most globally extensive terrestrial ecosystems in Earth history (DiMichele *et al.*, 2001).

Contributions to Modern and Ancient Tidal Sedimentology: Proceedings of the Tidalites 2012 Conference, First Edition. Edited by Bernadette Tessier and Jean-Yves Reynaud.

Extensive coal-mining activity in the U.S. and Europe has resulted in numerous exposures for study and analysis. Within the Illinois Basin area of the east-central U.S. (Fig. 1) mining of bituminous coal began circa 1820 and became an important industry by 1880. Today, it remains a key component of electrical-power generation worldwide. Due to the economic incentive to core-drill and excavate Pennsylvanian coal-bearing strata, we may know more about these terrestrial ecosystems than any others in Earth history. However, there remain many enigmas, such as upright palaeoforests, that have yet to be completely explained.

'Cyclothem' models

Pennsylvanian coal measures in the U.S. midcontinent contain lithostratigraphic units that in most cases have a very high degree of vertical repetition and lateral continuity. More precisely, many thin Pennsylvanian beds of underclay (palaeosols), coal, black-sheety shale and limestone exhibit great lateral persistence throughout and between major basins, such as the Appalachian Basin to the east and Forest City Basin to the west (Wanless & Weller, 1932; Archer & Greb 2011).

A generalised model of cyclical sedimentation, the 'cyclothem,' was initially developed in Illinois Basin (Weller, 1930, 1931) and subsequently applied to other areas, particularly in Kansas (Moore, 1935, 1964). Despite the seeming elegance of the cyclothem model, the concept has never been statistically or numerically verifiable (Zeller, 1964; Wilkinson, *et al.*, 2003). The model was used to define formations in Illinois during the 1940s and 1950s. Today, neither the Illinois nor the Kansas geological surveys retain the cyclothem concept as a part of formal lithostratigraphic nomenclature.

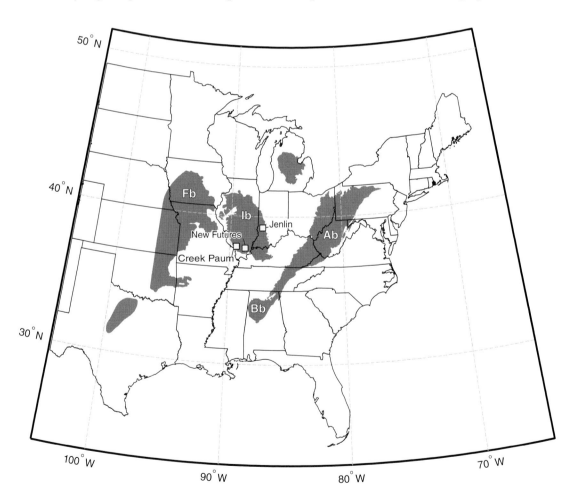

Fig. 1. Location of major Pennsylvanian coal basins in the eastern U.S. including the Appalachian Basin (Ab), Black Warrior Basin (Bb), Illinois Basin (Ib) and Forest City Basin (Fb). Specific sites within the Illinois Basin refer to active mines where upright trees are particularly common, well preserved and studied for this report.

As there are difficulties in defining and mapping 'cyclothems,' the concept should only be used informally (Kosanke *et al.*, 1960; Archer, 2008).

Coals

The stratigraphic succession encompassing the major economic coals of the Illinois Basin is the Desmoinesian Series of the Pennsylvanian System (Fig. 2). The Desmoinesian corresponds to the upper Moscovian and lower Kasimovian stages in the international time scale (Granstein *et al.*, 2012). Although coal beds ranging from Chesterian through Virgilian age have been mined in the Illinois basin, over 95% of resources and historic mining lie within the Desmoinesian Series. Within the Illinois Basin the major coals are of moderate thickness (typically 1 to 3 metres) but are regionally extremely extensive. Many coal beds can be traced with certainty from the Illinois basin into the Forest City basin (easternmost Kansas and environs) and probably into the northern Appalachian basin. In particular, the Herrin and Springfield coals are particularly widespread (Greb *et al.*, 2003).

Siliciclastic units

There two basic types of shale that directly overlie coal beds. This includes very black shale or shale that exhibits lighter-coloured shades of grey. The black, highly fissile phosphatic shale is commonly associated with thin limestone beds and both of these lithofacies can be very laterally continuous. This type of shale has been informally described as 'slate-like' or 'sheety' because it readily breaks into thin lamina and beds. Overall thicknesses of these units are rarely greater than 2 m and vertebrate and invertebrate faunas indicate widespread marine influences (Zangerl & Richardson, 1963). Some common fossils include the scallop *Dunbarella*, the inarticulate brachiopods *Lingula* and *Orbiculoidea*, occasional nautiloids and also fish remains, particularly sharks. Abundant conodonts provide a useful biostratigraphic framework. Within the subsurface, elevated amounts of radioactive elements yield a strong gamma-log response in the geophysical logs of black shale. Thus, the black shales are important for regional biostratigraphic and lithostratigraphic correlations. Many black shale units are so confidently correlated (e.g. Anna, Excello) that they carry the same name throughout the Illinois and Forest City basins.

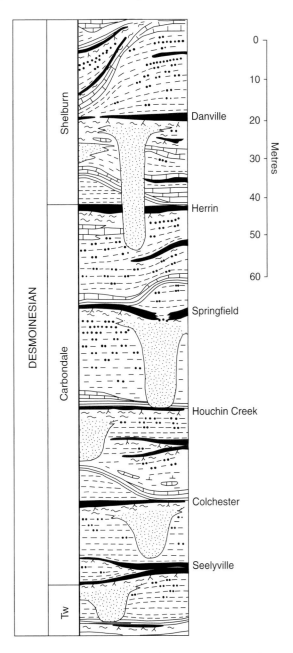

Fig. 2. Stratigraphic diagram showing positions of major coals and associated grey-shale wedges (tongue-like channels). Scale in metres.

At the other end of the shale spectrum, gray-shale units also include siltstone and sandstone. This lithofacies has been termed a 'gray shale wedge' (GSW) and, as compared to black shale, exhibits a much higher degree of lateral and vertical variability (Wanless, 1964). GSW facies are thickest adjacent to major river drainages that were contemporaneous with the widespread peat swamps (Hopkins, 1968; Gluskoter & Hopkins, 1970; Allgaier & Hopkins,

1975). GSWs are the most important factor in coal quality and thickness. Thicker coal seams, overlain by more than 6m-thick GSW, are lower in sulphur (1 to 2%) when compared to the usual levels of sulphur (3 to 5%) that occur in coal beds overlain directly by black shale. This pattern has been documented within several coal seams but most thoroughly for the Colchester, Springfield and Herbrin Coals (Hopkins 1968; Gluskoter & Simon, 1968, Gluskoter & Hopkins, 1970; Allgaier & Hopkins, 1975; Kvale & Archer, 1990).

Sequence stratigraphy

A prominent regional unconformity separates the older Mississippian System from the overlying Pennsylvanian strata in the Illinois Basin. A sub-Pennsylvanian unconformity is present in most places across the North American craton (Bristol & Howard, 1971, 1974).

GSW and the associated lateral lithostratigraphic variability were initially interpreted as the product of a terrestrial environment of deposition. This was related to the lack of obvious marine fossils and the common occurrence of well-preserved plant fossils. The progressive development of incised, valley-fill (IVF) models as well as tidal-estuarine models in the 1980s provided an alternative depositional model (Dalrymple et al., 1991, 1994, 2012; Archer & Kvale, 1993; Archer et al., 1994; Archer & Feldman, 1995; Nelson et al., 2002).

IVF models helped to explain the localised deposition within sub-Pennsylvanian valleys first noted by Bristol & Howard (1974) and also served to explain the GSWs initially described by Wanless (1964). Owing to concurrent palaeoglacial-eustatic flux, IVFs appear to be particularly common in the late Palaeozoic (Falcon-Lang, 2004; Falcon-Lang & DiMichele, 2010). Sites of abundant, well-preserved and diverse fossils, in many cases, can be related to IVF erosion and subsequent deposition (Feldman et al., 1993). Similar features termed 'incised channel fills' have also been described (Falcon-Lang et al., 2009).

OBSERVATIONAL DATA SUPPORTING CATACLYSMIC BURIAL

Recognition of tidal sedimentation within GSWs

In the late 1980s and early 1900s, cyclic ('tidal') rhythmites were described from GSWs exposed within coal mines in south-central Indiana (Kvale et al., 1989; Kvale & Archer, 1990; Archer & Kvale, 1993; Batemann, 1992) and north-eastern Illinois (Kuecher et al., 1990). These include essentially planar to low-angle, heterolithic rhythmites (Fig. 3A) that consist of low-angle, thinly interlaminated fine-grained sandstone and gray shale (Fig. 3B). Since their initial discovery, tidally influenced sedimentation has been recorded in many coal-bearing sections, including coals much older than those described herein; and from other coal basins in the eastern U.S. (Greb & Archer, 1995; Gastaldo et al., 2004a).

Where sediment accumulation was extremely rapid, the laminae within the rhythmites can exhibit a variety of small-scale cyclicity. The most common are neap-spring tidal bundles There are a number of analogues for such rhythmites and they have been documented within modern, hypertidal settings (Tessier et al., 1989; Dalrymple & Makino, 1989; Dalrymple et al., 1991, 2012; Tessier, 1993; Tessier et al., 1995; Archer, 2004, 2013). Ancient counterparts of tidal rhythmites exhibiting neap-spring cycles are also common throughout the geologic column (Archer, 1996a). Within inclined heterolith stratifications (IHS), sand-rich foresets are capped by thin mud drapes (Fig 3B).

Despite the strong imprint of tidal periodicities, geochemical evidence suggests that some deposition occurred within freshwater settings (Kvale & Mastalerz, 1998). The sparse fauna of GSW deposits, including pectinoid bivalves, linguloid brachiopods and the branchiopod *Leaia tricarinata*, are suggestion of fresh to brackish-water settings.

Due to the great extent of the supercontinent of Pangaea, palaeofluvial systems could easily have been of a similar magnitude, or larger, scale than the modern Amazonian system (Archer & Greb, 1995). Similar low salinity to freshwater conditions have been documented in modern hypertidal analogues of rhythmites (Archer, 2004, 2013). Today, the largest area of freshwater tidal flats is within the lower reaches of the Amazon River. Here, tidal effects extend for more than 1000 km inland and also extend out onto the continental shelf (Archer, 2005). In addition to the globe-spanning continent of Pangaea, the globe-spanning ocean of Panthalassa could have resonated strongly with the tide-raising forces of the moon and sun (Archer, 1996b). This could have resulted in the development of hypertidal conditions along the coastal areas of Pangaea.

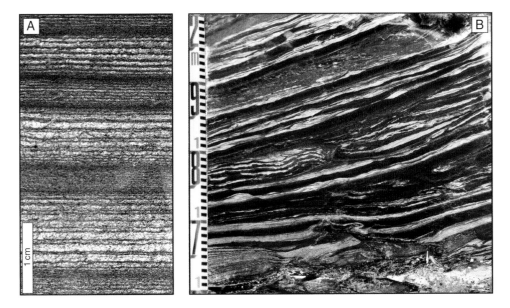

Fig. 3. (A) Cyclical, tidal rhythmite from Creek Paum Mine in southern Illinois. Neap-spring cycles and perigean-spring cycles are evident (Archer & Greb, 2011). (B) Close-up of mine wall showing inclined heterolithic stratification. Scale in cm and decimetres.

Upright plants, trees and forests

Upright forests of *in situ* trees have attracted a great deal of attention since the beginning of the formal study of Pennsylvanian geology and the number of mentions in the literature is very large (DiMichele & Falcon-Lang, 2011, 2012). Trees are so common in some mining regions that miners have given them nicknames such as 'kettlebottoms' and treat them as a serious hazard; the plug of rock within an upright fossil stump is prone to fall out of the mine roof (Chase & Sames, 1983). In the Illinois Basin such upright trees are commonly found almost exclusively within GSW facies. Large trees, ranging up to nearly 2 m in diameter, are often observed in the high walls of strip (opencast) mines. These trees seem to be concentrated along channel margins and the preserved stumps commonly attain vertical heights of 2 to 3 m. Upright, *in situ* trees and smaller plants (Fig. 4A) are common in GSW roof facies. Such in-place fossil plants are totally absent in mines with black shale or limestone roof facies. *In situ* fossil forests have been reported from a number of sites directly overlying coals (DiMichele & Nelson, 1989; DiMichele *et al.*, 2009). Within the Illinois Basin the more extensive entombed forest includes hundreds of upright trees that can be observed within single, underground mines (DiMichele & DeMaris, 1987; DiMichele *et al.*, 2009). Upright plants are also common above

laterally equivalent coals in the Appalachians (Chase & Sames, 1983; Gastaldo, *et al.*, 1995; DiMichele *et al.*, 1996; Gastaldo *et al.*, 2004b).

The most commonly encountered standing trees (Fig. 4A) are arborescent lycopsids or 'club-mosses', an extinct group of giant trees related to the modern, diminutive *Isoetes* (Bateman, 1992; Phillips & DiMichele, 1992; DiMichele & DeMaris, 1985; DiMichele & Phillips, 1994). Giant lycopsids were particularly prone to upright, *in situ* preservation because of their unique construction. The stems were supported by a thick rind of decay-resistant bark (Boyce *et al.*, 2010); such that water transport and support functions were entirely separated. This is quite different from modern seed-plant trees, where these functions are combined in the wood. As a consequence, the internal tissues of the ancient tree trunks decayed rapidly during or after tree maturation, producing a natural, cylindrical mould into which sediment could accumulate. The infilling sediments supported and preserved the hollow stump as cast and mould. The bark is commonly coalified. Lycopsid trees had an extensive, shallowly penetrating but tenacious rooting system, known as 'stigmaria'. These stout roots radiated horizontally up to at least 15 m from base of the trunk. Outrigger-like roots provided secure support for these tall (30 to 40 m) trees that often grew in soft

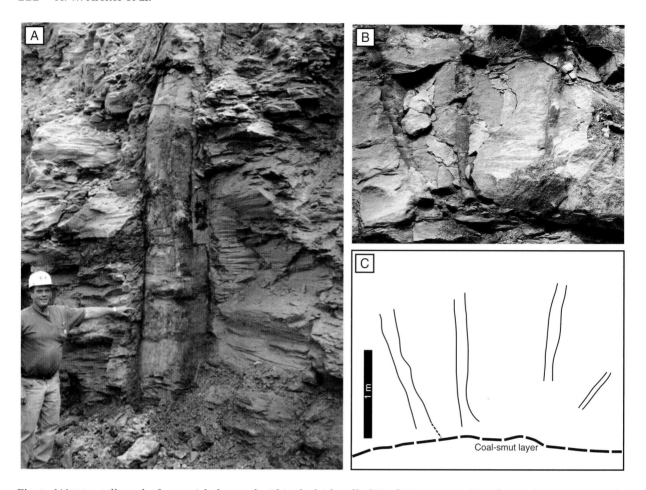

Fig. 4. (A) 3.3 m-tall trunk of an upright lycopod within the highwall of Creek Paum mine. Upright trunk is estimated to be 5.5 m in height. Photograph provided by Barry Sargent of Knight Hawk Coal. (B) Upright *Calamites* in shale overlying Murphysboro Coal in Jackson County, Illinois. (C) Tracing of B showing location of *Calamites* above a clastic-rich, coaly-shale layer ('coal smut').

substrates, such as uncompacted peat. Thus, lycopsids did not uproot and fall over like modern seed-plant trees. Despite several decades of mine-based research, none of the authors has ever seen an uprooted lycopsid trunk, nor have any of their field-experienced colleagues. Such trees would have stood upright for only a short period of time before decay eliminated the organic rind. Thus, all reported occurrences provide convincing evidence of rapid burial.

In addition to the lycopsids, smaller calamitean stems (Fig. 4B and C) are common and formed dense, monospecific stands (DiMichele *et al.*, 2009). This group is related to modern *Equisetum* and includes extant horsetails and scouring rushes. Calamiteans had, even during life, a hollow central stem. After death, the decay of tissue partitions within the stem permitted the entry of sediment. The result was upright preservation with a remaining thin coaly rind. Despite small stem diameters, calamiteans have been preserved to heights of several metres (DiMichele *et al.*, 2009) (Fig. 4B). Thus, a much smaller and architecturally very different plant also provides evidence of rapid burial within GSW facies.

Finally, much smaller scale but effectively *in situ* plants have also been recognised in association with upright lycopsid stumps (Gastaldo *et al.*, 2004a; DiMichele *et al.*, 2007). These smaller plants are generally represented by clumps of ground cover and by the stems and large foliar fragments of seed ferns and marattialean (true spore-bearing) ferns. Stems and large leaves of these plants do not form natural hollows during natural decay and would have been subject to rapid decay and degradation if exposed to oxidation. Thus, once the stems fell, rapid burial was necessary to isolate the plant remains from potential aerobic decay and destruction (Gastaldo, 2010). The famous Mazon Creek fauna of soft-bodied organisms from northern Illinois occurs in siderite concretions within a GSW overlying the Colchester Coal (Shabica & Hays, 1997). The preservation of such fossils demands rapid burial.

Thus, to conclude, a great variety of sizes of fossil plants have been preserved in place and encased within rhythmites. The plants are diverse and ranged from 30 m-tall lycopsids, several m-tall sphenopsids and smaller-scale ground foliage including ferns. Very high sedimentation rates, related to the local and regional development of hypertidal conditions, allowed the rapid creation of accommodation space and the entombing of plants. Finally, the grown and decay of Gondwanaland glaciations, particularly during periods of deglaciations also created abundant accommodation space for extended periods of deposition.

PROCESSES THAT CAN RAPIDLY INCREASE ACCOMMODATION SPACE

Peat compaction

A number of workers have tried to portray a palaeoenvironmental setting in which upright plants could be commonly preserved immediately above coal seams. It is widely accepted that the conversion of peat to coal involves some degree of compaction and dewatering. However, the timing of compaction is not well understood. Making a simplistic assumption that there was minimal internal compaction during peat accumulation, a 1 m-thick coal could have originally been a 10 to 20 m-thick peat. Assuming that approximately half of compaction occurred in the lower, buried portion of the peat pile, then the remaining compaction of the entire peat column could generate from 5 to 10 m of sediment accommodation space. This value falls within the measured heights of upright lycopsids discussed above. Compaction rates for various types of peats have been estimated to range as low as 1.4 : 1 up to rates of 30 : 1 (Ryer & Langer, 1980). Conversely, post-burial compaction, on the order of 3 : 1 at the most, was proposed by Wanless (1964) and Nadon (1998). Peat compaction could certainly generate some degree of accommodation space (Kvale & Archer, 1990), but whether or not the peat could compact in a short-enough timeframe is open for debate. Margins of coal beds commonly exhibit what appears to be significant

erosion (Fig. 5). This style of erosion suggests that significant peat compaction had already occurred prior to GSW deposition.

Occurrences of large, upright trees indicate that the peat would need to have compacted by several metres prior to the influx of the GSW. In addition, the IHS indicates that 2 to 4 m of water had to have been present so that the bar forms could migrate. Thus, a simple model invoking autocompaction alone does not seem to offer a viable explanation.

GSW rhythmites that directly overlie coal have a high degree of variability. In some parts of a mine, low-angle rhythmites are common (Fig. 6) and such rhythmites tend to contain relatively small upright plants. Close to the palaeochannels, GSW rhythmites are noticeably more thickly laminated, particularly in the lower parts of the section (Fig. 7). Inferred rates of vertical or lateral accretion would be decimetres or metres per month. Such rates commonly occur in modern hypertidal estuaries (Archer, 2013).

Despite the lingering uncertainties regarding the magnitude of Pennsylvanian peat compaction, this process, over regional scales, could provide a significant generation of accommodation space for the preservation of rapidly deposited, hypertidal lithofacies. Much more work on this topic will be necessary in order to understand the magnitudes and rates of both the coals as well as the enclosing sediments.

Faulting

Abrupt submergence caused by fault movements has been invoked as a mechanism to explain preservation of upright trees in the fossil record (Gastaldo *et al.*, 2004a). While examples can be cited from the modern record, investigations in the Illinois Basin disclose that tectonic movements played only a peripheral role in the burial and preservation of standing vegetation.

Similar to most of the Midcontinent and western United States, the Illinois Basin was tectonically active during the Pennsylvanian. This activity was part of the Ancestral Rocky Mountains (ARM) orogeny. Recurrent displacements along deep-seated strike-slip and high-angle reverse faults, with attendant folding of sedimentary cover, characterize this event. Most ARM structures in the Illinois Basin trend north to northwest and take the form of monoclines and belts of domes or anticlines at Pennsylvanian level. Chief among these are the Du Quoin Monocline, the La

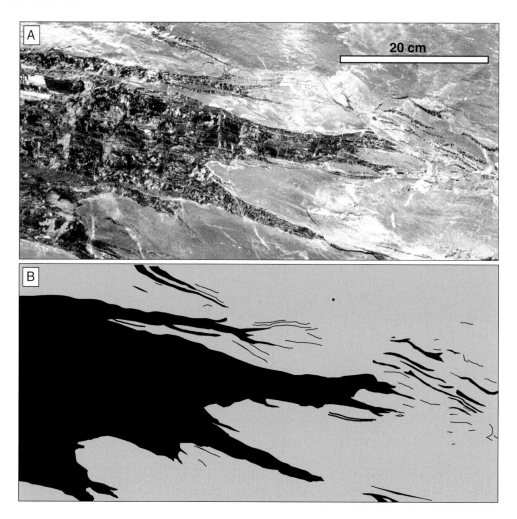

Fig. 5. (A) Frayed and abraded area of coal in an underground mine wall and (B) interpretation. Matrix sediment consists of basal grey-shale wedge lithofacies. The original peat must have been significantly compacted before the erosional event. This event was apparently penecontemporaneous with the deposition of the gray shale. Very thin coaly stringers are probably individual, coalified leaves and branches. Owing to the apparently resistant (felted) nature of the peat, the degree of erosion is remarkable and perhaps is related to event of significant energy, such as the passage of tidal bores.

Salle Anticlinorium and the Salem and Louden Anticlines (McBride & Nelson, 1999).

The structure of the Illinois Basin has been mapped in great detail, using the records of tens of thousands of exploration holes for oil, gas and coal along with seismic reflection and other geophysical surveys (Nelson, 1991; 1995; Nelson & Bauer, 1987; Kolata & Nelson, 1991). As a result, the structural framework of this basin is very thoroughly understood. Moreover, extensive and detailed mapping of faults and other structural disturbances has been carried out in underground coal mines (Krausse *et al.*, 1979; Nelson, 1981, 1983). Some of this underground mapping encompasses fields of fossil tree stumps; and the relationship of these fossils to regional features has been considered carefully.

On the grand scale, palaeochannels associated with GSW deposits and fossil trees tracked along trends of maximum basin subsidence. Locally, contemporaneous fault movements influenced channel trends. For example, the southern segment of the Walshville channel in the Herrin Coal parallels the eastern, downthrown side of the Du Quoin Monocline. *In situ* fossil stumps have been observed overlying the Murphysboro Coal in an area of south-western Illinois where thick coal and attendant GSW lie on the downthrown side of a monocline (Nelson *et al.*, 2011). A fossil forest above the Herrin Coal of east-central Illinois occurs on the downthrown side of an inferred north-east-striking basement fault (DiMichele *et al.*, 2007). However, the two best-known palaeochannels, the Walshville and the Galatia, each

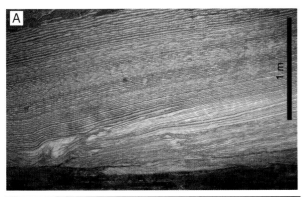

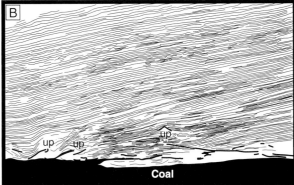

Fig. 6. (A) Mine-wall photograph and (B) interpretation (red lines) of cm-scale rhythmic bedding. Top of the Herrin Coal is exposed in the lower part of the image. Small upright plants ('up') are common along the top of the coal seam. In the lower left is an in-place, small tree trunk (rooted in underlying coal). Soft-sediment deformation around this trunk suggests the contemporaneous nature of plant growth and rhythmite deposition. This surface has been long-wall mined and thus has a slightly undulose surface produced by the cutter blades.

meandered for hundreds of kilometres across the basin are largely independent of mapped faults. In most of the mines where we have observed standing fossil trees, no tectonic disturbance of the coal and enclosing strata can be detected.

To summarise, rapid or sudden earth movements may have contributed to the burial of standing trees in isolated instances. However, the large majority of fossil forests in the Illinois Basin cannot be related to contemporaneous fault movements.

DYNAMICS OF DEGLACIATION

Applying Neogene models to Pennsylvanian Period settings

Dynamics of the deglaciation at the close of the Pleistocene are exceedingly complex and the study of global-climate changes is a very active

Fig. 7. (A) Mine-wall photograph and (B) interpretation of cross bedding and soft-sediment deformation. Planar sand beds (30 cm-thick) occur across the top the photograph. In places, these planar beds and other areas are obscured by white, limestone dust. This was sprayed on mine walls to reduce fire danger. Six individual sets of crossbeds are evident and these are separated by zones of slumping and extreme soft-sediment deformation.

area of on-going research. A number of calibrated data sets have been generated using proxy, palaeoclimatic information derived from a disparate variety of sources. These datasets, which include long ice cores, tree-ring measurements, submerged reef corals and speleothems, to name a few, provide very important constraints on the evaluation of climate-change and sea-level models (Milne *et al.*, 2005). Detailed understanding of Pleistocene processes can provide hypothetical models that can be compared to Pennsylvanian palaeoenvironments.

There are relatively direct linkages between global warming, deglaciation and sea-level changes. However, rapid changes in sea-level are not necessarily a direct response to climatically induced melting. Instead, periods of particularly rapid sea-level rise might relate much more directly to glacial hydrology (Shaw, 1989; Douglas, 1992; Alvarez *et al.*, 2011) than to simple, linear climate change.

Ice-rafted debris and Heinrich events

Within the late Pleistocene and Holocene, a number of Ice-Rafted-Debris (IRD) events have been recognised. IRD zones are found within cores taken from continental shelves and deep- ocean areas. These zones contain sand-sized sediment. As the background sedimentation was very fine grained, the occurrence of the thin, coarser zones is problematic. The coarser zones have been interpreted to be the result of armadas of icebergs that melted and dumped their load of coarser sediment.

Heinrich events are particularly well-developed IRD events and have an average duration of about 750 years (Maslin *et al.*, (2001). Roche *et al.* (2004) suggested that Heinrich event 4, which lasted for about 250 years, resulted in a 2 m sea-level rise. These events are related to global climatic fluctuations that coincide with the destruction of northern hemisphere glaciers. Heinrich events were produced by the release of a large volume of icebergs from tide-water glaciers. These events had an apparently abrupt onset that may have occurred within a few years (Maslin *et al.*, 2001). The Younger Dryas can be correlated with the youngest Heinrich Event (H0, see Fig. 8), which occurred at c. 12 ka. H1, the second Heinrich event occurred from 17 ka (Hemming, 2004) to 14 ka (Vidal, 1999). Vidal *et al.* (1999) discuss a linkage between such events in both the North and South Atlantic oceans.

Palaeomegafloods

The Last Glacial Maximum (LGM) was approximately 23 to 19 ka ago. The subsequent deglaciation included a great number of megafloods. The megafloods included the releases of meltwater from subglacial lakes as well as cataclysmic failure of large ice-based or morainal dams. To adequately describe the scale of these titanic discharges an oceanography term, the 'Sverdrup' unit, has been commonly applied. One Sverdrup (Sv) is equivalent to 10^6 m^3 s^{-1}. The entire average input of freshwater from the world's rivers into the ocean is approximately equal to 1 Sv. By comparison to modern large rivers, the freshwater

outflow of the Amazon River is approximately 0.3 Sv. The Meghna/Padma/Brahmaputra system and Chanjiang (Yangtze) have discharges of 0.05 Sv and 0.04 Sv, respectively. For comparison, the North American Mississippi River system has discharge of 0.02 Sv; thus it would require the simultaneous discharge of about 50 Mississippi-size rivers to equal 1 Sv.

In the following discussion of megafloods, we will first discuss the oldest and then proceed to the youngest events (Fig. 8). Following the LGM, some of the oldest megafloods were located in the region of Eurasia. Around 18 ka, large flows from Eurasian meltwater outbursts were spilling into the Aegean Sea (Baker, 2008).

Perhaps the best-studied series of megafloods occurred in the state of Washington in the northwestern U.S. The palaeofloods are commonly referred to as the 'Lake Missoula' floods or as the 'Spokane Flood.' Tremendous palaeofloods created an erosional landscape termed the 'Channelled Scablands' (Bretz, 1923; 1969). The Missoula floods were postulated to have been a single megaflood event (Bretz, 1969; Shaw *et al.*, 1999). Other workers have suggested multiple floods (Waitt, 1980). Benito & O'Conner (2003) suggested that some Lake Missoula discharges exceeded 3 Sv and that the largest single flood had a discharge near 10 Sv. This range of floods would have been equivalent to 10 to 33 times the discharges of the modern Amazon River system. Theoretical discharge models suggest that the lake could have been totally emptied into the ocean in about '100 days' (Shaw *et al.*, 1999, p. 608).

Within Arctic Canada, the Laurentide Ice Sheet formed a 3 km-thick dome over the Hudson Bay area approximately at 8.5 ka (Clark & Mix, 2002). During the on-going deglaciation, the moraine-dammed and ice-dammed megalakes included Lake Agassiz. As the thick ice sheet melted, approximately at 8.45 ka, megafloods emanating from this glacial megalake system included the most abrupt and widespread global-cooling event that has occurred over the last 100 ka (Clarke *et al.*, 2003, p. 923). Maximum discharges ranged from 5 to 10 Sv and the duration of the flow may have lasted 'less than one year' (Clarke *et al.*, 2003). The 8.2 ka cooling event is probably the combined result of these meltwater pulses. In Asia, megaflood discharges have been estimated to have been 11 Sv (Herget & Agatz, 2003). Very little is known about these systems at this time.

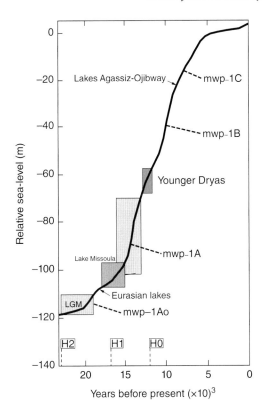

Fig. 8. Selected deglaciational events related to global changes in sea-level from 23 ka to the present (modified from Gornitz, 2009). Time range of the last glacial maximum (LGM) is based upon Clark & Mix (2002). Timings of meltwater-water pulses (mwp) are from Gornitz (2009). Megafloods related to large freshwater fluvio-glacial meltwater outbursts include: Eurasian lakes, outflows from Lake Missoula and discharges from the Lake Agassiz/Ojibway complex. Heinrich Events (H1 and H2) related to ice-rafted debris; H0 can be roughly equated to the Younger Dryas.

Melt-water Pulses (mwp)

Modern rates of sea-level rise have been estimated to average about 1.7 to 1.8 mm yr^{-1} (Douglas, 1992; Church *et al.*, 2004; Holgate & Woodworth, 2000). During the last deglaciation, short intervals exhibited rapid, sea-level rise and are referred to as 'melt-water pulses' ('mwp'). Even without a concurrent period of warming, rapid rises in sea-level could trigger: (1) destabilization of ice shelves, (2) collapse of continental-scale ice sheets and (3) initiate the release of tremendous volumes of glacial-megalake and subglacial meltwater (Weaver *et al.*, 2003). The mwp events were apparently very short and ranged from '10 to 10^3 years' (Gornitz, 2009, p. 887) and involved rapid changes in temperature and salinity of the oceans. Smaller scale modern events are termed jökulhlaups' and are

caused by subglacial melting or release of ice-dammed meltwater. Contemporary examples have been described as 'self-dumping glacial-dammed lakes' (Clague & Mathews, 1973, p. 501).

The most widely accepted meltwater pulses are designated as 'wmp-1A' and 'wmp-1B' (Fig. 8). These pulses have been discussed by many authors (see summary in Gornitz, 2009, table S1, p. 890). Event mwp-1A occurred from about 14.5 to 14.2 ka (Weaver *et al.*, 2003). Sea-level rose at rates exceeding 2 cm yr^{-1} or 20 m over a period of 1.0 ka (Fairbanks, 1989) whereas Weaver *et al.* (2003) estimated an average rate of rise of about 4 cm yr^{-1}. Based upon contemporary rates of sea-level rise of about 2 mm yr^{-1}, the estimated rates of flow for wmp-1A were 200 to 400 times quicker (Douglas, 1992; Holgate & Woodworth, 2008).

Other pulses, such as mwp-1Ao and mwp-1C (see Fig. 8) are far less widely noted. Event mwp-1Ao would be the oldest meltwater pulse and it ranged from about 19.2 ka to 19.0 ka (Yokoyama *et al.*, 2000; Clarke *et al.*, 2003). This pulse occurred within the youngest portion of the LGM. On the other hand, mwp-1C is the youngest major pulse and occurred at about 7.6 ka. This pulse caused a rise in sea-level of 4 cm yr^{-1} (Cronin *et al.*, 2007; Blanchon & Shaw, 1995). Older estimates for the timing of this pulse range from 8.2 ka (Törnqvist *et al.*, 2004) down to 7.6 ka (Blanchon & Shaw, 1995).

Comparison of Late Pennsylvanian and Late Cenozoic analogues

There are many similarities between Late Palaeozoic, coal-bearing rocks discussed herein and Late Cenozoic deglaciational events that have been compared and discussed herein. Both stratigraphic intervals occurred during extended periods of continental-scale cycles of glaciation and deglaciation. Based upon the observations discussed above, a typical Pennsylvanian grey-shale wedge (GSW) could have been initiated by a mwp event mechanically similar to Neogene events. A rapid rise in Pennsylvanian sea-level would have resulted in large, horizontal inland shifts of the coastline. The resultant transgression of an epicontinental basin, such as the low-slope Illinois Basin, would have quickly flooded coastal parts of widespread cratonic peats and rapidly progressed inland along riverine corridors. Initial generation of accommodation space could be related to peat compaction. Deposition of mud-rich laminae and mud-draped IHS, following the rapid compaction of the upper peat, resulted in slower and longer-term dewatering of the entire, laterally extensive peat bodies. Following this initial and internal peat compaction, sea-level rise related to a mwp could have allowed continued rapid deposition of tidal facies overlying the peat mires.

This scenario explains the initial and cataclysmic sedimentation that encased the upright plants. As the accommodation space was filled, there would be a shift to gradually decreasing sedimentation. Such observations are consistently seen within single mines and also in the repeatedly stacked 'cyclothems' that characterize the Illinois Basin coal measures. This depositional scenario also explains the common co-occurrence of thick coals and an overlying GSW. The thickest peats had a greater potential for internal dewatering and a higher potential for significant compaction.

The magnitude of the control of GSW deposition by mwp and Heinrich dynamics on the extremely flat landscape could have been very large. The botanical and sedimentological evidence demand rapid sedimentation within the lower GSW. The commonality of the upright trees, suggests that a synergistic combination of: (1) peat compaction, (2) mwp and Hienrich events and megafloods; and (3) rapid tidal sedimentation could have been a tipping point regarding the termination of peat accumulation. Over much longer periods of time and of a much smaller rate, a protracted regional to global transgression could have been driven by the processes related to Pangaean deglaciation.

CONCLUSIONS

The Late Palaeozoic encompasses the longest lasting glacial episodes that occurred during the Phanerozoic Eon. Within Pennsylvanian-age coals of the Illinois Basin a number of features can be interpreted to have been the result of coastal flooding and cataclysmic burial of mires by tidal-estuarine sedimentation. The dramatic flooding occurred during periods of active peat accumulation and fossil forests comprised of upright trees have been widely and spectacularly preserved.

The rapidity of the tidal-estuarine sedimentation was aided by a concurrent generation of sufficient accommodation. Several different processes have been put forth to explain the forests of upright trees. These include: (1) peat dewatering and compaction, (2) local to regional fault-related subsidence, or (3) dramatic, short-term rises in ancient sea-level changes related to glacio-eustacy. The notably and repetitious occurrences of GSW facies includes their deposition directly overlying major, economic coals. There is a very consistent stratigraphic occurrence of GSW facies that encased upright forests. These forests were growing in swamps that were subsequently converted to bituminous coal. The stratigraphic repetition indicates that unusual local or even regional factors were not the primary, underlying drivers. Invoking a global-scale, allogenic event of short duration appears to offer a more reasonable hypothesis.

Cenozoic glacial processes, which have occurred since the last LGM and throughout the ensuing deglaciation, indicate that sea-levels did not rise in a slow and uniform manner. During specific short-term intervals, melt-water pulses (mwp) resulted in very rapid and short-term rises of global sea-level. Some mwp events can be correlated to specific events of cataclysmic flooding caused by fluvio-glacial megafloods, such as those related to the megafloods released from Lake Missoula. It is suspected that similar large-scale mwp could have been common during Gondwana deglaciations. This simple mwp-based model could explain the common preservation of upright fossil forests within GSW that were originally rooted within the underlying coal beds.

ACKNOWLEDGEMENTS

Many persons and entities allowed access to and information regarding the underground and surface mines discussed in this manuscript. Without access to active mines, this research would not have been possible. Gregory M. Molinda (CDC/ NIOSH/PRL) provided some very useful photographic information from underground mines. WD acknowledges funding from the NMNU Small Grants Program to support fieldwork.

REFERENCES

Allgaier, **G.J.** and **Hopkins**, **M.E.** (1975) Reserves of the Herrin (No. 6) Coal in the Fairfield Basin in southeastern Illinois. *Illinois State Geol. Surv. Circular*, **489**, 31 pp.

Alvarez, **S.J.**, **Montoya**, **M.**, **Ritz**, **C.**, **Ramstein**, **G.**, **Charbit**, **S.**, **Dumas**, **C.**, **Nisancioglu**, **K.H.**, **Dokken**, **T.** and **Ganopolski**, **K.A.** (2011) Heinrich event 1: an example of dynamical ice-sheet reaction to oceanic changes. *Climate of the Past*, **7**, 1297–1306.

Archer, A.W. (2008) Cyclic sedimentation (cyclothem). In: *Encyclopedia of Paleoclimatology and Ancient Environments* (Ed. V. Gornitz), pp. 226–228, Springer, Dordrecht, The Netherlands.

Archer, A.W. (1996a) Panthalassa: paleotidal resonance and a global paleo-ocean seiche: *Paleoceanography*, **11**, 625–632.

Archer, A.W. (2004) Recurring assemblages of biogenic and physical sedimentary structures in modern and ancient extreme macrotidal estuaries. *J. Coastal Res.*, **43**, 4–22.

Archer, A.W. (1996b) Reliability of lunar orbital periods extracted from ancient cyclic tidal rhythmites. *Earth Planet. Sci. Lett.*, **141**, 1–10.

Archer, A.W. (2005) Review of Amazonian depositional systems. *Int. Assoc. Sedimentol. Spec. Publ.*, **35**, 17–39.

Archer, A.W. (2013) World's highest tides: Hypertidal coastal systems in North America, South America and Europe. *Sed. Geol.*, **284–285**, 1–25.

Archer, A.W. and Feldman, H.R. (1995) Incised valleys and estuarine facies of the Douglas Group (Virgilian): implications for similar Pennsylvanian sequences in the U.S. Mid-Continent. In: *Sequence stratigraphy of the mid-Continent* (Ed N. Hyne) Tulsa Geological Society, Tulsa, Oklahoma, pp. 119–140.

Archer, A.W., Feldman, H.R., Kvale, E.P. and Lanier, W.P. (1994) Pennsylvanian (Upper Carboniferous) fluvio- to tidal-estuarine coal-bearing systems: delineation of facies transitions based upon physical and biogenic sedimentary structures. *Palaeogeogr. Palaeoclimatol., Palaeoecol.*, **106**, 171–185.

Archer, A.W. and Greb, S.F. (1995) An Amazon-scale drainage in the early Pennsylvanian of Central North America: *J. Geol.*, **103**, 611–628.

Archer, A.W. and Greb, S.F. (2011) Hypertidal facies from the Pennsylvanian Period: Eastern and Western Interior Coal Basins, USA. In: *Principles of tidal sedimentology* (Eds R.A. Davis, Jr. and R.W. Dalrymple) pp. 421–436. Springer. The Netherlands.

Archer, A.W. and Kvale, E.P. (1993) Origin of gray-shale lithofacies ('clastic wedges') in U.S. midcontinental coal measures (Pennsylvanian): an alternative explanation. In: *Modern and Ancient Coal-Forming Environments* (Eds J.C. Cobb and B. Cecil), *Geol. Soc. Am. Spec. Pap.*, **286**, 181–192.

Baker, V. (2008) Late Quaternary megafloods. In: *Encyclopedia of paleoclimatology and ancient environments* (Ed V. Gornitz) pp. 506–507. Springer, The Netherlands.

Bateman, R.M. (1992) Evolutionary-developmental change in the growth architecture of fossil rhizomorphic lycopsids: scenarios constructed on cladistics foundations. *Biol. Rev.*, **69**, 527–597.

Benito, G. and O'Connor, J.E. (2003) Number and size of last-glacial Missoula floods in the Columbia River valley between the Pasco Basin, Washington and Portland, Oregon. *Geol. Soc. Am. Bull.*, **115**, 624–638.

Blanchon, P. and Shaw, J. (1995) Reef drowning during the last deglaciation: evidence for catastrophic seal-level rise and ice-sheet collapse. *Geology*, **39**, 4–8.

Boyce, C.K., Abrecht, M., Zhou, D. and Gilbert, P.U.P.A. (2010) X-ray photoelectron emission spectromicroscopic analysis of arborescent lycopsid cell wall composition and Carboniferous coal ball preservation. *Int. J. Coal Geol.*, **83**, 146–153.

Bretz, J.H. (1923) The Channeled Scablands of the Columbia Plateau. *J. Geol.*, **31**, 617–649.

Bretz, J.H. (1969) The Lake Missoula floods and the Channeled Scabland. *J. Geol.*, **77**, 505–543.

Bristol, H.M. and Howard, R.H. (1971) Paleogeographic map of the sub-Pennsylvanian Chesterian (upper Mississippian) surface in the Illinois Basin. *Illinois State Geological Survey Circular*, **458**, 16 p.

Bristol, H.M. and Howard, R.H. (1974) Sub-Pennsylvanian valleys in the Chesterian surface of the Illinois Basin and related Chesterian slump blocks. *Geol. Soc. Am. Spec. Pap.*, **148**, 315–335.

Chase, F.E. and Sames, G.P. (1983) Kettlebottoms: their relation to mine roof and support: *U.S. Bureau of Mines, Report of Investigations*, 8785, 12 p.

Church, J.A., White, N.J., Coleman, R., Lambeck, K. and Mitrovica, J. (2004) Estimates of the Regional Distribution of Sea-level Rise over the 1950–2000 Period. *J. Climate*, **17**, 2609–2625.

Clague, J.J. and Mathews, W.H. (1973) The magnitude of jökulhlaups. *J. Glaciology*, **12**, 501–504.

Clark, P.U. and Mix, A.C. (2002) Ice Sheets and sea-level of the Past Glacial Maximum. *Quatern. Sci. Rev.*, **21**, 1–7.

Clarke, G., Leverington, D., Teller, J. and Dyke, A. (2003) Superlakes, megafloods and abrupt climate change. *Science*, **301**, 922–923.

Cronin, T.M., Vogt, P.R., Willard, D.A., Thunell, R., Halka, J., Berke and M., Pohlman, J. (2007) Rapid sea-level rise and ice sheet response to 8, 200-year climate event. *Geophy. Lett.*, **34**, L20603.

Dalrymple, R.W., Boyd, R. and Zaitlin, B.A. (1994) Incised-valley systems: origin and sedimentary sequences: *SEPM Spec. Publ.*, **51**, 391 p.

Dalrymple, R.W., Mackay, D.A., Ichaso, A.A. and Choi, K.S. (2012) Processes, morphodynamics and facies of tide-dominated estuaries. In: *Principles of Tidal Sedimentology* (R.A. Davis Jr. and R.W. Dalrymple). Springer, pp. 79–107.

Dalrymple, R.W. and Makino, Y. (1989) Description and genesis of tidal bedding in the Cobequid Bay-Salmon River estuary, Bay of Fundy, Canada In: *Sedimentary Facies of the Active Plate Margin* (Eds A. Taira and F. Masuda) pp. 151–177. Terra Publishing, Tokyo.

Dalrymple, R.W., Zaitlin, B.A. and Boyd, R. (1991) Estuarine facies models: conceptual basis and stratigraphic implications. *J. Sed. Petrol.*, **62**, 1130–1146.

DiMichele, W.A. and DeMaris, P.J. (1985) Arborsecent lycopod reproduction and paleoecology in a coal-swamp environment of late Middle Pennsylvanian age (Herrin Coal, Illinois, U.S.A.). *Rev. Palaeob. Palynol.*, **44**, 1–26.

DiMichele, W.A. and DeMaris, P.J. (1987) Structure and dynamics of a Pennsylvanian-age *Lepidodendron* forest: colonizers of a disturbed swamp habitat in the Herrin (No. 6) Coal of Illinois. *Palaios*, **2**, 146–157.

DiMichele, W.A., Eble, C.F. and Chaney, D.S. (1996) A drowned lycopsid forest above the Mahoning coal (Conemaugh Group, Upper Pennsylvanian) in eastern Ohio, U.S.A. *Int. J. Coal Geol.*, **31**, 249–276.

DiMichele, W.A. and Falcon-Lang, H.J. (2011) Pennsylvanian 'fossil forests' in growth (To assemblages): origin, taphonomic bias and palaeoecological insights. *J. Geol Soc. London*, **169**, 585–605.

DiMichele, W.A. and Falcon-Lang, H.J. (2012) Calamitalean 'pith casts' reconsidered. *Rev. Palaeob. Palynol.*, **173**, 1–14.

DiMichele, W.A., Falcon-Lang, H.J., Nelson, W.J., Elrick, S. and Ames, P. (2007) Ecological gradients within a Pennsylvanian forest: *Geology*, **35**, 415–418.

DiMichele, W.A. and Nelson, W.J. (1989) Small-scale spatial heterogeneity in Pennsylvanian-age vegetation from the roof shale of the Springfield Coal (Illinois Basin). *Palaios*, **4**, 276–280.

DiMichele, W.A., Nelson, W.J., Elrick, S. and Ames, P.R. (2009) Catastrophically buried Middle Pennsylvanian *Sigillaria* and calamitean sphenopsides from Indiana, USA: What kind of vegetation was this? *Palaios*, **24**, 159–166.

DiMichele, W.A., Pfefferkorn, H.W. and Gastaldo, R.A. (2001) Response of Late Carboniferous and Early Permian plant communities to climate changes. *Annu. Rev. Earth and Planet. Sci.*, **29**, 461–487.

DiMichele, W.A. and Phillips, T.L. (1994) Paleobotanical and paleoecological constraints on models of peat formation in the Late Carboniferous of Euramerica. *Palaeogeogr., Palaeoclimatol, Palaeoecol.*, **106**, 39–90.

Douglas, B.C. (1992) Global sea-level rise, *J. Geophys. Res.*, **96**, 6981–6992.

Fairbanks, R.G. (1989) A 17,000-year glacio-eustatic sea-level record: influence of glacial melting rates on the Younger Dryas event and deep-ocean circulation. *Nature*, **342**, 637–642.

Falcon-Lang, H.J. (2004) Pennsylvanian tropical rainforests responded to glacial-interglacial rhythms. *Geology*, **32**, 689–692.

Falcon-Lang, H.J. and DiMichele, W.A. (2010) What happened to the coal forests during Pennsylvanian glacial phases? *Palaios*, **25**, 611–617.

Falcon-Lang, H.J., Nelson, W.J., Elrick, S., Looy, C.V., Ames, P.R. and DiMichele, W.A. (2009) Incised channel fills containing conifers indicate that seasonally dry vegetation dominated Pennsylvanian tropical lowlands. *Geology*, **37**, 923–926.

Feldman, H.R., Archer, A.W., Kvale, E.P., Cunningham, C.R., Maples, C.G. and West, R.R. (1993) A tidal model of Carboniferous Konservat Laggerstätten formation: *Palaios*, **8**, 485–498.

Gastaldo, R.A. (2010) Peat or no peat: why do the Rajang and Mahakam deltas differ? *Inter. J. Coal Geol.*, **83**, 162–172.

Gastaldo, R.A., H.W. Pfefferkorn and W.A. DiMichele (1995) Characteristics and classification of Carboniferous roof shale floras. In: *Historical perspectives of early Twentieth Century Carboniferous paleobotany in North America* (Eds P.C. Lyons, E.D. Morey and R.H. Wagner) *Geol. Soc. Am. Mem.*, **185**, 341–352.

Gastaldo, R.A., Stevanovic-Walls, I.M. and Ware, W.N. (2004a) Erect forests are evidence for coseismic base-level changes in Pennsylvanian cyclothems of the Black Warrior Basin, U.S.A. In: *Sequence stratigraphy, paleoclimate and tectonics of coal-bearing stratigraphy* (Eds J.C. Pashin and R.A. Gastaldo), *AAPG Studies in Geology*, **51**: 219–238.

Gastaldo, R.A., Stevanovic-Walls, I.M., Ware, W.N. and Greb, S.F. (2004b) Community heterogeneity of early Pennsylvania peat mires. *Geology*, **32**, 693–696.

Gluskoter, H.J. and Hopkins, M.E. (1970) Distribution of sulfur in Illinois coals. In: *Depositional environments in parts of the Carbondale Formation - Western and northern Illinois* (W.H. Smith, R.B. Nance and R.G. Johnson). *Illinois State Geol. Surv. Guidebook Series*, **8**, 89–95.

Gluskoter, H.J. and Simon, J.A. (1968) Sulfur in Illinois coals. *Illinois State Geol. Surv. Circ.*, **432**, 28 p.

Gornitz, V. (2009) Sea-level change, post-glacial. In: *Encyclopedia of paleoclimatology and ancient environments* (Ed. V. Gornitz) pp. 887–893. Springer, Dordrecht, The Netherlands.

Gradstein, F.M., Ogg, J.G., Schmitz, M.D. and Ogg, G.M. (2012) The geologic time scale. Elsevier, New York, 1176 p.

Greb, S.F. Andrews, W.M., Eble, C.F., DiMichele, W., Cecil, C.B. and Hower, J.C. (2003) Desmoinesian coal beds of the Eastern Interior and surrounding basins: The largest tropical peat mires in earth history. In: *Extreme depositional environments: Mega-end members in geologic time* (Eds M.A. Chan and A.W. Archer), *Geol. Soc. Am. Spec. Pub.*, **370**, 127–150.

Greb, S.F. and Archer, A.W. (1995) Rhythmic sedimentation in a mixed tide and wave deposit, eastern Kentucky, U.S.A. *J. Sed. Res.*, **B65**, 96–106.

Hemming, S.R. (2004) Heinrich events: massive Late Pleistocene detritus layers of the North Atlantic and their global climate imprint. *Rev. Geophys*, **370**, 127–150.

Herget, J. and Agatz, H. (2003) Modelling ice-dammed lake outburst floods in the Altai Mountains (Siberia) with HEC-RAS. In: *Palaeofloods, historical data and climatic variability: applications in flood risk assessment* (Eds V.R. Thorndycraft, G. Benito, N.M. Barriendos and M.S. Llast) Madrid, Proceedings of the PHEFRA Workshop, Centro de Ciencias Medioamientales pp. 177–181.

Holgate, S.J. and Woodworth, P.L. (2000) Evidence for enhanced coastal sea-level rise during the 1990s. *Geophys. Res. Lett.*, **31**, L07305.1-L07305.4.

Hopkins, M.E. (1968) Harrisburg (No. 5) Coal reserves of southeastern Illinois. *Illinois State Geol. Surv. Circ.* **431**, 25 p.

Kolata, D.R. and Nelson, W.J. (1991) Tectonic history of the Illinois Basin. *AAPG Mem.*, **51**, 25 p.

Kosanke, R.M., Simon, J.A., Wanless, H.R. and Willman, H.B. (1960) Classification of the Pennsylvanian strata of Illinois. *Illinois State Geol. Surv. Report of Investigations*, **214**, 84 p.

Krausse, H.-F., Damberger, H.H., Nelson, W.J., Hunt, S.R., Ledvina, C.T., Treworgy, C.G. and White, W.A. (1979) Roof strata of the Herrin (No. 6) Coal Member in mines of Illinois: their geology and stability: *Illinois State Geological Survey, Illinois Minerals Note*, **72**, 53 p.

Kuecher, G.M., Woodland, B.G. and Broadhurst, F.M. (1990) Evidence of deposition from individual tides and of tidal cycles from the Francis Creek Shale (host rock to the Mazon Creek Biota, Westphalian D (Pennsylvanian), northeastern Illinois. *Sed. Geol.*, **68**, 211–221.

Kvale, E.P. and Archer, A.W. (1990) Tidal deposits associated with low-sulfur coals, Brazil Fm. (Lower Pennsylvanian), Indiana. *J. Sed. Petrol.*, **60**, 563–574.

Kvale, E.P., Archer, A.W. and Johnson, H.J. (1989) Daily, monthly and yearly tidal cycles within laminated siltstones of the Mansfield Formation (Pennsylvanian) of Indiana. *Geology*, **17**, 365–368.

Kvale, E.P. and Mastalerz, M. (1998) Evidence of ancient freshwater tidal deposits. In: *Tidalites: Processes and Products* (Eds C.R. Alexander, R.A. Davis and V.J. Henry), *SEPM Spec. Pub.*, **61**, p. 95–107.

Maslin, J., Seidov, D. and Lowe, J. (2001) Synthesis of the nature and causes of rapid climate transition during the Quaternary. *Geophys. Monogr.*, **126**, 9–52.

McBride, J.H. and Nelson, W.J. (1999) Style and origin of mid-Carboniferous deformation in the Illinois Basin, USA. Ancestral Rocky Mountains deformation. *Tectonophysics*, **305**, 249–273.

Milne, G.A., Long, A.J. and Bassett, S.E. (2005) Modeling Holocene relative sea-level observations from the Caribbean and South America. *Quatern. Sci. Rev.*, **24**, 1183–1202.

Moore, R.C. (1964) Paleoecological aspects of Kansas Pennsylvanian and Permian cyclothems. In: *Symposium on Cyclic Sedimentation* (Ed D.F. Merriam) *Kansas Geol. Surv. Bull.*, **169**, 287–380.

Moore, R.C. (1935) Stratigraphic classification of the Pennsylvanian rocks of Kansas. *Kansas Geol. Surv. Bull.*, **22**, 256 p.

Nadon, G.C. (1998) Magnitude and timing of peat-to-coal compaction. *Geology*, **26**, 727–730.

Nelson, W.J. (1981) Faults and their effect on coal mining in Illinois. *Illinois State Geol. Surv. Circ.*, **523**, 45 p.

Nelson, W.J. (1983) Geologic disturbances in Illinois coal seams. *Illinois State Geol. Surv. Circ.* **530**, 52 p.

Nelson, W.J. (1995) Structural features in Illinois: *Illinois State Geological Survey, Bulletin* **100**, 144 p.

Nelson, W.J. (1991) Structural styles in the Illinois Basin. *AAPG Mem.*, **51**, 209–246.

Nelson, W.J. and Bauer, R.A. (1987) Thrust faults in southern Illinois Basin – result of contemporary stress? *Geol. Soc. Am. Bull.*, **98**, 302–307.

Nelson, W.J., Devera, J.A., Williams, L.M. and Staub, J.R. (2011) Bedrock geology of Oraville quadrangle, Jackson County, Illinois. *Illinois State Geological Survey, Illinois Geologic Quadrangle Map IGQ Oraville-BG*, 2 sheets, scale 1:24,000.

Nelson, W.J., Smith, L.B. and Treworgy, J.D. (2002) Sequence stratigraphy of the Lower Chesterian (Mississippian) strata of the Illinois Basin. *Illinois State Geological Survey Bulletin* **107**, 70 p.

Phillips, T.L. and DiMichele, W.A. (1992) Comparative ecology and life-history biology of arborescent lycopods in Late Carboniferous swamps of Euramerica. *Annals of the Missouri Botanical Garden*, **79**, 560–588.

Roche, D., Paillard, D. and Corgijo, E. (2004) Constraints on the duration and freshwater release of Heinrich event 4 through isotope modeling. *Nature*, **432**, 379–382.

Ryer, T.A. and Langer, A.W. (1980) Thickness change involved in the peat-to-coal transformation for a bituminous coal of Cretaceous age in central Utah. *J. Sed. Petrol.*, **50**, 987–992.

Shabica, C.W. and Hays, A.A. (1997) Richardson's guide to the fossil fauna of Mazon Creek. *Northeastern Illinois University, Chicago*, 308 p.

Shaw, J. (1989) Drumlins, subglacial meltwater floods and ocean responses. *Geology*, **17**, 853–856.

Shaw, J., Munro-Stasiuk, M., Sawyer, B., Beaney, C., Lesemann, J-E., Musacchio, A., Rains, B. and Young, R.R. (1999) The Channeled Scabland: back to Bretz? *Geology*, **27**, 605–608.

Tessier, B. (1993) Upper intertidal rhythmites in the Mont-Saint-Michel Bay (NW France): perspectives for paleoreconstruction. *Mar. Geol.*, **110**, 355–367.

Tessier, B., Archer, A. W., Lanier, W. P. and Feldman, H. R. (1995) Comparison of ancient tidal rhythmites (Carboniferous of Kansas, U.S.A.) with modern analogues (the Bay of Fundy and Mont-Saint-Michel, N.W. France). *Int. Assoc. Sedimentol. Spec. Publ.*, **24**, 259–271.

Tessier, B., Monfort, Yt ., Gigot, P. and Larsonneur, C. (1989) Enregistreme des tidaux en accretion verticale, ples en baie du Mont-Saint-Michele et dans la molasses marine Miocene du basin de Digne. *Bull. Soc. Geologie France*, **5**, 1029–1041.

Törnqvist, T.E., Bick, S.J., Gonzalez, J.L., van der Borg, I. and de Jong, A.F.M. (2004) Tracking the sea-level signature of the 8.2 ka cooling event: new constraints from the Mississippi Delta. *Geophys. Res. Lett.*, **31**, L23309.

Vidal, L., Schneider, R.R., Marchal, O., Bickert, T., Stocker, T.F. and Wefer, G. (1999) Link between the North and South Atlantic during the Heinrich events of the last glacial period. *Climate Dynam.*, **15**, 909–919.

Waitt, R.B., Jr. (1980) About 40 last-glacial Lake Missoula jökulhlaups through southern Washington. *J. Geol.*, **88**, 653–679.

Wanless, H.R. (1964) Local regional factors in Pennsylvanian cyclic sedimentation. In: *Symposium on cyclic sedimentation* (Ed. D.F. Merriam). *Kansas Geol. Surv. Bull.*, **169**, pt. 2, 593–606.

Wanless, H.R. and Weller, J.M. (1932) Correlation and extent of Pennsylvanian cyclothems. *Geol. Soc. Am. Bull.*, **43**, 1177–1206.

Weaver, A. J., Saenko, O.A., Clark, P.U. and Mitrovica, J.X. (2003) Meltwater Pulse 1A from Antarctica as a trigger of the Bølling-Allerød warm interval. *Science*, **299**, 1709–1713.

Weller, J.M. (1930) Cyclical sedimentation of the Pennsylvanian Period and its significance: *J. Geol.*, **38**, 97–135.

Weller, J.M. (1931) The conception of cyclical sedimentation during the Pennsylvanian Period. *Illinois State Geol. Surv. Bull*, **60**, 163–177.

Wilkinson, B.H., Merrill, G.K. and Kivett, S.J. (2003) Stratal order in Pennsylvanian cyclothems. *Geol. Soc. Am. Bull.*, **115**, 1068–1087.

Yokayama, Y., Lambeck, K., De Deckker P., Johnston, P. and Fifield, L.K. (2000) Timing of the Last Glacial Maximum from observed sea-level minima. *Nature*, **406**, 713–716.

Zangerl, R. and Richardson, E.S. (1963) The paleoecological history of two Pennsylvanian black shales. *Fieldiana Geol. Mem.*, **4**, 1–352.

Zeller, E.J. (1964) Cycles and psychology. In: *Symposium on cyclic sedimentation* (Ed. D.F. Merriam), *Kansas Geol. Surv. Bull.*, **169**, pt. 2, 631–636.

Tidal ravinement surfaces in the Pleistocene macrotidal tide-dominated Dong Nai estuary, southern Vietnam

TOSHIYUKI KITAZAWA* and NAOMI MURAKOSHI[†]

* Faculty of Geo-environmental Science, Rissho University, Kumagaya, 360-0194, Japan (E-mail: kitazawa@ris.ac.jp)
[†] Faculty of Science, Shinshu University, Matsumoto, 390-8621, Japan

ABSTRACT

Outcrop investigation of tidal ravinement surfaces (TRSs) in macrotidal, tide-dominated, estuary infill sediments deposited along the Dong Nai river in southern Vietnam, during the last interglacial period, identified two types of TRS: a bar-TRS and a wave/tidal ravinement surface (W/TRS). The bar-TRS was identified in the landward part of the estuary at the base of tidal sand-bar deposits. The W/TRS was identified in the seaward part of the estuary at the base of subtidal lag deposits. The bar-TRS is amalgamated with a sequence boundary and is overlain by a transgressive upward-coarsening succession. The W/TRS is amalgamated with a sequence boundary and the bar-TRS and is overlain by subtidal lag deposits, which are in turn overlain by an upward-fining highstand succession. In the intermediate part of the estuary profile, the bar-TRS is overlain by a transgressive upward-coarsening succession, which is in turn overlain by subtidal lag deposits.

Keywords: Tidal ravinement surface, tide-dominated estuary, macrotidal, tidal sand bar, Vietnam.

INTRODUCTION

Tidal ravinement surfaces (TRS: Allen, 1991; Allen & Posamentier, 1993) are important sequence stratigraphic boundaries in incised-valley systems because, in contrast to wave ravinement surfaces, they are developed only within the incised valleys and not on interfluves (Zaitlin et al., 1994). Although TRSs have been described in many estuaries, the term TRS is used commonly to describe amalgamated truncations, both those at the base of tidal inlet channels cutting barrier islands and those at the base of flood-tidal deltas in wave-dominated or wave and tide-dominated estuaries (e.g. Allen & Posamentier, 1993; Zaitlin et al., 1994).

In a transgressive tide-dominated estuary, a TRS is created by amalgamation of channel scours as a result of landward migration of tidal channels separating tidal sand bars (Dalrymple et al., 1992), which are elongate along the axis of a funnel-mouthed estuary. Outcrop-scale sedimentologic features of TRSs formed in tide-dominated estuaries are not well understood. Only a few examples have been described: Cretaceous TRSs in southern England (Wonham & Elliott, 1996; Yoshida et al., 2004); Eocene TRSs in Spitsbergen (Plink-Björklund, 2005); a Pleistocene TRS in southern Vietnam (Kitazawa, 2007); and Cretaceous TRSs in Wyoming–Utah, USA (Plink-Björklund, 2008). Recently, modern and ancient tide-dominated estuary infills, mainly interpreted from seismic profiles or borehole correlations, were reviewed by Tessier (2012), who considered the differences between TRSs in tide-dominated and wave-dominated estuaries. Recent studies of modern French incised-valley fills have shown that there can be a variety of sedimentary systems and TRS shapes in open, tide-dominated estuaries (Chaumillon, 2010; Tessier et al., 2010a, 2010b).

An important aim of this paper is to provide outcrop-scale descriptions of TRSs in Pleistocene

Contributions to Modern and Ancient Tidal Sedimentology: Proceedings of the Tidalites 2012 Conference,
First Edition. Edited by Bernadette Tessier and Jean-Yves Reynaud.
© 2016 International Association of Sedimentologists. Published 2016 by John Wiley & Sons, Ltd.

macrotidal tide-dominated estuary deposits on the basis of field observations rather than interpretations of seismic profiles or correlations of borehole data. By studying outcrop scale of TRS one can better understand the spatial distribution of sequence stratigraphic surfaces. By presenting a classification of TRSs on the basis of outcrop-scale features, the results of this study will contribute to the interpretation of the architectural development of other tidally influenced estuary deposits.

STUDY AREA AND LOCAL GEOLOGY

The study area is located along the Dong Nai River, southern Vietnam, near its junction with the Sai Gon River (Fig. 1). The Dong Nai River flows south through Bien Hoa and the Sai Gon River flows southeast through Ho Chi Minh City; both join the Nha Be River, which debouches into the South China Sea 50 km south of the study area. The floodplain of the Nha Be River system adjoins that of the Mekong River.

Middle to Upper Pleistocene deposits are exposed along the Dong Nai River. The Thu Duc Formation unconformably overlies the Ba Mieu Formation in this area (Fig. 2). The Ba Mieu Formation was deposited around marine isotope stage (MIS) 7 and the Thu Duc Formation during MIS 5–3 (Kitazawa *et al.*, 2006; Kitazawa *et al.*, in press). Kitazawa & Tateishi (2005) and Kitazawa (2007) suggested that these sequences represent a tide-dominated estuary and delta succession deposited in incised valleys. These formations are the only reported examples of Pleistocene tide-dominated estuary deposits seen in outcrop. The thickness of intertidal deposits indicates that the tidal range of the Thu Duc Formation was about 5 m (Kitazawa, 2007).

METHODS

The lower part of the Thu Duc Formation was investigated at 17 outcrops in the Nhon Trach and Phuoc Tan regions (Fig. 2). Outcrop-scale features and lateral changes of the erosional surfaces and the deposits were investigated along the palaeo-Dong Nai estuary. TRSs were classified on the basis of the characteristics of the deposits that overlie them, such as grain size, sedimentary structures and the presence of upward-coarsening/fining successions.

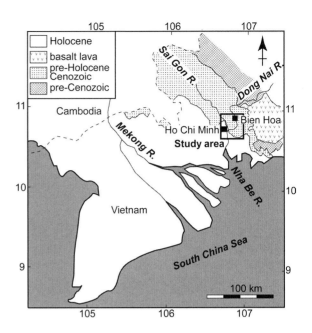

Fig. 1. Regional geologic map showing the location of the study area. Latitude and Longitude in degrees, North and East.

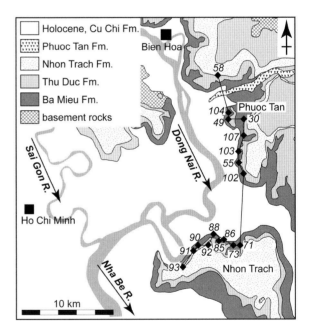

Fig. 2. Geologic map of the study area (modified from Kitazawa *et al.*, in press) showing outcrops investigated.

RESULTS

A correlation of the lithologic columns constructed from investigations at 17 outcrops is shown in Fig. 3. Figs 4, 5, 6 and 7 are photographs of investigated outcrops.

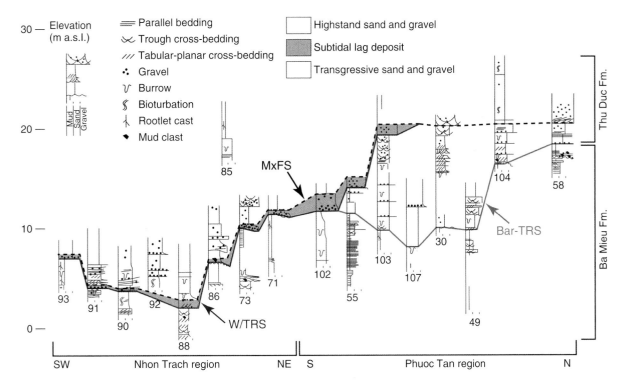

Fig. 3. Correlation of lithologic columns derived from outcrop investigations. See Fig. 2 for outcrop locations. MxFS: maximum-flooding surface.

Seaward part of the estuary: Nhon Trach region

In the Nhon Trach region, the Thu Duc Formation consists of a thick (more than 17 m) upward-fining succession from thick gravel to cross-bedded or bioturbated sand containing gravel, then to mud with rootlets (Fig. 3). The cross-bedded sands are bidirectional (Dominantly to the north and therefore a landward current) with mud drapes, indicating that tidal influence is present. It is interpreted to be a tidal sand bar that is commonly developed in the lower intertidal and subtidal zones of mesotidal or macrotidal estuaries and deltas (e.g. Meckel, 1975; Wright *et al.*, 1975; Dalrymple *et al.*, 1990). In tidal sand-bar deposits, a thin upward-fining succession is common and represents migration of the relief on the bar.

The oldest thick gravel unconformably overlies the Ba Mieu Formation (Fig. 4). The gravel consists of poorly sorted, massive or faintly bedded coarse sand and rounded gravel (commonly less than 3 cm diameter) with mud clasts. Laterally accreted channel-shaped lag beds were observed at several outcrops (Fig. 4B). Tidal or wave signatures were not found in the lag itself but the dominant palaeo-current direction in cross-bedded sand and gravel above the lag is to the north-east

(present landward) at outcrops 73, 90 and 91. It appears that deposition after the lag was affected strongly by flood tides. NE-SW oriented channels at outcrop 88 support a north-east current direction for the flood tides (Fig. 4B). The gravels are interpreted to be subtidal lags deposited by strong tidal currents at the bottom of major tidal channels and are sourced from older deposits in the 'subtidal erosional zone' of Kitazawa (2007) in the palaeo–Dong Nai estuary mouth adjacent to a tide-dominated shelf that has been eroded by tidal currents and storm waves (Dalrymple & Choi, 2007). The laterally accreted channel reflects major tidal channel migration.

The thick upward-fining succession is interpreted to represent regression of a tide-dominated delta during the highstand of the last interglacial period. The depositional environment changed from a subtidal erosional zone, to tidal sand bars and sand flats, then to muddy tidal flats and finally to salt marshes with mangrove vegetation. A maximum-flooding surface is recognised in the subtidal lag deposits. The erosional base of the gravel is interpreted to represent the TRS. The thickness between the highstand sand and gravel of the lowest subtidal lag (outcrop 88) and

Fig. 4. W/TRS in the Nhon Trach region. (A) W/TRS representing amalgamation of tidal channels that eroded tidal sand-bar deposits of the Ba Mieu Formation (outcrop 88). (B) Photograph and sketch of laterally accreted channel-shaped gravel beds interpreted as subtidal lag; enlargement of (A). Tidal channels were oriented NE-SW and migrated south-eastward. (C) Subtidal lag where tidal channel lags have converged; left side of (A). (D) W/TRS and subtidal lag unconformably overlying salt-marsh mud of the Ba Mieu Formation (outcrop 71).

the uppermost salt-marsh facies (outcrop 85) is about 16 m; this thickness represents the difference in elevation between the centre of the major tidal channel and its outermost borders in the estuary mouth during the period of maximum flooding.

Landward part of the estuary: Phuoc Tan region

Over most of the Phuoc Tan region, the lower Thu Duc Formation consists of an thick (more than 10 m) upward-coarsening succession from bioturbated, poorly sorted sand to cross-bedded sand and gravel (Fig. 5), overlain by a thick upward-fining

Fig. 5. Tidal sand-bar deposits (Phuoc Tan region). (A) Tidal sand-bar deposits coarsening upward from bioturbated, poorly sorted sand to cross-bedded sand and gravel (outcrop 30). The dominant palaeo current is eastward. (B) Enlargement of lower part of (A). Cross-bedded muddy sand with burrows. (C) Upward-coarsening sand of tidal sand bar (near outcrop 58). (D) Enlargement of lower part of (C). Poorly sorted cross-bedded sand and gravel.

succession similar to that in the Nhon Trach region. The basal poorly sorted sand unconformably over-lies salt-marsh mud of the Ba Mieu Formation (Fig. 6) but the basal lag is not always present and is thinner than that in the Nhon Trach region. Cross beddings in the upward-coarsening succession indicate that the dominant direction of sediment transport was landward. At several outcrops, the upward-coarsening succession is underlain by a thin upward-fining successions of cross-bedded sand with mud drapes and sand and mud beds showing flaser, wavy, or lenticular bedding (Figs 6D and 7). The upward-fining succession consists of tidal creek-fill and tidal flat deposits. In the southern part of the Phuoc Tan region, subtidal lag deposits unconformably overlie the Ba Mieu Formation (outcrop 102), as in the Nhon Trach region, or erodes poorly sorted deposits (outcrops 55 and 103).

Upward-coarsening successions are interpreted to represent the transgression of a tide-dominated estuary during the last interglacial period. The environment changed from tidal flats to tidal sand bars and then to a subtidal erosional zone.

The erosional base of the tidal sand bars is interpreted to represent a TRS and the base of the tidal flat deposits below the TRS as a transgressive surface. A maximum-flooding surface is recog-nised in the coarsest part of the sequence, that is, the uppermost part of the upward-coarsening succession.

The transgressive sand and gravel of tidal sand bar pinches out both landward (north) and sea-ward (south). The landward thinning is caused by a decrease of sediment accommodation space and the abrupt seaward pinchout reflects erosion by tidal channels in the subtidal erosional zone.

DISCUSSION

TRS classification

Fig. 8 shows a schematic longitudinal geologic section of a tide-dominated estuary derived from correlation of outcrop observations in the study area (Fig. 3). The TRS is laterally continuous (at least 20 km from south to north) and is therefore recognised as a regional stratigraphic boundary.

Fig. 6. Bar-TRS (Phuoc Tan region). (A) Bar-TRS without preservation of underlying transgressive deposits (outcrop 103). (B) Tidal sand-bar deposit overlying salt marsh mud with rootlets (near outcrop 30). (C) Bar-TRS underlain by early transgressive deposits unconformably overlying salt-marsh mud of the Ba Mieu Formation (near outcrop 58). TS, transgressive surface. (D) Early transgressive tidal flat deposit directly overlying the transgressive surface (TS) and underlying the bar-TRS; left side of (C).

TRS is commonly amalgamated with sequence boundary, so tidal erosion provides an important delineation of estuary infill. An incised valley formed by fluvial erosion during a sea-level lowstand is eroded and deepened by tidal channels during a subsequent marine transgression. The TRS in a tide-dominated estuary can be classified into two types on the basis of the deposits that

overlie them, which reflect the location within the estuary and the strength of tidal currents.

(1) A bar-TRS was developed in the Phuoc Tan region by landward migration during the early stage of transgression of tidal channels separating tidal sand bars in the central part of the estuary. The primitive bar-TRS may have been formed initially by intertidal channels in the most landward

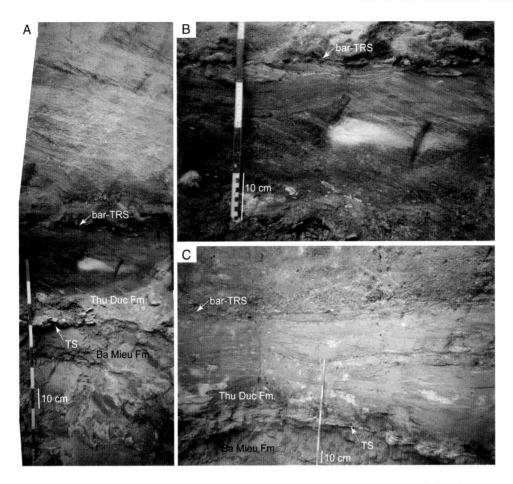

Fig. 7. Early transgressive deposits survived bar-tidal ravinement erosion (near outcrop 104). (A) Succession from salt marsh mud of the Ba Mieu Formation to early transgressive tidal creek fill and tidal flat deposits on the transgressive surface (TS), then to tidal sand-bar deposit overlying the bar-TRS. (B) Upward-fining succession of cross-bedded sand with mud clasts and mud drapes and sand and mud beds showing flaser or wavy bedding; enlargement of (A). (C) Flaser bedding beneath bar-TRS interpreted as early transgressive tidal flat deposits; correlated to sequence below bar-TRS in (B).

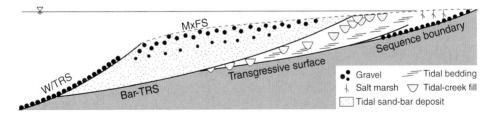

Fig. 8. Schematic longitudinal section of a tide-dominated estuary succession. Landward part of the section is from Kitazawa (2007). MxFS: maximum-flooding surface.

tidal sand bars but was subsequently more deeply eroded by channels separating tidal sand bars. The channels were filled with sand and gravel and amalgamated to form a laterally continuous stratigraphic boundary. At the beginning of the transgression, tidal flat deposits overlaid the transgressive surface, which was commonly eroded and

not preserved below the bar-TRS. In this case, the bar-TRS was amalgamated with the sequence boundary and transgressive surface.

(2) A W/TRS was developed in the Nhon Trach region by landward migration during the late stage of transgression of major tidal channels in the estuary mouth. Even for a tide-dominated estuary,

its seaward part is subject to wave influence owing to the unconfined and wide estuary mouth. We did not find shelf deposits or wave-influenced structures in the sediments overlying the W/TRS, but several channel lags converged to form a thick lag (Fig. 4) and the overlying sediments were deposited by flood tides. It is apparent that tidal currents had an important role in the erosion that formed the W/TRS but wave action also contributed. Erosion on the W/TRS removed most of the pre-existing transgressive deposits and gravel and mud clasts were preserved only as lag deposits because the tidal channels were deeper than the bar-TRS. In this case, the W/TRS was amalgamated with the sequence boundary, transgressive surface and bar-TRS. In the landward region, the transgressive upward-coarsening succession deposited on the bar-TRS escaped wave/tidal ravinement erosion and was overlain by subtidal lag deposits.

The two types of TRS represent different stages of transgression. A bar-TRS is formed on the sea floor by the first intrusion of inter-bar tidal channels and is preserved by the deposition of tidal sand bars. A bar-TRS can be renewed by deeper erosion to the base of previous tidal sand-bar deposits and then succeeded by deposition of new bar deposits. The bar-TRS corresponds to the base of the elongate tidal sand bar labelled C2 in the schematic section of a tide-dominated estuary, presented by Dalrymple *et al.* (1992; their fig. 14). A W/TRS is formed on the erosional sea floor of an estuary mouth adjacent to a tide-dominated shelf after landward migration of tidal sand bars. Tidal channels and wave action remove previous tidal sand-bar deposits (and more) and the sea floor lacks transgressive estuary deposits other than subtidal lag. The W/TRS corresponds to the base of the transgressive lag labelled C1 by Dalrymple *et al.* (1992; their fig. 14).

In the model of Dalrymple *et al.* (1992), transgressive tidal sand-bar deposits are thin in the mid-estuary region and have been completely removed in the outer estuary. Our study indicates that the aggradational rates of sands in the estuary were high and that tidal sand-bar deposits are thick (Fig. 8). Tidal sand-bar deposits pinch out seaward between the bar-TRS and W/TRS and are not subjected to W/TRS processes farther landward. Although transgressive tidal sand-bar deposits are abundant, for a large rise of sea level they would be overlain by shelf muds or erosionally removed

by shoreface ravinement, or by offshore tidal ravinement (Reynaud & Dalrymple, 2012).

The W/TRS concept can be applied to TRSs identified by other studies, where they are almost coincident with the maximum-flooding surface and the transgressive deposit overlying them is absent; examples include the Cobequid Bay–Salmon River estuary (Dalrymple & Zaitlin, 1994), the Seine estuary (Tessier *et al.*, 2010a) and the Mont-Saint-Michel estuary (Tessier *et al.*, 2010b). In a section of the Holocene Colorado River deposits shown by Meckel (1975), the transgressive succession is very thin, with poorly sorted sand containing shell fragments, mud clasts and pebbles. It represents subtidal lag deposit and its base can be recognised as the W/TRS.

For TRSs in wave-dominated or wave and tide-dominated estuaries, truncations at the base of tidal inlet channels and flood-tidal deltas (e.g. Allen & Posamentier, 1993; Zaitlin *et al.*, 1994) should be distinguished clearly from bar-TRSs and W/TRSs; a term such as 'inlet-TRS' would be appropriate. The tidal channels separating tidal sand bars are more widely spaced than the tidal inlet channels cutting barrier islands, as shown by Reinson (1992). As noted by Dalrymple & Zaitlin (1994), Kitazawa (2007) and Tessier (2012), the landward extents of bar-TRS and W/TRSs in tide-dominated estuaries are greater than those of an 'inlet-TRS' in a wave-dominated or wave and tide-dominated estuary.

CONCLUSIONS

TRSs in tide-dominated estuaries can be classified into two types based on outcrop investigation. A bar-TRS is defined by the base of a tidal sand bar in the landward part of the estuary and is overlain by a transgressive upward-coarsening succession. A W/TRS is defined by the base of subtidal lag deposits in the estuary mouth adjacent to erosional tide-dominated shelf. Both the bar-TRS and W/TRSs are amalgamated with both the sequence boundary and the transgressive surface over a wide area.

ACKNOWLEDGEMENTS

We are indebted to Drs Fujio Kumon, Koichi Hoyanagi, Kohki Yoshida and Katsura Ishida of Shinshu University and Dr. Masaaki Tateishi of

Niigata University for valuable discussions. We thank Drs Van Lap Nguyen and Thi Kim Oanh Ta of the Vietnam Academy of Science and Technology and Mr. Hoang Sam Nguyen for support and suggestions during the field investigations.

REFERENCES

Allen, G.P. (1991) Sedimentary processes and facies in the Gironde estuary: a recent model for macrotidal estuarine systems. In: *Clastic Tidal Sedimentology* (Eds D.G. Smith, G.E. Reinson, B.A. Zaitlin and R.A. Rahmani), *Bull. Can. Soc. Petrol. Geol., Mem.*, **16**, 29–400.

Allen, G.P. and Posamentier, H.W. (1993) Sequence stratigraphy and facies model of an incised valley fill: the Gironde Estuary, *France. J. Sed. Petrol.*, **63**, 378–391.

Chaumillon E., Tessier B. and Reynaud J.-Y. (2010) Stratigraphic records and variability of incised valleys and estuaries along French coasts. *Bull. Soc. Géol. Fr.*, **181**, 75–85.

Dalrymple, R.W. and Choi, K. (2007) Morphologic and facies trends through the fluvial–marine transition in tide-dominated depositional systems: A schematic framework for environmental and sequence-stratigraphic interpretation. *Earth-Sci. Rev.* **81**, 135–174.

Dalrymple, R.W., Knight, R.J., Zaitlin, B.A. and Middleton, G.V. (1990) Dynamics and facies model of a macrotidal sand-bar complex, Cobequid Bay-Salmon River Estuary (Bay of Fundy). *Sedimentology*, **37**, 577–612.

Dalrymple, R.W. and Zaitlin, B.A. (1994) High-resolution sequence stratigraphy of a complex, incised valley succession, Cobequid Bay – Salmon River estuary, Bay of Fundy, Canada. *Sedimentology*, **41**, 1069–1091.

Dalrymple, R.W., Zaitlin, B.A. and Boyd, R. (1992) Estuarine facies models: conceptual basis and stratigraphic implications. *J. Sed. Petrol.*, **62**, 1130–1146.

Kitazawa, T. (2007) Pleistocene macrotidal tide-dominated estuary–delta succession, along the Dong Nai River, southern Vietnam. *Sed. Geol.*, **194**, 115–140.

Kitazawa, T., Nakagawa, T., Hashimoto, T. and Tateishi, M. (2006) Stratigraphy and optically stimulated luminescence (OSL) dating of Quaternary sequence along the Dong Nai River, southern Vietnam. *Journal of Asian Earth Sciences*, **27**, 788–804.

Kitazawa, T., Murakoshi, N., Tateishi, M. and Hashimoto, T. (in press) OSL dating and tectonic movement of Quaternary deposits in the Ho Chi Minh City region, southern Vietnam. *Journal of Asian Earth Sciences.*

Kitazawa, T. and Tateishi, M. (2005) Geometry and preservation processes of tidal sand bar deposits of Middle Pleistocene macrotidal tide-dominated delta succession, southern Vietnam. *J. Sed. Soc. Japan*, **61**, 27–38.

Meckel, L.D. (1975) Holocene sand bodies in the Colorado delta area, northern Gulf of California. In: *Deltas: Models for Exploration* (Ed. M.L. Broussard), 239–265. Houston Geological Society, Houston.

Plink-Björklund, P. (2005) Stacked fluvial to tide-dominated estuarine deposits in high-frequency (fourth-order) sequences of the Eocene Central Basin, Spitsbergen. *Sedimentology*, **52**, 391–428.

Plink-Björklund, P. (2008) Wave-to-tide facies change in a Campanian shoreline complex, Chimney Rock Tongue, Wyoming-Utah, U.S.A. In: *Recent advances in models of siliciclastic shallow-marine stratigraphy* (Eds G.J. Hampson, R.J. Steel, P.M. Burgess and R.W. Dalrymple), *SEPM Spec. Publ.*, **90**, 265–291.

Reinson, G.E. (1992) Transgressive barrier island and estuarine systems. In: *Facies Models: Response to Sea Level Change* (Eds R.G. Walker and N.P. James) 3rd edn, 179–194. Geological Association of Canada.

Reynaud J.-Y. and Dalrymple R.W. (2012) Shallow-marine tidal deposits. In: *Principles of Tidal Sedimentology* (Eds R.A. Davis and R.W. Dalrymple) 335–369. Springer.

Tessier, B. (2012) Stratigraphy of tide-dominated estuaries. In: *Principles of Tidal Sedimentology* (Eds R.A. Davis and R.W. Dalrymple), 109–128. Springer.

Tessier, B., Billeaud, I. and Lesueur, P. (2010a) Stratigraphic organization of a composited macrotidal wedge: the Holocene infill of the Mont-Saint-Michel Bay. *Bull. Soc. Géol. Fr.*, **181**, 99–113.

Tessier, B., Delsinne, N. and Sorrel, P. (2010b) Holocene infilling of a tide-dominated estuarine mouth. The example of the macrotidal Seine estuary (NW France). *Bull. Soc. Géol. Fr.*, **181**, 87–98.

Wonham, J.P. and Elliott, T. (1996) High-resolution sequence stratigraphy of a mid-Cretaceous estuarine complex: the Woburn Sands of the Leighton Buzzard area, southern England. In: *Sequence Stratigraphy in British Geology* (Eds S.P. Hesselbo and D.N. Parkinson), *Geol. Soc. London Spec. Publ.*, **103**, 41–62.

Wright, L.D., Coleman, J.M. and Thom, B.G. (1975) Sediment transport and deposition in a macrotidal river channel: Ord River, Western Australia. In: *Estuarine Research* (Ed. L.E. Cronin), v. II, 309–322. Academic Press, New York.

Yoshida, S., Johnson, H.D., Pye, K. and Dixon, R.J. (2004) Transgressive changes from tidal estuarine to marine embayment depositional systems: The Lower Cretaceous Woburn Sands of southern England and comparison with Holocene analogs. *AAPG Bull.*, **88**, 1433–1460.

Zaitlin, B.A., Dalrymple, R.W. and Boyd, R. (1994) The stratigraphic organization of incised-valley systems associated with relative sea-level change. In: *Incised-valley Systems: Origin and Sedimentary Sequences* (Eds R.W. Dalrymple, R. Boyd and B.A. Zaitlin), *SEPM Spec. Publ.*, **51**, 45–60.

Tidally-modulated infilling of a large coastal plain during the Holocene; the case of the French Flemish Coastal plain

JOSÉ MARGOTTA*, ALAIN TRENTESAUX* and NICOLAS TRIBOVILLARD*

* University Lille 1 - UMR 8187, CNRS LOG, Villeneuve d'Ascq, France

ABSTRACT

The French Flemish coastal plain corresponds to the southern end of the extended lowlands that develop along the North Sea coast between Denmark and France. The aim of this study was to reconstruct the evolution of the Holocene deposits of the plain. It was achieved using core information from a large database combined with new cores and, for the first time in this area, the interpretation of very high-resolution seismic profiles along the coastal plain waterways. The results highlight the role of tides in the Holocene record despite the present-day wave-dominated coastal morphology. The definition of seven lithofacies and the determination of clay minerals, foraminifera and pollen assemblages led to the distinction of three main facies associations representative of the Holocene infill, all of them being dominated by tidal processes: subtidal sand facies, mud flats and marsh-swamp facies. They are grouped into two sedimentary units: A channel-belt unit at the base, overlain by an extended tidal flat unit infilling all the remaining available accommodation space. The distribution and geometry of these units are displayed thanks to several cross-sections covering most of the plain and allowing a pseudo-3D reconstruction. The stratigraphic organisation shows an aggradational channel-belt unit preserved landward, and a progradational tidal flat unit, preserved in the axis of the palaeoriver and in the seaward part of the plain. The arrangement of these Holocene units reveals the infilling process of the area due to the interaction of the rate of sea-level rise, accommodation space, marine sediment supply and the coastal hydrodynamic conditions dominated by a macrotidal regime and influenced by wave action.

Keywords: Holocene, coastal plain, tidal flats, facies model, VHR seismic reflection.

INTRODUCTION

Coastal environments (including deltas, estuaries, lagoons, strand plains and tidal flats) represent extremely variable and rapidly evolving sedimentary systems (Harris *et al.*, 2002). Many long-term geological processes (e.g. subsidence and isostasy) and short-term dynamics (e.g. tidal regime and waves) are reflected in their settings, controlling the geometry and type of sediments that accumulate in these zones (Davis & Hayes, 1984; Syvitski *et al.*, 2005).

In the coastal zones, sediment erosion, transport and deposition are closely related to the power of waves and tides (Swift & Thorne, 1991; Pendón *et al.*, 1998; Chaumillon *et al.*, 2010), which rule the coastal morphology and the arrangement of available sediments (Davis & Hayes, 1984). In addition, the impact of sea-level changes influences the evolution of coastal landforms. The measurement or prediction of these variables and their relative importance over the course of time provide the input parameters to determine the morphological changes of the coasts (Fitzgerald *et al.*, 2008). For the most recent history of the coastal areas, the effects of human influence should also be considered within the geomorphic change parameters (Reed *et al.*, 2009).

During the Quaternary, North-west Europe, as in most of the world (e.g. Voris, 2000), experienced

Contributions to Modern and Ancient Tidal Sedimentology: Proceedings of the Tidalites 2012 Conference,
First Edition. Edited by Bernadette Tessier and Jean-Yves Reynaud.
© 2016 International Association of Sedimentologists. Published 2016 by John Wiley & Sons, Ltd.

significant changes in its coastal zones linked to the successive glacial-interglacial cycles (Ehlers *et al.,* 2011). The most dramatic changes occurred in the landscape of the area linking the North Sea and the English Channel, with a marine connection in interglacial conditions and extensive continental shelves with a complex system of palaeovalleys that formed the 'Fleuve Manche' palaeoriver in glacial times (Gibbard, 1988; Lautridou *et al.,* 1999; Lericolais *et al.,* 2003). During these times, most of the sediment was transported to and by these rivers, with a possible wind influence for the finer material (Lefort, 2011).

Various factors interplayed in this change, such as tidal forcing (Scourse *et al.,* 2009), short-term glacio-hydro-isostatic adjustment (Lambeck, 2001) and long-term background tectonics (Hijma, 2009). In the Holocene, through the ensuing sea-level rise, glacial sediments were reworked by waves and currents to feed coastal areas. The general framework of coastal areas corresponds to Pleistocene palaeovalleys drowned by the fast sea-level rise converting the landscape into tidal basins (Baeteman *et al.,* 2002), in which changes occur locally in relation to the controlling factors, capable of shifting landward or seaward (Baeteman, 2010).

Recent studies carried out on estuaries and incised valleys along the French coasts (e.g. Chaumillon *et al.,* 2010), as well as on a part of the coastal lowlands of the southern North Sea (e.g. Baeteman, 1999; Beets & Van der Spek, 2000; Baeteman & Declerck, 2002), show the variety of stratigraphic architecture and morphological evolution throughout the Holocene. Due to its location in the Dover Strait, the Holocene deposits of the French Flemish Coastal Plain (FFCP) probably record the major change that occurred as the enclosed, landwardmost part of the Southern Bight of the North Sea that was flooded and turned to a high energy, megatidal seaway.

Many studies have been performed in the area (Sommé, 1977; Van der Woude & Roeleveld, 1985; Gandouin, 2003; Mrani-Alaoui, 2006; Meurisse-Fort, 2007), mostly focused on determining the lithofacies and ages of the Holocene infill. However, the configuration of this coastal infill remains open due to the complex spatial arrangement of these lithofacies and the lack of regional correlations covering the whole area. This paper, by compiling ancient data and with the help of new cores and seismic profiles, contributes to build a facies model of the Holocene deposits of

the FFCP. Extensive facies analysis, key features of depositional environments and the role of tides in the filling of this large embayment are discussed.

GEOGRAPHICAL AND GEOLOGICAL SETTING

The FFCP is a large coastal system located in northern France. It covers about 750 km^2 and forms the southern end of the vast continental southern North Sea coastal plain (Fig. 1). It is a low-lying region intensively affected by anthropogenic changes and susceptible to flooding by the sea. However, its triangular shape belongs to a heritage of Pleistocene glacial-interglacial cycles (Sommé *et al.,* 1999) and the most recent sea-level drop.

The FFCP has most of the characteristics of a simple and shallow coastal plain incised-valley system (Zaitlin *et al.,* 1994), incised during the Pleistocene (Sommé, 1977) and filled through subsequent rise of sea-level in the Holocene. The drainage system is rather limited, with only two rivers of importance: The Aa and Hem rivers, with 89 km and 25 km lengths, respectively. These rivers present a very low water discharge (10 m^{3}s^{-1} and 1.5 m^{3}s^{-1} respectively) (IFREMER, 1986) and they have been modified into the plain by canalisation and a dense network of artificial drainage (Fig. 1).

Currently, the FFCP is characterised by a semi-diurnal, macrotidal regime along the coast. Maximum tidal range varies from 7.3 m in Calais, to the west, and 6.5 m in Dunkerque, to the east (Fig. 1). Tidal currents flow parallel to the shoreline with velocities between 0.5 and 1 m s^{-1} (S.H.O.M, 1968; Cartier, 2011). They are mostly asymmetric (Clique & Lepetit, 1986; Cartier, 2011) with an eastward (i.e. flood) dominant direction (Grochowski *et al.,* 1993). The coast is under the control of predominantly south-west winds with a secondary direction from the north-east. Calais and Dunkerque, have a percentage of wave-height greater than 1 m of 32 and 19% respectively. Waves may attain up to 3 m during storms (Anthony *et al.,* 2010).

Published data by Denys & Baeteman (1995) indicates that sea-level on the FFCP has risen continuously during the Holocene (Fig. 2). The trend of sea-level rise shows a fast increase at the beginning of the Holocene (7 mm yr^{-1}) and two progressive decelerations at 7500 cal BP (2.5 mm yr^{-1}) and 5500 cal BP (0.7 mm yr^{-1}).

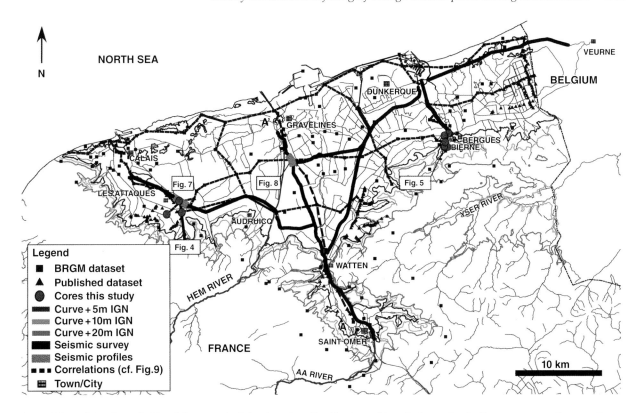

Fig. 1. Map of the French Flemish coastal plain showing distribution of boreholes, seismic lines, cross-sections and contour interval +5, +10 and +20 m NGF.

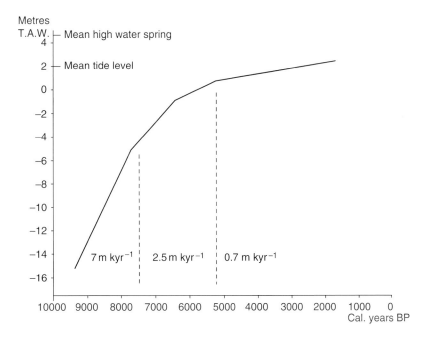

Fig. 2. Holocene sea-level curve in the FFCP. Adapted after Denys & Baeteman (1995). The Belgian datum (TAW) is 2.33 m lower than the French datum (NGF). The curve is based on the position of upper mean sea-level limit.

METHODS

This study is mainly based on the integration of core dataset and very-high resolution seismic profiles. In total, 10 undisturbed cores were collected using a percussion corer at 2 locations, Les Attaques and Bierne (Fig. 1). They were described following the classification of Reineck & Wunderlich (1968) and sampled for micropaleontology (foraminifera and pollen analysis), clay mineralogy and ^{14}C dating. Samples for radiocarbon dating were collected on peat and organic-rich layers. Ages were calibrated using the CALIB software (Stuiver & Reimer, 1986) in its 2011 revised version (Calib 6.1.0).

Core interpretation was compared and correlated with information from 385 boreholes from various available sources: the French National subsurface database of BRGM (BRGM, 2014) and boreholes from previous scientific studies along the plain (Bosch, 1975; Bootsman, 1977; Paris, 1977; Sommé, 1977; Sommé et al., 1994; Gandouin, 2003; Mrani-Alaoui, 2006), to obtain the regional sedimentary associations. As none of these 385 cores are still available, records, publications and data present in the databases were used. Thanks to the cores collected for the present study, some of them being located close to those in the database, it was possible to associate the different descriptions. This helps in making an efficient integration.

For the first time in the FFCP, seismic profiling was performed along the waterways. The configuration of these navigable channels determines the geometry of the profile network (Fig. 1). A total of 125 km of Very High Resolution (VHR) seismic profiles (around 80% of waterways open to navigation) were obtained using an IKB-Seistec Boomer profiler (Simpkin & Davis, 1993). This seismic device associated with a line-in-cone receiver is specially designed for shallow water environments.

A basic seismic processing with band-pass filtering and gain correction was implemented to enhance the seismic profiles. For interpretation, the seismic profiles were converted to depth using an average seismic velocity in saturated soft sediment of 1500 m s^{-1}, as has been carried out before by numerous authors (e.g. Dalrymple & Zaitlin, 1994; McGee, 1995; Bachrach, 1998; Saito et al., 1998; Novak, 2002; Lin et al., 2009). The interpretation was based on the seismic stratigraphic principles outlined by Mitchum et al. (1977) and Sangree & Widmier (1979).

FACIES ASSOCIATIONS

Facies analysis has been developed following classic facies models (Reineck & Singh, 1980; Dalrymple, 1992; Boyd et al., 1992). Based on the core description it was possible to distinguish seven lithofacies (From A to G) according to lithology, colour, sedimentary structures and micropalaeontological composition (Fig. 3). They were summarised and grouped into three facies associations along the sedimentary succession: 1) Subtidal sand facies association, 2) mud flat facies association; and 3) marsh-swamp facies association. Representative boreholes are shown in Figs 4 and 5.

Subtidal sands facies association

Sand-dominated lithofacies are exposed in all the cores described in the FFCP. In accordance with their stratigraphic position in cores, a distinction has been made within sand deposits outlined by their changes in sediment components, colour and the occurrence of shells.

In general, the most common sands of the plain are bluish grey, well-sorted, sub-rounded and very fine-grained. They look massive but present some slight uniform cross-lamination, which includes scarce discontinuous mud laminae and some shell fragments. Amongst the sporadic shell fragments, rests of *Cerastoderma edule* are distinguished (Fig. 3A). These sands are found in the lower part of the cores (Fig. 4).

The second type of sand presents typically light grey to brown colours (Fig. 3B). These sand beds are silty to medium-grained, rounded and moderate to well-sorted. They present a moderate abundance of peat fragments, some grains of glauconite and rock fragments, an abundance of shell fragments and occasional whole shells of *Cerastoderma edule*. These beds are characteristic of the upper part of the cores (Fig. 4), commonly eroding a thick peat layer. Sometimes they have been affected by soil formation.

Foraminifera assemblages found in this facies association show dominance of *Haynesina germanica* and include *Milliolinella subrotunda*, *Cribroelphidium excavatum*, *Triloculina trigonula* and *Haynesina depressula*. This preserved benthic-fauna strongly suggests subtidal settings and the influence of marine conditions (Culver & Banner, 1978; Wang & Chappell, 2001; Debenay et al., 2006).

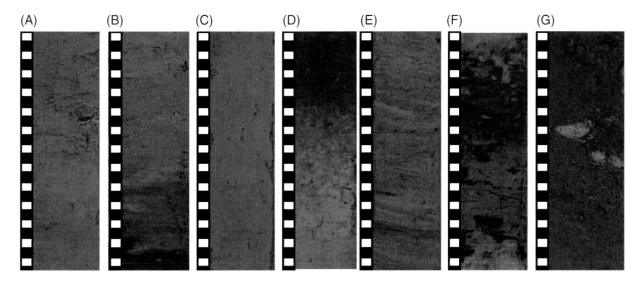

Fig. 3. Typical lithofacies displayed in the FFCP. A) General sand facies, B) Sand facies of the upper sedimentary units, C) General mud facies, D) Mud facies of the upper sedimentary beds and development of anthropized soil, E) Heterolithic facies characterized by wavy and flaser bedding, F) Organic-rich mud facies, bioturbated; and G) laminated peat layer with wood fragments. Scale in cm.

Clay mineralogy also shows a difference between both types of sands (Fig. 6). Grey-bluish sands present a quite homogeneous composition influenced by input from a marine source. These conditions are extended in mostly all the deposits of the Holocene succession. The link with a marine source is evidenced by the strong similarities of clay mineralogy observed in the current offshore throughout Calais and Dunkerque (Vicaire, 1991). Upper light grey/brown sands present smectite enrichment, similar to clay mineral composition observed in the Cretaceous rocks of the Artois hills (Deconinck et al., 1989; Deconinck & Chamley, 1995).

This change is interpreted as due to an increased influence of mainland source-rock for the clay fraction from the drainage basin of the Aa-Hem fluvial system in the uppermost sedimentary succession of the plain. Another possibility for the smectite enrichment would be the erosion of the chalk cliffs along the coast when the sea-level reached their feet. Nevertheless, the cliff retreat may have been not so important, as attested by the limited flint accumulation that had drifted toward the north-east in the Pierette fossil gravel bar described by Sommé in 1977 (Pierre, 2007).

Mud flat facies association

This facies association is frequent in the cores collected. It has been described as highly variable in thicknesses and stratigraphic positions. Commonly,

this facies is vertically accreted deposits, where the basal contact of each layer is generally sharp and the upper contact is gradational to the organic-rich sediments (Figs 4 and 5).

These deposits are characterised by grey bluish clayey-silts, either massive/structureless or with faint parallel laminations (Fig. 3C). In the upper part of the Holocene succession, the colour changes to light grey/brown and can be affected by the development of cultivated ground (Fig. 3D). Most of the time, the original fabric of the sediment had been modified by the intense root action. Some indistinct mottled texture is commonly observed. Scattered organic laminae and plant fragments have been identified. In addition, dispersed shell fragments of *Cerastoderma edule* and *Hydrobia ulvae* were found.

The benthic foraminifera found in this facies association are dominated by *Haynesina germanica*, accompanied by some variations in the presence of *Ammonia tepida, Cribroelphidium gerthi, Cribroelphidium williamsoni Milliolinella subrotunda* and *Bulliminella elegantissima*. This assemblage characterises intertidal mud flats (Culver & Banner 1978; Horton, 1999, Evans et al., 2001; Bernasconi & Cusminsky, 2009).

Mud flat facies association locally contains thin heterolithic beds (Fig. 3E) made of an alternation of greyish brown, light grey, well sorted, fine-grained sand and greyish brown mud, thin beds and laminae. These grade upward into mud-dominated deposits. Sedimentary structures in

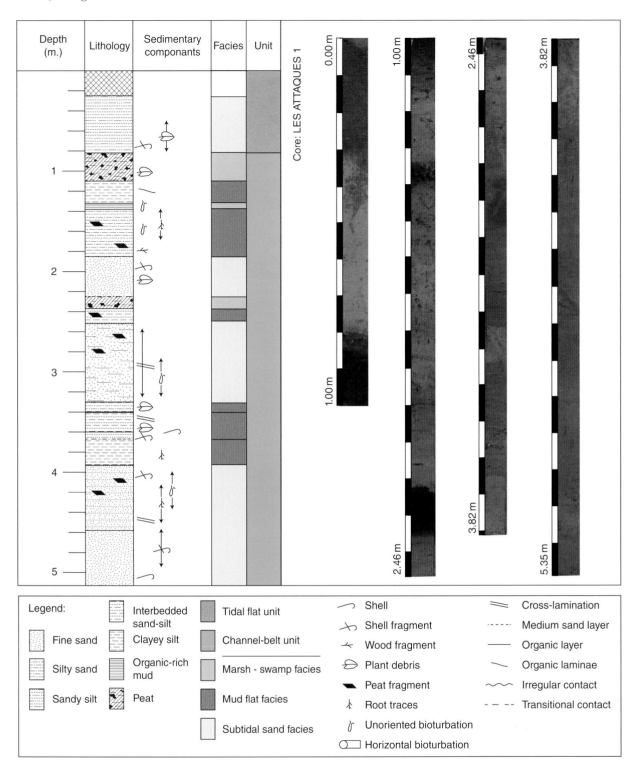

Fig. 4. Stratigraphic column of core Les Attaques 1. See location in Fig. 1.

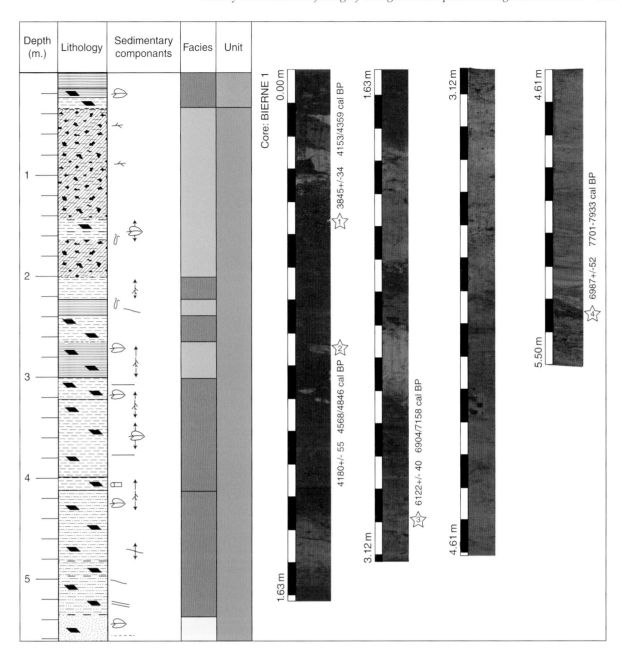

Fig. 5. Stratigraphic column of core Bierne 1. See location in Fig. 1. Numbers in stars along the core photo refer to levels that were sampled for ^{14}C dating. Same legend as Fig. 4.

the heterolithic beds include parallel and wavy bedding. In these typical tidal deposits, mud drapes correspond to slack water stages and sand beds to flood/ebb stages. (Reineck & Wundderlich, 1968; Weimer *et al.*, 1982).

Marsh-swamp facies association

This facies association consists of thin beds of organic-rich muds and peat layers gradational to the top of the mud flats facies (Figs 4 and 5). This facies indicates the evolution of salt marsh to freshwater marsh with peat accumulation due to total infilling of accommodation space. The organic-rich muds are gradational from brown-greyish clayey silts to brown/black organic-rich clays (Fig. 3F). They are characterised by concentration of plant debris and abundance of *Psilonichnus* and *Planolites* ichnotraces (Gingras *et al.*, 2007; 2012).

The organic-rich muds present a foraminifera's microfaunal content characterised by the presence

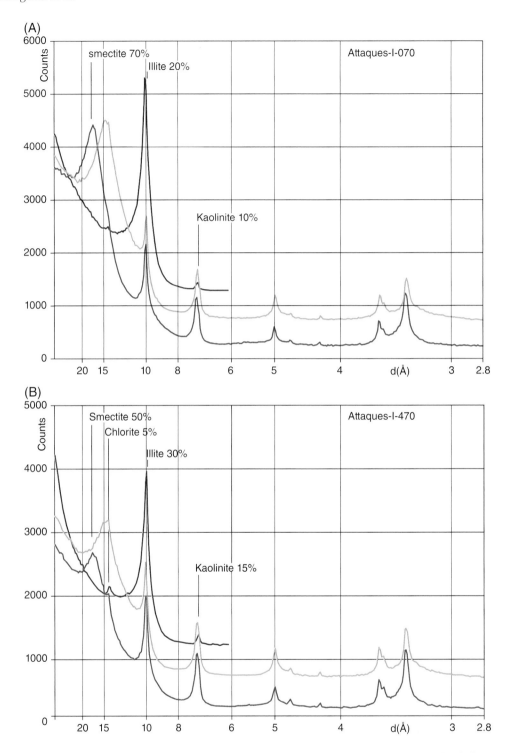

Fig. 6. Typical X-ray diffraction diagrams from Holocene deposits of the FFCP. For each diagram, the different colour lines represent the air-dried, glycolated and cooked analysis in green, blue and red, respectively. Quantification has been performed on the glycolated (blue line). A) Sample from les Attaques I borehole at a depth of 70 cm. B) Sample from les Attaques I borehole at a depth of 470 cm. Note the enrichment in smectite in the sample Attaques I-070, representative of the upper sedimentary succession.

of *Jadammina macrescens, Haynesina germanica, Cribroelphidium gerthi, Cribroelphidium williamsoni, Cribroelphidium margaritaceum, Miliammina fusca* and *Trochammina inflata*. Foraminifera assemblage is characteristic of brackish conditions (Alday *et al.*, 2013). A notable presence of agglutinated benthic-fauna suggests the development of supratidal environments (Culver & Banner, 1978; Wang & Chappell, 2001; Debenay *et al.*, 2006, Armynot du Châtelet *et al.*, 2008).

Peat layers present variable colours between black and brown-to-dark brown, respectively associated with textures ranging from sapric (amorphous) to fibric (fibrous, >67% fibre content, ASTM 1998). Fibric brown peats consist of undecomposed fibrous organic material; they can present an apparently laminated aspect and contain some roots, wood fragments and plants remains (Fig. 3G).

Pollen analysis shows dominance of *Alnus* and sedges (*Cyperaceae*) within the pollen association along the entire sampled interval. The presence of *Betula, Quercus* and *Gramineae* has also been noticed. Although their presence is scarce, it is important to indicate that pollen from aquatic plants as *Myriophyllum verticillatum* and *Potamogeton* were recognised. Pollen associations suggest the development of shrub swamps (CORINE Biotopes, 1997) overlying mud flats and marshes. Within these shrub swamps some wooded areas and constantly submerged zones (mostly related with freshwaters conditions) developed in a nearshore context. These types of wetlands were often found in the Holocene reconstitution of vegetation cover along the northern France region (Vergne, 2013).

SEISMIC STRATIGRAPHY

The seismic profiles image the uppermost part of the FFCP infilling succession. On average, the high-resolution seismic profiles are of bad quality due to operating in the waterways (shallowness, side reverberation, gas-rich sediments etc.). One clear reflector only, of good continuity and amplitude, can be found along a few profiles. This reflector was tied and correlated with sediment core analysis in order to calibrate seismic data and provide age and lithology to performed seismic interpretation (Fig. 7). This key marker corresponds to the limit between the pre-Holocene substrate and the Holocene deposits.

On the available seismic data, the pre-Holocene substrate forms the acoustic basement of all seismic

profiles. Besides some variability on depths around −10m to −20m NGF, it does not have remarkable geometric characteristics, structures or regular reflection patterns. The general trend is a gentle offshore dip with some limited slight incisions around the present-day Aa valley (Fig. 8). This substrate represents the land invaded by the rising sea-level, over which the Holocene infill is preserved.

The Holocene infill shows weak impedance contrasts and constant interferences (multiples). In spite of this, the internal pattern's configuration, terminations and the geometry of the reflectors allow the identification of two seismic units (Units A and B; Figs 7 and 8) that indicate the arrangement of the Holocene deposits and their spatial and temporal variations.

Seismic Unit A exhibits a cut-and-fill configuration lying upon the Pre-Holocene substrate. The seismic facies in Unit A is characterised by reflectors with low continuity and amplitude. Reflection patterns show oblique reflectors ranging from highly inclined to sub-horizontal and show channel-shape feature geometries. Indeed, this unit corresponds to the sand-dominated deposits widespread through the FFCP and largely forming the lower part of the Holocene sedimentary succession.

Seismic Unit B forms a drape overlying the pre-existing topography of unit A with gently onlapping, downlapping and concordant terminations (Fig. 8). The upper bounding surface coincides with the present-day canal floor. The seismic facies is characterised by regular reflectors with moderate amplitude and continuity. Internal configuration shows low angle, oblique parallel and subparallel reflectors indicating slightly dipping seaward reflectors as prograding foresets (Fig. 8). This is correlated with the mud-dominated facies frequently developed in the uppermost part of the Holocene infill and extended along the entire FFCP.

STRATIGRAPHIC FRAMEWORK

The stratigraphic organisation of the FFCP was reconstructed considering the detailed vertical facies succession deposited within the plain (Figs 4 and 5), the high-resolution seismic data (Figs 7 and 8) and regional association of cores and boreholes.

Based on the analysis of the depositional facies, their spatial distribution and comparison with the characteristics delineated by seismic facies, it is proposed that two sedimentary units constitute

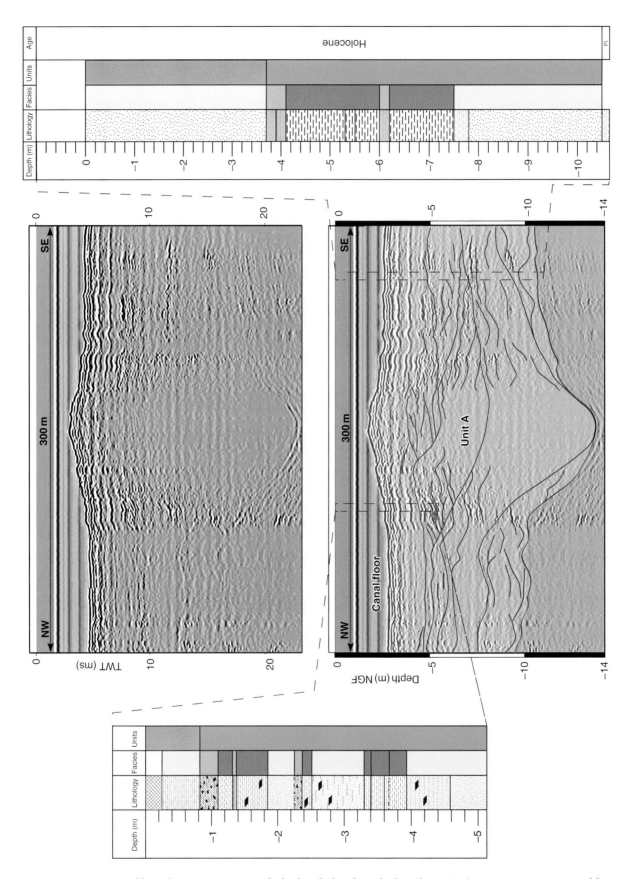

Fig. 7. VHR seismic profile and its comparison with the boreholes described in the FFCP (Core Les Attaques 1 and borehole 197) showing the Holocene sedimentary infilling. The bottom of borehole 197 reaches the pre-Holocene deposits. Core and borehole tops obtained from the surrounding surface at the edge of the canal are projected on the seismic profile. Same core legend as for Fig. 4.

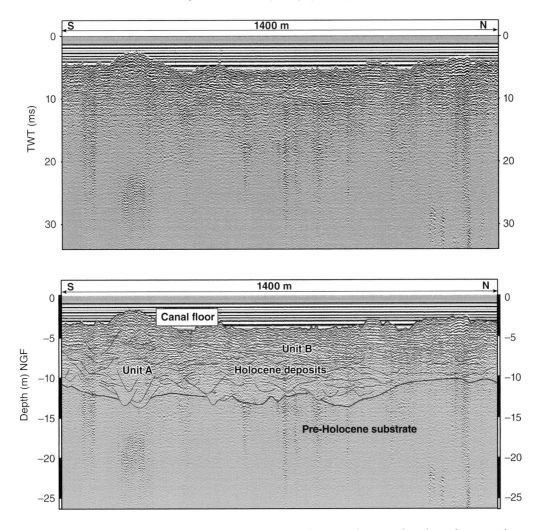

Fig. 8. Selected VHR seismic profile illustrating the interpretation of seismic facies within the Holocene sedimentary succession. Seismic facies A shows aggrading channelized geometries. Seismic facies B shows a sub-parallel geometry with slight northward progradation.

the regional Holocene infill of the FFCP: a lower channel-belt unit and an upper tidal flat unit. Their arrangement in time and space is a response to the rate of sea-level rise, accommodation space variations, sediment supply and the interactions of hydrodynamic processes. The stratigraphic architecture is presented in Fig. 9.

The channel-belt unit is the lower part of the Holocene infill. It unconformably overlies the pre-Holocene substratum, the top topography of which is interpreted as a sequence boundary that formed during the last sea-level drop and low-stand. This first unit is widespread along the entire area of the FFCP, being well-developed and preserved landward (Fig. 9). It presents mostly aggradational sedimentation patterns, dominated by subtidal sand facies with channelized features mainly developed along the axial network of the

Aa-Hem river system, as evidenced by the unit A on the seismic profiles (Fig. 8). Meanwhile, along the margins of the FFCP, mud flats and marsh facies were preserved, as well as subtidal sands facies derived from the small tributaries.

This unit represents the transgressive fill linked to the rapid sea-level rise during the Early Holocene (Walker *et al.,* 2012) in which Aa-Hem valley was flooded and the river system took the form of tidal channels, as the main way to fill the abundant accommodation space with sediments, mostly from marine input due to the limited contribution of the continental fluvial network. As the rate of sea-level rise slowed down around middle Holocene (Walker *et al.,* 2012), accommodation space was surpassed by sediment supply and mud flats and marsh facies were raised to an elevation favourable for peat development (Figs 4 and 5).

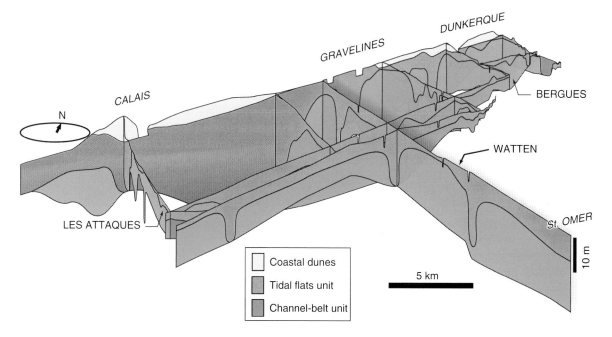

Fig. 9. Fence diagram giving a simplified regional picture of the 3D vertical and lateral stratigraphic relationships between Holocene sedimentary units of the FFCP. For location, see Fig. 1 (dotted lines).

The tidal flat unit overlies the channel belt unit along the FFCP except on certain distal or lateral areas where it reaches the pre-Holocene substratum (Fig. 9). This upper unit represents an important change in the depositional conditions of the FFCP area. Due to low accommodation space thick peat developed in the most internal zones. Siliciclastic sediment deposition occurs aligned in the major channels and seaward, indicating an infilling dominated by progradational trends. The passage to tidal flat conditions thus corresponds to the transition between the transgressive and the highstand system tracts, due to subtle changes in the topography; this does not correspond to a single surface. Seismic unit B can represent part of this prograding unit with seaward downlapping features (Fig. 8), especially in the outer segment of the valley. This change has occurred around 5500 cal BP (Denys & Baeteman, 1995) when the rate of sea-level rise decelerated again. This is in the range of the age found for other French incised valley systems (Chaumillon *et al.*, 2010).

In the axis of the Aa-Hem River system, tidal channels remain active, reworking previous deposits (from the channel-belt unit) and locally reaching the pre-Holocene sediments. Depositional facies of this tidal flat unit are represented by sand flat and tidal channel mostly in the central and seawards parts while mud flat facies develop on the flanks. These facies are primarily influenced by marine sediment supply. However, a contribution of the mainland reliefs through the river system and chalk cliff erosion is present, at least for the clay-size fraction, as attested by the clay analysis (Fig. 6).

COASTAL EVOLUTION

The reconstruction of large-scale development for the Holocene deposits of the FFCP clearly shows a sedimentary succession of transgressive and highstand deposits linked to the rate of sea-level rise and the balance of coastal dynamics (Fig. 10). It is proposed that the filling follows a two-fold evolution due to the interplay of deceleration in sea-level rise, accommodation space and the relative influence of tides and waves, which are well marked in the stratigraphic record. A palaeogeographic reconstruction is proposed (Fig. 11).

The shape of the initial palaeotopography is shown in Fig. 11A. The main feature of the pre-Holocene surface is the presence of the Aa-Hem system along the western side of the FFCP, which drains quite a large surface of relatively steep hills (Fig. 1). This is in opposition to the rest of the valley, east from Gravelines, as most of the water is

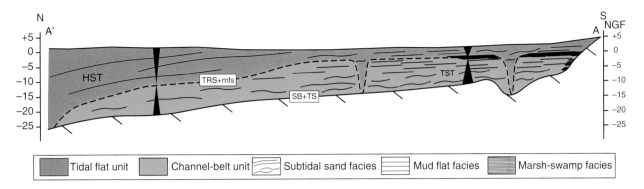

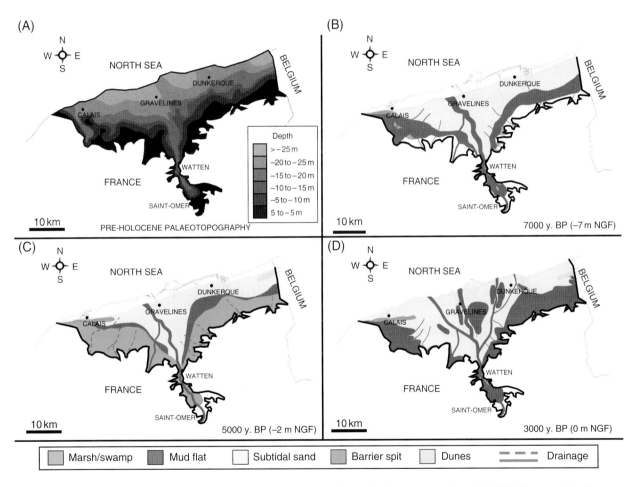

Fig. 10. Schematic longitudinal section (A-A' on Fig. 1) illustrating the regional sedimentary infilling and sequence stratigraphic arrangement of the FFCP along the axe of Aa river, the main river of the system. This synthetic section displays the arrangement of the distribution of facies associations prior the installation of the younger coastal dune system along the present-day coastline.

Fig. 11. Synthesis of the coastal system evolution of the FFCP. A) Initial palaeotopography. B, C, D) Palaeographical reconstructions revealing the environmental changes for three time intervals at 7000 y BP, 5000 y BP and 3000 y BP, i.e. before human modifications.

drained to the Ijzer (Fig. 1) and flows towards the Belgian side of the coastal plain (Liu *et al.*, 1992; Baeteman, 1999; Mathys, 2009). This will cause the tidal action to be amplified along the Aa-Hem valley axis, where the tidal prism mainly developed during the Holocene. The geometry of the valley is inherited from the glacial lowstand, it has a significant control in the accomodation space and development of facies distribution.

During the Early Holocene (Walker *et al.*, 2012) the rapid sea-level rise generated a high accommodation space in the drowned incised valley system. Sediments from offshore and especially from the major rivers, with Meuse and Rhine acting as sources (Hijma, 2009; Mathys, 2009), invaded the valley; the transgressive surface is amalgamated with the sequence boundary and sedimentation shift was landward along the valley. After a first stage of Holocene infilling (Fig. 11B), in the Middle Holocene (Walker *et al.*, 2012) from around 7500 cal BP, the sea-level rise experienced a first slowing down. This stage was greatly influenced by inherited pre-Holocene palaeotopography. Transgressive conditions are fully established and dominated along the entire area. Massive sediment load entering from the North Sea occurred (Anthony, 2000; Van der Molen & Van Dijck, 2000), being widespread and preserved inland, forming embayment geometry with estuarine conditions in the axial area.

Depositional conditions during this phase also present a rapidly increasing of tidal regime (Shennan *et al.*, 2000). Tidal currents have impacted upon the depositional setting in all parts of the FFCP. Borehole data illustrate the lateral energy gradient from high-energy-related sandy deposits in the axial Aa-Hem river system and muddy intertidal deposits along the flanks. It is assumed that this generalised impact of tidal dynamics is due to the macrotidal range combined with a very important tidal prism, allowing tidal currents to penetrate and be active further inland along the axis of this river system. Meanwhile, muddy and fine sand deposits also developed in the eastern part of the FFCP, where no connection with a fluvial network prevents the hegemony of subtidal sand facies.

In the second stage, towards the end of the Middle Holocene, from around 5500 cal BP, there was a change in the rate of sea-level rise; It slowed down again and low-energy conditions prevailed. Together with this slowing down, the ongoing sediment supply from the North Sea, influenced by the longshore sand transport northwards due to tides and waves (Anthony, 2000) surpassing the creation of accommodation space, the tidal prism

decreased and tidal energy moved away from the border sides of the plain. These conditions are conducive to the development of marshes along the border sides of the plain with a subsequent evolution to freshwater marshes with peat accumulation (Fig. 11C), as well as driving a coastal building seaward. However, tidal dynamics remained active in the main axis of the Aa river system, as attested to by deep channels observed in boreholes and punctual amalgamated cut-anfill facies observed along the axial seismic profile (Fig. 8). These first changes show the transition from transgressive to progradational patterns on this stage.

There were rapid geographical changes and landscape modifications during the evolution of this progradational stage along the FFCP. In the context of embayment filling and extended marsh depositions, the main tidal channels were still active and were probably narrow and deep (FitzGerald *et al.*, 2012). This gradually eroded the previous Holocene deposits, focussing the tidal ravinement in the active channels, as has occurred in the Mont-Saint-Michel bay (Tessier *et al.*, 2010a) and the Seine estuary (Tessier *et al.*, 2010b). Nevertheless, the details obtained from seismic profiles or from core data, as well as the lack of enough dating, do not allow us to conclude that this process erodes the pre-Holocene substratum.

Through this stage, tidal setting evolved and was progressively established in the whole FFCP, represented by a large tidal flat intertidal surface. Progressively, due to local subsidence from differential peat compaction (Van Asselen *et al.*, 2009) new available accommodation space was created along the extended areas where peat developed. Increasing tidal prism size due to easy erosion of previous Holocene deposits, combined with macrotidal range, presumed strong storm wave activity (Anthony *et al.*, 2010); and sediment supply from a marine source also interplayed to re-establish tidal flats landward. A palaeogeographic configuration of these changes is presented in Fig. 11D, around 3000 cal BP, showing the environmental settings at a time before human influence in the area (Thoen, 1986; Tys, 2007). Anthropization started to develop when Romans occupied the region. Dunes were fixed and, from the seventh century, lowlands were reclaimed. Auto-compaction induced by natural evolution or climatic changes (Allen, 2000), then drainage modification by anthropogenic activities (Paepe, 1960; Baeteman, 2005), must also be considered as other local variables that have had an impact in the definitive landscape of the coastal plain.

CONCLUSIONS

This study represents the first attempt to integrate VHR seismic reflection data to examine the Holocene sedimentary infill of the French Flemish Coastal Plain. The combination of VHR seismic profiles, cores and boreholes are used to reconstruct the regional sedimentary architecture and evolution of the Holocene deposits in respect to an incised-valley classic facies model.

Holocene deposits record the transition from transgressive to highstand system tracts. Rapid sea-level rise at the beginning of the Holocene formed a transgressive surface where the Holocene succession unconformably overlies pre-Holocene substrate. The Holocene fill can be subdivided in two major sedimentary units. The first an aggrading channel-belt unit well preserved inland in transgressive conditions, whilst the second unit consists of prograding tidal-flats with a better development seaward in the external part of the outer valley segment.

The Holocene sedimentary succession on the large-scale has responded to the interaction of the rate of sea-level rise with tidal current action and accomodation space. Rapid sea-level rise created a large accommodation space allowing marine sediments to invade far inland. From 7500 cal BP, slowing down of sea-level rise and the macrotidal influence appear to be more pronounced with a comprehensive transgressive infilling. After 5500 cal BP, the valley was filled, tidal prism size decreased and sediments began to prograde. Afterwards, renewed tidal sedimentation inland was favoured by peat compaction, inducing an increase of tidal prism in a low accommodation space.

The FFCP appears to be a sedimentary sink located in the southernmost part of the long continental North Sea coastal plain. Sediment supply from the North Sea was trapped inland, influenced by tidal transport and north-eastward longshore transport. Limited sediment budget was delivered from the Aa-Hem river system. Despite the continuous tidal influence along the Holocene and the current tidal conditions, the present-day coast resembles a wave-dominated coast due to man-influenced changes (e.g. partly by closure of the connections between coastal rivers and the North Sea) that induced modifications on the natural balance of the mechanisms acting along the coastal system.

ACKNOWLEDGEMENTS

The authors thank Wim Versteeg and the Renard Centre of Marine Geology for their collaboration in the seismic acquisition, Patrice Herbin and Christine Louvion from the Archaeological Service of the General Council of the Nord Department for their technical assistance during fieldwork and access to archaeological sites, Bernadette Tessier for her constructive help in the seismic data processing, Eric Armynot du Châtelet and Virginie Vergne, who collaborated efficiently in the micropalaeontological analyses, while Melesio Quijada gave us a precious technical support. Jean-Yves Reynaud made helpful remarks on a preliminary manuscript. Bernadette Tessier is also thanked as invited editor for her detailed review of the first submission. The International Association of Sedimentologists (IAS) and their Postgraduate Grant Scheme (PGS) gave financial support for radiocarbon dating.

REFERENCES

Alday, M., Cearreta, A., Freitas, M.C. and **Andrade, C.** (2013) Modern and late Holocene foraminiferal record of restricted environmental conditions in the Albufeira Lagoon, SW Portugal. *Geol. Acta*, **11** (1), 75–84.

Allen, J.R.L. (2000) Morphodynamics of Holocene salt marshes: a review sketch from the Atlantic and Southern North Sea coast of Europe. *Quatern. Sci. Rev.*, **19**, 1155–1231.

Anthony, E.J. (2000) Marine sand supply and Holocene coastal sedimentation in northern France between the Somme estuary and Belgium. In: *Coastal and Estuarine Environments: sedimentology, geomorphology and geo-archaeology* (Eds K. Pye and J.R.L. Allen), *Geol. Soc. London Spec. Publ.*, **175**, 87–97.

Anthony, E.J., Mrani-Alaoui, M. and **Héquette, A.** (2010) Shoreface sand supply and mid- to late Holocene aeolian dune formation on the storm-dominated macrotidal coast of the southern North Sea. *Mar. Geol.*, **276**, 100–104.

Armynot du Châtelet, E., Recourt, P. and **Chopin, V.** (2008) Mineralogy of agglutinated benthic foraminifera: implication for paleo-environmental reconstructions. *Bull. Soc. Géol. Fr.*, **179**, 583–592.

ASTM (1998) Standard practice for establishing allowable properties for visually-graded dimension lumber from Ingrade tests of full-size specimens. *Annual book of ASTM standards*. Wood. West Conshohocken, PA: American Society for Testing and Materials, **04.10**, 287–311.

Bachrach, R. (1998) *High resolution shallow seismic subsurface Characterization*. Unp. PhD thesis. Stanford University. California.155 p.

Baeteman, C. (2010) Geological Considerations on the Effect of Sea-level Rise on Coastal Lowlands, in Particular in Developing Countries. *Meded. Zitt. K.*

Acad. Overzeese Wet/Bull. Séancc. Acad. R. Sci. Outre-Mer, **56** (2), 195–207.

Baeteman, C. (1999) The Holocene depositional history of the Ijzer palaeo-valley (Western Belgian coastal plain) with reference to the factors controlling the formation of intercalated peat beds. *Geol. Belgica*, **2** (3–4), 39–72.

Baeteman, C. (2005) The Streif classification system: a tribute to an alternative system for organising and mapping Holocene coastal deposits .*Quat. Int.*, **133–134**, 141–149.

Baeteman, C. and **Declerck, P.-Y.** (2002) A synthesis of early and middle Holocene coastal changes in the Western Belgian lowlands. *Belgeo*, **2**, 77–107.

Baeteman, C., Scott, D.B. and **Van Strydonck, M.** (2002) Changes in coastal zone processes at high sea-level stand: a late Holocene example from Belgium. *J. Quatern. Sci.*, **17** (5–6), 547–559.

Beets, D.J. and **Van der Spek, A.J.F.** (2000) The Holocene evolution of the barrier and the back-barrier basins of the Belgium and the Netherlands as a function of late Weichselian morphology, relative sea-level rise and sediment supply. *Neth. J. Geosc. – Geol. Mijnbouw*, **79**, 3–16.

Bernasconi, E. and **Cusminsky, G.** (2009) Estudio paleoecológico de Foraminíferos de testigos del de Golfo Nuevo (Patagonia, Argentina). *Geobios*, **42**, 435–450 (in Spanish).

Bootsman, C. (1977) *De geologische opbown van de kustvlakte bij Ghyvelde (noord-Frankrijk)*. Study report. Instituut voor aardwetenschappen. Amsterdam, Vrije universiteit 74 p. (in Dutch).

Bosch, J.H.A. (1975) *Verlasg veldwerk Calaisis*. Study report. Instituut voor aardwetenschappen. Amsterdam, Vrije Universiteit. 41 p. (in Dutch).

Bout-Roumazeilles, V., Cortijo, E., Labeyrie, L. and **Debrabant, P.** (1999) Clay mineral evidence of nepheloid layer contribution to the Heinrich layers in the Northwest Atlantic. *Palaeogeogr., Palaeoclimatol., Palaeoecol.*, **146** (1–4), 211–228.

Boyd, R., Dalrymple, R.W. and **Zaitlin, B.A.** (1992) Classification of clastic coastal depositional environments. *Sed. Geol.*, **80**, 139–150.

BRGM (2014) http://infoterre.brgm.fr/

Cartier, A. (2011) *Evaluation des flux sédimentaires sur le littoral du Nord Pas de Calais : Vers une meilleure compréhension de la morphodynamique des plages macrotidales*. Unp. PhD thesis. Université du Littoral Côte d'Opale, Dunkerque (in French).

Chaumillon, E., Tessier, B. and **Reynaud, J.-Y.** (2010) Stratigraphic records and variability of incised valleys and estuaries along French coasts. *Bull. Soc. Géol. Fr.*, **181**, 75–85.

Clique, P.-M. and **Lepetit, J.-P.** (1986) De la Frontière belge à la baie de Somme. In: *Catalogue sédimentologique des côtes françaises*. Côte de la Mer du Nord et de la Manche, Coll. Direction Etudes & Recherches EDF, **61**, 11–133 (in French).

CORINE Biotopes (1997) *Types d'habitats français - ENGREF*, 175 p. (in French).

Culver, S.J. and **Banner, F.T.** (1978) Foraminiferal assemblages as flandrian palaeo-environmental indicators. *Palaeogeogr., Palaeoclimatol., Palaeoecol.*, **24**, 53–72.

Dalrymple, R.W. (1992) Tidal Depositional Systems. In: *Facies Models: Response to Sea Level Change* (Eds **R.G. Walker** and **N.P. James**). *Geol. Assoc. Canada*. St John's, Newfoundland. 195–218.

Dalrymple, R.W. and **Zaitlin, B.A.** (1994) High-resolution sequence stratigraphy of a complex, incised valley succession, Cobequid Bay - Salmon River estuary, Bay of Fundy, Canada. *Sedimentology*, **41**, 1069–1091.

Davis, R.A.J. and **Hayes, M.O.** (1984) What is a wave-dominated coast? *Mar. Geol.*, **60**, 313–329.

Debenay, J.-P., Bicchi, E., Goubert, E. and **Armynot du Châtelet, E.** (2006) Spatio-temporal distribution of benthic foraminifera in relation to estuarine dynamics (Vie estuary, Vendée, W France). *Estuar. Coast. Shelf Sci.*, **67**, 181–197.

Deconinck, J.-F. and **Chamley, H.** (1995) Diversity of smectite origins in late cretaceous sediments: example of chalks from Northern France. *Clay Mineral.*, **30**, 365–378.

Deconinck, J.-F., Holtzapffel, T., Robaszynski, F. and **Amédro, F.** (1989) Données minéralogiques, géochimiques et biologiques comparées dans les craies cénomaniennes et santoniennes du Boulonnais. *Geobios*, **11**, 179–188 (in French).

Denys, L. and **Baeteman, C.** (1995) Holocene evolution of relative sea level and local mean high water spring tides in Belgium – a first assessment. *Mar. Geol.*, **124**, 1–19.

Ehlers, J., Gibbard, P.L. and **Hughes, P.D.** (2011) Introduction. In: *Quaternary Glaciations – Extent and Chronology – A Closer Look* (Eds **J. Ehlers, P.L. Gibbard** and **P.D. Hughes**), **15**, 1–14. Elsevier.

Evans, J.R., Kirby, J.R. and **Long, A.J.** (2001) The litho- and biostratigraphy of a late Holocene tidal channel in Romney Marsh, southern England. *Proc. Geol. Assoc.*, **112**, 111–130.

Fitzgerald, D., Bynevich, I. and **Hein, C.** (2012) Morphodynamics and facies architecture of tidal inlets and tidal deltas. In: *Principles of Tidal Sedimentology* (Eds **Davis, R.A.** Jr, and **Dalrymple, R.W.**). Springer. 301–334.

Fitzgerald, D.M., Fenster, M.S., Argow, B.A. and **Buynevich, I.V.** (2008) Coastal Impacts Due to Sea-Level Rise. *Annu. Rev. Earth Planet. Sci.*, **36**, 601–647.

Gandouin, E. (2003) *Enregistrement paléoclimatique interdisciplinaire de la transgression holocène. Signature paléo-environnementale des Chironomidae (Diptères) du bassin de Saint-Omer (France)*. Unp PhD thesis, Université Lille 1, Lille, 246 p. (in French).

Gibbard, P.L. (1988) The history of great northwest European rivers during the past three millions years. *Phil. Trans. Roy. Soc. London*, **318**, 559–602.

Gingras, M.K., Bann, K.L., MacEachern, J.A., Waldron, W. and **Pemberton, S.G.** (2007) A conceptual framework for the application of trace fossils. In: *Applied Ichnology* (Eds J.A. MacEachern, K.L. Bann, M.K. Gingras and S.G. Pemberton). *SEPM Short Course Notes*, **52**, pp. 1–25.

Gingras, M.K., MacEachern, J.A. and **Dashtgard, S.** (2012) The potential of trace fossils as tidal indicators in bays and estuaries. *Sed. Geol.*, **279**, 97–106.

Grochowski, N.T.L., Collins, M.B., Boxall, S.R., Salomon, J.C., Breton, M. and **Lafite, R.** (1993) Sediment transport pathways in the Eastern English Channel. *Oceanol. Acta*, **16**, 531–537.

Harris, P.T., Heap, A.D, Bryce, S.M, Porter-Smith, R., Ryan, D.A. and Heggie, D.T. (2002) Classification of Australian clastic coastal depositional environments based upon a quantitative analysis of wave, tidal, and river power. *J. Sed. Res.*, **72** (6), 858–870.

Hijma, M.P. (2009) *River Valley to Estuary: The Early-Mid Holocene Transgression of the Rhine–Meuse valley, The Netherlands*. Unp. PhD thesis, Utrecht University, Utrecht, 192 p.

Horton, B.P. (1999) The contemporary distribution of inter-tidal foraminifera of Cowpen Marsh, Tees Estuary, UK: Implications for studies of Holocene sea level changes. *Palaeogeogr., Palaeoclimatol., Palaeoecol.*, **149**, 127–149.

IFREMER (1986) Le Littoral de la Région Nord-Pas de Calais qualité du milieu marin. Rapports Scientifiques Et Techniques De L'Ifremer. *Convention de Coopération Région Nord - Pas de Calais*, **3**, 136 p. (in French).

Lambeck, K. (2001) Glacial Crustal Rebound, Sea Levels and Shoreline. *Encyclopedia of Ocean Sciences* (Second Edition), 49–58.

Lautridou, J.-P., Auffret, J.-P., Baltzer, A., Clet, M., Lecolle, F., Lefebvre, D., Lericolais, G., Roblinjouve, A., Balescu, S., Carpentier, G., Descombes, J.-C., Occhietti, S., and Rousseau, D.-D. (1999) Le fleuve Seine, le fleuve Manche. *Bull. Soc. Géol. de Fr.*, **170**, 545–558 (in French).

Lefort, J.-P., Danukalova, G.A. and Monnier, J.-L. (2011) Origin and emplacement of the loess deposited in northern Brittany and under the English Channel. *Quat. Int.*, **240**, 117–127.

Lericolais, G., Auffret, J.-P. and Bourillet, J.-F. (2003) The Quaternary Channel river: seismic stratigraphy of its paleo-valleys and deeps. *J. Quatern. Sci.*, **18**, 245–260.

Lin, Y.-T., Schuettpelz, C.C., Wu, C.H. and Fratta, D. (2009) A combined acoustic and electromagnetic wave-based techniques for bathymetry and subbottom profiling in shallow water. *J. Appl. Geophys.* **68**, 203–218.

Liu, A.C., Missiaen, T. and Henriet, J.-P. (1992) The morphology of the top-Tertiary erosion surface in the Belgian sector of the North Sea. *Mar. Geol.*, **105**, 275–284.

Mathys, M. (2009) *The Quaternary geological evolution of the Belgian Continental Shelf, southern North Sea*. Unp. PhD Thesis, Ghent University, Ghent. 371 p.

McGee, T.M. (1995) High-resolution marine reflection pro-filing for engineering and environmental purposes. Part A: Acquiring analogue seismic signals. *J. Appl. Geophys.*, **33**, 271–285.

Meurisse-Fort, M. (2007) *Enregistrement haute résolution des massifs dunaires et impacts des tempêtes (Mer du Nord, Manche et Atlantique)*. Unp. PhD thesis, Université Lille 1, Lille, 299 p (in French).

Mitchum, R.M., Vail, P.R. and Sangree, J.B. (1977) Stratigraphic Interpretation of Seismic Reflection Patterns in Depositional Sequences. In: *Seismic stratig-raphy - Application to hydrocarbon exploration* (Ed. C.E. Payton), *AAPG Mem.*, **165**, 117–133.

Mrani-Alaoui, M. (2006) *Evolution des environnements sédimentaires holocènes de la plaine maritime fla-mande du Nord de la France: eustatisme et processus*. Unp. PhD thesis, Université du Littoral, Côte d'Opale, Dunkerque, 211 p (in French).

Novak, B. (2002) Early Holocene brackish and marine facies in the Fehmarn Belt, southwest Baltic Sea: depo-sitional processes revealed by high-resolution seismic and core analysis. *Mar. Geol.*, **189**, 307–321.

Paepe, R. (1960) La plaine maritime entre Dunkerque et la frontière belge. *Bull. Soc. Belg. Et. Géog.*, **29** ,47–66 (in French).

Paris, P. (1977) Verslag van het fysich-geografisch veldwerk in de omgeving van Calais, gedurende de Zomer van 1973. Study report, Vrije Universiteit, Amsterdam, 26 p. (in Dutch).

Pendón, J.G., Morales, J.A., Borrego J., Jimenez I. and Lopez M. (1998) Evolution of estuarine facies in a tidal channel environment, SW Spain: evidence for a change from tide- to wave-domination. *Mar. Geol.*, **60**, 43–62.

Pierre, P. (2007) Durée de l'évolution marine et recul hol-ocène d'un littoral à falaises, l'exemple du nord boulon-nais (France). *Quaternaire*, **18** (3), 219–231 (in French).

Reed, D.J., Davidson-Arnott, R.G.D. and Perillo, G.M.E. (2009) Estuaries, coastal marshes, tidal flats and coastal dunes. In: *Geomorphology and Global Environmental Change* (Eds **O. Slaymaker**, **T. Spencer** and **C. Embleton-Harman**), Cambridge University Press, 130–157.

Reineck, H.-E. and Singh, I.B. (1980) Depositional Sedimentary Environments. With Reference to Terrigenous Clastics. Second edition. Springer-Verlag, Berlin, Heidelberg, New-York, 549 pp.

Reineck, H.-E. and Wunderlich, F. (1968) Classification and origin of flaser and lenticular bedding. *Sedimentology*, **11**, 99–104.

Saito, Y., Katayama, H., Ikehara, K., Kato, Y., Matsumoto, E., Oguri, K., Oda, M. and Yumoto, M. (1998) Trans-gressive and highstand system tracts and post-glacial transgression, the East China Sea. *Sed. Geol.*, **122**, 217–232.

Sangree, J.B. and Widmier, J.M. (1979) Interpretation of depositional facies from seismic facies. *Geophysics*, **44**, 131–160.

Scourse, J., Uehara, K. and Wainwright, A. (2009) Celtic Sea tidal sand ridges, the Irish Sea Ice Stream and the Fleuve Manche Palaeotidal modelling of a transitional passive margin depositional system. *Mar. Geol.*, **259**, 102–111.

Shennan, I., Lambeck, K., Flather, R., Horton, B.P., McArthur, J., Innes, J., Lloyd, J.M., Rutherford, M. and Wingfield, R.T.R. (2000) Modelling western North Sea palaeogeographies and tidal changes during the Holocene. In: *Holocene Land-Ocean Interaction and Environmental Changes around the North Sea* (Ed. I. Shennan). *Geol. Soc. London*, **166**, 299–319.

S.H.O.M. (1968) Courants de marée dans la Manche et sur les côtes françaises de l'Atlantique. Paris, 287 p. (in French).

Simpkin, P. and Davis, M. (1993) For seismic profiling in very shallow water, a novel receiver. *Sea Technology*, **34**, 5.

Somme, J. (1977) *Les plaines du Nord de la France et leur bordure. Etude géomorphologique*. Unp. PhD thesis, Université Paris 1, Paris, 801 p. (in French).

Sommé, J., Munaut, A.V., Emontspohl, A.F., Limondin, N., Lefèvre, D., Cunat-Bogé, N. and Gilot, E. (1994) The Watten boring – an Early Weichselian and Holocene climatic and palaeoecological record from the French North Sea coastal plain. *Boreas*, **23**, 231–243.

Stuiver, **M.** and **Reimer**, **P.J.** (1986) A computer program for radiocarbon age calibration. Proceedings of the 12th International Conference. *Radiocarbon*, **28**(2B), 1022–1030.

Swift, **D.J.P.** and **Thorne**, **J.A.** (1991) Sedimentation on continental margins, I: A general model for shelf sedimentation. In: *Shelf sand and sandstone bodies: geometry, facies and sequence stratigraphy* (Eds **D.J.P. Swift**, **G.F. Oertel**, **R.W. Tillman** and **J.A. Thorne**), *Int. Assoc. Sedimentol. Spec. Publ.*, **14**, 3–1. Oxford, Blackwell.

Syvitski, **J.P.M.**, **Harvey**, **N.**, **Wolanski**, **E.**, **Burnett**, **W.C.**, **Perillo**, **G.M.E.** and **Gornitz**, **V.** (2005) Dynamics of the coastal zone. In: *Coastal fluxes in the Anthropocene. The Land-Ocean Interactions in the Coastal zone Project of the International Geosphere-Biosphere Programme* (Eds **C.J. Crossland**, **H.H. Kremer**, **H.J. Lindeboom**, **J.L. Marshall-Crossland** and **M.D.A.** Le **Tissier**), Berlin, Springer, 39–94.

Tessier, **B.**, **Billeaud**, **I.** and **Lesueur**, **P.** (2010a) Stratigraphic organisation of a composite macrotidal wedge: the Holocene sedimentary infilling of the Mont-Saint-Michel Bay (NW France). *Bull. Soc. Géol. Fr.*, **181**, 99–113.

Tessier, **B.**, **Delsinne**, **N.** and **Sorrel**, **P.** (2010b) Holocene sedimentary infilling of a tide-dominated estuarine mouth. The example of the macrotidal Seine estuary (NW France). *Bull. Soc. Géol. Fr.*, **181**, 87–98.

Thoen, **H.** (1986) Les Hommes et la Mer dans l'Europe du Nord-Ouest de l'antiquité à nos jours. L'activité des sauniers dans la plaine maritime flamande de l'Age du Fer à l'époque Gallo-romaine. *Revue du Nord-spécial hors série: Archéologie de la Picardie et du Nord de la France*, **1**, 23–46 (in French).

Tys, **D.** (2007) La formation du littoral flamand et l'intervention humaine. In Villes et campagnes en Neustrie: sociétés, économies, territoires, christianisation. In: *actes des XXVe Journées Internationales d'Archéologie Mérovingienne de l'AFAM*. Verslype, Laurent Pub. Montagnac, 211–219 (in French).

van Asselen, **S.**, **Stouthamer**, **E.** and van **Asch Th.W.J.** (2009) Effects of peat compaction on delta evolution: A review on processes, responses, measuring and modeling. *Earth-Sci. Rev.*, **92**, 35–51.

Van der Molen, **J.** and **Van Dijk**, **B.** (2000) The evolution of the Dutch and Belgian coasts and the role of sand supply from the North Sea. *Global Planet. Change*, **27**, 223–244.

Van der Woude, **J.D.** and **Roeleveld**, **W.** (1985) Paleoecological evolution of an interior coastal zone: the case of the Northern France coastal plain. *Bull. Assoc. Fr. pour l'étude du Quaternaire*, **22**, 31–39.

Vergne, **V.** (2013) Paysages forestiers du Nord de la France: approches géohistoriques et paléoécologiques. In: *La forêt domaniale du Nord-Pas-de-Calais. Bilan de 15 ans d'échanges Office National des Forêts - Conseil scientifique de l'environnement Nord – Pas-de-Calais*, 109–152 (in French).

Vicaire, **O.** (1991) Dynamique hydro-sédimentaire en mer du Nord méridionale (du cap Blanc-nez à la frontière belge). Unp. PhD thesis, Université Lille 1, Lille, 264 p (in French).

Voris, **H.K.** (2000) Maps of Pleistocene sea levels in Southern Asia: shorelines, river systems and time durations. *J. of Biogeogr.*, **27**, 1153–1167.

Walker, **M.J.C.**, **Berkelhammer**, **M.**, **Björck**, **S.**, **Cwynar**, **L**, **C.**, **Fisher**, **D.A.**, **Long A**, **J.** and **Weiss**, **H.** (2012) Formal subdivision of the Holocene Series/Epoch: a Discussion Paper by a Working Group of INTIMATE (Integration of ice-core, marine and terrestrial records) and the Subcommission on Quaternary Stratigraphy (International Commission on Stratigraphy). *J. of Quatern. Sci.*, **27** (7), 649–659.

Wang, **P** and **Chappell**, **J.** (2001) Foraminifera as Holocene environmental indicators in the South Alligator River, Northern Australia. *Quat. Int.*, **83–85**, 47–62.

Weimer, **R.J.**, **Howard**, **J.D.** and **Lindsa**, **D.R.** (1982) Tidal flats and associated tidal channels. In: *Sandstone Depositional Environments*. (Eds A. Scholle and D. Spearing). *AAPG Spec. Publ.* **M31**, 191–245.

Zaitlin, **B.A.**, **Dalrymple**, **R.W.** and **Boyd**, **R.** (1994) The stratigraphic organization of incised-valley systems associated with relative sea-level change. In: *Incised-valley Systems: Origin and Sedimentary Sequences* (Eds R.W. Dalrymple, R. Boyd and B.A. Zaitlin), *SEPM Spec. Publ.*, **51**, 45–60.

Sedimentology of a transgressive mixed-energy (wave/tide-dominated) estuary, Upper Devonian Geirud Formation (Alborz Basin, northern Iran)

MAHMOUD SHARAFI[†], SERGIO G. LONGHITANO[‡], ASADOLLAH MAHBOUBI[†], REZA MOUSSAVI-HARAMI[†] and HOSIEN MOSADDEGH[§]

[†] *Department of Geology, Faculty of Science, Ferdowsi University of Mashhad, Iran*
[‡] *Department of Sciences, University of Basilicata, Italy; e-mail: sergio.longhitano@unibas.it*
[§] *School of Earth Science, Kharazmi University, Tehran, Iran*

ABSTRACT

The Upper Devonian (Frasnian to Famennian) Geirud Formation exposed in the north-eastern margin of the Albortz Basin, Iran, records the development and the evolution of a mixed-energy (wave-tide dominated) estuary adjacent to an open coast. Four vertically-stacked groups of facies associations were detected in supporting this interpretation: (i) a river-dominated, bay-head delta zone, including basal fluvial channel belts, flood plains and tidally-influenced channels; (ii) a mixed-energy, central basin zone, consisting of sand-rich and mud-rich tidal flats, associated with tidal channels and bars; (iii) a wave-dominated, coastal and open-marine shelf zone, comprising siliciclastic marine deposits; and (iv) an open-marine shelf zone representing the uppermost stratigraphic interval of the Geirud Formation and consisting of carbonate offshore and bioclastic-rich deposits. The large-scale depositional architectures of the Geirud strata are exposed among three main investigated areas. The correlation of the three sections suggests that the Geirud Formation includes two major stratigraphic intervals: (i) a lowermost compound incised-valley fill; and (ii) an overlying transgressive-regressive interval. The latter also comprises shoreface and open-shelf sediments over the estuarine deposits, instead of coastal barrier or estuarine mouth sand deposits, which are usually considered as important depositional elements in the classical estuary facies models. The stratigraphic evolution reconstructed in this paper documents the infill of incised valleys during an important episode of relative sea-level rise and the ensuing overstepping of these valleys by an estuarine system which, in turn, became transgressed by siliciclastic shoreface and carbonate open marine facies.

Keywords: Albortz Basin, Upper Devonian, mixed-energy estuary, tidal influence, compound incised-valley fills, transgression.

INTRODUCTION

Amongst the most known and well-documented tide/wave-dominated depositional systems, transgressive fluvial-estuarine successions represent probably the most controversial type of sedimentary deposit to interpret. Although an impressive volume of scientific literature has documented estuaries in various settings (e.g. Dalrymple *et al.*, 1992; Simpson *et al.*, 2002; Boyd *et al.*, 1992, 2006; Chaumillon & Weber, 2006; Dalrymple, 2006; Dalrymple & Choi, 2007,

Chaumillon *et al.*, 2010, amongst others), their distinctive features and stratigraphic records are not always easy to decipher, either from outcrop and borehole datasets. Usually, estuarine successions occupy incised palaeo-valleys (Van Wagoner *et al.*, 1990; Boyd *et al.*, 1992; Allen & Posamentier, 1993, 1994; Zaitlin *et al.*, 1994; Dalrymple *et al.*, 1994; Dalrymple, 2006), which represent fluvially-eroded, elongate topographic lows, characteristically larger than single channels, across a regionally-mappable sequence boundary (Boyd *et al.*, 2006). Their infill

Contributions to Modern and Ancient Tidal Sedimentology: Proceedings of the Tidalites 2012 Conference, First Edition. Edited by Bernadette Tessier and Jean-Yves Reynaud.
261

accumulates during the subsequent base-level rise and it may often record multiple lowstand-driven incision episodes, resulting in compound incised valleys and complex stratigraphic models (e.g. Zaitlin *et al.*, 1994; Labaune *et al.*, 2010; Tesson *et al.*, 2011; Li & Bhattacharya, 2013; Tropeano *et al.*, 2013). The incised valley fills are thought to be rapidly overprinted by deltaic deposits, coastal barriers or estuarine mouth sands (e.g. Zaitlin *et al.*, 1994); whereas few case studies document estuaries transgressed by shoreface and open-marine strata.

Owing to their consequent predisposition to be good water, oil and gas repositories, the economic attention on continental to shallow-marine transgressive estuarine incised-valley fills has greatly increased in the last decades (Brown, 1993; Dolson *et al.*, 1991; Dalrymple *et al.*, 1994; Zaitlin *et al.*, 1994; Pulham, 1994; Buatois *et al.*, 2002; Boyd *et al.*, 2006; Chaumillon *et al.*, 2010).

When investigated in the subsurface, due to low-resolution seismic images or punctual, not-extensive well data, it is often difficult to obtain reliable reconstructions on the original depositional frameworks or the complex vertical/lateral facies relationships that characterise these successions (Dalrymple *et al.*, 1992). Moreover, estuarine and incised-valley fills are amongst the hardest successions to reconstruct from subsurface explorations because of their low width/depth ratio, limited lateral extent, ribbon geometry and the complex association of fluvial, tidal, wave and marine facies within them (Boyd *et al.*, 2006). Therefore, when exposed in outcrop, fluvial-to-marine sediments represent crucial key-case studies in order to compare their depositional architectures and internal heterogeneities with sealed, resource-bearing analogues.

The Devonian Geirud Formation in the Alborz Basin of northern Iran provides an excellent ancient analogue for fluvial-marine transitions accumulated within drowned palaeo-valleys and evolving basinward to coastal sediments deposited in a wider, open-marine setting. In this area, well-exposed outcrops allow detailed sedimentological analyses, the results of which complement the impressive number of previous studies focused on fluvial-open marine sediments from other localities around the world (Dalrymple *et al.*, 1992; 1994; Komatsu, 1999; Plink-Björklund, 2005; Folkestand & Satur, 2008; Fischbein *et al.*, 2009; Fröhlich *et al.*, 2010; Schwarz *et al.*, 2011, amongst others).

The focuses of the present study are: (i) to document the wide range of facies included in the Geirud Formation; (ii) to reconstruct the depositional features describing the original palaeoenvironmental setting; and (iii) to interpret the dynamic of a depositional system evolving from a fluvial-dominance to a tide/wave-dominance through time.

GEOLOGICAL AND STRATIGRAPHIC SETTING OF THE CENTRAL ALBORZ MOUNTAINS

The E-W trending Alborz mountains belt (the so-called 'Alborz zone') (Fig. 1A) is one of the most investigated geological-structural zones of northern Iran (Stöcklin, 1974; Annells *et al.*, 1975, 1977; Huber & Eftekhar-nezhad, 1978a, 1978b; Vahdati Daneshmand, 1991; Şêngör, 1990; Alavi, 1996; Axen *et al.*, 2001; Jackson *et al.*, 2002; Allen *et al.*, 2003; Guest *et al.*, 2006; Zanchi *et al.*, 2006, amongst others). The Alborz and the adjacent central Iranian sector (Fig. 1A) represent the remnants of the Early Palaeozoic passive margin of Gondwana, which underwent an important rifting phase during the Ordovician to Silurian time span (Stöcklin, 1968; Stöcklin1974; Berberian & King, 1981; Şêngör *et al.*, 1988; Şêngör, 1990; Saidi & Akbarpour, 1992). The Ordovician-Silurian rifting phase was followed by renewed continental shelf deposition from the Middle Devonian to the Middle Triassic. During this timespan, Iran was located in the northern margin of Gondwana, along the southern border of the Palaeo-Tethys Ocean (Golonka, 2007; Bagheri & Stamfli, 2008). The Gondwanan affinities of Iran are clearly indicated by Silurian ostracods (Hairapetian *et al.*, 2011), Early Ordovician, Silurian and Late Devonian palynomorphs (Ghavidel-Syooki, 1995; Ghavidel-Syooki *et al.*, 2011). Furthermore, structural similarities between the Pre-Cambrian basement of Central Iran and the Pan-African orogeny also support the peri-Gondwanan location of Iran (Nadimi, 2007; Hairapetian *et al.*, 2011).

During the Triassic, following the northward subduction of the Palaeotethys, the Albortz margin collided with the Eurasian plate forming a mega-tectonic suture which is presently exposed in the northern side of the Alborz Mountains (Fig. 1B) (Stampfli *et al.*, 1991; Şêngör & Natal'in, 1996; Guest *et al.*, 2006). Compressional tectonics continued up to the late Cretaceous-Palaeocene and

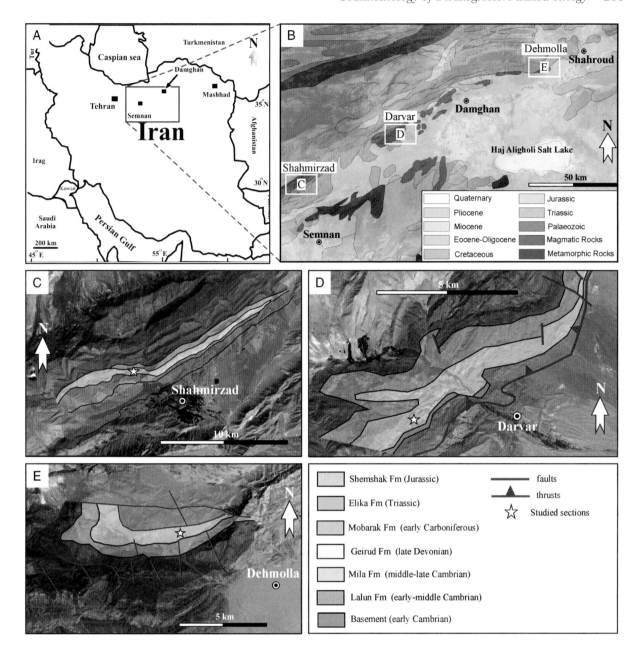

Fig. 1. (A) General map of Iran showing the location of the studied area. (B) Geology of the Alborz Mountains comprised between Semnan, Damghan and Shahrud, with the indication of the three studied areas (modified from the Geological map of Iran at scale 1:2.500.000). (C, D, E) Geological sketches of the studied areas around Shahmirzad, Darvar and Dehmolla, superimposed on Google Earth images.

was replaced by an extensional phase during the Eocene. From the Miocene to Quaternary, this crustal zone was affected by compressional to transpressional events, responsible for the exhumation of the orogeny and consequent present-day denudation (Guest *et al.*, 2006).

The Alborz area was an extensive continental shelf during the Middle Devonian to the Middle Triassic, when the Alborz-Central Iran block collided with Eurasia, generating siliciclastic sediments in continental to transitional zones and carbonates in shallow-marine to deeper-marine settings (Fig. 2) (Stöcklin, 1974; Clark *et al.*, 1975; Berberian & King, 1981; Shahrabi, 1991; Alavi, 1996; Stampfli *et al.*, 2001; Rezaeian, 2008). One of the most interesting and complex

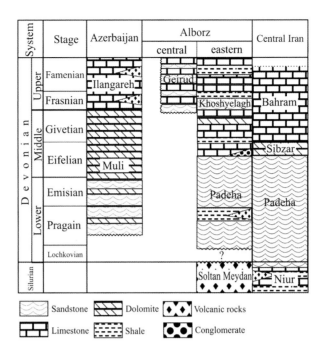

Fig. 2. Correlation chart of the Silurian and Devonian deposits of the Alborz Mountains and central Iran (modified from Wendt *et al.*, 2005).

stratigraphic sequences is represented by the Upper Devonian Geirud Formation.

General stratigraphy of the Geirud Formation

The 240 to 370 m-thick Geirud Formation (Assereto, 1963) represents a vertically-continuous succession including dark calcareous shale, siltstone, sandstone and fossiliferous limestone, extensively exposed in the central Alborz Mountains (Fig. 1C, D and E). It rests unconformably on the upper Cambrian-lower Ordovician Milla Formation and is overlain by the lower Carboniferous Mobarak Formation (Fig. 3A) (Ghavidel-Syooki, 1994). A major stratigraphic hiatus separates the Milla and Geirud Formations (Fig. 3B) and extends from the Late Ordovician through Silurian and Early-Middle Devonian time, possibly equivalent with the Caledonian orogeny (Ghavidel-Syooki, 1995). The Late Devonian (Frasnian-Famennian) age of the Geirud Formation is well constrained on the basis of brachiopod (Bozorgnia, 1964), palynomorph (Ghavidel-Syooki, 1995) and goniatite (Dashtban, 1995) fossil remains.

Previous studies interpreted the Geirud Formation as the record of continental and shallow-marine depositional environments (Gaetani,

1965; Assereto, 1966; Bozorgnia, 1973; Stampfli, 1978; Alavi-Naini, 1993; Lasemi, 2001), developed after a major relative sea-level cycle which affected the Palaeotethys sea during the Middle-Late Devonian (Lasemi, 2001; Wendt *et al.*, 2005).

The internal subdivision of the Geirud Formation has been debated in the past. Assereto & Gaetani (1964) firstly divided it into four members, whereas Stepanov (1971) referred the formation *stricto sensu* to the lowermost member only, including the other units in the overlying Mobarak Formation. This latter stratigraphic identification was generally accepted in the most recent studies (e.g. Stöcklin & Setudenia, 1991; Wendt *et al.*, 2002; 2005).

In detail, the Geirud Formation comprises terrestrial shales, quartzites, red sandstones and red conglomerates, passing laterally and upward to sandstones, containing large-scale tabular cross-stratification and to shallow-marine to fully-marine dolomites, calcarenites and deep-marine red shales (Ueno *et al.*, 1997). These lithologies are well exposed along the southern flank of the Alborz Mountains, in the central-western sector of the Alborz Basin (Fig. 1A), between the cities of Semnan, Damghan and Sharud (Fig. 1B).

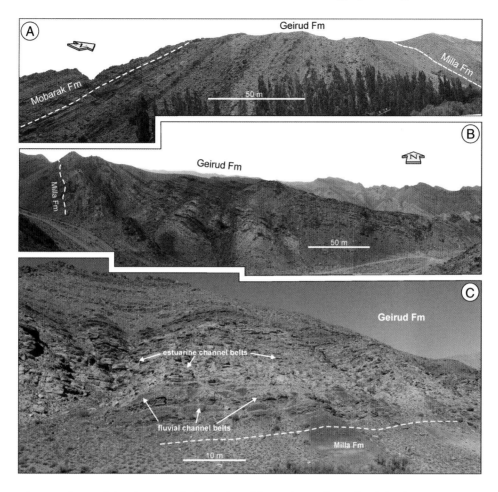

Fig. 3. Panoramic overviews of the three studied stratigraphic sections where the Geirud Formation was investigated. (A) The Shahmirzad, (B) Darvar and (C) Dehmolla sections.

DATASET AND METHODS

The Geirud Formation is today tectonically deformed to form a flank of an extensive anticline (Fig. 3A and B). Three main outcrops were chosen to acquire the stratigraphic and sedimentological data reported in this study; from west to east, they are: the Shahmirzad (Fig. 3A), Darvar (Fig. 3B) and Dehmolla (Fig. 3C) sections (Fig. 1C, D and E for their location).

The preservation of the outcrops allowed detailed architectural analysis. Bed dimensions, geometry, lithology and sedimentary structures were measured and recorded in outcrop from each bed. Physical sedimentological (texture, composition, mechanical structures, contacts, geometry, palaeocurrent directions, etc.) and ichnological attributes, as well as vertical trends and stacking patterns were examined to define and interpret facies and facies associations (Tables 1 and 2). The

three sections (Shahmirzad, Darvar and Dehmolla; Fig. 3) were thus correlated by using the main lithostratigraphic elements and bounding surfaces (Fig. 4). This dataset provided a basis for the interpretation of sedimentary processes, depositional environments and systems.

RESULTS

The sedimentological dataset documented for the Geirud Formation allowed the recognition of four groups of facies associations (Table 1), each recording a distinctive depositional zone. These groups, including genetically-related facies associations, occur in different stratigraphic positions along the Geirud Formation. The three outcrops chosen for our observations (Fig. 3A to C) expose the best-preserved strata belonging to the Geirud Formation. Their differ slightly in the general

Table 1. Letter symbols used to codify the lithofacies recognised in the studied deposits.

Groups	Facies associations	Facies	Depositional environments	Depositional zones
1	A1	Gmm, Gt, St, Sp, Sm Sl	fluvial channel belt	river-dominated bay-head delta zone
	A2	Fm Sl	inter-channel flood-plain	
	A3	St, Sp Shl	transitional fluvio-tidal channel	
2	B1	Fm, Wr, Sm, Shl Cr	tidal flat	mixed-energy, central basin zone
	B2	St, Sp Sl	tidal channel	
	B3	Sp, Hb, Shl, Wr Cr	longitudinal tidal bar	
3	C1	Fm, Shl HCS	offshore-transition	wave-dominated, siliciclastic coastal and open-marine shelf zones
	C2	St, Sp HCS	wave-dominated shoreface	
4	D1	Fm Shl	offshore	carbonate open-marine shelf zone
	D2	Fm, HCS, St Cr	nearshore	

Table 2. Groups of facies associations, facies and interpreted depositional environments and zones detected within the Geirud Formation (for letter symbols see Table 2).

Letter symbol	Description	Biota and bioturbation	Sedimentary process
Gmm	Matrix-supported, poorly sorted gravel organized into structureless beds with sharp bases and tops	Absent	Hyper-concentrated debris flow accumulated during high-flow regime of unidirectional currents
Gt	Clast-supported, well-sorted gravel organized into trough cross strata with sharp bases and tops	Absent	Deposition of bedload during low flow regime of unidirectional currents
Se-Ss	Coarse to very coarse grained, very poorly sorted sand with scattered granules, organized into cross strata containing gravel lags	Absent	Rapid deposition within chute or scours excavated into cohesive muddy substrates
Sp	Moderately-sorted to well-sorted sandstone organized into planar cross strata with sharp bases and tops	Sporadic bivalve shell debris	Migration of 2D dunes under low-energy flow regime
Sl	Low-angle cross-laminated, moderately-sorted to well-sorted sandstone	Absent	Washed-out sand flows occurring between subcritical and supercritical currents
Shl	Horizontally-laminated, moderately-sorted to well-sorted sandstone	Absent	Currents flowing under low-flow regime
Sm	Moderately-sorted to well-sorted structureless (massive) sandstone	Absent or containing rare bioclasts (bivalves)	Rapid deposition by high-energy flow or post-deposition modification (e.g. collapse of previously formed beds)
St	Trough cross-stratified, moderately-sorted to well-sorted sandstone, with gravel lags often occurring at the base of the troughs	Rare shell debris, bivalves, brachiopods	Migration of 3D dunes under low-flow regime
Wr	Symmetric ripples	Absent	Deposition under oscillatory wave motion in shallow-marine environments
Cr	Asymmetric ripples	Moderate bioturbation in places	Migration of ripples under unidirectional currents
Hb	Fine-to-medium grained, moderately-sorted sandstone containing foresets with opposite directions of migration (herringbones)	Absent or concentrated in intervals within neap lamina cycles	Bi-directional tidal currents
HCS	Medium-to-coarse grained, moderately-sorted sandstone containing hummocky and swaley cross-lamination	Shell debris, bivalves, brachiopods, echinoderm. Rare bioturbation in the sandstone strata, intense bioturbation in the finer-grained interfaces	Oscillatory waves generated after high-energy storms and rapidly vanishing

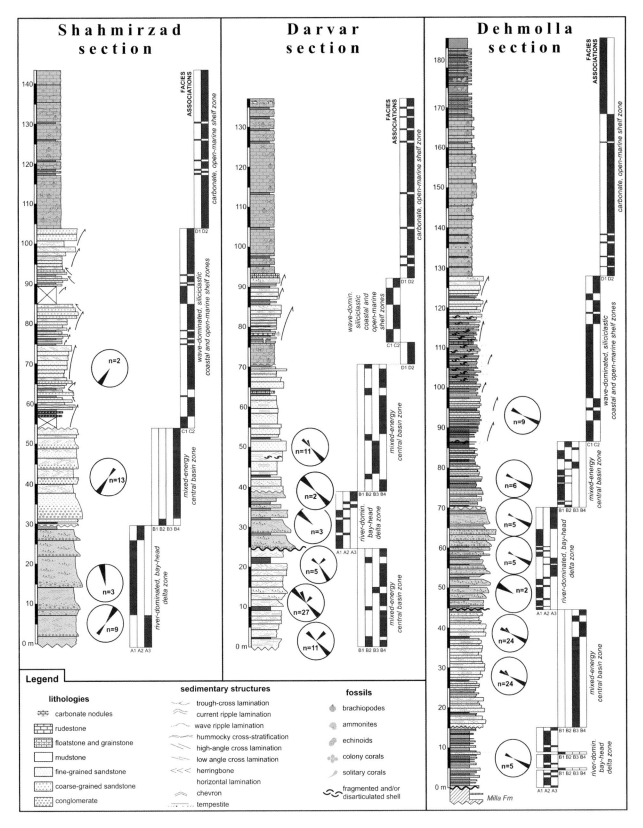

Fig. 4. Stratigraphic-sedimentological logs measured in the three studied sections and facies association panels (see Table 2 for facies association letter symbols).

thickness of the outcropping succession, from 140 to 180 m, but their relative position across the investigated area (Fig. 1B) is strategic in order to detect along-strike facies changes and lateral overall geometries of the main bounding stratigraphic surfaces formed in this part of the Alborz Basin during the Devonian.

Group 1 (facies associations A1 to A3): river-dominated bay-head delta zone

This group includes three facies associations: Fluvial channel belt deposits (A1), Flood-plain deposits (A2) and tidally-influenced fluvial channel deposits (A3).

Facies association A1: Fluvial channel belt deposits

Description: This association comprises six facies: *Gmm, Gt, St, Sp, Sm* and *Sl* (see Table 2 for an explanation of the letter symbols used), which are generally organized into composite, thinning-upward and fining-upward, lenticular-shaped strata sets (Fig. 5A and B). Each strata set shows highly-varying thicknesses, ranging from a few centimetres to a few metres, and is often encased within mudstones pertaining to the facies association A2 (Fig. 5C and D). Sediments consist of reddish pebble and granule conglomerates, passing upwards to coarse to fine pebbly sandstones (Fig. 5D). Internally, individual strata are 0.4 to 1.2 m-thick, have erosional concave-up bases and are normally-graded. Basal conglomeratic facies consist of matrix-supported (*Gmm*) to clast-supported (*Gt*) pebble-grade to granule-grade (2 to 8 cm in diameter) deposits, containing sub-angular to sub-rounded elements made up of quartz, chert and a minor amount of mud clasts (Fig. 5E and F). Conglomerates are locally structureless or are diffusely cross-stratified (Fig. 5F). These facies evolve upwards to red-coloured and pink-coloured, medium-grained and fine-grained sandstone, well-sorted to medium-sorted. Sandstone strata range in thickness from a few centimetres to 1 m. Internally, they show trough and planar cross-lamination (facies *St* and *Sp*) indicating a NW-directed, palaeoflow direction (Fig. 5G and H) or they are devoid of any sedimentary structure (*Sm*). The top of each fining-upward strata set is commonly characterised by low-angle and horizontal lamination (facies *Sl*).

Interpretation: The fining-upward and thinning-upward strata sets described in the facies association A1 represent a multi-storey channel-fill belt accumulated in a braided-type fluvial setting, possibly characterised by a high-gradient relief (Bridge, 2003; 2006). As observable in many modern analogue depositional systems, uni-directional water currents are usually concentrated in sinuous and interlaced channels which may expand and shift position through a combination of vertical incision and lateral migration (Collinson, 1996; Miall, 1996). During high-energy flow stages (Steel & Thompson, 1983), single floods occur as high-density debris flows transporting coarse-grained bed-load material which scour the underlying finer and cohesive deposits (Nemec & Steel, 1984; Miall, 1996; Nouidar & Chellai, 2001; Fabuel-Perez *et al.*, 2009; Fröhlich *et al.*, 2010), generating the concave-up erosional surfaces observed at the base of each strata set. As the flood vanes, the coarsest and heaviest bedload accumulates, producing a gravel basal lag comparable to the conglomeratic facies *Gmm* and *Gt* described at the base of each strata set. The coexistence of sub-rounded and sub-angular clastic elements is a feature commonly described in such debris flow deposits (e.g. Nemec & Kazancy, 1999), where well rounded pebbles are frequently subject to intraparticle impacts during the mass transport, producing an *in-situ* brecciation and consequent sub-angular elements. The flanks of the incising channel may collapse during this stage, accumulating structureless strata at the base, similar to those of facies *Sm*. During the subsequent high-energy stage, the main channels may migrate in other sectors of the braided river system (Steel & Thompson, 1983; Nemec & Kazanci, 1999). Consequently, basal lags are buried rapidly by sand-size sediments reworked by traction currents of different energy strength. The resulting deposits can be identified in the described facies *St*, *Sp* and *Sl*, whose internal cross-lamination records the progressive decrease of the fluvial flood energy and the consequent channel-filling by bedforms indicative of changes in flow regime (Miall, 1996). These bedforms were possibly 3D, sinuous-crested dunes migrating in the deeper part of the active channels (Miall, 1996; Bridge, 2003, Fabuel-Perez *et al.*, 2009; Fröhlich *et al.*, 2010; Schwartz *et al.*, 2011). Planar and low angle cross-lamination at the top of fining-upward sequences indicate the migration of low-relief

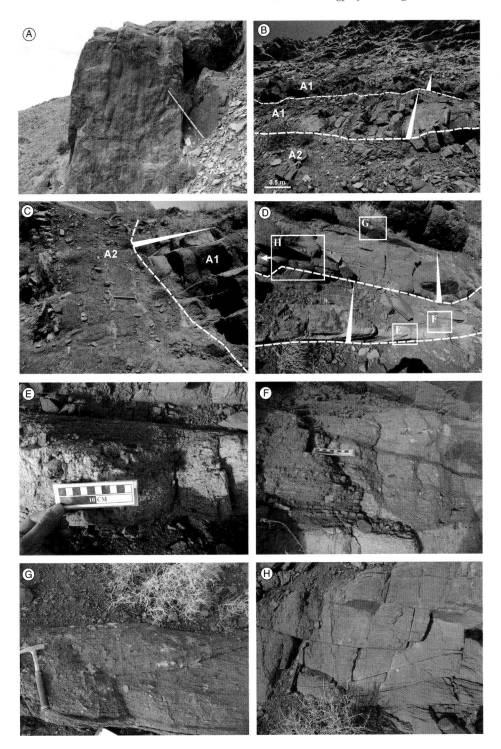

Fig. 5. Outcrop photographs of the sedimentary deposits belonging to the facies associations A1, A2 and A3. (A) Structureless, reddish sandstone of facies association A1. (B) Erosive-based, vertically-stacked lenticular strata of the facies association A1 encased in mudstone deposits of facies association A2. (C) Detail of the erosive base of a river channel-fill cutting the underlying heterolithic mudstone-sandstone strata of the facies association A2. (D) Channel-fill sandstones are often normally-graded (white arrows). (E) Detail of a pebbly sandstone at the base of a channel-fill strata. (F) Low-angle cross-lamination in reddish sandstones. (G) High-angle cross-lamination comprised within stratasets (H) of different dimensions.

bed waves (bedload sheets) on upper-stage plane beds in the shallow areas of active channels during falling flow stages (Miall, 1996; Bridge, 2003; 2006). The different thicknesses of the fining-upward strata sets observed within this facies association can be related to channel belts of various sizes, recording streams of different energy and width. Small-scale channel belt deposits are in fact encased within mudstones, indicating that they record overflood episodes which activate peripheral parts of the fluvial systems (Plint & Browne, 1994; Miall, 1996; Bridge, 2003; Schwartz *et al.*, 2011).

Facies association A2: Flood-plain deposits

Description: This association, occurring as intercalations within the previously described deposits (Fig. 5B and C), comprises two facies: *Fm* and *Sl*. Facies *Fm* consists of red and pink, claystones, locally mixed to a small amount of siltstones, forming 0.5 to 2 m-thick, massive to laminated tabular strata. Locally, these deposits are intercalated with facies *Sl* which consists of very fine sandstone (Fig. 5C). Subaerial exposure evidences, such as pedogenized soils or root casts, associated with burrows and desiccation cracks, were seldom observed in this facies association.

Interpretation: This association is thought to be the product of settling from suspension during flooding events generated by overbank flows, which are quite frequent in a floodplain environment of braided river systems (Miall, 1996; Bridge, 2003). Mudstone and fine sandstone strata thus represent deposition under overbank flows of weaker energy or the result of settling accumulation in ephemeral pools comprised between bars (Weissmann *et al.*, 2011). The scarcity of any subaerial markers in the topmost levels of these intervals suggests very frequent flooding episodes, as well as quick lateral migration of the channel belts, which are very common features in a dynamic braided river system developed in a humid, discharge-active climatic regime.

Facies association A3: Transitional fluvio-tidal channel deposits

Description: This association represents a sand-rich succession which overlies the fluvial facies associations A1 and A2 through sharp or grading contacts (Fig. 6A). Facies association A3 includes three recurrent facies: *St*, *Sp* and *Shl*. Individual

sedimentary units range from 1 to 5 m in thickness, have sharp or erosional bases (Fig. 6A, B and C) locally marked by rip-up mudstone clasts and show frequent fining-upward grain-size trends. Sediments consist of basal coarse to fine-grained sandstone containing sparse glauconitic grains, showing high-angle to medium-angle cross lamination and forming trough cross strata (*St*) (Fig. 6B). Cross lamination consists of laminasets, internally characterised by bundles (Fig. 6B and C) and bounded by discontinuity surfaces which have been distinguished into three hierarchical orders (I to III in Fig. 6B), on the basis of the thickness of the laminasets they delimit (Fig. 6C). This facies often contains trough cross stratification passing upwards to planar cross lamination (*Sp*) and to horizontal lamination (*Shl*) (Fig. 6D). Foresets show NE-SW (Sharmirzad) and NW-SE (Dehmolla) palaeocurrents, appear totally deprived of mudstone intervals or drapes (Fig. 6E) and locally are organized into herringbone cross stratification (Fig. 6F). Bioturbation is rarely present and consists of sporadic inclined burrows attributable to *Palaeophycus* and *Thalassinoides* trace fossils.

Interpretation: Fining-upward strata sets composing facies association A3 are interpreted as fluvial channel-fills developed in a region with low-gradient relief and adjacent to a coastal marine environment. Cross stratification observed in these sandstone strata are interpreted as the result of the complex interaction between upper-stage and lower-stage current flows (Bridge, 2006). Laminae bundles, hierarchically-different discontinuity surfaces and herringbone structures (Fig. 6E), which are all features commonly attributed to the action of flood and ebb tidal currents in transitional or coastal environments (Boyd *et al.*, 2006), indicate the tidal impact on fluvial discharge and the influence of flood tidal currents within the river stream that usually occur during low-stage fluvial current regime (Shanley *et al.*, 1992; Spalletti, 1996; Yoshida, 2000; Fischbein *et al.*, 2009; Bhattacharya & Bhattacharya, 2006; Martinius & Gowland, 2010). Discontinuity surfaces of different hierarchical order also suggest that sediment accumulation occurred according to episodes of deposition under the effect of traction currents which decelerated, possibly during slack water periods or low-energy, neap tides in the coastal areas. The absence of the mud drapes within the foresets thus suggests a transport capacity which is still sufficient to move the

Fig. 6. Outcrops showing sediments of the facies association A3. (A) Sandstone cross-stratasets of facies association A3 lying on erosive base. Internally, sandstone tidal channel-fill consists of vertically-stacked cross strata, including cross lamination (see line drawing in B). (C) Detail of the previous outcrop, showing different hierarchies (I, II, III) of discontinuity surfaces. (D) Trough cross lamination and (E and F) herringbone cross lamination are also present in this lithofacies (see Table 1 for facies letter symbols).

fine-grained fraction perennially in suspension. General medium-energy conditions are also supported by little glauconite content and low bioturbation (Dashtgard *et al.*, 2010; Sharafi *et al.*, 2012a, 2012b; Sharafi *et al.*, 2013). Facies association A3 is thus considered as the filling of transitional, fluvio-tidal channels located seaward of the limit of tidal action of the alluvial plain (Dalrymple *et al.*, 1992; Ichaso & Dalrymple, 2006), which is known as 'tidal limit' in tidally-dominated estuaries (Boyd *et al.*, 2006; Dalrymple & Choi, 2007). Also, the increasing thickness of the channel-fill strata compared to the average channel size of the facies association A1, indicates an increased transport capacity which is common in tide-influenced estuaries (Dalrymple *et al.*, 2012).

Group 2 (Facies associations B1 to B3): mixed-energy, central basin zone

This group includes three facies associations: Tidal flat deposits (B1), tidal channel deposits (B2) and longitudinal tidal bar deposits (B3). The group 2 occurs in the lowermost stratigraphic intervals of the Darvar and Dehmolla sections but it recurs again upwards, in the middle intervals of all the three logs (Fig. 4).

Facies association B1: Tidal flat deposits

Description: Facies association B1 is exposed in limited outcrops in the studied area and overlies the deposits of facies association A2 (Fig. 7). Sediments of facies association B1 form 0.1 to 0.5 m-thick tabular strata sets, characterised by no evident bed-thickness or vertical grain-size trends (Fig. 8A). This association mostly consists of heterolithic sandstone/mudstone couplets, with mudstone strata progressively decreasing upwards (Fig. 8B). The association contains five main facies: *Fm*, *Wr*, *Shl*, *Sm* and *Cr*. Mudstone strata of facies *Fm* form 0.30 to 1.5 m-thick finely laminated-flat strata and are generally structureless or indistinctly laminated (Fig. 8B). They are intercalated with thinly-bedded, less than 5 cm-thick, very fine sandstone and siltstone deposits, forming massive (*Sm*) and indistinctly laminated (*Shl*) strata (Fig. 8C). Bioturbation is absent. Locally, bioclastic debris occurs as bivalve-rich and brachiopod-rich thin intercalations (Fig. 8D). Strata of this facies dominate in the lowermost interval of the association but they progressively thin upward, replaced by prevailing sandstone facies (Fig. 8B). Facies *Wr* is made up of very fine-grained and medium-grained sandstone organized into strata commonly less than 10 cm-thick and containing flaser structures (Fig. 8E). Strata contain symmetrical, WNW-ESE-oriented ripple cross lamination (Fig. 8E). These deposits alternate with centimetres-thick to decimetres-thick fine sandstone strata, which include plain-parallel lamination (*Shl*) and asymmetrical, ESE-pointing ripple cross-lamination (*Cr*) (Fig. 8F). Facies association

B1 typically grades upward to coarser-grained sediments of association B3. Coupled in strata sets, these two associations form coarsening-upward and thickening-upward sequences up to 6 m-thick. B2 deposits can also be erosionally overlain by facies association B1 (Fig. 7).

Interpretation: Heterolithic deposits of facies association B1 are interpreted as the record of a tidal flat environment, which commonly is thought to be an important physical component of a larger estuarine-type depositional system or open-coast of mesotidal to macrotidal seas (Van Straaten, 1961; Klein, 1985). Mud derives from fall-out of suspended fine-grained sediments, whereas sandstone facies result from occasional incursions of traction currents flowing under lower-flow regime (Shanley *et al.*, 1992; Dalrymple *et al.*, 1992; Nouidar & Challai, 2001; Fabuel-Perez *et al.*, 2009). Mudstone-dominated facies occurring in the lowermost stratigraphic interval records upper tidal flat deposits, which are commonly characterised by dominant muddy sediments accumulated from suspension (Reading & Collinson, 1996). Mudstone beds represent deposition during slack water stages (Shanley *et al.*, 1992; Fabuel-Perez *et al.*, 2009). Symmetrical ripples observed in the facies *Wr* (Fig. 8E) suggest oscillatory currents generated by waves, possibly induced by the wind on the water surface (Shanley *et al.*, 1992; Fabuel-Perez *et al.*, 2009). Mid-tidal flat facies are indicated by strata characterised by balanced volumes of mud and sand deposits, reflecting roughly equal periods of suspension and bedload deposition. Thus, bed geometries change from lenticular through wavy and flaser (Fig. 8E and F) as the proportion of mud decreases upwards (Reineck & Wunderlich, 1968). This zone generates vertically stacked laminites or tidal

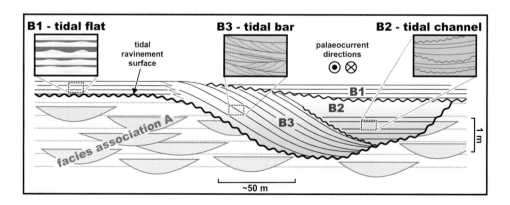

Fig. 7. Conceptual along-strike reconstruction of the geometric lateral relationships between facies associations B1, B2 and B3.

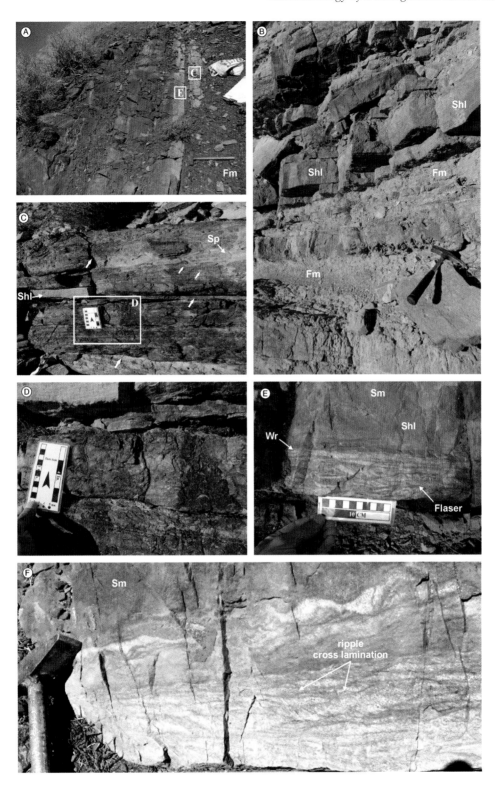

Fig. 8. Deposits of facies association B1. (A) Heterolithic mudstone-sandstone stratasets interpreted as the record of tidally-generated rhythmites accumulated in a sand-dominated tidal flat environment. (B) Mud-dominated heterolithic facies pass upward to sand-dominated strata couplets. (C) Fine-grained sandstone strata are often characterised by: (D) Bioclastic debris deriving from bivalve and brachiopod rests; (E) symmetrical, wavy cross lamination associated with plain-parallel (*Shl*) cross lamination; (F) flaser cross lamination alternating asymmetrical ripples and massive (*Sm*) fine-grained sandstone strata (see Table 1 for facies letter symbols).

rhythmites, possibly recording neap-spring cycles (Tessier *et al.*, 1989; Tessier, 1993; Dalrymple *et al.*, 1991). Sand-rich facies in the uppermost interval of the association (Fig. 8B) record low-tidal flat deposits, where current-generated sand bedforms such as asymmetrical and symmetrical ripples predominate. The low bioclastic content and the absence of bioturbation suggest extremely-stressed conditions caused after periodic and/or significant salinity fluctuations (De Mowbray, 1983; Frey & Howard, 1986; Nouidar & Chellai, 2001; Buatois *et al.*, 2002; Savrda & Nanson, 2003; Mángano & Buatois, 2004; Gingras *et al.*, 2012).

Facies association B2: Tidal channel deposits

Description: The deposits belonging to this facies association erosionally overlie sediments of associations B3, whereas they are overlain by facies association B1 (Fig. 9A). Facies association B2 includes three facies: *St*, *Sp* and *Sl*. Sediments mostly consist of fine-grained and medium-grained sandstone, with subordinate muddy sandstone, containing abundant gravel-size skeletal fragments and glauconitic grains. Strata are vertically stacked to form 1.5 to 3 m-thick, often amalgamated bed sets having internal erosional surfaces and no evident vertical grain-size trends (Fig. 9B). Strata sets have erosional basal surfaces (Fig. 9C), characterised by bivalve, brachiopod and echinoderm skeletal concentrations, whose fragments also occur throughout the entire bed thickness but with a lower abundance. Internal sedimentary structures are trough, tabular and low-angle cross-lamination (facies *St*, *Sp* and *Sl*) (Fig. 9D), locally showing single mud drapes. Reactivation surfaces are often visible separating cross-laminated sandstone intervals. Palaeocurrents indicate NW-SE-oriented dominant trends, associated with subordinate NW-oriented values (Fig. 4).

Interpretation: The erosional bases with associated lags, the dominance of traction-generated cross strata and the absence of wave-generated structures, indicate that sand deposition occurs in tidally-dominated channel-fills or swatchways (Fig. 7) cross-cutting tidal bars (facies association B3) across a middle estuary zone (Dalrymple & Rhodes, 1995; Elliott, 1986; Yoshida, 2000;

Fig. 9. (A) Tidal channel/bar deposits of facies associations B2 and B3. (B) Tidal channel facies are organized into amalgamated strata sets with internal erosional surfaces and no evident vertical grain-size trends. (C) Strata sets have erosional basal surfaces. (D) Internal sedimentary structures are trough, tabular and low-angle cross-bedding (see Table 1 for facies letter symbols).

Nouidar & Chellai, 2001; Dalrymple & Choi, 2007; Folkestand & Satur, 2008; Dalrymple *et al.*, 2012). Tidal currents are suggested by the occurrence of foreset bundles and reactivation surfaces in the finer-grained sandstone (Visser, 1980; Smith, 1988; Dalrymple *et al.*, 1990; Folkestand & Satur, 2008; Longhitano *et al.*, 2012a). The amalgamation or the erosional stacking of the internal units indicate that the tidal channel filling process was roughly continuous or characterised by repeated rapid depositional and erosional episodes. Shell and glauconitic clasts indicate a close association with marine environments through the action of landward-directed, flood tidal currents.

Facies association B3: Longitudinal tidal bar deposits

Description: This facies association is composed of five facies: *Sp*, *Hb*, *Shl*, *Wr* and *Cr*. Sediments consist of fine-grained to coarse-grained, moderately-sorted to well-sorted sandstone, containing abundant glauconitic grains and skeletal fragments, organized into strata sets with highly-varying thicknesses, ranging from 3 up to 12 m. Strata sets have concave-up geometry and are bounded at the bases and tops by master erosional surfaces (I) with 1.5° to 4° of inclination in the same direction of the general sediment transport (Fig. 10A). Internally, II-order and III-order discontinuity surfaces form high-angle cross-lamination (*Sp*), containing foresets (IV order surfaces) with angular and tangential geometry, often characterised by bidirectional, opposite (herringbone), NW-SE-trending palaeocurrent directions (*Hb*) associated with normally-oriented (NE-directed) foresets (Fig. 10B). In the upper bed intervals, horizontal lamination (*Shl*) and locally wave and current ripples (*Wr*, *Cr*) are present (Fig. 10C and D). Rare bioturbation structures (*Skolithos* and *Ophiomorpha nodosa*) occur sparsely within the strata. The deposits belonging to this association are laterally adjacent with, and erosionally overlain by, sediments of facies association B2, *i.e.* tidal channel deposits (Fig. 7 and 10E).

Interpretation: The stratal geometries and internal organisation of this facies suggest primary deposition of sand-size sediment under the effect of tidal currents. The sedimentary structures and low bioturbation of sandstone packages suggest the accumulation of tidal bars and ripples in a stressed subaqueous setting (Yoshida, 2000;

Nouidar & Chellai, 2001; Fabuel-Perez *et al.*, 2009). Rare mudstone drapes indicate that the finest fraction was transported perennially in suspension; which is common in high-energy environments. Their depositional architectures and lateral relationships with the above-described deposits allow us to interpret these facies as the record of longitudinal tidal sand bars migrating across an estuarine environment (Dalrymple *et al.*, 1992; Allen & Posamentier, 1994; Yoshida, 2000; Folkestand & Satur, 2008; Tessier, 2012). The bioturbation and glauconite suggest that these deposits accumulated in the most external part of a tidally-dominated estuary, across the transition to the open-marine environment, where sedimentation is influenced by wave and tidal processes (Plink-Björklund & Steel, 2006; Dalrymple & Choi, 2007; Such *et al.*, 2007; Schwarz *et al.*, 2011). Tidal bars develop in shallow-water marine environments, as can be observed in many modern depositional systems (Dalrymple *et al.*, 1990; Nio & Yang, 1991; Yoshida *et al.*, 2001). The direction of migration of tidal bars occurs orthogonally or at high angle with respect to the main tidal current direction (e.g. Dalrymple & Rhodes, 1995; Olariu *et al.*, 2011; 2012). In an estuarine setting, longitudinal tidal bars are adjacent to distributary tidal channels, which migrate during their evolution over the tidal bars deposits, generating the master surfaces of erosion observed as major architectural element (I in Fig. 10B) within the cross-stratified deposit of the facies association B3. Internal discontinuity surfaces are due to: local changes in the flow direction (II), reactivation of bedform migration after slack-water stages (III) and foreset accretion (IV). The vertical stacking of tidal bars may produce compound sandbodies, which are very common features in transgressive, tidally-dominated successions (Dalrymple *et al.*, 1990; Longhitano *et al.*, 2010; 2012b; Olariu *et al.*, 2011; Reynaud & Dalrymple, 2012; Longhitano, 2013).

Group 3 (Facies associations C1 to C2): wave-dominated, siliciclastic coastal and open-marine shelf zones

Sediments included within 'group 3' comprise two siliciclastic facies associations which are closely associated, forming coupled stratal units (Fig. 11A). From the base to the top, these units consist of: offshore-transition deposits (C1) and wave-dominated shoreface deposits (C2).

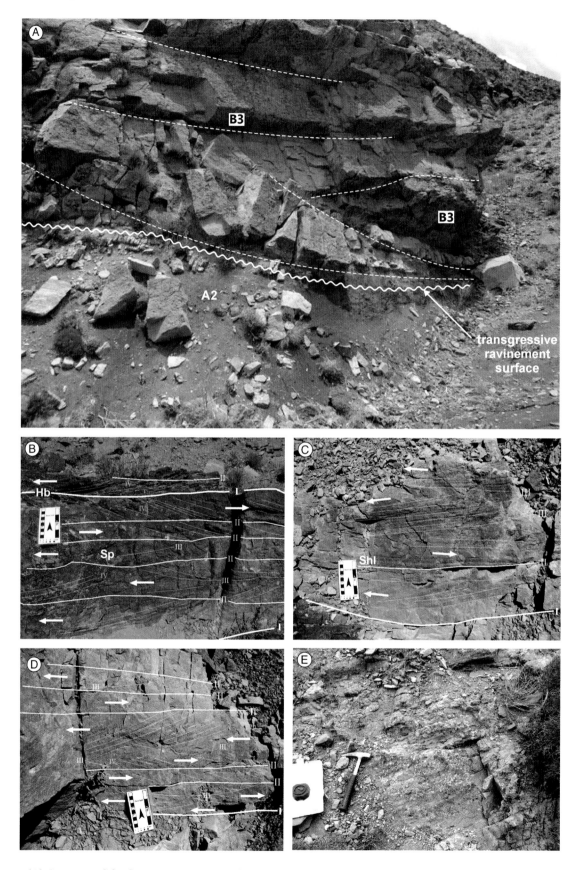

Fig. 10. (A) Outcrops of the facies association B3. (B) Stratasets separated by master erosional surfaces (I) are interpreted as tidal sand bars. (C) Internally, they contain II-order, III-order and IV-order discontinuity surfaces, comprising foresets with angular and tangential geometry and with bidirectional, opposite palaeocurrent directions (D). (E) Bioturbation structures derived from *Skolithos* and *Ophiomorpha nodosa* (see Table 1 for facies letter symbols).

Fig. 11. (A) Vertically-stacked (tectonically-inclined) stratasets separated by sharp surfaces and interpreted as parasequences of the facies associations C1 and C2. (B) Offshore-transition mudstone and fine-grained sandstone of facies association C1, containing *Thalassinoides, Ophiomorpha, Palaeophycus, Planolites, Rhizocorallium*, fugichnia trace fossils (C). (D) Thickening-upward (white arrow) sandstone strata of facies association C2, including high-angle cross lamination (E) and hummocky cross lamination (F) (see Table 1 for facies letter symbols).

Facies association C1: Offshore-transition deposits

Description: The deposits belonging to this facies association form tabular lithosomes with thickness ranging from 1 m to 2–3 m (Fig. 11B) and characterised by sharp basal surfaces of erosion overlain by gravel-size, skeletal fragments, mostly derived from bivalves, brachiopods and echinoderms. Facies included in this association are *Fm, Shl* and *HCS*. Sediments consist of highly-bioturbated and often structureless 0.5 to 1 m-thick mudstone strata (*Fm*), containing 10 to 30 cm-thick sandstone intercalations showing plane-parallel lamination (*Shl*) (Fig. 11B). Often, such strata sets include thicker and coarser sandstone intervals, having erosional basal contacts and thicknesses of 50 to 70 cm, internally characterised by hummocky cross-stratification (*HCS*). A number of different trace fossils, including *Thalassinoides, Ophiomorpha, Palaeophycus, Planolites, Rhizocorallium* and *fugichnia* occur, are preserved as interface structures (Fig. 11C).

Interpretation: The deposits of this facies association indicates a subaqueous depositional environment dominantly characterised by sedimentation occurring from fall-out of fines moved in suspension where, intermittently, traction currents occurred transporting sandy sediments as low-energy regime turbidity flows. Occasionally, this environment was affected by powerful oscillatory flows, possibly generated by vanishing storm-driven currents, which produced erosion of the more proximal sandy deposits (*i.e.*, shoreface) and subsequent rapid deposition seaward due to the decreasing of the flow energy as the environment returned under fair-weather conditions. Such processes thus suggest an offshore-transition environment (Nouidar & Chellai, 2001; MacEachern et al., 2007; Sharafi et al., 2010; Fröhlich et al., 2010; Schwarz et al., 2011), occupying the distal and deeper part of a wave-dominated coastal system (Walker & Plint, 1992).

Facies association C2: Wave-dominated shoreface deposits

Description: This association grades over the deposits belonging to the previous facies association C1 (Fig. 11A) and consists of coarsening-upward stratal units up to 10 m-thick composed of facies *St, Sp* and *HCS* (Fig. 11D). Sediments are moderately to well-sorted, very fine sand grading-upward to coarse sand and granule-size bioclastic sand, often containing gravel strata in the uppermost interval of the unit. From the base to the top of each stratal unit, sediments are organized into a 2 to 5 m-thick interval which consists of 50 to 80 cm-thick sandstone strata, often amalgamated, comprising trough and planar cross lamination (*St, Sp*) (Fig. 11E) and rare hummocky cross-stratification (*HCS*) (Fig. 11F). Gravel-size skeletal fragments, mostly derived from bivalves, brachiopods and echinoderms, are abundant and randomly distributed or concentrated in patches.

Interpretation: The vertical facies transition, the upward changes of internal architectures, as well as the relationships with the underlying offshore-transition strata that characterises each stratal unit are features consistent with shoreface regressive sequences described in a number of wave-dominated modern and ancient coastal settings (Fisher & McGowen, 1967; Plint, 2000; Hampson & Storms, 2003; Swift et al., 2008; Varban & Plint, 2008; Hampson, 2010). The entire facies composing these stratal units appear dominated by current-generated and wave-generated sedimentary structures, which commonly identify lower to upper shoreface environments located across the fair-weather wave base level (Clifton, 1969; Clifton et al., 1971; Elliott, 1986; Walker & Plint, 1992; Hampson & Storms, 2003).

Group 4 (Facies associations D1 to D2): carbonate open-marine shelf zone

This group of facies associations occupies the uppermost stratigraphic interval of the studied succession and includes offshore deposits (D1) and bioclastic nearshore deposits (D2).

Facies association D1: offshore carbonate deposits

Description: Sediments of facies association D1 overlie the previous shoreface deposits through a sharp contact. This association comprises two main recurrent facies: *Fm* and *Shl*. Sediments dominantly consist of dark-grey mudstones (*Fm*) which are typically structureless but also, locally, mudstones containing well-distinct horizontal lamination (shales) occur. This facies shows rare intercalations of fine-grained sandstone and siltstone 10 to 20 cm-thick (Fig. 12A). Scarce skeletal elements (bivalves, brachiopods) are present in the mudstone facies.

Fig. 12. (A) Thinly-bedded laminated black shales of facies association D1, intercalated with packstone and grainstone of facies association D2. (B) Brachiopod and echinoderm fragments. (C) Colonizing corals *Syringopora* intercalated to fine-grained facies. (D, E) Hummocky cross-stratification, characterised by internal cross lamination. (F) *Arenicolites, Protovirgularia (pr), Diplocraterion, Palaeophycus, Thalassinoides, Chondrites* and *Helminthopsis* trace fossils.

Interpretation: This association records an open-marine, offshore depositional environment characterised by sediment accumulation that occurred from fall-out of suspended fines below the storm wave base (Reading & Collinson, 1996; Fröhlich *et al.*, 2010). The structureless or laminated aspect of the mudstone strata suggests deposition under slightly different conditions of sediment rate or different degrees of biological homogenization. The less-frequent fine sandstone and siltstone intercalations record weak turbidity currents flowing in a most distal part of a shelf, possibly generated after episodes of major energy (storms) affecting the nearshore coastal areas (Fröhlich *et al.*, 2010).

Facies association D2: Bioclastic nearshore deposits

Description: The deposits belonging to the facies association D2 form a 40 to 45 cm-thick strata set which sharply overlies the previously-described units and is overlain by the strata of the Mobarak Formation. Facies association D2 includes four main facies: *Fm*, *HCS*, *St* and *Cr* (Fig. 12A). The basal sediments of this stratigraphic interval are dominantly composed of black to dark grey mudstone and wackestone containing 20 cm-thick black shale intercalations (*Fm*). Upwards, sediments consist of dominant packstone and grainstone, forming 20 to 40 cm-thick strata (Fig. 12A). The bioclastic content mostly includes *Syringopora* corals (Fig. 12B). Subordinate bivalves, gastropods, ammonites, solitary and colony corals, bryozoans, trilobites, benthic foraminifers and ostracods also occur, showing various degrees of fragmentation. Admixed to these fossil associations, peloids, intraclasts and quartzite material were also observed. In the lowermost interval, colonizing corals (*Syringopora*) intercalated to fine-grained facies were found (Fig. 12C). In the uppermost, coarsening-upward interval of this facies association, strata are internally characterised by the presence of hummocky cross-stratification (*HCS*) (Fig. 12D and E), associated to local trough cross-lamination (*St*) and current ripple lamination (*Cr*). Coarse-grained 20 to 30 cm-thick intervals often occur, characterised by normal-graded textures and undulate erosive bases. Biogenic structures include *Arenicolites*, *Protovirgularia*, *Diplocraterion*, *Palaeophycus*, *Thalassinoides*, *Chondrites* and *Helminthopsis* (Fig. 12F), mostly displaying interface lifestyle.

Interpretation: The deposits of facies association D2 records a vertical transition from basal deep-marine carbonate sediments that progressively become even more shallowing upward. This trend is indicated by the increasing overall grain size of the bioclastic-rich facies, associated to sedimentary structures, such as 'tempestite' layers, that unequivocally record the influence of the wave base on to the sediments and, therefore, the evolution from offshore to offshore-transition/ shoreface environments through time. The lowermost, deeper interval suggests the occurrence of periods when anoxic conditions predominated in the substrate, recorded by the black shale intercalations. This oxygen-deficient substrate was sporadically colonized by *Chondrites*, probably indicating short periods of increasing oxygen concentration (Bromley & Ekdale, 1984; Ekdale & Mason, 1988; Schwarz *et al.*, 2011). The local occurrence of *in-situ* colony corals indicates that patch reefs of limited extension developed in this environment, built up by small colonies of tabulate and rugose corals, associated with brachiopods, in a shelf environment. Brachiopod-dominated and echinoderm-dominated fossil assemblages observed in the uppermost interval suggest shallower marine conditions.

Depositional zones of the Geirud Formation

The facies associations described and interpreted in the previous sections represent the record of distinct depositional zones belonging to a mixed, wave-dominated and tide-dominated estuary adjacent seaward to an open marine coast (Fig. 13). The four groups of lithofacies associations interpreted in the previous sections record sediment accumulation across a coastal area, whereas their internal features represent along-strike facies heterogeneities. No correlative data were detected in a perpendicular cross section, due to the lack of exposed sections because of the tectonic deformation of the Geirud strata. However, a possible along-dip reconstruction was inferred by using some of the most common depositional models for wave-dominated estuaries (i.e. Zaitlin *et al.*, 1994; inset in Fig. 14).

The correlation of Fig. 14 shows abrupt vertical changes from estuarine bay-head deltaic lithofacies evolving upward to estuarine central basin deposits and to shallow-marine and open-marine sediments. The various depositional areas (Table 1) record: (i) a river-dominated bay-head delta zone, which includes basal fluvial channel belts (f.a. A1), flood plains (f.a. A2) and transitional fluvio-tidal channels (f.a. A3); (ii) a mixed energy, central basin zone, which consists of tidal flats (f.a. B1), associated with tidal channels (f.a. B2) and longitudinal tidal bars (f.a. B3); (iii) a wave-dominated, siliciclastic coastal zone made up of marine deposits including offshore-transition (f.a. C1) and wave-dominated shoreface (f.a. C2) environments; and (iv) a carbonate open-marine shelf zone representing the uppermost

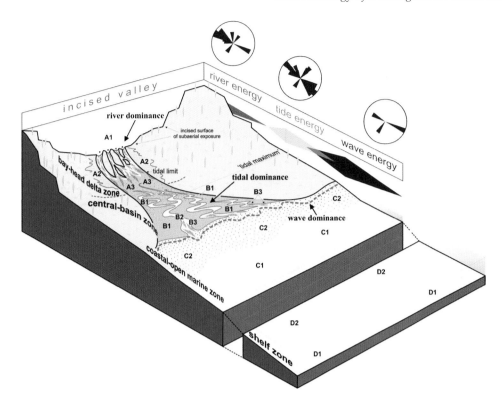

Fig. 13. Palaeogeographic reconstruction of the estuarine depositional system for the deposits of the Geirud Formation at the time of initial transgression. The system was subject to a mix of different energetic factors: fluvial flows dominated in the innermost bayhead deltaic zone, tidal currents flowed bi-directionally within the central basin zone, whereas waves reworked sediments along the coastal zone. Distally, shoreface deposits merged into shelf carbonates (letter symbols: A1 = fluvial channel belt deposits; A2 = flood-plain deposits; A3 = tidally-influenced fluvial channel deposits; B1 = tidal flat deposits; B2 = tidal channel deposits; B3 = tidal bar deposits; C1 = offshore-transition deposits; C2 = wave-dominated shoreface deposits; D1 = offshore carbonate deposits; and D2 = nearshore carbonate deposits).

stratigraphic interval of the Geirud Formation and constituted by offshore carbonate (f.a. D1) and bioclastic-rich nearshore (f.a. D2) deposits (Fig. 13).

Main stratigraphic surfaces of the Geirud formation

The stratigraphic correlation obtained across the three sections of Shamirzad, Darvar and Dehmolla (Fig. 14) allows us to reconstruct the main stratigraphic surfaces along a transect-oriented parallel to the southern Albortz basin-margin, which was part of the larger Devonian Tethys.

The correlation (Fig. 14) highlights the complex geometry of the underlying bounding surfaces of the studied stratigraphic succession, including the basal unconformity that separates the Geirud Formation from the underlying Milla Formation (sb_1 in Fig. 14) and the topmost erosional surface

with the overlying Mobarak Formation (sb_3 in Fig. 14).

The overall shape of the basal unconformity sb_1 points out a palaeo-relief, highly visible from a satellite view (Fig. 15A). In the western sector of the study area (Fig. 15A), some of the deepest W-shaped valleys which characterise this surface differ in elevation from the adjacent highs of about 200 to 300m. The overlying terrestrial sediments show along-strike rapid thickness changes and indicate continental-to-marine strata aggradation as the result of the filling a pre-existent morphology along this margin of the Albortz Basin during the Devonian.

Another relevant unconformity is represented by the sb_2 surface (Fig. 14; see also Fig. 15A and B). This erosional surface cuts somewhat the underlying deposits, showing 10 to 20m-deep incisions (Fig. 15A). However, its lateral geometry and the volume of underlying sediment eroded suggest

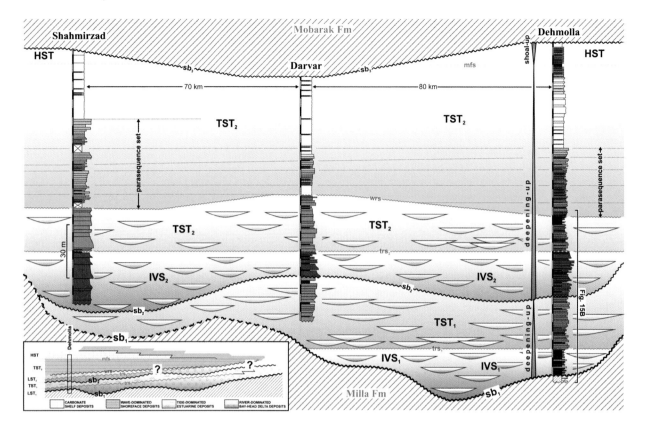

Fig. 14. Sequence stratigraphic correlation panel of the Geirud Formation. This correlation was obtained by using the coastal deposits of the group 3 (C1 and C2) as general key bed-set, as they show an overall tabular geometry (see Fig. 15A) laterally recognisable either at outcrop-scale and at kilometric (seismic-line)-scale (letter symbols: *sb* = sequence boundary; *trs* = tidal ravinement surface; *wrs* = wave ravinement surface; *mfs* = maximum flooding surface; *IVF* = incised-valley fill; *TST* = transgressive systems tract; *HST* = highstand systems tract). The inset in the bottom-left corner shows an inferred along-dip reconstruction (inspired after Zaitlin *et al.*, 1994) of the various systems tracts detected in the main panel (question marks are in sectors with no stratigraphic data).

a lower rank unconformity if compared with the basal master surface sb_1.

Other important stratigraphic boundaries observable in the Geirud Formation are flat erosional surfaces (i.e. *trs* and *wrs* in Fig. 14), laterally continuous for several kilometres (Fig. 15A). The overlying sediments are represented by estuarine, tidally-dominated deposits (i.e. tidal channels) or coastal, wave-dominated deposits (i.e. shoreface) (Fig. 15B).

Minor stratigraphic surfaces occur in the intermediate interval of the studied succession and are represented by sharp contacts bounding the bases of each sub-littoral parasequence (facies associations C1 and C2). They are commonly characterised by shell remains and pebble concentrations associated to glauconitic clasts. On these surfaces, offshore-transition lithofacies occur as basal deposits of coarsening-upward parasequences.

DISCUSSION

The stratigraphic arrangement of the four groups of facies associations indicates the stacking of different depositional systems which record: (i) compound incised-valley fills in the lowermost stratigraphic interval and (ii) transgressive–regressive systems in the strata of the middle-uppermost interval of the Geirud Formation.

The palaeogeographic scenario reconstructed for this margin of the Albortz Basin during the transgression possibly consisted of a series of river valleys separated by interfluves (Fig. 13). Within these depressions, a tidally-influenced estuarine setting developed adjacent to a wave-dominated coast.

As observed in many modern analogues (e.g. the Gironde), river flows, tidal currents and waves can interact or separately dominate different parts of the same estuary, leading to the development of

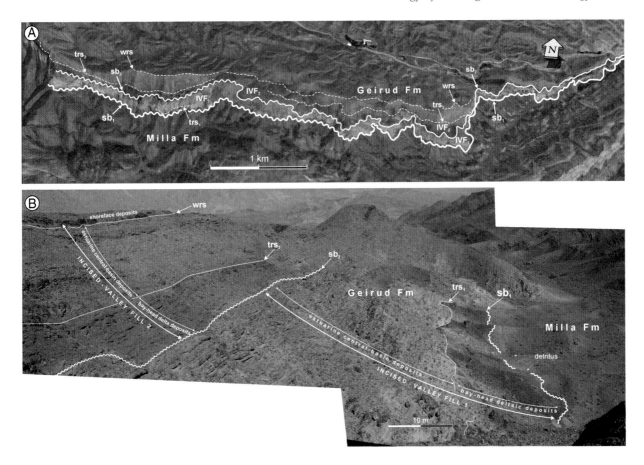

Fig. 15. (A) Satellite overview of the Geirud Formation near the Dehmolla area. The main stratigraphic surfaces, as well as the basal lithofacies, are laterally continuous for several kilometres. (B) Detail of the lowermost stratigraphic interval of the Geirud Formation (Dehmolla Section) interpreted as a compound incised-valley fill.

a mixed-energy depositional system (Allen & Posamentier, 1994; Féniès & Tastet, 1998; Lericolais *et al.*, 2001).

Sedimentary processes within a mixed-energy estuary

During the Upper Devonian, the southern margin of the palaeo-Tethys was characterised by a wide coastal zone (Stöcklin, 1974; Berberian & King, 1981; Berberian, 1983; Alavi, 1996; Stampfli *et al.*, 2001; Rezaeian, 2008), where sediments were distributed across a transitional (from continental to open-marine) coastal area (Fig. 13). The sedimentary environments detected from the studied stratigraphic succession concur to interpret the main depositional systems as a mixed, wave-dominated and tide-dominated estuary (Reinson, 1992; Dalrymple *et al.*, 1992; Boyd *et al.*, 2006; Tessier, 2012), passing seaward to an open coast and a marine shelf system.

At the time of initial transgression in the Geirud area, fluvial systems passed seawards to estuarine settings and, in turn, to a wave-dominated coast and an open-marine shelf (Fig. 13). In this mixed-energy coastal setting, wave, river and tidal processes coexisted and the relative intensity of these processes varied spatially through the system at any time (Yoshida *et al.*, 2007). As documented in the modern mixed, tide/wave-dominated Gironde Estuary (Central Bay of Biscay, SW France), wave energy is highest along the open coastline or near the estuarine mouth and decreases markedly farther landwards because of frictional dissipation and sheltering (Allen, 1991; Allen & Posamentier, 1993).

In the Geirud basal stratigraphic interval, fluvial deposits (f.a. A3) suggest the existence of a bay-head deltaic zone. According to modern analogues (e.g. Dalrymple *et al.*, 1992; Dalrymple & Choi, 2007; Boyd, 2010; Dalrymple, 2010; Tessier, 2012) in this zone, river currents decrease in strength and in relative influence on the sediments in a seaward direction because of: (i) the decreasing gradient of the river streams, (ii) the progressive enlargement of the alluvial plain cross-sectional area, as the river approaches the sea and (iii) the transition from single to multiple distributary and tidal channels (Boyd *et al.*, 2006; Dalrymple & Choi, 2007).

However, in contrast with the facies models of mixed energy estuaries (e.g. Chaumillon *et al.*, 2010), the Geirud terrestrial lithofacies do not indicate 'classical' bay-head deposits but rather braidplain environments and the absence of a real delta. This is suggested as a possible circumstance in the case of small river systems, where fluvial channels at the river mouths are virtually continuous with the tidal channels of the estuarine central-basin zone. Consequently, sediments continue to be transported off the bay-head zone, without producing any relevant accumulation (Chaumillon *et al.*, 2010).

Tidal currents also play an important role on sediment distribution in the estuary central basin zone in the form of landward-directed (flood) and seaward-directed (ebb) flows, but their strength and the relative size of tidal bedforms (e.g. dunes, bars), depends on the tidal excursion on which this zone is subject (e.g. micro-tidal, meso-tidal, macro-tidal or mega-tidal). In modern analogues, the central basin zone is also the coastal area where the maximum tidal-current speeds occur (Zaitlin *et al.*, 1994), near the place where the distributary channels bifurcate (Dalrymple & Makino, 1989; Archer, 2013). This area is referred to as the 'tidal maximum' (Dalrymple, 2006; Dalrymple & Choi, 2007).

In the Geirud strata, wave-generated facies appear volumetrically important, indicating that the wave action was crucial at the seaward end of the system, possibly characterised by large, open-water fetch. Wave energy at the bed increased landward from the shelf toward a shallower inshore profile, reaching a maximum at the mouth of the estuary (Chaumillon *et al.*, 2008). Due to the open-mouth character of many present-day systems, wave energy possibly penetrated some distance into the estuary but frictional dissipation in shallow water caused wave-generated currents to decrease in strength in a landward direction. Consequently, wave-dominated strata of the Geirud formation constitute an isolated volume of facies, spatially well-separated from the rest of the fluvial/tidal facies.

The continuous mixing of wave-generated and tide-generated facies, as well as the 'size' of the various bedforms interpreted from the suite of cross stratified facies recognised in the three sections thus indicate that this system was subject to a moderate tidal range (i.e. mesotidal <4 m), resulting in a balance between wave penetration at the mouth and tidal bi-directional currents.

The outer estuary is commonly characterised by (ebb/flood) tidal deltas and associated coastal barriers with inlets in mixed-energy and wave-dominated estuarine mouths (e.g. Roy *et al.*, 1980; Roy, 1984; Zaitlin & Schultz, 1984, 1990; Boyd & Honig, 1992; Dalrymple *et al.*, 1992; Plink-Björklund, 2005). These lithofacies were not detected in the Geirud strata, where estuarine tidally-dominated lithofacies are overlain by shoreface wave-dominated lithofacies through a wide wave ravinement surface (Fig. 14). This important facies lack can be explained with the following alternative hypothesis: (i) tidal deltaic, barrier and inlet deposits were totally removed by the erosional wave action during the ensuing marine transgression; (ii) the tidal regime was not able to produce such tidally-driven features, as the mixed energy estuary merged seaward directly into a sandy shoreface without any relevant barrier complex; and (iii) the Geirud system was a shore-normal, open-ended estuary which, according to the Reinson's classification (1992) was deprived of any relevant sediment accumulation at the mouth, due to this estuarine morphologies typical of coastlines subjected to a mesotidal regime.

Sequence stratigraphic evolution of the Geirud Formation

In a sequence-stratigraphic framework, the Geirud Formation represents a 3rd-order(?) depositional sequence, delimited at the base and top by two regional-scale sequence boundaries. However, within this interval, internal, higher-frequency sequences can be observed, suggesting that the filling of previous lowstand incision occurred during two minor relative sea-level cycles, leading to the building of a so-called 'compound incised-valley fill' (Boyd *et al.*, 2006). Instead, the rest of the overlying deposits record an important phase of marine transgression and an ensuing normal regression, after the cessation of the relative sea-level rise.

The Geirud compound incised-valley fill

The lowermost 45 m-thick stratigraphic interval of the Geirud Formation includes a compound incised-valley fill (Fig. 14). It consists of a major basal sequence boundary (sb_1 in Fig. 14) which represents a regional-scale unconformity, laterally continuous for several kilometres and recognisable from the satellite view (Fig. 15A). This surface shows incisions up to 200 to 300 m-deep and represents the landscape profile of a series of adjacent incised valleys excavated during a major sea-level lowstand on the underlying sediments belonging to the Milla Formation. The overlying deposits, which occur in the lowermost strata of the Darvar and Dehmolla sections (Figs 14 and 15B), record fluvial-estuarine aggrading strata (IVS_1) accumulated during a first early transgressive phase. As the transgression continued, fully-estuarine strata of the facies associations B1 to B3 (TST_1) retrograded over a sharp surface interpreted as a tidal ravinement surface (trs_1 in Fig. 14). The $IVS_1 + TST_1$ strataset appears not fully preserved because it is re-incised by a new erosional surface (sb_2 in Figs 14 and 15A). This younger unconformity represents a new surface of subaerial exposure incised onto the previous generation of incised-valley fills during an ensuing higher-frequency sea-level fall and thus, it can be regarded as a minor sequence boundary. This higher-order unconformity represents the base for the aggradation of a new generation of incised-valley fills (IVS_2 in Fig. 14), which evolved upwards to facies associations whose internal features are comparable with the underlying older incised-valley fill (Fig. 15B).

As documented in many facies models and case studies (e.g. Gustason *et al.*, 1986; Gustason *et al.*, 1988; Reinson *et al.*, 1988; Krystinik, 1989; Wood & Hopkins, 1989, 1992; Krystinik & Blakeney-DeJarnett, 1994; Bowen & Weimer, 1997; 2003; Ardies *et al.*, 2002; Zaitlin *et al.*, 2002; Leckie *et al.*, 2005; Tropeano *et al.*, 2013), estuaries are commonly re-established in the same location during subsequent sea-level cycles, leading to multiple cut-and-fill events in the sedimentary record. Therefore, incision of previous deposits during relative sea-level lowstand episodes can occur repeatedly, as the relative sea-level lowstand phase can be punctuated by a series of higher-order/amplitude cycles responsible of the recurring erosion of previously-deposited strata and ensuing infilling (Boyd *et al.*, 2006). The resulting 'compound fill' therefore records multiple cycles of incision and deposition resulting from fluctuations in base level and is therefore punctuated by one or more sequence boundaries in addition to the main, lower-order sequence boundary at the base of the incised valley (Boyd *et al.*, 2006).

The Geirud transgressive-regressive complex

The 120 m-thick stratigraphic interval of the Geirud Formation overlying the compound incised-valley fill (Fig. 14) records a complete transgressive-regressive cycle of relative sea-level oscillation. This interval rests on a second tidal ravinement surface (trs_2 in Fig. 14) on which estuarine facies associations developed. The upward stratigraphic interval is represented by shoreface strata overlying the previous deposits through a wave ravinement surface (*wrs* in Fig. 14) and accumulated during the prosecution of the transgression in this part of the basin.

The associations *C1* and *C2* form vertically-stacked stratal units which are bounded by sharp basal surfaces (Fig. 11A) marked by concentrations of residual bioclastic-rich material. These surfaces can be interpreted consistently as flooding surfaces, bounding single parasequences, each recording transgressive/regressive episodes of sedimentation in wave-dominated coastal settings (e.g. Messina *et al.*, 2007). Transgression may have occurred as a long-lasting relative sea-level rise punctuated by minor periods of sea-level still-stands. During these momentary episodes of shorter duration, the sediment accumulation rate overpassed the rate of accommodation space, generating phases of normal regressions and resulting in shoaling-upward parasequences (Fig. 11A). The ensuing rapid rise of the relative sea-level was thus recorded in the condensed horizons at the base of each unit (e.g. Chiarella *et al.*, 2012; Chiarella & Longhitano, 2012). This vertical pattern is consistent with the classical parasequence definition (Van Wagoner *et al.*, 1987; Einsele *et al.*, 1991), or R-type cycles (Zecchin, 2007). Their accumulation, usually referred to middle-shelf to outer-shelf settings, can also occur in coastal areas characterised by low-gradient topography (e.g. Chiarella *et al.*, 2012; Chiarella & Longhitano, 2012). During rapid transgressive phases, sediment supply is relatively low but increases during the ensuing relative sea-level still-stands, producing normal regression strata and progradation of nearshore facies (Hampson & Storms, 2003). Such architecture, which is common of transgressive parasequence sets described in many coastal, wave-dominated successions (e.g. Van Wagoner *et al.*, 1990; Swift *et al.*, 1991; Posamentier & Allen, 1993; Helland-Hansen & Martinsen, 1996; Coe, 2003; Storms & Hampson, 2005; Messina *et al.*, 2007; Chiarella & Longhitano,

2012), suggests that the transgression was punctuated by temporarily arrests of the relative sea-level, during which the overpass of the sediment accumulation rate on the accommodation space generated normal regressions.

Coastal strata are stratigraphically overlain by shelf limestones (Fig. 14) indicating deeper open-marine conditions reached during the late stage of the transgression. Around the intermediate interval of such topmost bioclastic-rich interval, the vertical 'return' from offshore to nearshore deposits, confirmed by the occurrence of repeated 'tempestite' layers, suggests a possible 'turnabout zone' or maximum flooding surface (*mfs* in Fig. 14), recording the definitive end of the relative sea-level rise and the ensuing progradation of coastal carbonates during the highstand phase.

CONCLUSIONS

Sedimentary facies documented in the upper Devonian Geirud Formation, in central-northern Iran, provided a background for the reconstruction of the various depositional environments recorded in this part of the Alborz Basin and their stratigraphic evolution during a complex phase of marine transgression. The followings are the results of this work:

1. The Geirud Formation, from 240 to 370 m in thickness, is organized into stratal units that revealed four groups of facies associations, recording vertical (=temporal) changes from continental and open-marine environments.
2. These environments are represented by: (i) fluvial-dominated, bay-head deltaic deposits, including fluvial channel conglomerates, flood plain fines and tidally-influenced channelized sandstone; (ii) central estuarine deposits, consisting of sand-dominated and mud-dominated tidal flats, associated with sandstone tidal channel and bars; (iii) siliciclastic offshore-transition and shoreface sandstone deposits; and (iv) carbonate open-marine shelf deposits, represented by offshore and bioclastic-rich nearshore wackestones and packstones.
3. These facies associations record the sedimentation of a mixed-energy, wave-dominated and tide-dominated estuary passing seaward to an open-marine nearshore shelf. This complex of laterally-adjacent depositional systems infilled a system of incised valleys and retrograded on the northern margin of the Alborz Basin during

a nearby-continuous episode of marine transgression.
4. The stratigraphic distribution of the tidally-influenced/dominated facies and their volume compared to the bulk of the Geirud Formation also suggest that the system developed under a mesotidal range at least, to enable to generate landward propagation of tidal currents and resulting bedforms.
5. The present case study documents an unusual depositional scenario for a mixed-energy estuary, for the following reasons:
   i. the lack of a real bay-head delta, which is instead represented by a braidplain-type riverine system of small size;
   ii. the absence of relevant tidal deltas and associated inlets, possibly caused by the mesotidal regime acting at the coastline;
   iii. consequent shallow transgressive ravinement surfaces, because of the lack of tidal inlets that usually produce constriction and relatively powerful tidal flows, indicating that the estuary entrance was possibly a zone of bypass rather than an environment of sediment accumulation;
   iv. the occurrence of shoreface sandstones in the outer environments, instead of coastal barriers or tidal inlet deposits. This feature, which is uncommon in the well-known facies models of estuaries, possibly indicates the dominance of shoaling waves acting across a gently-inclined, inshore coastal profile.
6. The overall stratigraphy of the studied succession suggests that sediments accumulated after one major relative sea-level change, which was punctuated by minor, lower-amplitude base-level fluctuations. This interplay generated a compound incised-valley fill documented in the lowermost stratigraphic interval and a multiple, transgressive parasequence set observed in the intermediate stratigraphic interval of the Geirud Formation. Estuarine-to-shoreface strata thus evolved upwards to shelf limestone accumulated during a final highstand phase.

REFERENCES

Alavi, M. (1996) Tectonostratigraphic synthesis and structural style of the Alborz mountain system in northern Iran. *J. Geodynamics*, **21**, 1–33.

Alavi-Naini, M. (1993) Paleozoic stratigraphy of Iran. In: *Treatise on the Geology of Iran* (Ed. H. Hushmandzadeh), *Geol. Surv. Iran*, Tehran, **5**, 1–492.

Allen, G.P. (1991) Sedimentary processes and facies in the Gironde estuary; a recent model for macrotidal estuarine systems. In: *Clastic Tidal Sedimentology* (Eds D.G. Smith, G.E. Reinson, B.A. Zaitlin and R.A. Rahmani), *Can. Soc. Pet. Geol. Mem.*, **16**, 29–40.

Allen, G.P. and Posamentier, H.W. (1993) Sequence stratigraphy and facies model of an incised valley fill: the Gironde Estuary, France. *J. Sed. Petrol.*, **63**, 378–391.

Allen, G.P. and Posamentier, H.W. (1994) Transgressive facies and sequence architecture in mixed tide- and wave-dominated incised-valley: example from the Gironde Estuary, France. In: *Incised Valley Systems: Origin of Sedimentary Sequences.* (Eds R.W. Dalrymple, R. Boyd, B.A. Zaitlin), *SEPM Spec. Publ.*, **51**, 225–240.

Allen, M., Ghassemi, M.R., Shahrabi, M. and Qorashi, M. (2003) Accommodation of late Cenozoic oblique shortening in the Alborz range, northern Iran. *J. Struct. Geol.*, **25**, 659–672.

Annells, R.N., Arthurton, R.S., Bazley, R.A. and Davies, R.G. (1975) Geological quadrangle map of Iran, Qazvin and Rasht sheet: Tehran, Geological Survey of Iran, scale 1:250,000.

Annells, R.S., Arthurton, R.S., Bazley, R.A.B., Davies, R.G., Hamedi, M.A.R. and Rahimzadeh, F. (1977) Geological map of Iran, Shakran sheet 6162: Tehran, Geological Survey of Iran, scale 1:100,000.

Archer, A.W. (2013) World's highest tides: Hypertidal coastal systems in North America, South America and Europe. *Sed. Geol.*, **284–285**, 1–25.

Ardies, G.W., Dalrymple, R.W. and Zaitlin, B.A. (2002) Controls on the geometry of incised valleys in the Basal Quartz unit (Lower cretaceous), Western Canada Sedimentary Basin. *J. Sed. Res.*, **72**, 602–618.

Assereto, R. (1966) Explanatory notes on the geological map of upper Djadjerud and Lar valleys (Central Elburz, Iran). *Istituto di Geologia dell'Università di Milano, G*, **232**, 1–86.

Assereto, R. (1963) The Paleozoic formations in central Elburz (Iran) (preliminary note). *Rivista Italiana di Paleontologia e Stratigrafia*, **69**, 503–543.

Assereto, R. and Gaetani, M. (1964) Nuovi dati sul Devoniano della catena dell'Imam Zadeh Hashim (Elburz Centrale -Iran). *Rivista Italiana di Paleontologia e Stratigrafia*, **70**, 631–636.

Axen, G.J., Stockli, D.F., Lam, P., Guest, B. and Hassanzadeh, J. (2001) Implications of preliminary (U-Th/He cooling ages from the central Alborz Mountains, Iran. *Geol. Soc. Am.*, Abstracts with Programs, **33**, 7, 257.

Bagheri, S. and Stampfli, G.M. (2008) The Anarak, Jandaq and Posht-e-Badam metamorphic complexes in central Iran: New geological data, relationships and tectonic implications. *Tectonophysics*, **451**, 123–155.

Berberian, M. (1983) The southern Caspian: A compressional depression floored by a trapped, modified oceanic crust. *Can. J. Earth Sci.*, **20**, 163–183.

Berberian, M. and King, G.C.P. (1981) Towards a paleogeography and tectonic evolution of Iran. *Can. J. Earth Sci.*, **18**, 210–265.

Bhattacharya, H.N. and Bhattacharya, B. (2006) A Permo-Carboniferous Tide-storm interactive system: Talchir Formation, Raniganj Basin, India. *J. Asian Earth Sciences*, **27**, 303–311.

Bowen, D.W. and Weimer, P. (2003) Regional sequence stratigraphic setting and reservoir geology of Morrow incised-valley sandstones (lower Pennsylvanian), eastern Colorado and western Kansas. *AAPG Bull.*, **87**, 781–815.

Bowen, D.W. and Weimer, P. (1997) Reservoir geology of incised valley sandstones of the Pennsylvanian Morrow Formation, Southern Stateline Trend, Colorado and Kansas. In: *Shallow Marine and Non-Marine Reservoirs—Sequence Stratigraphy, Reservoir Architecture and Production Characteristics* (Eds K.W. Shanley and B.F. Perkins). SEPM, Gulf Coast Section, Eighteenth Annual Research Conference, 55–66.

Boyd, R. (2010) Transgressive wave-dominated coasts. In: *Facies models 4* (Eds N.P. James and R.W. Dalrymple). *Geol. Assoc. Can.*, St. John's, pp 265–294.

Boyd, R., Dalrymple, R. and Zaitlin, B.A. (1992) Classification of clastic coastal depositional environments. *Sed. Geol.*, **80**, 139–150.

Boyd, R., Dalrymple, R.W. and Zaitlin, B.A. (2006) Estuary and incised valley facies models. In: *Facies Models Revisited* (Eds H.W. Posamentier and R.G. Walker). *SEPM Spec. Publ.*, **84**, 171–234.

Boyd, R. and Honig, C. (1992) Estuarine sedimentation on the eastern shore of Nova Scotia. *J. Sed. Pet.*, **62**, 569–583.

Bozorgnia, F. (1964) Microfacies and microorganisms of Paleozoic through Tertiary sediments of some parts of Iran. *National Iranian Oil Company, Tehran-Iran*, **158**, 1–22.

Bozorgnia, F. (1973) Paleozoic foraminiferal biostratigraphy of central and east Alborz Mountains, Iran. *National Iranian Oil Company, Geological Laboratories*, **4**, 1–185.

Bridge, J.S. (2006) Fluvial facies models: Recent developments. In: *Facies Models Revisited* (Eds H.W. Posamentier and R.G. Walker). *SEPM Spec. Publ.*, **84**, 85–170.

Bridge, J.S. (2003) *Rivers and Floodplains: Forms, Processes and Sedimentary Record.* Blackwell Publishing, Oxford.

Bromley, R.G. and Ekdale, A.A. (1984) Trace fossil preservation in flint in the European chalk. *J. Paleontol.*, **58**, 298–311.

Brown, L.F., Jr. (1993) Seismic and Sequence Stratigraphy: Its Current Status and Growing Role in Exploration and Development (course notes): New Orleans Geological Society, Short Course No. 5.

Buatois, L.A., Mangano, M.G., Alissa, A. and Carr, T.R. (2002) Sequence stratigraphic and sedimentologic significance of biogenic structures from a Late Paleozoic marginal- to open-marine reservoir, Morrow Sandstone, subsurface of southwest Kansas, USA. *Sed. Geol.*, **152**, 99–132.

Chaumillon, E., Bertin, X., Falchetto, H., Allard, J., Weber, N., Walker, P., Pouvreau, N. and Woppelmann, G. (2008) Multi time-scale evolution of a wide estuary linear sandbank, the Longe de Boyard, on the French Atlantic coast. *Mar. Geol.*, **251**, 209–223.

Chaumillon, E., Tessier, B. and Reynaud, J.-Y. (2010) Stratigraphic records and variability of incised valleys and estuaries along French coasts. *Bull. Soc. Géol. Fr.*, **181**, 75–85.

Chaumillon, E. and Weber, N. (2006) Spatial variability of modern incised valleys on the French Atlantic Coast: Comparison between the Charente and the Lay–Sèvre incised-valleys. In: *Incised Valleys in Time and Space* (Eds R.W. Dalrymple, D.A. Leckie and R.W. Tillman). *SEPM Spec. Publ.*, **85**, 57–85.

Chiarella, D. and Longhitano, S.G. (2012) Distinguishing depositional environments in shallow-water mixed, bio-siliciclastic deposits on the basis of the degree of heterolithic segregation (Gelasian, southern Italy). *J. Sed. Res.*, **82**, 969–990.

Chiarella, D., Longhitano, S.G., Sabato, L. and Tropeano M. (2012) Sedimentology and hydrodynamics of mixed (siliciclastic-bioclastic) shallow-marine deposits of Acerenza (Pliocene, Southern Apennines, Italy). *Italian Journal of Geosciences*, **131**, 136–151.

Clark, G.C., Davies, R.G., Hamzepour, G. and Jones, C.R. (1975) Explanatory text of the Bandar-e-Pahlavi quadrangle map, 1:250,000. *Geological Survey of Iran*, Tehran, Iran. pp.198.

Clifton, H.E. (1969) Beach lamination: nature and origin. *Mar. Geol.*, **7**, 553–559.

Clifton, H.E., Hunter, R.E. and Phillips, R.L. (1971) Depositional structures and processes in the non-barred high-energy nearshore. *J. Sed. Petrol.*, **41**, 651–670.

Coe, A.L. (Ed.) (2003) *The Sedimentary Record of Sea-Level Change*. Cambridge University Press, Cambridge, 288 pp.

Collinson, J.D. (1996) Alluvial sediments. In: *Sedimentary environments: processes, facies and stratigraphy* (Ed. H.D. Reading), third edition (Blackwell, Oxford), 37–82.

Dalrymple, R.W. (2006) Incised valleys in time and space: an introduction to the volume and an examination of the controls on valley formation and filling. In: *Incised valleys in time and space* (Eds R.W. Dalrymple, D.A. Leckie and R.W. Tillman), *SEPM Spec. Publ.*, **85**, 5–12.

Dalrymple, R.W. (2010) Introduction to siliciclastic facies models. In: *Facies models* (Eds N.P. James and R.W. Dalrymple). *Geol. Assoc. Can.*, St. John's, 59–72.

Dalrymple, R.W., Boyd, R. and Zaitlin, B.A. (1994) History of research, types and internal organization of incised valley systems: introduction to the volume. In: *Incised-Valley Systems: Origin and Sedimentary Sequences* (Eds R.W. Dalrymple, R. Boyd and B.A. Zaitlin). *SEPM Spec. Publ.*, **51**, 1–10.

Dalrymple, R.W. and Choi, K. (2007) Morphology and facies trends through the fluvial marine transition in tide-dominated depositional systems: a schematic framework for environmental and sequence stratigraphic interpretation. *Earth Sci. Rev.*, **81**, 135–174.

Dalrymple, R.W., Knight, R.J., Zaitlin, B.A. and Middleton, G.V. (1990) Dynamics and facies model of a macrotidal sandbar complex, Cobequid bay — Salmon river estuary (Bay of Fundy). *Sedimentology*, **37**, 577–612.

Dalrymple R.W., Mackay D.A., Ichaso A.A. and Choi K.S. (2012) Processes, Morphodynamics and Facies of Tide-Dominated Estuaries. In: *Principles of Tidal Sedimentology* (Eds R.A. Davis and R.W. Dalrymple), Springer, pp. 79-107.

Dalrymple, R.W. and Makino, Y. (1989) Description and genesis of tidal bedding in the Cobequid Bay-Salmon River estuary, Bay of Fundy, Canada. In: *Sedimentary*

Facies of the Active Plate Margin (Eds A. Taira and F. Masuda). *Terra Scientific*, Tokyo, 151–177.

Dalrymple, R.W., Makino, Y. and Zaitlin, B.A. (1991) Temporal and spatial patterns of rhythmites deposition on mud flats in the macrotidal Cobequid Bay–Salmon River Estuary. Bay of Fundy. In: *Clastic Tidal Sedimentology* (Eds D.G. Smith, G.E. Reinson, B.A. Zaitlin and R.A. Rahmani). *Mem. Can. Soc. Petrol. Geol.*, **16**, 137–160.

Dalrymple, R.W. and Rhodes, R.N. (1995) Estuarine dunes and bars. In: *Geomorphology and Sedimentology of Estuaries* (Ed. G.M.E. Perillo), pp. 359–422. *Dev. Sedimentol.* 53, Elsevier Science, New York.

Dalrymple, R.W., Zaitlin, B.A. and Boyd, R. (1992) Estuarine facies models: conceptual basis and stratigraphic implications. *J. Sed. Petrol.*, **62**, 1130–1146.

Dashtban, H. (1995) Upper Devonian goniatites (Famennian) from central Alborz. *GeoSciences, Scientific Quarterly Journal*, **4**, 36–43.

Dashtgard, S.E., MacEachern, J.A., Frey, S.E. and Gingras, M.K. (2010) Tidal effects on the shoreface: Towards a conceptual framework. *Sed. Geol.*, **279**, 42–61.

De Mowbray, T. (1983) The genesis of lateral accretion deposits in recent intertidal mudflat channels. Solway Firth, Scotland. *Sedimentology*, **30**, 425–435.

Dolson, J., Muller, D., Everts, M.J. and Stein, J.A. (1991) Regional paleogeographic trends and production, Muddy Sandstone (Lower Cretaceous), Central and Northern Rocky Mountains. *AAPG Bull.*, **75**, 409–435.

Einsele, G., Ricken, W. and Seilacher, D. (1991) Cycles and events in stratigraphy: basic concepts and terms. In: *Cycles and Events in Stratigraphy* (Eds G. Einsele, W. Ricken and A. Seilacher). Springer-Verlag, Berlin, 1–19.

Ekdale, A. and Mason, T. (1988) Characteristic trace fossil assemblages in oxygen poor sedimentary environments. *Geology*, **16**, 720–723.

Elliott, T. (1986) Siliciclastic shorelines. In: *Sedimentary Environment and Facies* (Ed. H.G. Reading). Blackwell, Oxford, 155–188.

Fabuel-Perez, I., Redfern, J. and Hodgetts, D. (2009) Sedimentology of an intra-montane rift-controlled fluvial dominated succession: The Upper Triassic Oukaimeden Sandstone Formation, Central High Atlas, Morocco. *Sed. Geol.*, **218**, 103–140.

Féniès, H. and Tastet, J.P. (1998) Facies and architecture of an estuarine tidal bar (the Trompeloup bar, Gironde Estuary, SW France). *Mar. Geol.*, **150**, 149–169.

Fischbein, S.A., Joeckel, R.M. and Fielding, C.R. (2009) Fluvial-estuarine reinterpretation of large, isolated sandstone bodies in epicontinental cyclothems, Upper Pennsylvanian, northern Midcontinent, USA and their significance for understanding late Paleozoic sea-level fluctuations. *Sed. Geol.*, **216**, 15–28.

Fisher, W.L. and McGowen, J.H. (1967) Depositional systems in the Wilcox Group (Eocene) of Texas and their relationship to occurrence of oil and gas. *Trans. Gulf-Coast Assoc. Geol. Soc.*, **17**, 105–125.

Folkestand, A. and Satur, N. (2008) Regressive and transgressive cycles in a rift-basin: Depositional model and sedimentary partitioning of the Middle Jurassic Hugin Formation, Southern Viking Graben, North Sea. *Sed. Geol.*, **207**, 1–21.

Frey, R.W. and Howard, J.D. (1986) Mesotidal estuary sequences, a perspective from the Georgia Bight. *J. Sed Petrol.*, **56**, 911–924.

Fröhlich, S., Petitpierre, L., Redfern, J., Grech, P., Bodin, S. and Lang, S. (2010) Sedimentological and sequence stratigraphic analysis of Carboniferous deposits in western Libya: Recording the sedimentary response of the northern Gondwana margin to climate and sea-level changes. *J. Afr. Earth Sci.*, **57**, 279–296.

Gaetani, M. (1965) Brachiopods and molluscs from Geirud Formation, Member A (Upper Devonian and Tournaisian). *Riv. Ital. Paleontol. Stratigr.*, **71**, 679–770.

Ghavidel-Syooki, M. (1994) Biostratigraphy and paleobiogeography of some Paleozoic rocks at Zagros and Alborz Mountains. *Iran Geol. Surv. Publ.*, 168 pp.

Ghavidel-Syooki, M. (1995) Palynostratigraphy and palaeogeography of a Palaeozoic sequence in the Hassanakdar area, Central Alborz Range, northern Iran. *Rev. Palaeobot. Palynol.*, **86**, 91–109.

Ghavidel-Syooki, M., Hassanzadeh, J. and Vecoli, M. (2011) Palynology and isotope geochronology of the Upper Ordovician–Silurian successions (Ghelli and Soltan Maidan Formations) in the Khoshyeilagh area, eastern Alborz Range, northern Iran; stratigraphic and palaeogeographic implications. *Rev. Palaeobot. Palynol.*, **164**, 251–271.

Gingras, M.K., MacEachern, J.A. and Dashtgard, S.E. (2012) The potential of trace fossils as tidal indicators in bays and estuaries. In: Modern and ancient depositional systems: perspectives, models and signatures (Eds S.G. Longhitano, D. Mellere and R.B. Ainsworth). *Sed. Geol.*, Special Issue, **279**, 97–106.

Golonka, J. (2007) Phanerozoic paleoenvironment and Paleolithofacies maps. *Late Paleozoic. Geologia*, **33**, Zeszyt 2, 145–209.

Guest, B., Axen, G.J., Lam P.S. and Hassanzadeh, J. (2006) Late Cenozoic shortening in the west-central Alborz Mountains, northern Iran, by combined conjugate strike-slip and thin-skinned deformation. *Geosphere*, **2**, 35–52.

Gustason, E.R., Ryer, R.A. and Odland, S.K. (1986) Unconformities and facies relationships of the Muddy Sandstone, northern Powder River Basin, Wyoming and Montana. *AAPG Bull.*, **70**, 1042–1068.

Gustason, E.R., Wheeler, D.A. and Ryer, T.A. (1988) Structural control on paleovalley development, Muddy Sandstone, Powder River Basin, Wyoming. *AAPG Bull.*, **72**, 871–880.

Hairapetian, V., Mohibullah, M., Tilley, L.J., Williams, M., Miller, C.G., Afzal, J., Ghobadi Pour, M. and Hejazi, S.H. (2011) Early Silurian carbonate platform ostracods from Iran: A peri-Gondwanan fauna with strong Laurentian affinities. *Gondwana Research*, **20**, 645–653.

Hampson, G.J. (2010) Sediment dispersal and quantitative stratigraphic architecture across an ancient shelf. *Sedimentology*, **57**, 96–141.

Hampson, G.J. and Storms, J.E.A. (2003) Geomorphological and sequence stratigraphic variability in wave-dominated, shoreface-shelf parasequences. *Sedimentology*, **50**, 667–701.

Helland-Hansen, W. and Martinsen, O.J. (1996) Shoreline trajectories and sequences: description of variable depositional-dip scenarios. *J. Sed. Res.*, **66**, 670–688.

Huber, H. and Eftekhar-nezhad, J. (1978a) Geological map of Iran, sheet no. 1, northwest Iran: Tehran, National Iranian Oil Company, scale 1:1,000,000.

Huber, H. and Eftekhar-nezhad, J. (1978b) Geological map of Iran, sheet no. 2, north-central Iran: Tehran, National Iranian Oil Company, scale 1:1,000,000.

Ichaso, A. and Dalrymple, R.W. (2006) Abstracts with Programs. On the Geometry of Tidal-meanders. *Geol. Soc. Am.*, **7**, 38, 186 pp.

Jackson, J., Priestley, K., Allen, M. and Berberian, M. (2002) Active tectonics of the South Caspian Basin. *Geophys. J. Int.*, **148**, 214–245.

Klein, G.D. (1985) Intertidal flats and intertidal sand bodies. In: *Coastal Sedimentary Environments*, (Ed. R.A. Davis), 2nd edn., Springer-Verlag, New York, 187–224.

Komatsu, T. (1999) Sedimentology and sequence stratigraphy of a tide- and wave dominated coastal succession: the Cretaceous Goshoura Group, Kyushu, southwest Japan. *Cretaceous Res.*, **20**, 327–342.

Krystinik, L.F. (1989) Morrow formation facies geometries and reservoir quality in compound valley fills, central State Line area, Colorado and Kansas (abstract). *AAPG Bull.*, **73**, 375–389.

Krystinik, L.F. and Blakeney-Dejarnett, B.A. (1994) Sedimentology of the Upper Morrow Formation in eastern Colorado and western Kansas Morrow sandstone (Pennsylvanian) of southeastern Colorado and Kansas. In: *Unconformity–Related Hydrocarbons in Sedimentary Sequences* (Eds J.C. Dolson, M.L. Hendricks and W.A. Wescott), Rocky Mountain Association of Geologists, Reprint, RMAG Sandstone Reservoirs of the Rocky Mountains, 167–180.

Labaune, C., Tesson, M., Gensous, B. Parize, O., Imbert, P. and Delhaye-Prat, V. (2010) Detail architecture of a compound incised valley and correlation with forced regressive wedges : exemples of Late Quaternary Têt and Agly rivers, western Gulf of Lions, Mediterranean Sea, France. *Sedimentary Geology*, **223**, 360–379.

Lasemi, Y. (2001) Facies analysis, depositional environments and sequence stratigraphy of the Upper Pre-Cambrian and Paleozoic rocks of Iran (in Persian). *Iran Geol. Surv. Publication*, 1–180.

Leckie, D.A., Wallace-Dudley, K.E., Vanbeselaere, N.A. and James, D.P. (2005) Sedimentation in a low-accommodation setting: non-marine (Cretaceous) Mannville and marine (Jurassic) Ellis Groups, Manyberries field, southeastern Alberta. *AAPG Bull.*, **88**, 1391–1418.

Lericolais, G., Berné, S. and Féniès, H. (2001) Seawards pinching out and internal stratigraphy of the Gironde incised valley on the shelf (Bay of Biscay). *Mar. Geol.*, **175**, 183–197.

Li, Y. and Bhattacharya, J.P. (2013) Facies-Architecture study of a stepped, forced regressive compound incised valley in the Ferron Notom Delta, Southern Central Utah, U.S.A. *J. Sed. Res.*, **83**, 206–225.

Longhitano S.G. (2013) A facies-based depositional model for ancient and modern, tectonically-confined tidal straits. *Terra Nova*, **25**, 446–452.

Longhitano, S.G., Chiarella, D., Di Stefano, A., Messina, C., Sabato, L. and Tropeano, M. (2012a) Tidal signatures in Neogene to Quaternary mixed deposits of southern Italy straits and bays. In: *Modern and ancient depositional systems: perspectives, models and signatures* (Eds S.G. Longhitano, D. Mellere and R.B. Ainsworth). *Sed. Geol.*, Special Issue, **279**, 74–96.

Longhitano, S.G., Mellere, D., Steel, R.J. and Ainsworth, R.B. (2012b) Tidal Depositional systems in the Rock Record: a Review and New Insights. In: *Modern and ancient depositional systems: perspectives, models and signatures* (Eds S.G. Longhitano, D. Mellere and R.B. Ainsworth). *Sed. Geol.*, Special Issue, **279**, 2–22.

Longhitano, S.G., Sabato, L., Tropeano, M. and Gallicchio, S. (2010) A mixed bioclastic/siliciclastic flood-tidal delta in a micro-tidal setting: depositional architectures and hierarchical internal organization (Pliocene, southern Apennines, Italy). *J. Sed. Res.*, **80**, 36–53.

MacEachern, J.A., Bann, K.L., Pemberton, S.G. and Gingras, M.K. (2007) The ichnofacies paradigm: high-resolution paleoenvironmental interpretations of the rock record. In: *Applied Ichnology* (Eds J.A. McEachern, K.L. Bann, M.K. Gingras, S.G. Pemberton). *SEPM*, Short Course Notes, **52**, Tulsa, 27–64.

Mángano, M.C. and Buatois, L.A. (2004) Ichnology of carboniferous tide-influenced environments and tidal flat variability in the North American midcontinent. In: *The Application of Ichnology to Palaeoenvironmental and Stratigraphic Analysis* (Ed. D. McIlroy). *Geol. Soc. London Spec. Publ.*, **228**, 157–178.

Martinius, A.W. and Gowland, S. (2010) Tide-influenced fluvial bedforms and tidal bore deposits (Late Jurassic Lourinhã Formation, Luisitanian Basin, Western Portugal). *Sedimentology*, **58**, 285–324.

Messina, C., Rosso, A., Sciuto, F., Di Geronimo, I., Nemec, W., Di Dio, T., Di Geronimo, R., Maniscalco, R. and Sanfilippo, R. (2007) Anatomy of a transgressive systems tract revealed by integrated sedimentological and palaeoecological study: the Barcellona P.G. Basin, northeastern Sicily, Italy. In: *Sedimentary Processes, Environments and Basins - A Tribute to Peter Friend* (Eds G. Nichols, C. Paola, E.A. Williams). *Int. Assoc. Sedimentol. Spec. Publ.*, **38**, 367–399.

Miall, A.D. (1996) *The Geology of Fluvial Deposits: Sedimentary Facies, Basin Analysis and Petroleum Geology.* Springer, Berlin.

Nadimi, A. (2007) Evolution of the Central Iranian basement. *Gondwana Research*, **12**, 324–333.

Nemec, W. and Kazanci, N. (1999) Quaternary colluvium in west-central Anatolia: sedimentary facies and palaeoclimatic significance. *Sedimentology*, **46**, 139–170.

Nemec, W. and Steel, R.J. (1984) Alluvial and coastal conglomerates: their significant features and some comments on gravelly mass-flow deposits. In: *Sedimentology of Gravels and Conglomerates* (Eds E.H. Koster and R.J. Steel). *Memoir*, **10**, 1–31.

Nio, S.D. and Yang, C.S. (1991) Diagnostic attributes of clastic tidal deposits: a review. In: *Clastic Tidal Sedimentology* (Eds D.G. Smith, G.E. Reinson, B.A. Zaitlin and R.A. Rahmani), *Mem. Can. Soc. Petrol. Geol.*, **16**, 3–28.

Nouidar, M. and Chellai, E.H. (2001) Facies and sequence stratigraphy of an estuarine incised-valley fill: Lower Aptian Bouzergoun Formation, Agadir Basin, Morocco. *Cretaceous Res.*, **22**, 93–104.

Olariu, I.M., Olariu, C., Steel, R.J., Dalrymple, R.W and Martinius, A.W. (2011) Anatomy of a laterally migrating tidal bar in front of a delta system: Esdolomada Member, Roda Formation, Tremp-Graus Basin, Spain. *Sedimentology*, **59**, 356–378.

Olariu C., Steel, R.J., Dalrymple, R.W. and Gingras, M. (2012) The sedimentological, ichnological and architec-

tural characteristics of compound dunes in a tidal seaway, the Lower Baronia sandstones (Lower Eocene), Ager Basin, Spain. In: *Modern and ancient depositional systems: perspectives, models and signatures* (Eds S.G. Longhitano, D. Mellere and R.B. Ainsworth). *Sed. Geol.*, Special Issue, **279**, 134–155.

Plink-Björklund, P. (2005) Stacked fluvial and estuarine deposits in high-frequency (4th-order) sequences of the Eocene Central Basin, Spitsbergen. *Sedimentology*, **52**, 391–428.

Plink-Björklund, P. and Steel, R. (2006) Incised valleys on an Eocene coastal plain and shelf, Spitsbergen e part of a linked shelf-slope system. In: *Incised Valleys in Time and Space* (Eds R.W. Dalrymple, D.A. Leckie and R.W. Tillmann). SEPM Spec. Publ., **85**, Tulsa, 281–308.

Plint, A.G. (2000) Sequence stratigraphy and palaeogeography of a Cenomanian deltaic complex: the Dunvegan and lower Kaskapau formations in subsurface and outcrop, Alberta and British Columbia, Canada. *Can. Bull. Petrol. Geol.*, **48**, 43–79.

Plint, A.G. and Browne, G.H. (1994) Tectonic event stratigraphy in a fluvio/lacustrine, strike-slip setting: the Boss Point Formation (Westphalian A), Cumberland Basin, Maritime Canada. *J. Sed. Res.*, **64**, 341–364.

Posamentier, H.W. and Allen, G.P. (1993) Variability of the sequence stratigraphic model: effects of local basin factors. *Sed. Geol.*, **86**, 91–109.

Pulham, A.J. (1994) The Crusiana field, Llanos basin, eastern Colombia: high resolution sequence stratigraphy applied to late Paleocene–Early Oligocene, estuarine, coastal plain and alluvial clastic reservoirs. In: *High Resolution Sequence Stratigraphy: Innovations and Applications* (Ed. S. Johnson). University of Liverpool, Liverpool, England, 63–68.

Reading, H.G. and Collinson, J.D. (1996) Clastic Coasts. In: *Sedimentary Environments: Processes, Facies and Stratigraphy* (Ed. H.G. Reading). Third ed. Blackwell Science, Oxford, 154–231.

Reineck, H.E. and Wunderlich, F. (1968) Classification and origin of flaser and lenticular bedding. *Sedimentology*, **11**, 99–104.

Reinson, G.E. (1992) Transgressive barrier island and estuarine systems. In: *Facies models: Response to Sea Level change* (Eds R.G. Walker and N.P. James). *Geol. Assoc. Can.*, St. Johns, Newfoundland, Canada, 179–194.

Reinson, G.E., Clark, J.E. and Foscolos, A.E. (1988) Reservoir geology of Crystal Viking field, Lower Cretaceous estuarine tidal channel-bay complex, south-central Alberta. *AAPG Bull.*, **72**, 1270–1294.

Reynaud, J.-Y. and Dalrymple, R.W. (2012) Shallow-marine tidal deposits. In: *Principles of Tidal Sedimentology* (Eds R.A. Davis Jr and R.W. Dalrymple), Springer, New York, 335–370.

Rezaeian, M. (2008) Coupled tectonics, erosion and climate in the Alborz Mountains, Iran. PhD thesis, University of Cambridge; 219 p.

Roy, P.S. (1984) New South Wales estuaries: their origin and evolution. In: *Coastal Geomorphology in Australia* (Ed. B.G Thom). Academic Press, New York, pp. 99–121.

Roy, P.S., Thom, B.G. and Wright, L.D. (1980) Holocene sequences on an embayed high-energy coast: an evolutionary model. *Sed. Geol.*, **26**, 1–19

Saidi, A. and Akbarpour, M.R. (1992) Geological Map of Iran. 1:100,000 series, sheet No. 676, Kiyasar. Geological Survey of Iran, Tehran.

Savrda, C.E. and Nanson, L.L. (2003) Ichnology of fair-weather and storm deposits in an upper Cretaceous estuary (Eutaw Formation, western Georgia, USA). *Palaeogeogr. Palaeoclimatol. Palaeoecol.*, **202**, 67–83.

Schwartz, R.K., O'Brien, T.J., Barber, D.E., Ness, J.B. and Weislogel, A.L. (2011) Braided channel system in the Paleogene Beaverhead intermontane basin: A longitudinal segment in the paleo-Missouri headwater system of southwest Montana [abs]: 2011 Annual GSA, Minneapolis. *GSA Abstracts with Programs*, **43**, 431.

Schwarz, E., Veiga, G.D., Spalletti, L.A. and Massaferro, J.L. (2011) The transgressive infill of an inherited-valley system: The Springhill Formation (lower Cretaceous) in southern Austral Basin, Argentina. *Mar. Petrol. Geol.*, **28**, 1218–1241.

Şêngör, A.M.C. (1990) A new model for the late Paleozoic-Mesozoic tectonic evolution of Iran and implications for Oman. In: *The geology and tectonics of the Oman region* (Eds M.P. Searle and A.C. Ries). *Geol. Soc. London*, 797–831.

Şêngör, A.M.C., Altiner, D., Cin, A., Ustaomer, T. and Hsu, K.J. (1988) Origin and assembly of the Tethyside orogenic collage at the expense of Gondwana Land. In: *Gondwana and Tethys* (Eds M.G. Audley-Charles and A. Hallman). *Geol. Soc. London Spec. Publ.*, **37**, 119–181.

Şêngör, A.M.C. and Natal'in, B.A. (1996) Paleotectonics of Asia: Fragments of a synthesis. In: *The tectonic evolution of Asia* (Eds A. Yin and M. Harrison). Cambridge, Cambridge University Press, 486–640.

Shahrabi, M. (1991) *Geological map of Iran*. 1:250,000 series, Gorgan, Geol. Surv. Iran, Tehran.

Shanley, K.W., McCave, P.J. and Hettinger, R.D. (1992) Tidal influence in Cretaceous fluvial strata from Utah, USA: a key to sequence stratigraphic interpretation. *Sedimentology*, **39**, 905–930.

Sharafi, M., Ashuri, M., Mahboubi, A. and Moussavi-Harami, R. (2012a) Stratigraphic application of *Thalassinoides* ichnofabric in delineating sequence stratigraphic surfaces (Mid-Cretaceous), Kopet-Dagh Basin, northeastern Iran. *Palaeoworld*, **21**, 202–216.

Sharafi, M., Ashuri, M., Mahboubi, A., Moussavi-Harami, R. and Nadjafi, M. (2010) Sequence stratigraphy of the Aitamir Formation (Albian–Cenomanian) in Sheikh and Bi-bahreh synclines in the west Kopet-Dagh Basin. *Journal of Science*, University of Tehran, **35**, 201–211.

Sharafi, M., Mahboubi, A. and Moussavi-Harami, R. (2012b) The relation between glauconitization and calcite cementation with the relative sea level changes in the mixed siliciclastic-carbonate sediments of Aitamir Formation (Mid-Cretaceous), Kopet-Dagh Basin. *Investigation of Stratigraphy and Sedimentology*, **48**, 19–36.

Sharafi, M., Mahboubi, A., Moussavi-Harami, R., Ashuri, M. and Rahimi, B. (2013) Sequence stratigraphic significance of sedimentary cycles and shell concentrations in the Aitamir Formation (Albian–Cenomanian), Kopet-Dagh Basin, northeastern Iran. *J. Asian Earth Sci.*, **67–68**, 171–186.

Simpson, E.L., Dilliard, K.A., Rowell, B.F. and Higgins, D. (2002) The fluvial-to-marine transition within the post-rift Lower Cambrian Hardyston Formation, Eastern Pennsylvania, USA. *Sed. Geol.*, **147**, 127–142.

Smith, D.G. (1988) Tidal bundles and mud couplets in the McMurray Formation, northeastern Alberta, Canada. *Bull. Can. Petrol. Geol.*, **36**, 216–219.

Spalletti, L. (1996) Estuarine and shallow marine sedimentation in the upper Cretaceous e lower tertiary west e central Patagonian Basin (Argentina). In: *Geology of Siliciclastic Shelf Seas* (Eds M. de Batist and P. Jacobs). *Geol. Soc. London*, Spec. Publ., **117**, 81–93.

Stampfli, G.M. (1978) Etude geologique generale de l'Elburz oriental au S de Gonbad-e-Qabus Iran N-E. Theses presentee a la Faculte des Sciences de l'Universite de Geneve, 1–329.

Stampfli, G.M., Borel, G.D., Cavazza, W., Mosar, J. and Ziegler, P.A. (2001) Palaeotectonic and palaeogeographic evolution of the western Tethys and Peri-Tethyan domain (IGCP Project 369), *Episodes*, **24**, 222–228.

Stampfli, G.M., Marcoux, J. and Baud, A. (1991) Tethyan margins in space and time. *Palaeogeogr. Palaeoclimatol. Palaeoecol.* **87**, 373–409.

Steel, R.J. and Thompson, D.B. (1983). Structures and textures in Triassic braided stream conglomerates ('Bunter' Pebble Beds) in the Sherwood Sandstone Group, North Staffordshire, England. *Sedimentology*, **30**, 341–367.

Stepanov, D.L. (1971) Carboniferous stratigraphy of Iran. Sixieme Congres de Stratigraphie et de Geologie du Carbonifere, Sheffield 11th to 16th September 1967, **4**, 1505–1518. Publishing Company 'Ernest van Aelst'; Maastricht.

Stöcklin, J. (1974) Northern Iran: Alborz Mountains. In: *Mesozoic-Cenozoic orogenic belts; data for orogenic studies; Alpine-Himalayan orogens* (Ed. A.M. Spencer). *Geol. Soc. London Spec. Publ.*, **4**, 213–234.

Stöcklin, J. (1968) Structural history and tectonics of Iran: a review. *AAPG Bull.*, **52**, 1229–1258.

Stöcklin, J. and Setudehnia, A. (1991) Stratigraphic Lexicon of Iran. *Geol. Surv. Iran Reports*, **18**, 1–376.

Storms, J.E.A. and Hampson, G.J. (2005) Mechanisms for forming discontinuity surfaces within shoreface-shelf parasequences: sea level, sediment supply or wave regime? *J. Sed. Res.*, **75**, 67–81.

Such, P., Buatois, L.A. and Mangano, M.G. (2007) Stratigraphy, depositional environments and ichnology of the lower paleozoic in the Azul Pampa area e Jujuy Province. *Rev. Asoc. Geol. Argentina*, **62**, 331–344.

Swift, D.J.P., Parsons, S.B. and Howell, K.A. (2008) Campanian continental and shallow marine architecture in a eustatically modified clastic wedge: Mesaverde Group, Wyoming, USA. In: *Recent Advances in Models of Siliciclastic Shallow-Marine Stratigraphy* (Eds G.J. Hampson, R.J. Steel, P.M. Burgess and R.W. Dalrymple), *SEPM Spec. Publ.*, **90**, 473–490.

Swift, D.J.P., Phillips, S. and Thorne, J.A. (1991) Sedimentation on continental margins: V. Parasequences. In: *Shelf Sand and Sandstone Bodies-Geometry, Facies and Sequence Stratigraphy* (Eds D.J.P. Swift, G.F. Oertel, R.W. Tillman, J.A. Thorne). *Int. Assoc. Sedimentol. Spec. Publ.*, **14**, 153–187.

Tessier, B. (2012) Stratigraphy of Tide-Dominated Estuaries. In: *Principles of Tidal Sedimentology* (Eds R.A. Davis, Jr. and R.W. Dalrymple). Springer, New York, 621 p.

Tessier, B. (1993) Upper intertidal rhythmites in the Mont-Saint-Michel Bay (NW France): Perspectives for paleo-reconstruction. *Mar. Geol.*, **110**, 355–367.

Tessier, B., Monfort, Y., Gigot, P. and Larsonneur, C. (1989) Enregistrement des cycles tidaux en accrétion verticale, adaptation d'un outil de traitement mathématique. Exemples en baie du Mont-Saint-Michel et dans la molasse marine miocène du bassin de Digne. *Bull. Soc. Géol. France*, **V5**, 1029–1041.

Tesson, M., Labaune, C., Gensous, B., Suc, J-P., Melinte-Dobrinescu, M., Parize, O., Imbert, P. and Delhaye-Prat, V. (2011) Quaternary 'compound' incised valley in a microtidal environment, Roussillon Continental Shelf, Western Gulf of Lions, France. *J. Sed. Res.*, **81**, 708–729.

Tropeano, M., Cilumbriello, A., Sabato, L., Gallicchio, S., Grippa, A., Longhitano, S.G., Bianca, M., Gallipoli, M.R., Mucciarelli, M. and Spilotro, G. (2013) Surface and subsurface of the Metaponto coastal plain (Gulf of Taranto - southern Italy): present-day *vs.* LGM-landscape. *Geomorphology*, Special Issue, **203**, 115–131.

Ueno, K., Watanabe, D., Igo, H., Kakuwa, Y. and Matsumoto, R. (1997) Early Carboniferous Foraminifera from the Mobarak Formation of Shahmirzad, Northeastern Alborz Mountains, Northern Iran. In: *Late Paleozoic foraminifera, their biostratigraphy, evolution and paleoecology and the mid-Carboniferous boundary* (Eds C.A. Ross, J.R.P. Ross and P.L. Brenckle). *Cush-man Foundation for Foraminiferal Research, Special Publication*, **36**, 149–152.

Vahdati Daneshmand, F. (1991) *Amol: Geological quad-rangle map of Iran: Tehran*, Geol. Surv. Iran, scale 1:250,000.

Van Straaten, L.M.J.U. (1961) Sedimentation in tidal flat areas. *Alberta Soc. Petrol. Geol.*, **9**, 203–226.

Van Wagoner, J.C., Mitchum, R.M., Campion, K.M. and Rahmanian, V.D. (1990) Siliciclastic sequence stratigra-phy in well logs, cores and outcrops; concepts for high-resolution correlation of time and facies. *AAPG, Methods in Exploration series*, **7**, Tulsa.

Van Wagoner, J.C., Mitchum, R.M., Posamentier, H.W. and Vail, P.R. (1987) An overview of sequence stratigraphy and key definitions. In: *Atlas of Seismic Stratigraphy, volume 1* (Ed. A.W. Bally). *AAPG, Studies in Geology*, **27**, 11–14.

Varban, B.L. and Plint, A.G. (2008) Sequence stacking patterns in the Western Canada foredeep: influence of tectonics, sediment loading and eustasy in deposition of the Upper Cretaceous Kaskapau and Cardium Formations. *Sedimentology*, **55**, 395–421.

Visser, M.J. (1980) Neap–spring cycles reflected in Holocene subtidal large-scale bedform deposits: a preliminary note. *Geology*, **8**, 543–546.

Walker, R.G. and Plint, G.A. (1992) Wave- and storm-dominated shallow marine systems. In: *Facies models: Responses to sea level change* (Eds R.G. Walker and N.P. James). *Geol. Assoc. Can.*, St. John's, Newfoundland, 219–238.

Weissmann, G, Hartley, A. and Nichols, G. (2011) Alluvial facies distributions in continental sedimentary basins - distributive fluvial systems. In: *River To Rock Record: The Preservation Of Fluvial Sediments And Their Subsequent Interpretation* (Eds S. Davidson, S. Leleu and C. North), *SEPM*, **79**, pp. 327–355.

Wendt, J., Kaufmann, B., Belka, Z., Farsan, N. and Karimi Bavandpur, A. (2002) Devonian/Lower Carboniferous stratigraphy, facies patterns and palaeogeography of Iran. Part I. Southeastern Iran. *Acta Geologica Polonica*, **52**, 129–168.

Wendt, J., Kaufmann, B., Belka, Z., Farsan, N. and Karimi Bavandpur, A. (2005) Devonian/Lower Carboniferous stratigraphy, facies patterns and palaeogeography of Iran, Part II. Northern and central Iran. *Acta Geol. Pol.*, **55**, 31–97.

Wood, J.M. and Hopkins, J.C. (1989) Reservoir sandstone bodies in estuarine valley fill: Lower Cretaceous Glauconitic Member, Little Bow Field, Alberta, Canada. *AAPG Bull.*, **73**, 1361–1382.

Wood, J.M. and Hopkins, J.C. (1992) Traps associated with paleo-valleys and interfluves in an unconformity bounded sequence: Lower Cretaceous Glauconitic Member, southern Alberta, Canada. *AAPG Bull.*, **76**, 904–926.

Yoshida, S. (2000) Sequence and facies architecture of the upper Blackhawk Formation and the Lower Castlegate Sandstone (Upper Cretaceous), Book Cliffs, Utah, USA. *Sed. Geol.*, **136**, 239–276.

Yoshida, S., Jackson, M.D., Johnson, H.D., Muggeridge, A.H. and Martinius, A.W. (2001) Outcrop studies of tidal sandstones for reservoir characterization (Lower Cretaceous Vectis Formation Isle of Wight, southern England). Sedimentary Environments Offshore Norway - Palaeozoic to Recent. *NPF Special Publication*, **10**, 233–257.

Yoshida, S., Steel, R.J. and Dalrymple, R.W. (2007) Changes in depositional processes; an ingredient in a new gen-eration of sequence stratigraphic models. *J. Sed. Res.*, **77**, 447–460.

Zaitlin, B.A., Dalrymple, R.W. and Boyd, R. (1994) The stratigraphic organization of incised valley systems associated with relative sea-level change. In: *Incised Valley Systems: Origin and Sedimentary Sequences* (Eds R.W. Dalrymple, R. Boyd and B.A. Zaitlin). *SEPM Spec. Publ.*, **51**, 45–60.

Zaitlin, B.A. and Shultz, B.C. (1984) An estuarine-embayment fill model from the Lower Cretaceous Mannville Group, west-central Saskatchewan. In: *Mesozoic of Middle North America Scott* (Eds D.F. Scott and D.J. Glass). *Can. Soc. Petrol. Geol. Mem.*, **9**, 455–469.

Zaitlin, B.A. and Shultz, B.C. (1990) Wave-influenced estu-arine sand body, Senlac heavy oil pool, Saskatchewan, Canada. In: Sandstone Petroleum Reservoirs (Eds J.H. Barwis, J.G. McPherson and J.R.J. Studlick). New York, Springer-Verlag, 363–387.

Zaitlin, B.A., Warren, M.J., Potocki, D., Rosenthal, L. and Boyd, R. (2002) Depositional styles in a low accommo-dation foreland basin setting: an example from the Basal Quartz (Lower Cretaceous), southern Alberta. *Bull. Can. Petrol. Geol.*, **50**, 31–72.

Zanchi, A., Berra, F., Mattei, M., Ghassemi, M.R. and Sabour, J. (2006) Inversion tectonics in central Alborz, Iran. *J. Struct. Geol.*, **28**, 2023–2037.

Zecchin, M. (2007) The architectural variability of small-scale cycles in shelf and ramp clastic systems: the controlling factors. *Earth-Sci. Rev.*, **84**, 21–55.

Sedimentary facies and Late Pleistocene-Holocene evolution of the northern Jiangsu coast and radial tidal ridge field, South Yellow Sea, China

YONG YIN*[†‡], PEIHONG JIA*[†‡] and QING LI[†‡]

[†] *The Key Laboratory of Coast & Island Development, School of Geographic & Oceanographic Sciences, Nanjing University, Hankou Rd.22, Nanjing, 210093, P. R. China (yinyong@nju.edu.cn)*
[‡] *Present address: Key Laboratory of Coast and Island Development (Nanjing University), School of Geogarphic and Oceanographic Sciences, Xianlin Ave. 163, Nanjing, 210093, P. R. China*
* *Corresponding authors: Yong Yin (yinyong@nju.edu.cn), Peihong Jia (jiaph@nju.edu.cn)*

ABSTRACT

In the radial tidal ridge field, northern Jiangsu coast of the South Yellow Sea, Late Pleistocene-Holocene glacial-eustatic sea-level fluctuations have formed a set of particular tidal genetic strata. Sedimentary facies associations are attributed to tidal flat, tidal ridge, tidal channel and creek infilling, flood plain environments as well as tidal ravinement and channel base lag. Five sedimentary units were delineated according to facies assemblages and major facies change boundaries. The facies assemblages indicate that two stacked aggradational tide-dominated units (U1 and U2), were formed during MIS 3 (Late Pleistocene). The sand source from the palaeo-Yangtze River mainly contributed to the formation of thick aggradational sequences in these units. The unit U2 features tidal channels characterized by extensive lateral migration. The tidal environment during the Late Pleistocene was similar to the present littoral tidal flat. During the Last Glacial Maximum when sea-level fell to -150 m on the South Yellow Sea shelf, the study area became terrestrial exposed and flood plain deposits accumulated (unit U3). A prominent transgressive surface occurs at the boundary between the Late Pleistocene deposits and the Holocene succession. A tidal ravinement surface separates the Holocene strata into a transgressive unit (U4, mainly composed of tidal flat facies) and a tidal sand ridge unit (U5). The formation of tidal sand ridges corresponds to the sand input from the abandoned Yellow River Delta after AD 1128. Glacio-eustatic sea-level changes and tidal erosion controlled the evolution of the Late Pleistocene-Holocene succession, while tidal current fields and sand supply from nearby deltas have played a key role in the RTR system formation.

Keywords: Pleistocene, Holocene, tidal sand ridge, tidal flat, cores, seismic, sediment supply, sea-level change, South Yellow Sea.

INTRODUCTION

Tidal sedimentary systems along the coast of the Yellow Sea and East China Sea have been studied intensively (Yang, 1989; Liu *et al.*, 1989; Nio & Yang, 1991; Liu, 1997; Wang *et al.*, 1999; Li *et al.*, 2001; Hori *et al.*, 2001; Jin & Chough, 2002; Jin *et al.*, 2002; Choi & Dalrymple, 2004; Choi & Kim, 2006; Liu *et al.*, 2007; Liu *et al.*, 2010). However, the evolution of tide-dominated coastal environments with large supply of sediment, like the northern Jiangsu coast of western South Yellow Sea has not received much attention. Unlike most studied tide-dominated coastal areas, the northern Jiangsu coast is influenced by two major river mouths: the Yangtze River to the south and the abandoned Yellow River to the north (Fig. 1). Hence, sediment supply from nearby deltas has played an important role in the development of the sedimentary system since the Late Pleistocene.

The South Yellow Sea is a shallow, semi-closed, epicontinental sea between China and the Korean

Contributions to Modern and Ancient Tidal Sedimentology: Proceedings of the Tidalites 2012 Conference,
First Edition. Edited by Bernadette Tessier and Jean-Yves Reynaud.

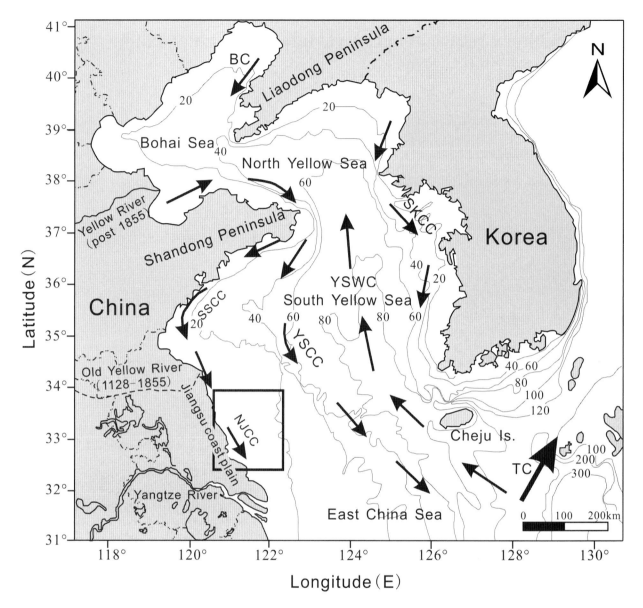

Fig. 1. Location of the study area (square: The northern Jiangsu coast and offshore area) on the western coast of the South Yellow Sea. The contour lines indicate the water depth in metres. YSCC, Yellow Sea Coastal Current; YSWC, Yellow Sea Warm Current; TC, Tsushima Current; SSCC, South Shandong Coastal Current; NJCC, North Jiangsu Coastal Current; SKCC, South Korean Coastal Current; BC, Bohai Current.

peninsula, with an average water depth of 46 m (Fig. 1). The western coast of the South Yellow Sea within the Jiangsu Province is composed of wide, low gradient tidal flats with an average slope of between 1.6/1000 and 1.75/1000. The coast is associated with a unique radial tidal ridge (RTR) system along the nearshore to offshore areas. The RTR developed under a convergent-divergent tidal current field and radiate perpendicular or at a high angle to the coastline (Wang,

2002). The RTR system covers an area of 2047 km² above mean sea-level and includes 70 tidal ridges of different sizes and tidal channels situated between 0 and 25 m water depth.

Previous works mostly focused on the sediment distribution, seismic stratigraphy, hydro- and sedimentary dynamics of the RTR (Ren, 1986; Fu & Zhu, 1986; Yang, 1989; Wang, 2002). The Late Pleistocene-Holocene evolution of the coastal plain and the RTR field and the influences of sediment supply from nearby deltas through time are still not well understood due to the lack of sufficient borehole and seismic data.

In this study we present sedimentary facies, units and stratal organisation from multi-site sediment core data (Fig. 2). The objectives of the study are to: (1) reconstruct the long-term evolution of the northern Jiangsu coast and its RTR system; and (2) to evaluate the factors, which controlled the formation of the RTR system.

GENERAL SETTINGS

Coastal geomorphology

The northern Jiangsu coast includes a low-elevation coastal plain, a stretch of broad tidal flats and a prominent underwater fan-shaped RTR system (Fig. 2). The coastal plain was influenced by the relict Yellow River course to the north and the Yangtze River to the south. Before the Yellow River shifted to its present course, the river drained into the South Yellow Sea at the Jiangsu coast between AD 1128 and AD 1855. In this period the river deposited between 6000 and 7000×10^8 t of sand, which resulted in the formation of a vast delta (Zhang & Wang, 1991). The palaeo-Yangtze River also supplied a large quantity of sand to the northern Jiangsu coast whilst entering the South Yellow Sea between Qiang Port and Lüsi Port at around 40 ka BP (MIS3; Sun *et al.*, 2015). After the Yangtze River migrated southward to its present course it left a vast delta plain south of the northern Jiangsu coastal plain.

The elevation of the coastal plain decreases from south-east and north toward the lowest central part. Before humans moved into this part of the plain, the area was composed of paralic swamps and lagoon plains. Due to Holocene sea-level fluctuations, a succession of four chenier ridges, with maximum height of 5 m, was formed in the coastal plain.

The progradational tidal flat of the northern Jiangsu coast has an average width of 4 to 5 km and a mean slope of $1/10^{-3}$ (Wang, 2012). North of Qiang Port the tidal flats have a width of 15 to 27 km with a mean slope of $0.15/10^{-3}$ (Fig. 2). The tidal flat is composed of coarse silt, sandy silt and silty sand with small amounts of clay. It progrades seaward at an average rate of 200 m yr^{-1} and accretes by 4 to 5 cm yr^{-1} (Zhang *et al.*, 1984). The tidal flat is characterized by tidal creeks and channels, which account for 10% of the total tidal flat area (Zhang & Wang, 1991). Tidal creeks and channels are conduits for sand transportation, connecting with sub-tidal channels or directly flowing to the sea at the mouth (Fig. 3). The lengths of the tidal creeks and channels range between hundreds of metres to 20 km. Tidal creeks above the mean high tidal level have a dendritic shape, with a length of a few hundreds to thousands of metres and a width of few metres to dozens of metres. Due to fine-grained sediment accumulation, the channels experience head withering, narrowing and shallowing as the tidal flat progrades. The tidal channels between mean high and mid-tidal levels are more sinuous, deeper and they migrate quicker than the creeks of the high tidal flat. Their width ranges from few metres to tens of metres when plants colonize the bank; and this increases to tens of metres up to hundreds of metres with banks free of plants. The tidal channels between mid-tidal and lower-tidal levels present a trumpet shape, with a width of few hundred metres to 1000 m, connecting with the sub-tidal sea at the mouth. They are straighter than those of the mid-tidal flat but still meander fast (Chen, 2001). On average, tidal creeks and channels may migrate laterally between 130 m and 230 m yr^{-1} causing tidal flats to be reworked strongly. The tidal channel deposits may account for 60 to 80% of highly reworked tidal flats (Zhang *et al.*, 1984).

The RTR system extends from mean low-tidal level to a water depth of about 30 m. Ridges are separated by sub-tidal channels and lined by tidal flats. Individual sand ridges are 5 km to 100 km long and 2 km to 25 km wide. Most ridges have a broad and flat top crest and protrude 5 to 20 m above the adjacent channel floor. Shallow seismic profiles across the sand ridge crests show large-scale very low-angle inclined layers, which dip towards the sides of the ridges. This suggests that ridges experience lateral accretion both during flood and ebb tides. The RTR system can be subdivided into three parts, the northern group, north-eastern group and southern group according to their elongation and magnitude (Fig. 2).

The northern ridges are elongated in a NNW–SSE to N–S direction. Dong Sha is the largest of the ridges with a maximum length of 58.8 km in a NNW–SSE direction and a maximum width of 30 km in a NE direction (Figs 2 and 3). Its highest elevation is 5.8 m above zero metres of local chart datum. The surface area above mean sea-level was about 775 km^2 in 1979. The area had decreased to

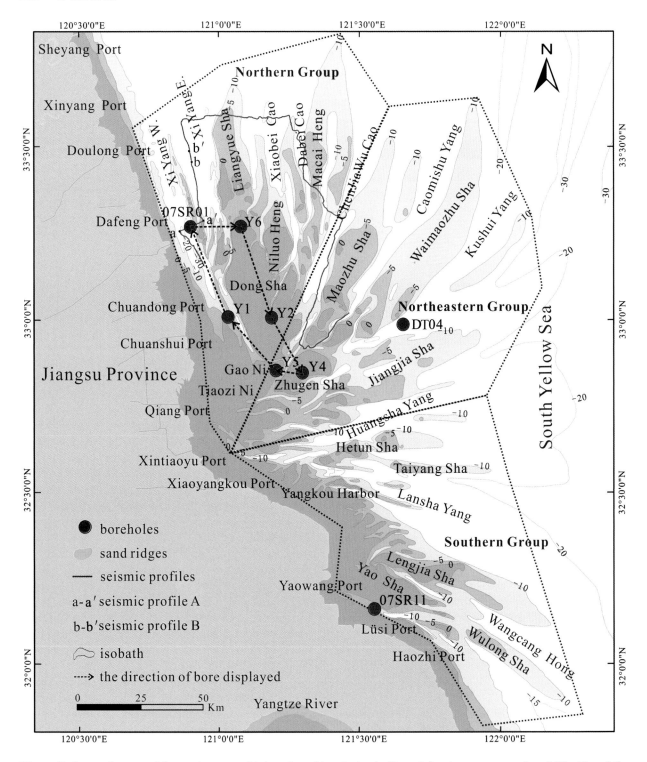

Fig. 2. Bathymetric map of the study area, with location of boreholes (collected for the present study, cf. Fig. 7) and the seismic lines (collected in the 1990s by Wang, 2002; Figs 9 and 10 for profiles A and B, respectively). Colour scale: yellow: land; green: intertidal area; blue: subtidal. Water depth is in metres. Note that in Chinese, 'Sha' and 'Heng' mean ridge and shoal, 'Yang', 'Hong' and 'Cao' mean tidal channel and 'Ni' means flat.

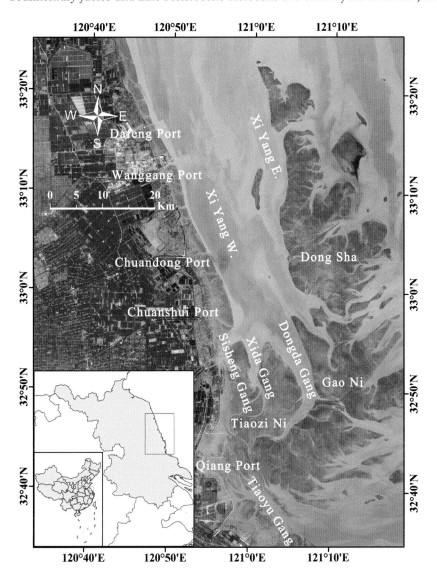

Fig. 3. Aerial photograph of the littoral tidal flat of northern Jiangsu coast showing the creek and channel pattern on the tidal flats; and one of the most prominent ridges of the RTR system, the Dong Sha ridge. Note that 'Gang' means tidal creek and channel in Chinese.

$578 \, km^2$ in 2006 because of erosion induced by strong tidal currents (Zhang, 2013). The Xi Yang tidal channel is located west of Dong Sha. The length of the channel is 90 km and it extends in a NNW direction (Figs 2 and 3). The water depth in the channel increases from 4.0 m at its south-eastern end to 14.8 m at the channel mouth. When delineated by the 15 m bathymetric contour, the Xi Yang channel has a length of 55 km and a width of 3 to 4 km. According to sea charts from 1979 and 2006, the channel bottom has been eroded into a rugged morphology with swales more than 20 m deep.

The north-eastern ridges are the largest ones; for example, Waimaozhu Sha ridge has a length of 115 km (Fig. 2). These well-developed ridges are related to sand dispersal from offshore to onshore by flood tides and *vice versa* by ebb tides. Between 1975 and 2009 the Waimaozhu Sha ridge extended 30 to 40 km northward, at an average rate of 0.88 to 1.18 km yr^{-1} (Zhang, 2013).

The southern area of the RTR system is composed of smaller ridges than those of the northern and north-eastern parts, probably due to reduced sediment supply from the abandoned Yellow River delta located to the north (Wang & Zhang, 1998).

Tidal regime

A radial tidal current regime appears in the offshore area of the northern Jiangsu coast, with rectilinear current characteristics (Zhu & Chang, 2000). The tidal regime in offshore areas is regular semi-diurnal. On the littoral area, due to deformation induced by sea bottom-morphology, the M_2 tide becomes prominent and the tidal regime is irregular semidiurnal. Currents converge to the direction of Qiang Port during flood (Fig. 4A), while they diverge from this place in a fan-like pattern of 150° arc during ebb (Fig. 4B). Constrained into the channels, the tidal currents have rectilinear characteristics. In the Xi Yang tidal channel, the flood and ebb currents reach mean maximum velocity of 1.24 m s⁻¹ and 1.20 m s⁻¹, respectively. At the Xi Yang hydrological survey station the maximum velocity of the flood can reach up to 2.15 m s⁻¹ and the ebb tide 1.80 m s⁻¹. At most hydrological survey stations, the flood tide has a higher velocity and a shorter duration than the ebb tide.

The mean tidal ranges on the northern Jiangsu coast vary between 1.5 and 5 m (Table 1). The macrotidal region is located on the apex zone of the RTR system, at Xintiaoyu Port, with a maximum tidal range of 5.68 m. The tidal range decreases toward the north and south. To the north, at the abandoned Yellow River mouth, the tidal range is 1.76 m and at the Lüsi Port it is 3.84 m. The distribution of spring tidal range is similar to the mean tidal range. The spring tidal range reaches up to 8.08 m at the Yangkou Harbor and 9.62 m at Xintiaoyu Port. No palaeo-tidal regime data from MIS3 has been reconstructed but a two-dimensional tidal model has shown that tidal currents on the Yellow Sea have generally been semi-diurnal throughout the postglacial stages, with an *F*-ratio (index of diurnal inequality) and S_2/M_2 ratio (index of spring-neap cycle)

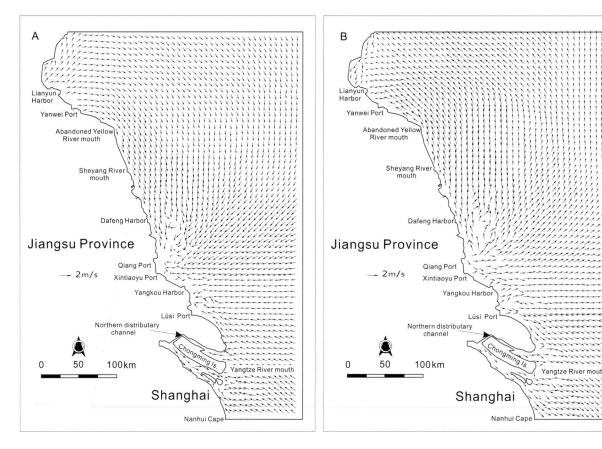

Fig. 4. The convergent-divergent tidal current field of the South Yellow Sea western coast and offshore areas. (A) Maximum current velocity during flood tide. (B) Maximum current velocity during ebb tide (modified after Yang, 2010).

Table 1. Mean and spring tidal ranges along the northern Jiangsu coast. Tidal range decreases from the convergent area of the RTR toward both north and south (See Fig. 4 for location of hydrological survey stations).

Location	Mean tidal range/m	Spring tidal range/m
Lianyun Harbour	3.69	6.11
Yanwei Port	3.24	5.39
Abandoned Yellow River Mouth	1.76	3.13
Sheyang River mouth	2.15	4.16
Dafeng Harbour	3.56	6.64
Qiang Port	4.84	7.85
Xintiaoyu Port	5.68	9.62
Yangkou Harbour	4.61	8.08
Lüsi Port	3.84	6.21
Northern distributary channel of Yangtze River	3.04	5.95

similar to present (Uehara & Saito, 2003). When sea-level was 45 m below present-day level, around 10 ka BP, the Yangtze estuary extended south-east towards the East China Sea for about 400 km while the East China Sea and Yellow Sea, north of 28°N, was a semi-enclosed embayment. The spatial distribution of M_2 tides was different from today and no amphidrome was present in the Yellow Sea (Uehara *et al.*, 2002). After sea-level reached the Holocene maximum at 6 ka BP and the Yangtze estuary opened up to the South Yellow Sea, a radial tidal flow pattern, similar to present, originated at the estuary mouth (Uehara *et al.*, 2002; Zhu, 2002). After 3.8 ka BP, the radial tidal flow field in the north of the Yangtze estuary gradually developed towards its present pattern (Zhu, 2002).

Surface sediment distribution

The RTR system is dominated by sand and silt deposits, with minor clay content (Fig. 5). The sediment is composed of 3.5 to 92.6% sand, 7.3 to 89.4% silt and 0.1 to 10.6% clay.

The distribution of grain size is closely related to the RTR's topography and hydrodynamic conditions. The northern part of the RTR is dominated by silt and sandy silt, with minor clayey silt (Fig. 5). The silt and sandy silt are distributed in the nearshore area of the abandoned Yellow River Delta. Sediments become coarser offshore and to the south. The central part is mainly composed of sand (fine and very fine-grained sand) and silty sand (Zhang, 2013). Fine-grained and

very fine-grained sand is mostly distributed on the sand ridges, whilst silty sand are associated with tidal channels. The distal parts of the channels are composed of clayey silt and silty clay. Debris of calcium concretions, originating probably from the underlying strata, have been found in tidal channels where pre-Holocene deposits outcrop at the sea floor. The southern part is mostly composed of silty sand and sandy silt, with minor clayey silt and sand. Sands are likely to occur on the crest of the ridges and sandy silt and clayey silt in the tidal channels. The silty sands mostly occur in the nearshore area but become finer to the south as well as seaward. The surface sediment distribution shows a trend with fine-grained sediments occurring in both the north and south; and coarse-grained sediments in the central part. This distribution pattern implies that the north and south parts are mostly influenced by fine sediment input from the abandoned Yellow River Delta and present-day Yangtze River Delta, while the sediments in the central part are mostly winnowed by tidal currents and waves.

Since the 1980s, the sediments on the ridge crests were still dominated by very fine-grained sand. However, in the nearshore area, the amount of silty material has increased, replacing the former sand deposits. On the outreaches of the RTR system, the clayey sediment areas have decreased and were replaced by silty sands. This suggests an increase of hydrodynamics, such as wave action, during the last 30 years, maybe in relation to sea-level rise.

MATERIALS AND METHODS

Cores were recovered from the ridges and channels in water depths between 0 and 15.4 m in order to reconstruct the stratigraphy and evolution of the northern Jiangsu coast and offshore area (Fig. 2). Cores, 75 mm in diameter, were obtained by using rotary drilling devices with a recovery rate of 60 to 70%. The cores were split lengthwise and the sediment described in terms of colour, lithology, texture and structures, microfossils, bioturbation and sedimentary breaks. Sediment colour was assigned on the basis of the Munsell Color Chart. One half of the cores was used for sampling and the other half was stored and archived in the Key Laboratory of Coast and Island Development of MOE, Nanjing University, for future reference.

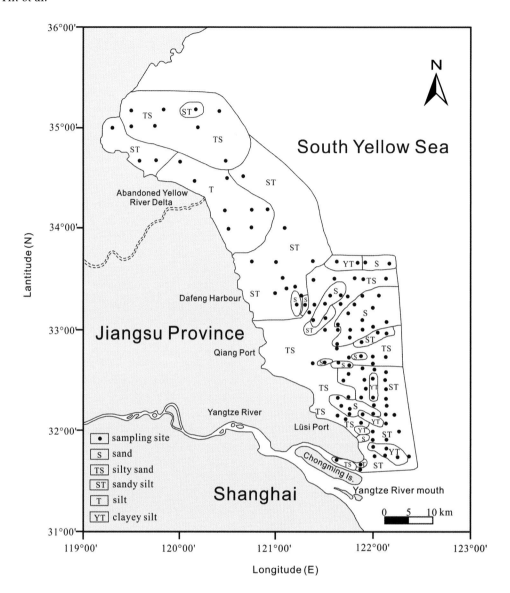

Fig. 5. Distribution of surface sediments in the nearshore and offshore zones of the Jiangsu coast (modified after Zhang, 2013).

Grain size analyses were performed on samples collected at 5 cm intervals. After removal of organic matter and calcium carbonate, grain size analyses were measured using a Malvern Mastersizer 2000 laser particle size analyser. The data were summarized using the method of Folk & Ward (1957).

Foraminifera samples were collected at 40 cm intervals and examined by the Nanjing Institute of Geology and Palaeontology (Chinese Academy of Sciences), in order to obtain additional information on depositional environments.

Twenty-six samples were collected for ¹⁴C dating. AMS (accelerator mass spectrometry) analyses were undertaken on bivalve and gastropoda shells as well as plant debris by Beta Analytic (Miami, FL, USA) and the School of Archaeology and Museology (Peking University, China). The ¹⁴C ages were calibrated by Beta Analytic and by using Calib Rev. 5.0.1 (Stuiver *et al.*, 2005). A Delta-R of −80±60 was applied as local reservoir correction. Few ages were too old and fell beyond the endpoints of the calibration curve (Table 2).

Table 2. ^{14}C age data. Note that the first 20 samples were analysed and calibrated by Beta Analytic, USA, and the last 6 by the School of Archaeology and Museology, Peking University, China (calibration using Stuiver *et al.*, 2005).

Code No. (Bata)	Sample No.	Down core depth/m	Materials	Conventional ages/yr B.P.	Calendar ages/cal yr B.P.	
					Intercept	Range/2σ
346527	Y1-04	13.93	Molluscan shell	700±30	310	420–280
346533	Y1-10	49.45	Snail	41,370±660	44,660	45,540–43,740
346536	Y1-13	55.05	Snail	39,240±500	43,060	43,960–42,390
346538	Y1-15	58.15	Snail	>42,650	—	—
342052	Y2-02	14.05	Snail	870±30	530	650–450
342053	Y2-03	56.54	Snail	40,900±530	44,420	45,120–43,570
342054	Y4-01	3.78	Molluscan shell	650±30	390	490–250
342056	Y4-03	14.65	Molluscan shell	770±30	480	550–320
342059	Y4-06	23.51	Oyster	7520±30	8030	8190–7920
342066	Y5-02	3.08	Molluscan shell	520±30	260	400–50
342068	Y5-04	5.85	Plant debris	560±30	280	430–120
342069	Y5-05	10.12	Molluscan shell	690±30	420	510–270
342070	Y5-06	15.24	Molluscan shell	790±30	490	600–390
342071	Y5-07	17.26	Molluscan shell	880±30	530	650–460
342072	Y5-08	19.48	Molluscan shell	6180±30	6710	6880–6540
342077	Y6-02	5.81	Molluscan shell	160±30	AD Post 1950	AD Post 1950
342078	Y6-03	16.16	Molluscan shell	6760±40	7370	7470–7230
342079	Y6-04	21.04	Oyster	6880±40	7450	7570–7350
342084	Y6-09	47.80	Molluscan shell	43,130±680	—	—
342085	Y6-10	52.74	Molluscan shell	>43,500	—	—
BA090468	07SR01-02-C	0.12	Shell	3920±35	3990	4215–3830
BA090470	07SR01-17-C	3.45	Organic bearing mud	32,940±130	36,845	37,425–36,510
BA090474	07SR01-51-C	15.37	Shell	35,495±140	39,750	40,070–39,405
BA090476	07SR01-56-C	18.13	Shell	42,185±320	45,235	45,740–44,705
BA090478	07SR01-62-C	20.20	Shell	42,645±615	45,590	46,655–44,600
BA090479	07SR01-73-C	23.26	Shell	>43,000	—	—

Finally, additional information is provided for the present article thanks to high resolution seismic reflection data collected in the RTR area in the 1990s (Wang, 2002).

RESULTS

Sedimentary facies, interpretation and chronology

Based on lithology, particle size, sedimentary texture and structure and fossil content, six sedimentary facies associations have been distinguished (Fig. 6):

1. Tidal flat facies association: this facies association occurs in the lower to mid part of all cores as well as in the bottom of the upper part of sediment cores Y1, Y4 and Y6 (Fig. 2 for core location; and Fig. 7). The tidal flat facies association

consists of olive grey to dark grey coarse silt, very fine-grained sand and silty clay, which appear mostly as 'sand-mud' couplets. The couplets sometimes show systematic vertical thickness variations typical of tidal rhythmites (Dalrymple & Makino, 1989). Tidal bedding structures such as lenticular, flaser and wavy bedding are commonly observed (Fig. 6). Sediment facies are commonly organized in fining-up successions, with flaser bedding or massive sand in the lower part, silt and mud wavy bedding in the middle part and silt and mud lenticular bedding in the top. Burrows are more common in mud and mud-silt bedding than in massive sand. Bidirectional cross-laminations locally occur in flaser and wavy bedding, recording the action of both flood and ebb currents. Nearshore to inner shelf foraminifera species *Ammonia beccarii, Ammonia annectens*, with low abundance and low diversity,

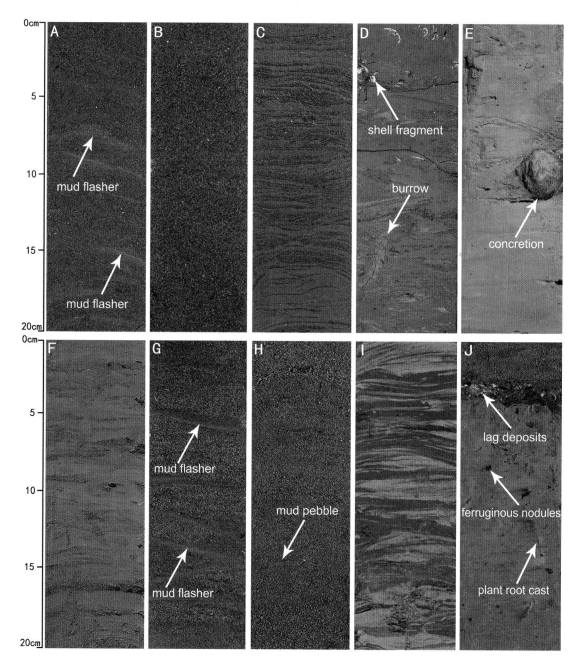

Fig. 6. Photographs of representative sedimentary facies from the Late Pleistocene to Holocene strata described in the cores (Fig. 2 for location). (A) Fine-grained sand with flaser bedding, lower-intertidal flat (borehole Y5, 54.4 m deep). (B) Massive fine sands, lower-intertidal flat (borehole Y5, 50.9 m deep). (C) Silt-mud wavy bedding, mid-intertidal flat (borehole Y2, 34.28 m deep). (D) Silt-mud lenticular bedding with shell fragments and burrows, upper intertidal flat (borehole Y6, 47.8 m deep). (E) Mud with vague horizontal laminae, upper intertidal flat (borehole Y2, 44.2 m deep). The greenish blue colour on the surface probably indicates terrestrial exposure. A calcium carbonate concretion is embedded into the mud. (F) Mud-dominated lenticular bedding, upper intertidal flat (borehole Y6, 13.3 m deep). (G) Flaser bedding facies, sand ridge bottom (borehole Y5, 17.0 m deep). (H) Massive fine-grained sand with embedded mud pebbles and chips, top of sand ridges (borehole Y2, 0.17 m deep). (I) Inclined Heterolithic Stratification (IHS) facies with bidirectional foreset laminae, tidal channel infilling (borehole 07SR01, 4.0 m deep). (J) Stiff mud with dark brown ferriginous mottles and plant root cast (palaeosol), flood plain deposits. The sharp erosional surface on top of the mud succession is the Holocene transgressive surface, overlain by a transgressive lag (borehole Y5, 19.45 m deep). All core sections are 20 cm long.

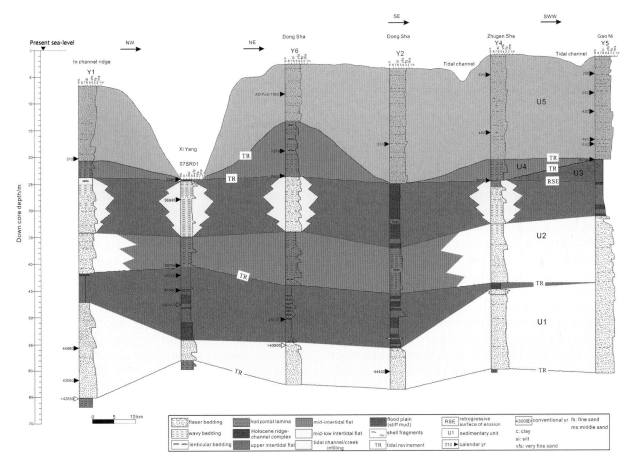

Fig. 7. Correlation diagram between collected cores (location in Fig. 2). Five units (U1, U2 U3, U4 and U5) are identified on the basis of facies interpretation and ¹⁴C ages. Each unit is separated by erosional surface formed during transgression and sea-level fall. See Fig. 2 for location of the Xi Yang channel, Dong Sha and Zhugen Sha tidal sand ridges; and Gao Ni flat.

characterize this facies association. Littoral to estuary bivalve and gastropod species, such as *Potamocorbulaamurensis* (Schrenck) and *Nassarius* (*Phrontis*) *caelatulus* Wang, are present as well. The tidal flat facies association is subdivided into three main facies: (1) lower intertidal flat facies are composed of light olive grey very fine-grained and fine-grained sand with minor coarse silt. It is fairly to poorly sorted and contains massive sand and flaser bedding (Fig. 6A and B). Individual successions of lower intertidal facies have thickness ranging from 1.5 to 6.5 m. Abundant shell fragments, quartzose granules, debris of calcium carbonate concretions and mud pebbles are embedded in the facies that rests through an erosional contact on underlying strata, generally represented by upper tidal facies; (2) Mid-intertidal flat facies are composed of light olive grey to olive grey interlayers of fairly to poorly sorted clayey silt

and silty clay, as well as small amounts of fairly to poorly sorted very fine-grained silt interca- lated with clayey silt. Individual successions of mid-intertidal facies have thicknesses ranging from 1.9 to 4.3 m. Sediment structures are repre- sented by wavy and lenticular bedding (Fig. 6C). Ripple cross bedding is occasionally bidi- rectional laminated. Carbonized plant debris is concentrated in places as lenticular-like thin layers. Burrows are occasionally present; (3) Upper intertidal flat facies consist of olive grey to dark grey silty clay, intercalated with len- ticular or millimetre-thick layers of clayey silt (Fig. 6D, E and F). The thickness of individual successions of this facies ranges from 1.8 to 4.4 m and is often more or less eroded by overlying deposits. Straight and sinuous burrows are common. Calcium carbonate concretions occur in places where temporary exposure and mete- oric water leaching took place (Fig. 6E). Dark organic mottles, interpreted as cast of plant roots, occur occasionally.

2. Tidal ridge facies association: this facies association is observed in the upper part of all cores except in core 07SR01 (Fig. 7). It is composed of light olive grey to brownish grey fairly sorted very fine-grained and fine-grained sand intercalated with thin layers (1 to 6mm-thick) of yellowish brown silty clay and clayey silt (average proportion of sand 79.2%, silt 17.5% and clay 3.4%). Flaser bedding and massive sand are commonly observed and organized as coarsening up successions ranging in thickness between 1 and 4m (Fig. 6G and H). Mollusk fragments and carbonized plant debris are concentrated in places. Debris of calcium carbonate and ferro-concretion remains deposited as granule-grade particles (2 to 4mm in diameter) are present. A composite foraminifera assemblage of inner shelf species represented by *Ammonia beccarii* and *Ammonia annectens* and mid-shelf species represented by *Bolivinarobusta* and *Buliminamarginata* characterizes this facies association. The assemblage has a higher abundance and diversity in comparison to the tidal flat facies association. In addition, nearshore species of bivalves and gastropods are observed.

3. Tidal channel and creek infilling facies association: this facies association is mainly present in the middle part of the boreholes (Fig. 7). It is organized in 1 to 5m-thick fining-up successions. The successions are dominated by wavy and lenticular bedding composed of millimetre-thick to centimetre-thick brownish-grey silt-clay, interlayered with coarse silt. Cyclic variation in thickness and a systematic change in sedimentary structures, from wavy-bedding in the lower part to lenticular-bedding in the upper part are assigned to inclined heterolithic stratification (IHS) (Fig. 6I). Shell layers representing channel base lags are locally present. Bi-directional laminae are well developed and contorted bedding occurs in places. The degree of bioturbation is about 15% with 5mm long burrows, inclined or perpendicular to bedding planes. A high abundance but low diversity of littoral to nearshore foraminifera species such as *Ammonia beccarii, Noniongrateloipi, N. schwageri, Ammonia pauciloculata, Cribrononion poeyanum* typify this facies association. The tidal creeks in the lower part of cores, e.g. at 50.6m depth in borehole Y2 and 44.6m depth in borehole Y4 are much smaller than those in the middle part of cores (Fig. 7). Their thickness is less than 70cm. The infillings display slightly inclined silt-clay and clay interlayers and distinct contact with underlying deposits. The up-fining infilling succession, IHS structures accompanied by basal channel lags, indicate that this facies is related to tidal channel infilling.

4. Flood plain facies association: this association is present only in the sediment core Y5 and is composed of yellow-brown clay and silty clay, with millimetre-thin horizontal and wavy laminae (Fig. 7). It appears as a uniform layer with no grading change from bottom to top, but with sharp upper and lower contacts. These deposits are completely devoid of marine faunal fossils. The sediments contain abundant pedogenic features such as argillans, dark and reddish mottles, calcium carbonate concretions, microcracks and casts of plant roots (Fig. 6J). The dark and reddish mottles are most probably analogous to the mottled fabric of palaeosols. The calcium carbonate concretions indicate subaerial exposure and downward leaching extracted from the upper soil horizons. All these features suggest a terrestrial floodplain environment prone to palaeosol formation. ^{14}C ages provided by different authors (Zheng, 1999; Qin *et al.*, 2004; Li & Wang, 1998; Li *et al.*, 2001) indicate that the flood plain deposits recognised as stiff mud (palaeosoil) in the Yangtze River delta and northern Jiangsu coast plain areas were formed between 25 and 15 ka BP.

5. Ravinement lag facies association: this facies association is a few centimetres thick and is observed in sediment cores Y4 and Y5 (Fig. 7) immediately above an erosional surface that truncates flood plain or upper intertidal deposits. It is characterized by brownish grey to grey very fine-grained sand, with abundant pebble-grade shell fragments and calcium carbonate concretion clasts (Fig. 6J). The overlying sediments are tide-dominated facies. Shell fragments provide cal. ^{14}C ages of ca 6700 and 8000 cal yr BP (Fig. 7, Table 2). The authors suggest that these lag deposits are due to tidal reworking during the Holocene transgression.

6. Channel base lag facies association: This facies association has been observed in the topmost layer of sediment core 07SR01, i.e. the bottom of the modern tidal channel Xi Yang (Fig. 2). It consists of a 12cm-thick layer of moderate yellowish-brown silt with dark yellowish-brown

silty clay and obscure cross-bedding. Large amounts of oyster shell fragments together with rock granules and 3 mm-long calcium carbonate clasts are present. Dating of shell fragments yield ages between 4215 and 3830 cal. yr BP (Table 2). These facies that rest on a distinct erosional boundary are typical of channel base lag deposits due to strong tidal current reworking.

Core stratigraphy

Stratal architecture is illustrated in a reconstructed core correlated section constructed with borehole Y1 to borehole Y5 and borehole 07SR01 (Fig. 7). According to facies evolution, surfaces and radiocarbon ages, five sedimentary units, U1, U2, U3, U4 and U5, are recognised, from base to top. All are bounded at their base by an erosional surface of at least regional extent.

The three basal sedimentary units, U1, U2 and U3, are Late Pleistocene in age (Fig. 7). The unit U1 includes a complete tidal flat succession from lower to upper intertidal flat in boreholes Y1, Y4 and 07SR01. The thickness of U1 ranges from 16.4 m in borehole Y4 to a maximum of 23 m in borehole Y1. The lowest part of the lower intertidal flat facies is missing in boreholes Y2 and Y6, probably because the drills did not reach it. The upper intertidal flat facies has a thickness of 15 m in borehole 07SR01, 11.2 m in borehole Y6 and 10.1 m in borehole Y2. The facies then diminishes to 5.3 m and 1.2 m in borehole Y1 and borehole Y4, respectively, and finally pinches out in borehole Y5, probably due to erosion before deposition of the overlying strata. Tidal creek infilling is present in this unit and occurs at a depth of 50.6 m in borehole Y2 and 44.6 m in borehole Y4 (Fig. 7). The terrestrial exposure features such as reddish colour, calcium carbonate concretions and plant rootlets in the upper tidal flat facies suggest a minor sea-level fall during that period of MIS3. This took place around 40,000 cal yr BP, according to ^{14}C dates (Table 2). Unit U2 is also composed of tidal flat deposits, including tidal channels and creeks. The thickness of the unit ranges from 15.4 to 22.8 m (Fig. 7). The boreholes Y2 and Y5 display an upper intertidal flat interval ranging in thickness from 12 to 6.2 m, respectively, and thinning to 1.2 m in borehole Y4. The upper part of U2 along the boreholes 07SR01, Y1, Y4 and Y6 is represented by tidal channel infilling facies. The thickness of this interval

is 15.0 m, 10.0 m, 10.1 m and 8.2 in boreholes 07SR01, Y6, Y1 and Y4, respectively. The tidal channel in the borehole 07SR01 reaches the upper part of the U1.

The unit U3 is composed of flood plain deposits. The unit has a limited spatial distribution and is only present in the borehole Y5, with a thickness of 4.3 m (Fig. 7). This poor extension of U3 suggests that the unit was deeply eroded, probably during the Holocene transgression.

The upper two units, U4 and U5, represent the Holocene strata (Fig. 7). U4 + U5 thickness varies between 17 to 23 m. In borehole 07SR01, drilled into the Xi Yang tidal channel, U4 and U5 are condensed in the decimetre-thick channel lag. The basal surface of U4 and U5 is clearly erosional. The unit U4 is composed of mid-intertidal to upper-intertidal flat deposits. According to the ^{14}C age of 8030 ± 30 cal yr BP found in borehole Y4, the deposition of U4 records the early stage of the Holocene transgression (Fig. 7 and Table 2). U4 is not laterally continuous and its thickness varies from 10.2 m in borehole Y6 to 3.9 m in borehole Y4 and 2.7 m in borehole Y1. The unit is missing in all other boreholes (Fig. 7). The top of U4 is eroded by an undular surface with a relative relief height between 3 and 10 m. This erosional surface, which separates U4 and U5, is interpreted to have formed as a result of tidal reworking during the Holocene marine transgression prior to the sand ridge formation.

The unit U5 is composed of ridge facies association in boreholes Y1 to Y5 (Dong Sha, Zhugeng Sha and Gao Ni ridges; Fig. 7). U5 is 11 to 21 m-thick and consists of five to seven coarsening-up elementary sequences, 3 to 4 m-thick (Fig. 7). Each sequence evolves from flaser bedding to massive sand. The authors relate these sequences to the lateral migration of the sand ridges. Prior to the formation of the ridges, the tidal currents strongly reworked the underlying strata and removed the Early Holocene deposits partly (boreholes Y1, Y4 and Y6) or completely (boreholes Y2, Y5). The authors assume that, locally, the tidal erosion also reworked the top of the Pleistocene unit U2.

DISCUSSION

Long term evolution

Radiocarbon ages from boreholes indicate that the U1 and U2 strata deposited during the MIS3 (Table 2; Fig. 7). The study area was a tide-dominated

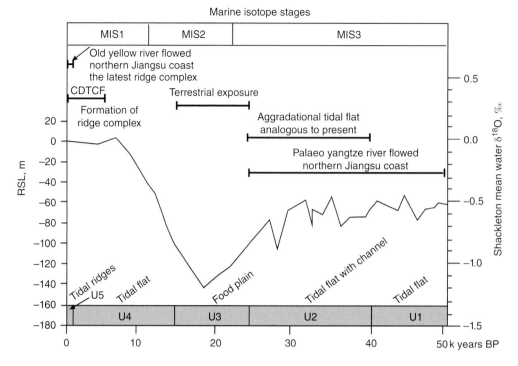

Fig. 8. The main stages of evolution of the northern Jiangsu coast and RTR field, correlated with global sea-level curve for the past 60 ka (from Shackleton, 2000). U1 to U5: sedimentary units; MIS: Marine Isotopic Stage; CDTCF: convergent-divergent tidal current field.

coastal environment during that time (Figs 7 and 8). U1 and U2 may correspond to stage 3.3 and stage 3.1 respectively, according to the ^{14}C dates. The study area experienced two phases of transgression due to sea-level rise and a regression in between during these time intervals. The tidal flat units U1 and U2 are significantly thicker than would be expected of a normal progradational succession. For example, U2 may reach a thickness up to 23 m. Upper intertidal flat successions commonly reach 10 m, and lower to mid intertidal flat intervals 5 to 10 m (Fig. 7). One explanation is that the formation of the intertidal flats occurred in the presence of significant in-place accommodation. Stacked aggradational sequences also occur in a sediment succession cored in the western offshore of the Korean peninsula (Jin *et al.*, 2002). The sequence III that these authors described (borehole YSDP 105), equivalent to the unit U2 identified herein, is 25 m-thick. This sequence III that developed in a tidal flat environment, with creeks and channels, is of an aggradational nature. Given that the amplitude of the RSL fluctuations that occurred during the MIS3 was of the order of 20 m (Fig. 7, Shackleton, 2000) it can be assumed that a balance existed between RSL

rise and sediment supply in order to allow the observed vertical aggradation of tidal-flat successions. The source materials were probably supplied by the palaeo-Yangtze River when it flowed into the South Yellow Sea. The drilled core 07SR11 (Fig. 2) from the nearshore of Lüsi port indicates that the palaeo-Yangtze River reached the South Yellow Sea during the MIS3 (Sun *et al.*, 2015). Furthermore, subsidence probably combined with glacio-eustatic sea-level rise to allow such an aggradation of tidal flats.

During the Last Glacial Maximum (LMG), the sea-level on the South Yellow Sea shelf fell by about 150 m below the present day level and, as a result, the shelf was exposed. The palaeo-Yellow River was thought to enter the South Yellow Sea to the north of the RTR field (Liu *et al.*, 2010), then to flow along the central trough to the north of Cheju Island (Fig. 1), and finally to enter the East China Sea. The northern branch of the palaeo-Yangtze River entered the South Yellow Sea south of the RTR field, flowed ESE to finally terminate in the Okinawa Trough on the shelf break at 31°N (Geng, 1981). The study area was a broad flood plain (Unit U3), associated with the palaeo-Yellow and palaeo-Yangtze rivers.

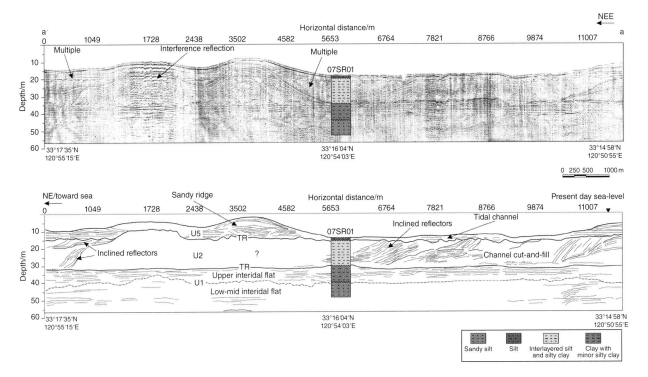

Fig. 9. Seismic profile A (after Wang, 2002), across Xi Yang tidal channel (see Fig. 2 for location). In U2, the channel infilling geometries feature long-distance lateral migration, comparable to present-day tidal channels. U3 and U4 are probably completely eroded by the marine erosion induced by the Holocene transgression and therefore not present. TR: tidal ravinement surface.

The earliest Holocene transgression record in the study area, according to the ^{14}C date from the borehole Y4, may date back to 8000 cal yr BP (Table 2; Figs 7 and 8). At this time the sea-level reached -10 m below present but the marine transgression in the central Yellow Sea may have commenced prior to about 11.5 ka (Chough *et al.*, 2004). The postglacial transgression was associated with strong tidal reworking, resulting in multiple unconformities in the core records. Tidal erosion started when the Holocene transgression inundated the study area. Before the sandy ridges started to develop, the strong tidal currents eroded most of the Holocene transgressive deposits (U4), as well as parts of the Late Pleistocene strata (U2). As a result of the erosion, a relict morphology with maximum relative relief of 10 m remained. When the old Yellow River flowed into the South Yellow Sea and delivered great amounts of sands to the study area ca. 1000 years ago, the tidal currents began to rework this newly available stock of sediment. The reworking moulded the deposits into a radial configuration under a convergent-divergent tidal current regime.

The Late Pleistocene channel infilling

The Late Pleistocene channel structures described in U2 have previously been interpreted as buried mid-Holocene sand ridges on seismic profiles (Wang, 2002). The new ^{14}C ages indicate that the infilling deposits probably developed in the late part of MIS3 between ca. 40,000 and 37,000 cal yr BP (Table 2, Figs 7 and 8). The core stratigraphy shows that the channel fill displays a fining-upward succession preserved in a regressive succession (Fig. 7).

On the seismic profiles of Wang (2002), the channel-like configuration and internal clinoform reflectors representing lateral accretion make it easy to distinguish the cut-and-fill structures from the present-day ridges (Figs 9 and 10). The internal structure of the Late Pleistocene channels usually shows a concave-up shape, while modern ridges show a convex-up shape. Furthermore, the modern sand ridges generally display a flat, hummocky shape with an internal low angle cross bedding structure, dipping in one or two directions. The Late Pleistocene channels show stacked

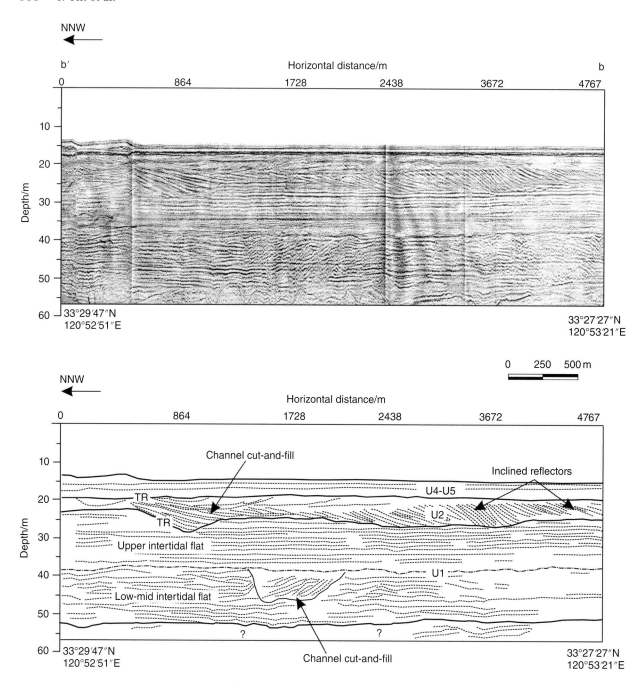

Fig. 10. Seismic profile B (after Wang, 2002) parallel to Xi Yang tidal channel (Fig. 2 for location). Stacked cut-and-fill geometries are visible in U2, which suggest lateral migration over long distance for these tidal channels. TR: tidal ravinement surface.

cut-and-fill features. The basal shell layer with a ¹⁴C age of 39,750 cal yr BP (Fig. 7, Table 2) represents the lowest channel base lag in U2, and may date the onset of channel functioning, although the shells in the lag must be older than the channel.

The tidal origin, rather than fluvial, of the Late Pleistocene channels is supported by the occurrence of abundant foraminifera and nearshore gastropod fossils in the infilling successions. The low angle dipping cross-sets with alternating thicker and thinner packages of muds and silts (Hughes, 2012) distinguished on seismic profiles and in cores also support the tidal origin for these channels (Figs 6I, 9 and 10).

The present-day tidal creeks and channels on the northern Jiangsu coast may represent analogues for the Late Pleistocene channels. The present-day tidal flat is wide and with well-developed tidal creek and channel system (Fig. 3). The creeks or channels, especially those on the lower to mid tidal flats, are meandered and migrate actively. This is a common phenomenon on progradational tidal flats. The main factors controlling the migration of creeks and channels include tide in-flowing volume into the channels in the late phase of ebb tide, the different routes of flood and ebb tides and the runoff from land and storm surges (Zhang & Wang, 1991). A direct control is channel bank stability. Indeed, the silty nature of the northern Jiangsu tidal flats favours active channel migration, especially during storm surges.

The satellite and aerial images allow quantifying the migration of tidal channels. The middle and end portions of the Sisheng Gang on the low tidal flat (Fig. 3) reached between 1973 and 1978 a migration rate of 0.25 km yr^{-1} and 0.5 km yr^{-1}, respectively. The rate for the middle portion increased to 1 km yr^{-1} between 1978 and 1980 (Zhang & Wang, 1991). Maximum channel migration rates were as high as 9.55 km yr^{-1} (Chen *et al.*, 2004).

The authors believe that the abundant migration structures in the Late Pleistocene unit U2 indicate that the tidal environment during the late MIS3 was similar to the present-day environment which is characterized by progradational and aggradational tidal flats. The great amount of sand supplied during that time to the study area by the palaeo-Yangtze River running into the South Yellow Sea in the southern part of the RTR field between Qiang Port and Lüsi Port may have induce tidal flat progradation. In addition, the sandy nature of U2 tidal flats favoured active and long distance channel migration as illustrated on the seismic line B (Fig. 10).

The unit U1 contains fewer and smaller tidal creeks than U2 (Figs 7, 9 and 10), although the borehole DT04 (not described in this study, Fig. 2 for location) shows in U1-equivalent strata, sixteen stacked tidal creek infilling sequences, metric in thickness, with well-developed IHS. On the seismic line B (Fig. 10) a tidal channel 680 m wide and 8 m deep is clearly imaged in U1, but the distance of lateral migration is apparently not as long as in U2. This basically suggests that the U1 and U2 tidal flat environments were not comparable.

Formation of sand ridges

The formation of the sand ridges has been discussed in several studies. Wang (2002) thought that the tidal ridges were the product of the Holocene transgression and that the ridges developed synchronously with the sea-level rise. Chen *et al.* (1995) suggested that the formation of the ridges dates back 7 ka to 6.5 ka BP and was caused by sediment input from the Yangtze River. Li *et al.* (2001) argued that the sand ridge system formed after the Holocene transgression maximum, although the numerical modelling indicated that a convergent-divergent tidal current field had set up in the early Holocene (Zhu, 1998). Li *et al.* (2001) further pointed out that two sand ridge systems exist, i.e. a sub-surface system buried underneath the modern coastal plain, which probably formed between 4200 and 4500 yr BP, and the active system which formed after 2000 yr BP. Finally, Liu & Xia (2004) concluded that the sand ridge morphology was shaped in the last 2000 years, as a result of the debouching of the old Yellow River.

Two primary conditions are required for the formation of the RTR; a convergent-divergent tidal current field and a large amount of sediments. Numerical modelling shows that during the Holocene transgression maximum a radial tidal flow pattern, similar to the present-day one, was established in the Yangtze estuary, which was widely open to the Yellow and East China Seas (Uehara *et al.*, 2002). The tides and tidal currents were gradually enhanced in the north of Yangtze Estuary after 3.8 ka BP and finally evolved to the current convergent-divergent tidal current field in the study area (Zhu, 1998).

The abandoned Yellow River Delta to the north of the RTR system played an important role concerning sand supply. Before the river course became abandoned in AD 1855 it flowed partially to the northern Jiangsu coast in the period between AD 1128 and AD 1494, then fully between AD 1494 and AD 1855. The coastline around the abandoned Yellow River mouth prograded 15 km between AD 1128 and AD 1578, reaching an average annual rate of 33 m (Li, 1991). The coastline progradation rate increased to 267 m yr^{-1} resulting in 74 km of coastline progradation between AD 1579 and AD 1855. During this period the old Yellow River carried annually 1.2×10^9 t of sediments to the Jiangsu coast (Li, 1991). The sands were partly trapped in the river mouth, forming deltaic mouth bars and shoals. However, parts of the sand were transported southward to the RTR system by flood tides via the main tidal channel, e.g. the Xi Yang channel (Fig. 2). The south moving suspended sediments were pumped by tides to construct the tidal flat landward and to pile up the ridges seaward. Due to a large sand input, the tidal flat prograded quickly, whilst previous isolated small underwater ridges grew fast, coalesced and finally merged into a big island, e.g. the Dong Sha sand ridge. Hence, as a result of the huge sand supply delivered to the study area by the 16th to 19th centuries, the growth of sand ridges accelerated, leading underwater ridges to emerge rapidly since the 16th century (Zhang & Chen, 1992).

After the Yellow River returned to the Bohai Bay (post-1855, Fig. 1), the abandoned delta mouth bars and shoals, disappeared within 30 years, due to wave erosion. The eroded sands were transported southward, further promoting the development of sand ridges. At present, the sediment source from the abandoned Yellow River mouth is almost cut off. Tidal currents continuously shape and rework the ridge-channel morphology, increasing the channel to ridge amplitude. The contribution of the modern Yangtze River to the RTR system construction can be considered as small since the suspended materials it supplies can move northward as far as Lüsi port (Fig. 2) during the summer monsoon season (Yang, 2010). Although the rate is obviously lower than before, sand ridge growth is still continuing. The authors believe that the materials eroded from the RTR system and *in situ* from the channels play an important role in maintaining the current sedimentation in the study area.

The radiocarbon ages indicate that the formation of the tidal ridges occurred since 600 years ago (Fig. 7), although the radial tidal current field existed since the last 6 ka. One explanation is that the large amounts of sediment supply began when the old Yellow River flowed in the area of the northern Jiangsu coast, i.e. since in the 12th century. Another explanation is that the ridges are morphodynamically active, e.g. the Dong Sha ridge migrated eastward with an average rate of 75 m a^{-1} from 1963 to 1979 (Zhang & Chen, 1992). The ridges of the Jiangsu coast area are obviously more mobile than other tidal ridges such as those in the North Sea (Gao & Collins, 2014). Active migration might rework the ridge deposits in a relatively short time, so that no sediment older than

a few hundreds of years is preserved in the ridges. Besides tidal currents, storms during the summer season can occasionally rework a whole sand ridge. Small-scale ridges, especially, are easily eroded during storm tides, contributing to the non-preservation of ancient deposits.

CONCLUSION

Long-term evolution of the northern Jiangsu coast and RTR field indicates that the Northern Jiangsu coast has been a tide-dominated environment since the Late Pleistocene, except during the LGM. Two stacked aggradational tidal flat units, U1 and U2, developed during the MIS3. The characteristics of these tidal flat units, especially U2, demonstrate that the environment was comparable to the present, with high sediment input. The palaeo-Yangtze River that entered the South Yellow Sea to the south of the RTR field was the main supplier of sediments to the area. The tidal channels associated with the tidal flat succession in U2 have a large spatial distribution and probably migrated extensively. Rapid migration was possible due to the silty nature of the tidal flat deposits and the loose and unstable channel banks. Furthermore, increased tidal volume during storm surges may also have accounted for rapid and extensively channel migration. Before the Holocene transgression, the study area experienced a terrestrial exposure when sea-level fell to -150 m during the LGM. A vast flood plain developed between the palaeo-Yellow River to the north and palaeo-Yangtze River to the south.

When the sea inundated the study area at around 8 ka BP the study area was transformed back into tidal flat environment. From ca. 6 ka BP, a convergent-divergent tidal current field pattern set up and radial tidal ridges probably started to develop. From the 12th century, prominent ridges emerged as the result of large amounts of sediment delivered by the palaeo-Yellow River, which discharged immediately north of the study area. After the Yellow River returned to Baohai Bay in AD 1885, the mouth bars of the abandoned delta were eroded by waves, enhancing sediment supply and consequently favouring the RTR growth. At present, the strong tidal currents are still shaping the ridge and channel morphology, causing the relative relief between ridge crests and channel floors to increase. Active migration and occasional complete reworking by storm waves probably explain why the age of some ridges never exceeds a few hundred years.

ACKNOWLEDGEMENTS

This research was supported by the National Key Basic Research Program of China (Grant no. 2013CB956500), National Natural Science Foundation of China (Grant no. 40776023) and the public welfare project of the State Oceanic Administration, People's Republic of China (Grant no. 201005006). We thank Dr Robert Dalrymple and Dr Alain Trentesaux for their reviews and constructive comments on the manuscript. We also thank Dr. Bernadette Tessier, Dr. Mikkel Fruergaard and Dr. Bernd Wünnemann for assistance in correcting and improving the manuscript.

REFERENCES

Chen, B.Z., Li, C.X. and Ye, Z.Z. (1995) Post glacial sedimentation and environmental evolution on northern flank of Yangtze River delta, east China. *Acta Oceanol. Sinica*, **17**, 64–75.

Chen, C.J. (2001) Change in tidal creek after mud flat being enclosed in the middle coast in Jiangsu province. *Mar. Sci. Bull.*, **20**, 71–79 (in Chinese with an English abstract).

Chen, J., Feng, W.B. and Zhang, R.S. (2004) Stability of tidal creek system of Tiaozini sand shoals in the north Jiangsu coast. *Sci. Geogr. Sinica*, **24**, 94–100 (in Chinese with an English abstract).

Choi, K.S. and Dalrymple, R.W. (2004) Recurring tide-dominated sedimentation in Kyonggi Bay (west coast of Korea): similarity of tidal deposits in Late Pleistocene and Holocene sequences. *Mar. Geol.*, **212**, 81–96.

Choi, K.S. and Kim, S.P. (2006) Late Quaternary evolution of macrotidal Kimpo tidal flat, Kyonggi Bay, west coast of Korea. *Mar. Geol.*, **232**, 17–34.

Chough, S.K., Lee, H.J., Chun, S.S. and Shin, Y.J. (2004) Depositional processes of late Quaternary sediments in the Yellow Sea: a review. *Geosci. J.*, **8**, 211–264.

Dalrymple, R.W. and Makino, Y. (1989) Description and genesis of tidal bedding in the Cobequid Bay–Salmon River estuary, Bay of Fundy, Canada. In: *Sedimentary facies of the active plate margin* (Eds A. Taira, F. Masuda), Terra Scientific, Tokyo, pp 151–177.

Folk, R.L. and Ward, W. (1957) Brazos River bar: A study in the significance of grain size parameters. *J. Sed. Petrol.*, **27**, 3–26.

Fu, M.Z. and Zhu, D.K. (1986) The sediment sources of the offshore submarine sand ridge field of the coast of Jiangsu Province. *Journal of Nanjing University (Natural Sciences Edition)*, **22**, 536–544 (in Chinese with an English abstract).

Gao, S. and Collins, M.B. (2014) Holocene sedimentary systems on continental shelves. *Mar. Geol.*, **352**, 268–294.

Geng, X.S. (1981) The submarine buried valley system on eastern China continental shelf. *Mar. Sci.*, **2**, 21–26 (in Chinese).

Hori, K., Saito, Y., Zhao, Q.H., Cheng, X.R., Wang, P.X., Sato, Y. and Li, C.X. (2001) Sedimentary facies of the tide-dominated paleo-Changjiang (Yangtze) estuary during the last transgression. *Mar. Geol.*, **177**, 331–351.

Hughes, Z.J. (2012) Tidal channels on tidal flats and marshes. In: *Principles of Tidal Sedimentology* (Eds R.A. Davis and R.W. Dalrymple). Springer Dordrecht Heidelberg London, New York, 269–300.

Jin, J.H. and Chough, S.K. (2002) Erosional shelf ridges in the mid-eastern Yellow Sea. *Geo-Mar. Lett.*, **21**, 219–225.

Jin, J.H., Chough, S.K. and Ryang, W.H. (2002) Sequence aggradation and system tracts partitioning in the mid-eastern Yellow Sea: roles of glacio-eustasy, subsidence and tidal dynamics. *Mar. Geol.*, **184**, 249–271.

Li, C.X. and Wang, P.X. (1998) *Late Quaternary Stratigraphy of the Changjiang Delta*. China Science Press, Beijing, 222pp (in Chinese with an English abstract).

Li, C.X., Zhang, J.Q., Fan, D.D. and Deng, B. (2001) Holocene regression and the tidal radial sand ridge system formation in the Jiangsu coastal zone, east China. *Mar. Geol.*, **173**, 97–120.

Li, Y.F. (1991) The development of the abandoned Yellow River delta. *Geogr. Res.*, **10**, 29–39 (in Chinese with an English abstract).

Liu, J., Saito, Y., Kong, X.H., Wang, H., Wen, C., Yang, Z.G. and Nakashima, R. (2010) Delta development and channel incision during marine isotope stages 3 and 2 in the western South Yellow Sea. *Mar. Geol.*, **278**, 54–76.

Liu, Z.X. (1997) Yangtze shoal-a modern tidal sand sheet in the northwestern part of the East China Sea. *Mar. Geol.*, **137**, 321–330.

Liu, Z.X., Berné, S., Saito, Y., Yu, H., Trentesaux, A., Uehara, K., Yin, P., Liu, J.P., Li, C.X., Hua, G.H. and Wang, X.Q. (2007) Internal architecture and mobility of tidal sand ridges in the East China Sea. *Cont. Shelf Res.*, **27**, 1820–1834.

Liu, Z.X., Huang, Y.C. and Zhang, Q.A. (1989) Tidal current ridges in the southwestern Yellow Sea. *J. Sed. Petrol.*, **59**, 432–439.

Liu, Z.X. and Xia, D.X. (2004) *Tidal sands in the China seas*. China Ocean Press, Beijing, 222 pp.

Nio, S.D. and Yang, C.S. (1991) Sea-level fluctuations and the geometric variability of tide-dominated sandbodies. *Sed. Geol.*, **70**, 161–193.

Qin, J.G., Wu, G.X., Zheng, H.B. and Li, C.X. (2004) Fossil signs in the uppermost hard clay of the Yangtze delta and adjacent continental shelves. *Mar. Geol. & Quatern. Geol.*, **24**, 11–18 (in Chinese with English abstract).

Ren, M.E. (1986) *Comprehensive Investigation of Coastal Zone and Tidal Flat Resources, Jiangsu Province*. China Ocean Press, Beijing, 517 pp (in Chinese).

Shackleton, N.J. (2000) The 100,000-year Ice-Age cycle identified and found to lag temperature, carbon dioxide, and orbital eccentricity. *Science*, **289**, 1897–1902.

Stuiver, M., Reimer, P.J. and Reimer, R.W. (2005) CALIB 5.0.1 [WWW program and documentation] (<http://calib.qub.ac.uk/>).

Sun, Z.Y., Li, G. and Yin Y. (2015) The Yangtze River deposition in southern Yellow Sea during Marine Oxygen Isotope Stage 3 and its implications for sea-level changes. *Quatern. Res.*, **83**, 204–215.

Uehara, K. and Saito, Y. (2003) Late Quaternary evolution of the Yellow/East China Sea tidal regime and its impacts on sediments dispersal and seafloor morphology. *Sed. Geol.*, **162**, 25–38.

Uehara, K., Saito, Y. and Hori, K. (2002) Paleotidal regime in the Changjiang (Yangtze) Estuary, the East China Sea, and the Yellow Sea at 6 ka and 10 ka estimated from a numerical model. *Mar. Geol.*, **183**, 179–192.

Wang, Y. (2012) *Regional oceanography of China Seas-Marine geology.* China Ocean Press, Beijing, 676 pp.

Wang, Y. (2002) *Study on the radial tidal sandy ridge system on South Yellow Sea.* China Environmental Science Press, Beijing, 433pp.

Wang, Y.P. and Zhang, R.S. (1998) Geomorphology responses on dynamics pattern of Jiangsu offshore sandbanks, eastern China. *Marine Sciences*, **3**, 43–47 (in Chinese with an English abstract).

Wang, Y., Zhu, D.K., You, K.Y., Pan, S.M., Zhu, X.D., Zou, X.Q. and Zhang, Y.Z. (1999) Evolution of radiative sand ridge field of the South Yellow Sea and its sedimentary characteristics. *Science in China (Series D)*, **42**, 97–112.

Yang, C.S. (1989) Active, moribund and buried tidal sand ridges in the East China Sea and the southern Yellow Sea. *Mar. Geol.*, **88**, 97–116.

Yang, Y.Z. (2010) Numerical study on suspended sediment flux on Radial Tidal Sandy Ridge System, South Yellow Sea, China. The doctoral thesis of Hehai University.

Zhang, C.K. (2013) *Jiangsu coastal and offshore comprehensive investigation and evaluation report.* Science Press, Beijing, 372–385 (in Chinese).

Zhang, G.D., Zhu, J.C., Wang, Y.Y. and Wang, H.Z. (1984) Study on modern tidal creeks and channels on north Jiangsu tidal flat. *Acta Oceanol. Sinica*, **6**, 223–234 (in Chinese).

Zhang, R.S. and Chen, C.J. (1992) Evolution of Jiangsu offshore banks (radial offshore tidal sands) and probability of Tiaozini sands merged into mainland. China Ocean Press, Beijing, 124pp (in Chinese with an English abstract).

Zhang, R.S. and Wang, X.Y. (1991) Tidal creek system on tidal mud flat of Jiangsu province. *Acta Geogr. Sinica*, **46**, 195–206 (in Chinese with an English abstract).

Zheng, X.M. (1999) Aeolian deposition and environment in Changjiang Delta and extending sea areas. East China Normal University Press, Shanghai, 174 pp (in Chinese with an English abstract).

Zhu, Y. and Chang, R. (2000) Preliminary study of the dynamic origin of the distribution pattern of bottom sediments on the continental shelves of the Bohai Sea, Yellow Sea and East China Sea. *Estuar. Coast. Shelf. Sci.*, **51**, 663–680.

Zhu, Y.R. (1998) Sediment dynamics study on the origin of the radial sand ridges in the southern Yellow Sea. PhD thesis, Department of Marine Geology and Geophysics, Tongji University, No.98002.

Zhu, Y.R. (2002) A numerical simulation study on evolution process of the tide and tidal current on the continental shelves of the Bohai Sea, Yellow Sea and East China Sea since the last deglaciation. *Journal of Ocean University of Qigngdao*, **32**, 279–286 (in Chinese with an English abstract).

Facies, architecture and stratigraphic occurrence of headland-attached tidal sand ridges in the Roda Formation, Northern Spain

KAIN J. MICHAUD* and ROBERT W. DALRYMPLE[†]

* Petrel Robertson Consulting Ltd., Suite 500, 736 – 8th Avenue, S.W. Calgary, AB, T2P 1H4, Canada
[†] Department of Geological Sciences and Geological Engineering, Queen's University,
Kingston, ON, K7L 3N6, Canada

ABSTRACT

The Eocene Roda Formation in Northern Spain is composed of two 3rd-order, regressive-transgressive cycles that are in turn composed of at least 18 stacked progradational deltaic parasequences. Within this paper, trangressive tidal ridges that overlie and locally incise into six of these parasequences are described and documented, three in the lower, Roda member, and three in the upper, Esdolomada member. Grain size, bedding-plane geometry, sedimentary structures, bioturbation and fossils were documented and these observations lead to the division of the ridge deposits into five facies. Ridges can be either erosively or gradationally based, with the nature of the basal contact changing spatially within a single ridge, with a gradational base being most common along the offshore flank and down-current (north-westerly) tip of the ridges. Gradationally based parts of ridges in the Roda member pass upward from bioturbated Facies 5 to rippled Facies 4 and then into one of the cross-bedded facies. Ridge growth appears to have involved an evolution from deposition on their landward 'lee' side to accretion on their seaward 'stoss' side, relative to the locally dominant westerly directed 'ebb' current. Isopach maps, palaeocurrent data and stratigraphic position indicate that the tidal sands in the Roda Formation are headland-attached ridges that formed from the transgressive reworking of deltaic protuberances. They are elongated approximately parallel to the palaeoshoreline, which is coincident with basin-wide NW-SE oriented tidal currents. Ridges are not present throughout the Roda Formation, occurring only at the tip of those parasequences that prograded at least 2 km into the basin. This indicates that the deltaic headland to which the ridges were attached had to grow to a certain size in order to cause sufficient local acceleration of the tidal currents.

Keywords: Transgressive, headland, ridge, bar, Roda, Esdolomada, Spain.

INTRODUCTION

In modern shallow-marine settings, tidal currents are often an effective agent of sea floor reworking. That effect is enhanced where the shoreline is transgressing, as the shelf receives little new sediment because of up-system trapping in estuaries. Modern tidal ridges produced during transgressive reworking are being documented in increasing detail as geophysical techniques improve. A wide array of studies around the British Isles and the East China and Yellow seas have investigated the large-scale architecture, grain-size trends, surface bedforms and flow around such ridges. Specific examples include: the Shambles Bank, southern U.K. (Berthot & Pattiaratchi, 2006); Kwinte Bank and Middelkerke banks, southern North Sea (Berné et al., 1994; Mathys, 2009); south-west Florida Inner Shelf (Davis et al., 1993); Milne Bank, Prince Edward Island, Canada (Shaw et al., 2008); Helwick Sandbank, Southern Wales (Schmitt et al., 2007); Kaiser Bank, Celtic Sea (Reynaud et al., 1999); Longe de Boyard Sandbank, French Atlantic coast (Chaumillon et al., 2008); Snouw and Braek banks,

Contributions to Modern and Ancient Tidal Sedimentology: Proceedings of the Tidalites 2012 Conference,
First Edition. Edited by Bernadette Tessier and Jean-Yves Reynaud.

Dunkerque area, Northern France (Tessier *et al.*, 1999); Dutch Coast (Van de Meene, 1996); East Bank, North Sea (Davis & Balson, 1992); Norfolk Banks, North Sea (Horillo-Caraballo & Reeve, 2008); and ridges offshore from the Changjiang delta, East China Sea (Yang, 1989). Despite this extensive literature, the detailed nature of the facies comprising such banks remains poorly constrained because of the difficulty in sampling using cores and grab samples.

Facies models are typically built around high quality, modern examples. The facies models that are used to identify ancient tidal ridges are based largely on early work on the tidal shelf seas around Great Britain (summarized in Reynaud & Dalrymple, 2012). These models predict that the deposits of tidal ridges should typically be sandy and composed of the deposits of simple and compound dunes that accumulated on one or both sides of the ridge, generating lateral-accretion architecture (Dalrymple, 2010). However, numerous recent studies report a wider range of sedimentary structures, architectures and flow patterns that indicate the deposits within tidal ridges can be variable and complex (Berné *et al.*, 1998; Reynaud *et al.*, 1999; Jin & Chough, 2002; Schmitt *et al.*, 2007; Shaw *et al.*, 2008). Due to the difficulty in studying these deposits in the modern environment, ancient examples must also be documented to strengthen our understanding of tidal-ridge deposits.

To address the need for additional ancient examples, detailed documentation of six headland-associated tidal sand ridges within the Eocene Roda Formation is presented here. The Roda Formation has been the subject of numerous sedimentological, palynological and structural-geology studies (Nio, 1976; Nio & Yang, 1991; Puigdefàbregas, 1992; Vincent, 2001; Torricelli, 2006; Leren *et al.*, 2010; Olariu *et al.*, 2012). Much of this previous work has focussed on the steep-fronted, coarse-grained deltas of the lower, Roda member that are well exposed along the sides of the Isábena River valley (Joseph *et al.*, 1993). By comparison, relatively little detailed work has been done on the tidal deposits that are variably interpreted to overlie or inter-finger with the deltaic deposits (Nio, 1976; Lopez-Blanco *et al.*, 2003; Tinterri *et al.*, 2007; Martinius, 2012; Olariu *et al.*, 2012).

The Roda Formation is composed of two 3rd-order, regressive-transgressive cycles that are in turn composed of at least 18 stacked, high-frequency deltaic parasequences. Here, we document the presence of tidal ridges immediately overlying six of these parasequences, three in the lower Roda member and three in the overlying Esdolomada member. By analysing all of the tidal ridges, our aim is to describe: 1) the range of preserved facies within such ridges; 2) their internal architecture; and 3) their stratigraphic occurrence, in order to understand better the general setting in which they formed. As several ridges are documented and analysed, it is also possible to evaluate the degree and causes of variation between ridges.

GEOLOGIC SETTING

The deposits that are the subject of this study crop out in an erosional window along the northern flank of the Tremp-Graus Basin (Fig. 1), which formed during early subsidence of the south Pyrenean foreland basin (Burbank *et al.*, 1992). The Tremp-Graus Basin originated as a thrust-top or piggyback basin situated between two southward directed thrusts, the Boixols thrust to the north and the Montsec thrust to the south. The basin was asymmetrical with a steeper northern margin, from which most of the sediment infill was sourced, and a gentler southern flank (Martinius, 2012). Deposition within the basin documents some of the earliest stages of foreland-basin sedimentation in the Pyrenees.

The Tremp-Graus Basin contains a thick succession of Late Cretaceous to Oligocene deposits (Fig. 2). Initial subsidence during the Cretaceous created accommodation that was filled with strata deposited in fluvial and lacustrine environments of the Tremp Formation (Nio, 1976). As the rate of base-level rise increased into the Palaeocene and Eocene, the basin was flooded, creating a shallow warm sea approximately 25 km in width (narrowing to the east) and 90 to 100 km in length (Plaziat, 1981; Verges & Burbank, 1996). Past studies focusing on the palaeo-fauna reveal that the water depth in the Roda area reached approximately 80 m, but was typically only 10 to 30 m, with possible water temperatures near 20 °C (Martinius, 1995; Molenaar & Martinius, 1996; Torricelli *et al.*, 2006). Clear, shallow, warm marine water allowed shallow-marine organisms to thrive, resulting in thick calcareous deposits. Early limestones (the Alveolina and La Puebla limestones; Fig. 2) are composed of red algae-coral reefs in

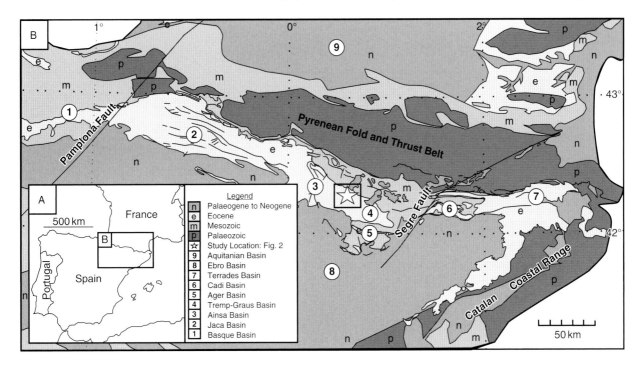

Fig. 1. (A) General location map and (B) more detailed geologic map showing the main sedimentary basins formed in the southern foreland area of the Pyrenean Fold and Thrust Belt. The deposits under investigation, marked by a star in the inset box, were deposited in the Eocene Tremp-Graus Basin (4) which was closed to the east as a result of uplift associated with the transverse Segre Fault and opened westward into deeper water in the Basque Basin (1) and the proto-Atlantic Ocean.

the north, with carbonate ramp facies to the south. They inter-finger with basinal marls of the Serraduy Formation that accumulated in deeper water (but <60 to 80 m deep) to the south and west (Martinius, 1995; Torricelli *et al.*, 2006).

Periodic coarse clastic deposition occurred along the northern margin of the basin as rivers transported sediment from rising highlands in the north-east that are associated with the active Boixols thrust. The first influx of coastal sands in the area began during the Roda Formation, which accumulated over an interval of approximately 1 My, from ~53 to 52 Ma based on biostratigraphic data (Torricelli, 2006) or from ~53.4 to 52.4 Ma based on magnetopolarity data (Lopez-Blanco *et al.*, 2003) (Fig. 2). The Roda Formation is composed of a series of 5 to 25 m-thick progradational sandstone tongues that are commonly capped with fossil-rich mudstones, carbonate units and bioclastic lags (Fig. 3). These sandstone-mudstone cycles are themselves stacked in two larger-scale (150 m-thick) progradational-retrogradational wedges that record two longer-term (i.e. 3rd-order) progradation-transgression cycles (Fig. 3; Sequences 1 and 2 *sensu* Michaud (2011) following the approach

advocated by Galloway (1989)), which are separated by a thick shale succession in which the El Villar Limestone is interpreted to represent the maximum flooding surface. The progradational portion of the lower wedge is classically defined as the Roda member, whereas the upper wedge forms the bulk of the Esdolomada member of the Roda Formation. The stacking patterns of the higher-order sandstone-mudstone alternations within the longer-term cycles allow these sequences to be subdivided into systems tracts (Fig. 3), with the maximum regressive intervals corresponding to the lowstand systems tracts. Both members are well-exposed along the sides of the Isábena River valley (Figs 4 and 5) and are locally exposed in smaller gullies cutting into the valley wall, thus allowing for good 3D coverage.

The Roda Formation is overlain by the 10 to 30 m-thick Morillo Limestone, which is in turn overlain by sandstones and conglomerates of the San Esteban Formation. In the study area the San Esteban Formation is composed of amalgamated channel sandstones and conglomerates of fluvial origin; equivalent deposits to the south have been interpreted as deltaic (Van Eden, 1970).

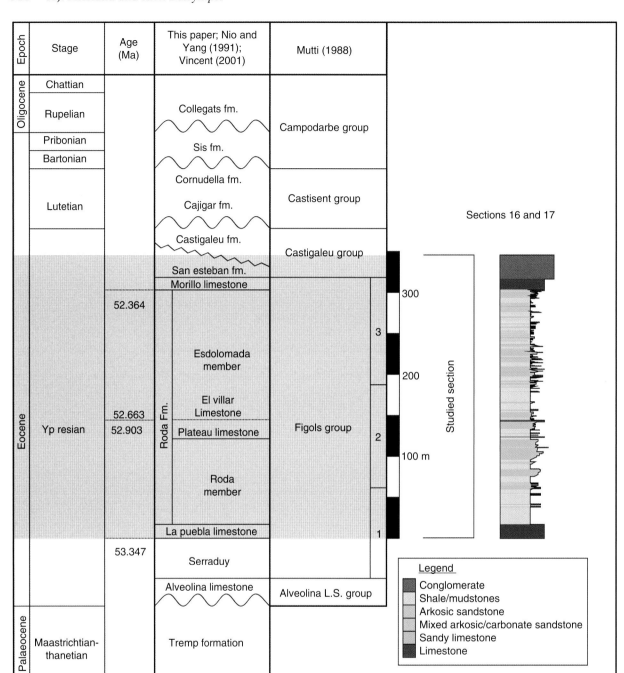

Fig. 2. Stratigraphic terminology and representative measured section (see location of measured section in Fig. 5) through the study interval. The generalized terminology of Mutti (1988) works well at a basin-wide scale, whereas the more detailed subdivisions proposed by Nio & Yang (1991) match the depositional succession in the study area best and are used here. Ages from Lopez-Blanco *et al.* (2003).

METHODOLOGY

Outcrop investigation of the Eocene Roda Formation in the Isábena River valley, which is located in the western part of the Tremp-Graus Basin (Figs 4 and 5), took place in two field seasons totalling over two months. Most previous studies in the area have focused largely on the lower Roda member. This study examined both the lower (Roda) and upper (Esdolomada) members, throughout the study area. Twenty-one stratigraphic sections totalling more than 3 kilometres

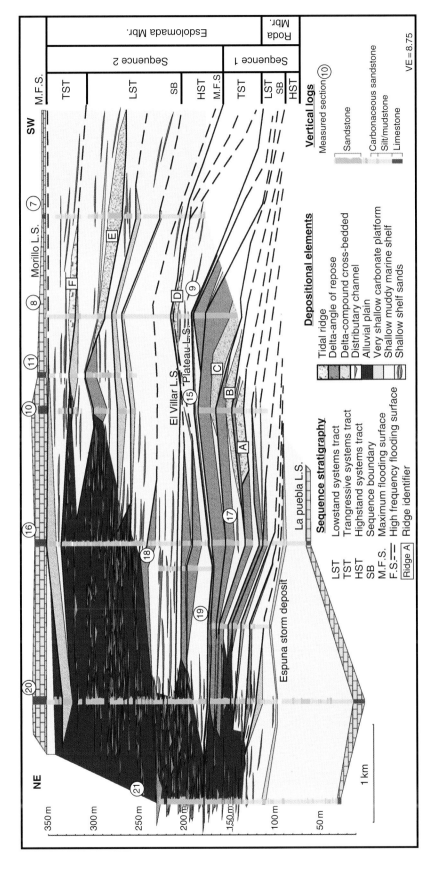

Fig. 3. Dip-oriented cross-section of the Roda Formation showing the location of Ridges A to F. Circle labels indicate the measured sections (see Fig. 5 for location). Square labels indicate the tidal ridges. Sequence-stratigraphic subdivision of the succession is shown in the right-hand column; see Michaud (2011) for additional discussion. Flooding surfaces (solid blue where certain; dashed where approximate) highlight parasequence tops and the end of local deltaic progradation. The tidal ridges overlie these flooding surfaces and formed during high-frequency transgressions.

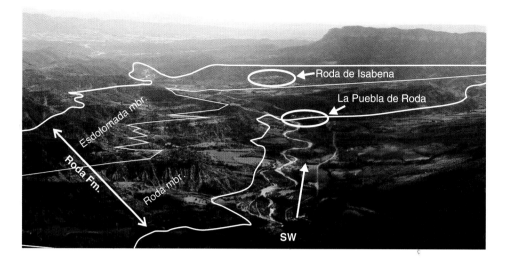

Fig. 4. View looking south-west along the Isábena River valley, showing the outcrop belt of the Roda Formation. Local towns are circled and identified. Photograph is taken from above the town of Serraduy, looking in the general direction of deltaic progradation. See Fig. 5 for town locations and scale.

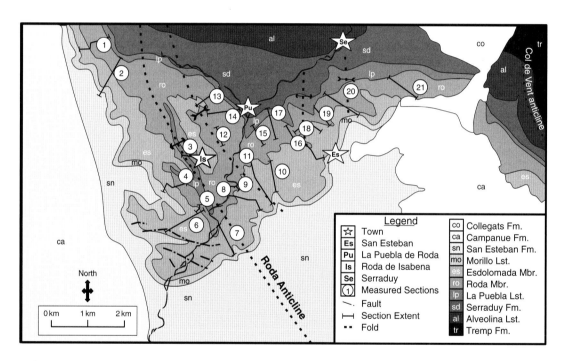

Fig. 5. Geologic map of the study area in the Isábena River valley (refined from Tinterri (2007) using high resolution satellite images; folds indicated as in Lopez-Blanco *et al.* (2003)). See regional location in Fig. 1 and panoramic photo in Fig. 4. Locations of measured sections are marked 1 to 21 and are indicated on Fig. 3. The strata under investigation are the Roda (ro) and Esdolomada (es) members of the Roda Formation.

in length were logged at a centimetre scale. The vertical sections have horizontal separations ranging from less than 100 metres to 2 kilometres. The area between sections was also examined, photographed and mapped. Special attention was given to tracing bounding surfaces in an effort to correlate the vertical logs with confidence and to understand horizontal stratal transitions. The data

acquired include high-resolution photographs of outcrops and detailed observations of sedimentary structures, palaeocurrent orientations, thickness of units, bioturbation diversity and intensity (*sensu* Bann *et al.*, 2004), fossilized organisms (cf. Martinius, 1995), grain-size trends, sedimentary structures and the attitude of bedding planes that define the architecture of sediment packages.

STRATIGRAPHIC ORGANIZATION

Previous studies have shown that the Roda Formation is comprised of repeated deltaic successions that prograded toward the south-south-west. Detailed mapping of the Roda Formation documents the presence of at least 18 progradational cycles within the study area, each of which is composed of channelized sands and conglomerates in the north-east that pass basinward into extensive sheets of coarse-grained to medium-grained deltaic sand that ultimately pinch out into silty mudstones as grain size decreases into the basin (Fig. 3). At the distal end of several of these progradational tongues, well sorted sandstones form 10 to 25 m-thick lenses that are focus of this study. Before documenting these sandstone lenses in more detail, we summarize the characteristics of the deposits with which they are associated.

The channelized sands and conglomerates in the north-east consist of upward fining, 1 to 9 m-thick discontinuous sandstone lenses containing common pebbles and boulders (Fig. 6). The sandstones contain cross beds, peaty horizons, common plant and wood fossils, rhizolith traces and rare shelly debris. Burrows include *Ophiomorpha, Thalassinoides, Planolites, Gastrochaenolites* and *Teredolites*, but the bioturbation intensity (*sensu* Bann *et al.*, 2004) is commonly 0 to 2 (unburrowed to weakly burrowed) and only locally reaches 5 to 6. These sandstones are interbedded with poorly exposed dark purple to brown, planar-bedded mudstone. Plant debris, peat and coal laminations are common and rhizoliths are present but rare. The sandstones are interpreted as brackish to fully marine channel-bar deposits, whereas the mudstones appear to be marginal marine-overbank deposits.

The progradational, deltaic sandstone sheets are 5 to 20 m-thick and are composed of gradationally based coarsening-upward arkosic sandstone. These sandstone sheets contain inclined stratification (Fig. 7) that dips in a south-westerly (i.e. seaward) direction, with the sandstones passing gradationally outward and downward into fine-grained mudstones with intercalated carbonate horizons (Fig. 8A). These inclined surfaces display two different bedding styles: i) simple 5 to 20 m-thick, near-angle-of-repose cross beds (Gilbert-style bedding; Fig. 7A and B) and ii) low-angle inclined cross-bedded sandstone in which 10-30 cm-thick cross-bed sets migrate obliquely down erosive bed-set boundaries (master-bedding surfaces) that dip at 5 to 15° forming large-scale compound cross beds (Fig. 7C). Gilbert-style bedding is more common within the coarser-grained sandstones that contain gravel and cobbles, whereas compound cross-bedding is typically found in lower coarse to medium-grained sandstones. Trace fossils are rare except on bounding surfaces, with *Ophiomorpha* and *Thalassinoides* being the only recognizable traces. Bioturbation intensity is typically 0 to 2. These sandstones are interpreted as brackish delta mouth-bar and delta-front deposits (Lopez-Blanco *et al.*, 2003; Tinterri, 2007).

These deltaic sandstones are typically capped by a prominent surface that is intensely burrowed (bioturbation intensity 3 to 6), which is overlain by a 1 m-thick lag that contains a variable mixture of arkosic sandstone and carbonate fossil fragments, with the carbonate content ranging between 10 to 90%. Carbonate grains include foraminifera, echinoderms, gastropods and bivalves with rare hermatypic corals (Fig. 8B to E). The arkosic sandstone component is typically coarse grained. Bioturbation intensity is typically 5 to 6 with *Thalassinoides, Rosselia, Ophiomorpha* and *Scolicia* as the dominant trace fossils. The higher diversity and ichnospecies composition of this trace-fossil assemblage indicate that sedimentation took place in an unstressed shallow-marine environment. The abrupt increase in carbonate content and bioturbation on the delta top indicates a decrease in sedimentation rate; such deposits are thus interpreted as representing flooding surfaces caused by at least local transgression.

These bioclastic lags are typically overlain by fossiliferous, planar-laminated mudstones that contain gastropods and large benthic foraminifera, with common bryozoa and trace amounts of crab debris, stingray barbs, echinoderms, nautiloids, corals and bivalves. This faunal composition indicates that these shales accumulated in a fully marine environment (Martinius & Molenaar, 1991; Martinius, 1995; Molenaar & Martinius, 1996), which reinforces the interpretation that the deltaic sandstones are capped by a flooding surface.

Five of the progradational cycles are overlain and locally incised by 10 to 25 m-thick, well sorted sandstone lenses (Fig. 3: sandbodies A through E); a 6th sandstone lens is fully encased in shale (Fig. 3: sandbody F). These isolated sandstone bodies are composed of well sorted fine to medium-grained, cross-bedded and ripple-laminated sandstone, with little or no mud, either as discrete beds or laminae, or admixed into the sand by burrowing. This indicates that the water was clean, which

Fig. 6. Channel deposits in the proximal part of the Roda Formation. (A) Sandy point bar, draped by a muddy abandoned-channel-fill deposit that contains several isolated sandstone beds (white outlines at right). (B) Aggradational, coarse-grained channel-fill deposit (white outline) with numerous internal lags, that interfingers with muddy overbank deposits at left. Scale bar is 150 cm long and is marked in 10 cm intervals.

in turn implies that the river supply of sediment was much less than that which occurred during progradation of the deltaic sandstone tongues. The deposits contain a diverse suite of burrows, with bioturbation intensities ranging from 0 to 5. The burrow types (see below) and high diversity indicate that these sandstones are marine in origin with little or no stress due to reduced salinity, which implies that river input had decreased significantly. These features, combined with their stratigraphic position overlying progradational deltaic sandbodies, demonstrate that they are transgressive in origin. Whether or not these transgressions are local or regional cannot be determined within the confines of the study area. Their lenticular geometry suggests that they are sand ridges, an interpretation that will be discussed further below.

TRANSGRESSIVE RIDGE FACIES

Within the Roda Formation, six progradational cycles are overlain by elongate, lenticular sandbodies, herein interpreted as transgressive tidal sand ridges. There are three in the Roda member (with a possible fourth to the south-west, where the member disappears beneath younger strata) and an additional three in the Esdolomada member (Fig. 3). The facies within the ridges are differentiated by grain size, bedding-plane geometry, sedimentary structures, bioturbation and body fossils (Table 1). Within the cross-bedded deposits that comprise the body of most of the ridges, the relative dip directions of the individual cross beds and the master bedding planes between them (i.e. the architectural element present) were used as the primary distinguishing criteria.

Fig. 7. Delta mouth-bar and delta-front deposits. (A and B) Steep-fronted, Gilbert-type deltas. (C) Lower-angle delta-front foresets that are internally cross-bedded, with a compound cross-bedded, forward-accretion architecture. The small step at the base of the photo represents the local horizontal; inclined master bedding is clearly visible. The internal cross beds cannot be seen from this distance. Figure circled for scale in (A) and (B); scale bar (circled in C) is 150 cm long and is marked in 10 cm intervals.

Facies 1

Description: This facies consists of 100 to 500 cm-thick, high-angle cross-beds composed of fine to lower coarse, well-sorted, arkosic sandstone (Table 1). Smaller sets can stack to form packages up to 500 cm thick. Foreset beds within the large cross-beds are 2 to 20 cm thick and dip at up to 30° (Fig. 9A). Cyclic thickening and thinning of fore-set beds suggest the presence of crude tidal bundles, but regular thickness variations are absent.

Low-angle erosion surfaces incise the cross beds locally. Bedset boundaries are typically horizontal but some are inclined, although without a preferred dip direction, and show incision into underlying cross-beds; overall, the bed-sets display vertical aggradation. Some mudstone drapes are present and are concentrated in the toesets and bottomsets of the cross-beds but mudstone layers are thin (less than a few mm) and sporadic in their occurrence.

Fig. 8. (A) Photograph in the distal, south-west part of the Roda Formation (measured section 4; see location in Fig. 5) illustrating cyclic alternation of carbonate grainstones (resistant) and shelf muds (recessive). Marine deposition is clearly illustrated by the presence of fossilized crabs (B), corals, echinoderms, large benthic foraminifera, stingray barbs, bryozoa and gastropods (B to E).

Table 1. Facies within the tidal ridges of the Roda Formation.

Facies	Description	Ichnology	Interpretation
1	Well sorted fine to lower coarse arkosic arenite composing 1 to 5 m-thick cross-beds forming packages up to 5 m-thick. Contains crude tidal bundles and rare mudstone drapes.	Bioturbation intensity ranges from 1 to 5 but is lower (2) in the foresets and higher in the bottomsets (3 to 5). Foresets contain fugichnia, *Rosselia*, *Cylindrichnus*, *Asterosoma*, *Planolites*, *Thalassinoides*, *Ophiomorpha*, *Teichichnus* and cryptic bioturbation. Bottomsets commonly contain the preceding burrows as well as less commonly found *Chondrites*, *Skolithos*, *Rusophycus*, *Arenicolites*, *Scolicia*, *Bergaueria*, *Protichnites* and *Macaronichnus*.	Lee-side deposits of large to very large dunes or the steep side of an elongate tidal bar with strongly asymmetric currents. *Cruziana* Ichnofacies indicates normal-marine conditions and a rapid sedimentation rate decreasing downward, away from the bedform lee face.
2	Well sorted upper fine to medium and rarely lower coarse angular arkosic sandstone composing 10 to 300 cm-thick cross-bed sets dipping mostly up (but at a slight angle to) master-bedding surfaces. Contains cyclic tidal-bundle sets, double mud drapes and cyclic mudstone/sandstone alternations. Total thickness of this facies reaches 20 m.	Bioturbation intensity of 1 to 5 with lower values in foresets (1 to 3). Contains abundant *Rosselia*, *Asterosoma*, *Thalassinoides*, *Planolites*, *Palaeophycus*, *Polykladichnus*, *Rusophycus*, *Cruziana*, *Ophiomorpha*, *Teichichnus*, *Asteriacites*, *Chondrites* and cryptic bioturbation.	Formed by dunes migrating up the stoss side of compound dunes or tidal ridges. Bedform migration was slower than in Facies 1. Currents were tidal in origin and the *Cruziana* Ichnofacies indicates normal-marine conditions. Dune height indicates a potential water depth of 12 to 18 m.
3	Well sorted upper fine to medium-grained angular sandstone forming cross-bed sets dipping in the same direction as master-bedding surfaces. Contains herring-bone cross-stratification, tidal bundles, mud laminations and grain-size striping.	Bioturbation intensity ranges from 0 to 5 with no difference between foresets and bottomsets. Contains *Rosselia*, *Arenicolites*, *Teichichnus*, *Scolicia*, *Planolites*, *Skolithos*, *Cruziana*, *Ophiomorpha*, *Thalassinoides*, *Psammichnites* and cryptic bioturbation.	Formed by dunes migrating down the lee face of a compound dune or tidal ridge. Currents were tidal in origin. Higher stress than Facies 1 and 2 possibly related to higher suspended-sediment concentration and/or lower salinity due to minor river influence.
4	Poorly sorted, angular, fine to medium-grained sandstone forming 1 to 3 cm-thick cross-laminated sets stacking to form packages up to several metres in thickness. Interbedded with Facies 5.	Bioturbation intensity ranges from 3 to 6. Contains *Thalassinoides*, *Planolites*, *Phycosiphon*, *Rusophycus*, *Asteroides*, *Chondrites*, *Teichichnus*, *Skolithos*, *Psammichnites*, *Palaeophycus* and cryptic bioturbation.	Formed by current ripples migrating in the troughs of compound dunes or tidal ridges. Current speeds slower than Facies 1 to 3.
5	Well sorted, planar laminated, fine grained grey siltstones and mudstones forming packages from 10 to 300 cm-thick. Contains flame structures, mudstone-sandstone rhythmites, paired mud drapes and bioclastic lags composed of bivalves, gastropods and large benthic foraminifera.	Bioturbation intensity ranges from 2 to 4. Contains orange-red oxidized *Thalassinoides*, *Rosselia* and *Asterosoma*.	Muddy sediment that accumulated in the troughs of dunes and tidal sand ridges. Shell lags within thicker accumulations indicate prolonged periods of weak currents punctuated by periods of accelerated currents to concentrate the shells.

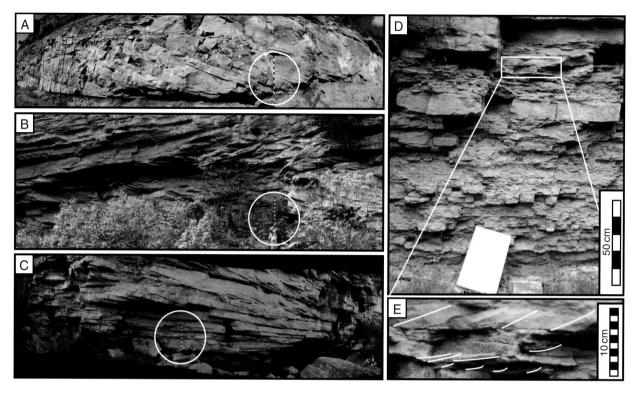

Fig. 9. Sandy ridge facies with 1.5 m stick marked at 10 cm intervals (circled) for scale in A to C. (A) Facies 1 composed of large, angle-of-repose foresets. (B) Facies 2 composed of cross-beds migrating obliquely up gently inclined, erosive set boundaries. (C) Facies 3 composed of cross-beds migrating obliquely down gently inclined, erosive set boundaries. (D) Coarsening-upward packages of ripple cross-lamination (Facies 4) passing upward into increasingly cross-bedded sands (Facies 3). (E) Close-up of Facies 4 sandstones composed of bioturbated current ripples dipping obliquely down shallowly dipping master bedding surfaces.

In the foresets, bioturbation consists predominantly of fugichnia, *Rosselia, Cylindrichnus, Asterosoma, Planolites, Thalassinoides, Ophiomorpha, Teichichnus* and cryptic bioturbation (Fig. 10). Bottomsets commonly contain the preceding list of burrows but also contain the less commonly found trace fossils *Chondrites, Skolithos, Rusophycus, Arenicolites, Scolicia, Bergaueria, Protichnites* and *Macaronichnus*. The bioturbation index ranges from 1 to 5 but is lower (BI 2) within the foresets, with higher bioturbation indices in the toesets and bottomsets of the cross-beds. *Skolithos* and *Rosselia* burrows reach lengths of up to 2 m.

Interpretation: These high-angle cross-beds were formed by avalanching on the lee side of either large to very large dunes or the steep (lee) side of an elongate tidal ridge. Evidence of tidal activity is widespread within the sandstone bodies being described (Nio, 1976; Lopez-Blanco *et al.*, 2003; Tinterri *et al.*, 2007; Martinius, 2012; see also below) and the large cross beds of Facies 1 are, therefore, interpreted to be the result of strongly asymmetric sediment transport by tidal currents. The paucity of clear tidal sedimentary

structures in Facies 1 appears to be a result of bedform size, as Facies 1 bedforms are large and would have migrated very slowly. As a result, the tidal bundles that might be present in Facies 1 would be thin and difficult to recognize. Low-angle reactivation surfaces are sporadically developed and irregularly spaced and may thus have been formed by smaller over-riding dunes.

The diverse suite of trace fossils belonging to the *Cruziana* Ichnofacies indicates the presence of normal-marine conditions, with a rapid sedimentation rate as indicated by the low bioturbation index and angle-of-repose foresets. Two ichnocoenoses are present. The primary suite (consisting of fugichnia, *Rosselia, Cylindrichnus, Asterosoma, Planolites, Thalassinoides, Ophiomorpha, Teichichnus* and cryptic bioturbation, with BI values of 1 to 3) occurs in the foresets (Fig. 10B) and represents behaviours that are adapted to high depositional rates (MacEachern *et al.*, 2005c). The second ichnocoenose (consisting of *Planolites, Chondrites, Skolithos, Rusophycus, Arenicolites, Scolicia, Protichnites* and *Bergaueria*) occurs in the toeset and bottomset strata and

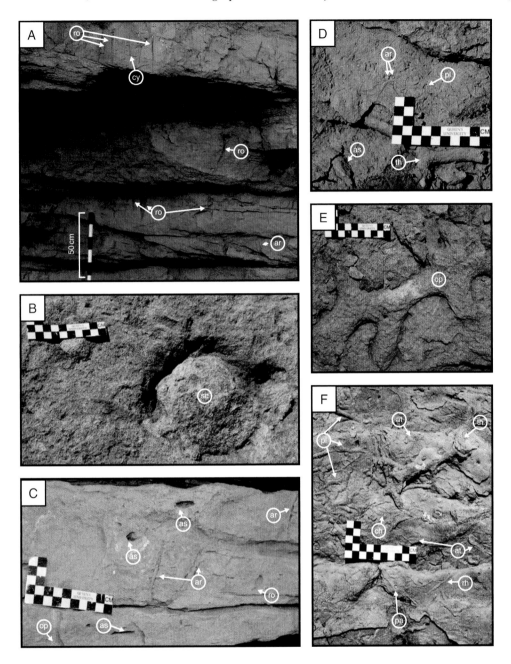

Fig. 10. Trace fossils common to the sandy portion of tidal ridges (Facies 1 to 4). (A and C) *Rosselia* (ro) and *Asterosoma* (as) are very abundant and well preserved. (B) Echinoid traces (*Scolicia* (sc)) are abundant on bedding planes and can reach BI6 in some locations. (C to D) *Asterosoma* (as), *Arenicolites* (ar), *Planolites* (pl) and *Ophiomorpha* (op) are also quite common. (E to F) Large *Ophiomorpha* (op), *Palaeophycus* (pa), *Chondrites* (ch), *Asteroides* (at) and *Planolites* (pl) are also moderately abundant. (A to E) are taken from cross-bedded Facies 1 to 3 whereas (F) is from current-rippled Facies 4.

reflects slower rates of sedimentation as evidenced by a higher level of bioturbation (BI 3 to 5).

Facies 2

Description: This facies consists of upper fine to medium and rarely lower coarse, well sorted angular arkosic sandstone in which 10 to 300 cm-thick

cross-bed sets climb obliquely up gently inclined (1 to 10°) erosive bed-set boundaries that are herein referred to as master-bedding surfaces (Fig. 9B; Table 1). Individual foreset beds in the cross-beds are 1 to 2 cm thick, 50 to 300 cm long in the current-parallel direction and dip at 30 to 35°. The master-bedding surfaces bounding cross-beds (erosional set boundaries) can be traced for over 100 m up-dip. Cyclic variations in bundle thicknesses and in the abundance of mud drapes, many of which are doublets, are present within the cross beds (Fig. 11A). Packages of this facies can reach 20 m in thickness.

The trace-fossil suite includes abundant *Rosselia, Asterosoma, Thalassinoides, Planolites, Palaeophycus, Polykladichnus, Rusophycus, Cruziana, Ophiomorpha, Teichichnus, Asteriacites, Chondrites* and cryptobioturbation (Figs 10 and 11B). *Rusophycus, Thalassinoides* and *Palaeophycus* are dominant on sandy foresets and at the erosive set boundaries between cross-beds, whereas *Rosselia* is the dominant vertical trace, reaching lengths exceeding one metre, cross cutting up to 3 or 4 cross-bed sets. Bioturbation intensity partly obscures bedding along set boundaries and in muddier intervals within the cross beds (BI 1 to 5), whereas sandier foresets are less intensely bioturbated (BI 1 to 3).

Interpretation: This facies was formed by dunes that migrated obliquely up inclined surfaces, forming upstream-accreting architectural elements. The inclined master bedding planes can represent the stoss side of either compound dunes or tidal ridges. Deposition was rapid but the bedforms migrated less rapidly than those that formed Facies 1, as indicated by the higher degree of bioturbation. As in Facies 1, the ichnogenera present within Facies 2 constitute two ichnocoenoses: an archetypal *Cruziana* Ichnofacies that occurs in the toeset and bottomset deposits and an impoverished, high-sedimentation-rate *Cruziana* Ichnofacies that occurs in the foresets. *Asteriacites* (a five armed echinoderm trace) suggests that normal-marine to slightly brackish-water conditions existed (Turner & Meyer, 1980).

Well-developed tidal bundling and cyclic mud draping indicate that tidal currents were responsible for deposition. The bioturbation level is highest in the muddier neap-tide deposits and least in the thicker spring-tide bundles (Fig. 11B). If the Leclair & Bridge (2001) approach to depth reconstruction is used, cross-bed sets are approximately 1/3 of the height of their ancestral dune. Average preserved cross-bed thickness of approximately 2 metres indicates that dunes ~6 m in height formed the deposits. Water depth, in turn, is typically 6 to 10 times the height of subaqueous dunes (Bridge & Tye, 2000; Leclair & Bridge, 2001), which would indicate a possible water depth of 36 to 60 metres. However, this analysis probably over-estimates depths as dune

climbing is apparently occurring; thus, the preserved sets are probably closer to actual dune height. If the thickest cross-bed sets (2 to 3 m) are taken as being nearly the height of the dune that formed them, then water depth would be closer to 12 to 18 m.

Facies 3

Description: Facies 3 is composed of compound cross-bedded upper fine to medium-grained sandstone that is well sorted and angular (Table 1). Cross-bed foresets dip in the same direction as their set boundaries, which themselves have dips of 1 to 10° (Fig. 9C), thus defining forward-accretion, descending architectural elements. Individual foresets range in height from 10 to 100 cm and typically dip at 30°. Dip orientation is generally unidirectional, but occasional reversals exist in the form of sets of ripple cross-lamination that climb up the foresets of the larger cross bed (i.e. they are not counter-flow ripples). Master-bedding surfaces are also periodically interrupted by higher-angle erosive (reactivation) surfaces. Mud laminations are present on foresets and in bottomsets and grain-size striping was observed within some cross beds. Cyclic thickening and thinning of tidal bundles is present. The trace-fossil suite is uniform with no difference between foresets and bottomsets, the common traces being *Rosselia, Arenicolites, Teichichnus, Scolicia, Planolites, Skolithos, Cruziana, Ophiomorpha, Thalassinoides, Psammichnites* and cryptobioturbation (Fig. 10). Bioturbation index ranges from 0 to 5. The total thickness of this facies is 2 to 10 m.

Interpretation: These deposits were formed by simple dunes that migrated down the lee face of a larger compound dune or tidal ridge. Tidal bundles and mud drapes (Fig. 11A and B) indicate that tidal currents were the predominant sediment-transporting agent. Reactivation surfaces are common in tidal environments where the subordinate current causes partial bedform reversal (Klein, 1970).

While the forward-accretion architecture of Facies 3 is similar to that seen in the cross-bedded mouth-bar deposits within some of the progradational sandstone tongues, the grain size is typically markedly finer than that of the mouth-bar deposits. More significantly, the trace-fossil suite within Facies 3 is much more marine in character than those within the delta deposits. The trace-fossil assemblage is similar to Facies 1

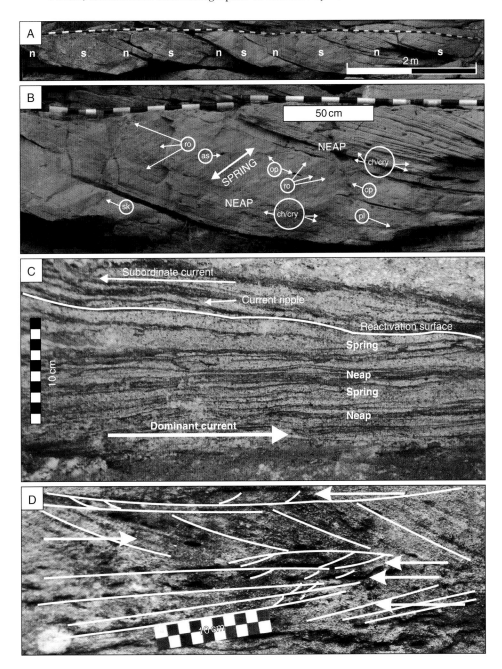

Fig. 11. Tidal features and bioturbation in sandy tidal-ridge deposits. (A) Facies 2 in Ridge C: spring (s) and neap (n) deposition is indicated by cyclic sandstone-mudstone development. (B) Close-up of neap-spring cyclicity. Intense and diverse bioturbation occurs within muddy foresets (neap-tide deposits), whereas less bioturbation occurs within sandier foresets (spring-tide deposits). Trace fossils present are *Rosselia* (ro), *Ophiomorpha* (op), *Asterosoma* (as), *Skolithos*, (sk), *Chondrites* (ch) and crytobioturbation (cry). (C) Facies 3 and 4 in Ridge F. Cyclic mud drapes and reverse current ripples migrating up dune foresets. (D) Herringbone cross-stratification in Ridge F.

and 2 (proximal *Cruziana* Ichnofacies), but is slightly less diverse, indicating the presence of a small stress. The presence of volumetrically more mud than in Facies 1 suggests that this stress might be related to a periodically higher suspended-sediment concentration and/or periodically lower salinity. The single ichnocoenose present in Facies 3 differs from the ichnological character of Facies 1 and 2, which had high-energy foresets and low-energy bottomsets. It appears that Facies 3 had relatively equal stresses on the foresets and bottomsets of the dunes.

Facies 4

Description: Facies 4 is composed of 1 to 3 cm-thick cross-laminated sets of poorly sorted, angular, fine-grained to medium-grained sandstone (Figs 10D and 11D; Table 1) that stack to form cosets that reach several metres in thickness. Dip direction is largely unidirectional, but reversals are present. Facies 4 has a higher mudstone percentage than Facies 1 to 3 and weathers recessively. Facies 4 occurs between occurrences of Facies 1 to 3 and is interbedded with the muddy Facies 5 in the lower part of the ridges and compound dunes.

The bioturbation signature includes *Thalassinoides, Planolites, Phycosiphon, Rusophycus, Asteroides, Chondrites, Teichichnus, Skolithos, Psammichnites, Palaeophycus* and cryptobioturbation (Fig. 10F). The bioturbation index ranges from 3 to 6.

Interpretation: These deposits were formed by current ripples that migrated on and around compound dunes and tidal ridges. The finer grain size than Facies 1 to 3, combined with the presence of ripples instead of dunes, indicates current speeds were slower. The trace-fossil suite is similar to the sandy facies; however, the bioturbation index is typically higher, with bioturbation commonly partially to completely obliterating physical sedimentary structures.

Facies 5

Description: Facies 5 is composed of well sorted, planar-laminated, fine grained grey siltstones and mudstones (Fig. 12; Table 1). The thickness of Facies 5 ranges from 10 cm to about 2 to 3 m. Due to poor exposure, the structures within Facies 5 are commonly difficult to determine. However, centimetre-thick mudstone-sandstone rhythmites, paired mud drapes and flame structures are visible in some heterolithic occurrences. Carbonate lags composed of Arcidae bivalves, various *Turritella* gastropods and large benthic foraminifera are common within the muds and are typically a few centimetres thick (Fig. 12D). Facies 5 typically correlates as the distal equivalent to Facies 2 cross-beds and Facies 4 cross-laminated sands (Fig. 12). Locally, the higher energy Facies 2 erodes into Facies 5. Where the contact is gradational, the mud percentage typically increases outward from the toe of Facies 2 cross-beds from ca. 10% to 100% over a distance of approximately 10 m.

The bioturbation signature includes *Thalassinoides, Rosselia* and *Asterosoma*, all of which are oxidized orange-red in the dark grey-blue shale. Burrows are difficult to discern, as is bioturbation intensity. The degree of continuity of horizontal shelly beds indicates the bioturbation index is most probably within the range of 2 to 4.

Interpretation: These muddy sediments accumulated in the troughs of large simple and compound dunes and adjacent to the tidal sand ridges. Thinner occurrences (2 to 50 cm thick) appear to be associated with the troughs of simple and compound dunes, whereas mudstones several metres in thickness were deposited alongside and/or on top of the larger tidal sand ridges. Shell lags within these thicker ridge-associated mudstones indicate periods of colonization by marine bivalves, punctuated by periods of accelerated currents which eroded the muds and concentrated the shells.

FACIES DISTRIBUTIONS, ISOPACH PATTERNS AND PALAEOCURRENTS

All of the tidal sandstone bodies have a large-scale lenticular geometry (Fig. 3). Isopach maps of these bodies were created by mapping measured thicknesses along outcrop exposures and extrapolating into covered areas using palaeocurrents and master-bedding trends (Fig. 13). Dimensions are best constrained to the north-west; the south-easterly extent is less well known because the deposits extend into the subsurface. Outcrop orientation is typically perpendicular to ridge length, allowing the width of the bodies to be determined with reasonable confidence, but the lengths are less certain. Thus, we estimated total ridge length and south-easterly extent using the typical dimensions and length/width ratios of similar tidal ridges reported by Off (1963; see also Wood, 2003).

Isopach data from the six tidal sandstone bodies show maximum thicknesses of 25 to 30 m and widths of up to 2 km, which implies that lengths are a minimum of 8 to 10 km (Fig. 13). Ridges B, D, E and F are noticeably asymmetric with the steep side facing to the south-west (i.e. offshore). The other ridges are partially covered by younger sediments or vegetation, such that the degree of asymmetry cannot be determined accurately. The crests of all ridges are oriented north-west to south-east, approximately at right angles to the south-westerly progradation direction of the delta lobes and essentially parallel to the average shoreline

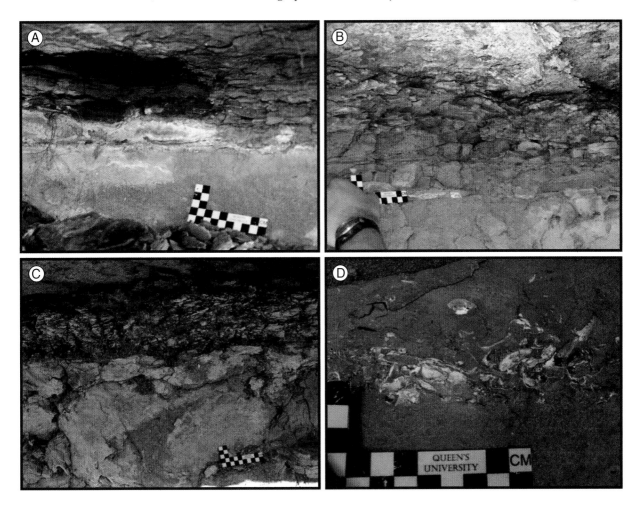

Fig. 12. Muddy Facies 5. Within sandy Facies 1 to 3, muds are thin with sparse bioturbation (A). Moving downward into the underlying mudstones (from A through D), muddy layers between the sandstones thicken (B) and bioturbation intensity increases, as a low-diversity assemblage of oxidized sand-filled tubular burrows becomes abundant. Below sandy Facies 1 to 3, mudstones are thick (C) and carbonate lags composed of large benthic foraminifera, gastropods and shells from the family Arcidae are found locally (D).

trend for the deltaic bodies. Palaeocurrent directions, as measured using dune cross-bedding, are typically nearly parallel to the crest (Fig. 13), although there is considerable scatter, presumably because of the three-dimensional nature of the dunes, the potential for the dunes to be oblique to the flow (cf. Dalrymple & Rhodes, 1995), the veering of currents across the ridge crest and the resulting bending of the dune crests relative to the ridge axis, and the difficulty of measuring foreset dips accurately on surfaces with only a small amount of three-dimensional relief.

Ridges A to C in the Roda member contain equal proportions of Facies 1 to 3, whereas Ridges D to F in the Esdolomada member have sandy

components composed of Facies 2 to 4 and an elevated proportion of the muddy Facies 5 (Fig. 14).

TIDAL RIDGE ARCHITECTURE

In order to illustrate the internal organization of facies within the sandstone bodies, two particularly well-exposed examples (Ridges B and F; Fig. 3) are described in detail. Ridge B contains the same facies as Ridges A and C but is different from Ridges D to F which lack Facies 1 and contain an increased percentage of Facies 4 (Fig. 14). Ridge F is then described in order to document the range of variability in ridge deposits present within the

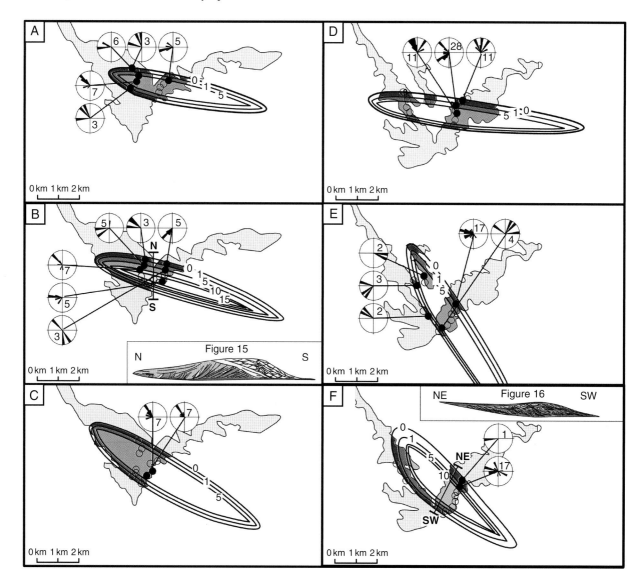

Fig. 13. Isopach maps (thickness in metres) with palaeocurrent rose diagrams (segments give the direction toward which the current moved; numbers indicate number of measurements) for tidal ridges A to F (see Fig. 3 for stratigraphic locations). (A to C) Ridges in the Roda member. (D to F) Ridges in the Esdolomada member. Grey and coloured portions indicate where the member and ridge outcrop, respectively, and circles represent measurement locations. Crests are typically oriented NW-SE with preserved palaeocurrents towards the north-west and south-east, although with considerable scatter. Deltaic deposits which lie to the north-east of these ridges have relatively uniform palaeocurrents towards the south-west, making them distinct from the ridges. Ridge lengths are drawn to fall within typical dimensions of tidal ridges (Off, 1963; Wood, 2003) but the south-easterly extent is unconstrained because the ridges are not exposed there.

Roda Formation. (A detailed description of Ridge D can be found in Olariu *et al.* (2012)).

Ridge B

Three genetically distinct units can be recognized within Ridge B (Fig. 15A). They are well exposed and are differentiated on the basis of facies, palaeocurrent orientation and the dip direction of the

master-bedding surfaces relative to the WNW-ESE ridge-crest orientation (Fig. 13B).

Unit 1

Description: Unit 1 constitutes the nucleus of the ridge and is composed of Facies 1 to 5 that show master-bedding dips to the north-north-east (Fig. 15A, B and C). The base is sharp to the north

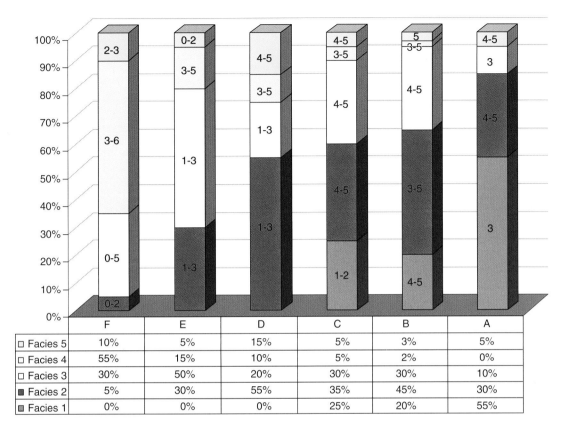

	F	E	D	C	B	A
□ Facies 5	10%	5%	15%	5%	3%	5%
□ Facies 4	55%	15%	10%	5%	2%	0%
□ Facies 3	30%	50%	20%	30%	30%	10%
■ Facies 2	5%	30%	55%	35%	45%	30%
■ Facies 1	0%	0%	0%	25%	20%	55%

Fig. 14. Facies composition of Ridges A to F. Bioturbation index is indicated by numbers 0 to 6 within each column. Facies 1 is only present in Ridges A to C, whereas elevated percentages of Facies 4 and 5 are found in Ridges D to F.

and truncates delta-front sandstones, whereas to the south it is increasingly gradational, with the mudstone percentage decreasing upward over a vertical distance of 2 m. The typical succession in this area is as follows: 0 to 1 m of Facies 5 mudstones that grade upward into 1 to 2 m of Facies 4 rippled sandstone, which is then overlain by 10 to 20 m of Facies 1 to 3 sandstones. Unit 1 thickens northward toward the ridge crest where it reaches 20 m in thickness.

Facies 1 deposits comprise approximately 50% of Unit 1. Large, 3 to 10 m-high, sandy foresets dip to the north, commonly at 30°. Low-angle erosive surfaces (interpreted as reactivation surfaces) are present within Facies 1 and are mantled by Facies 2 or 3 deposits. Facies 3 deposits comprise approximately 30% of Unit 1. Cross-bed sets range in thickness from 10 to 200 cm and migrate down erosive north-dipping surfaces, which are inclined at 1 to 15°. Facies 2 deposits comprise approximately 10% of Unit 1. Small, 10 to 50 cm, cross-beds climb obliquely up steep 10 to 15° master-bedding surfaces. Facies 2 is overlain by Facies 3 and is underlain by either Facies 1 or 3 (Fig. 15A and C).

Interpretation: The NW-SE tidal currents flowed nearly parallel to a WNW-ESE-oriented sand-ridge crest (cf. Fig. 13B). The locally dominant north-west-directed current preferentially flowed along and obliquely up the southern side of the ridge, as indicated by the preferred dip direction of the cross bedding within this unit, whilst the locally subordinate south-east-directed current was directed along and obliquely up the northern flank (Fig. 13B). During the dominant tidal-current phase, sediment was eroded from the gently dipping stoss (southern) side of the ridge and moved over and along the crest as small dunes (Facies 3). As water depth increased to the north of the ridge crest, sediment was deposited as the large cross beds of Facies 1. The subordinate tidal current reworked a thin layer of sand on the northern face of the ridge into south-easterly migrating dunes (Facies 2), which are locally preserved on the steep northern side. Variability in the preservation of Facies 2 on reactivation surfaces may reflect short-term, yearly to decadal, fluctuations in tidal-current speed.

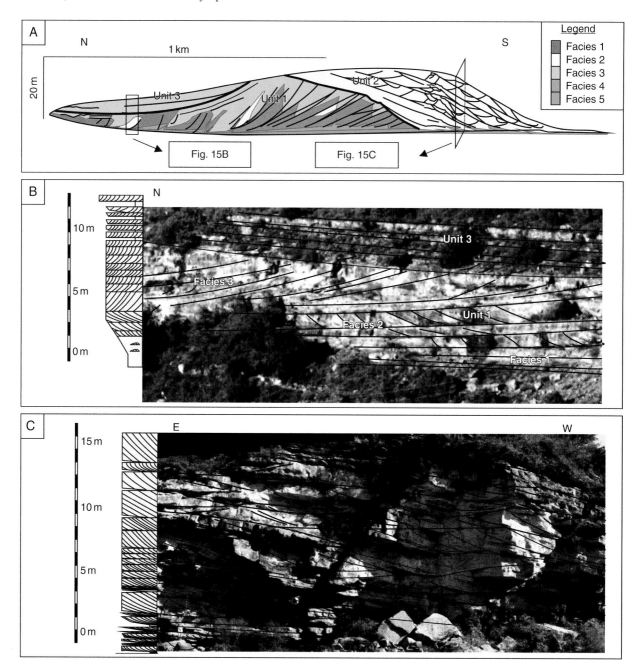

Fig. 15. (A) Oblique cross-section of Ridge B with master-bedding surfaces in grey (see Fig. 3 for stratigraphic location and Fig. 13 for geographic location) showing the distribution of facies and the locations of the photos in (B) and (C). (B and C) Panoramic photos with vertical logs (measured on the left edge of photos) illustrate facies and grain-size trends within the ridge. The photographs and logs are correlated with the facies illustrated in (A). The original, non-annotated version of photos can be found in Michaud (2011).

Unit 2

Description: Unit 2 is found on the southern flank of the ridge and is composed of Facies 2, 3, 4 and 5 with master-bedding planes that dip consistently toward the southwest (Fig. 15). Similar to Unit 1, it thickens toward a WNW-ESE-oriented ridge crest

where it reaches 25 to 30 m in thickness. The base of Unit 2 in the south is marked by 1 to 4 m of mudstone with the content of siltstone and sandstone (reaching upper-medium sand size) increasing upward. Immediately above the first sandstone, bioturbated sandstones are inter-bedded with centimetre-thick mudstones which become thinner and decrease in abundance upward. The internal architecture is dominated by Facies 2 with its upstream-accreting cross-beds. Unlike the rare occurrences of Facies 2 in Unit 1, in which the cross beds were relatively small (10 to 50 cm-thick sets), the sets in Facies 2 in Unit 2 are 2 to 3 m thick and make up at least 90% of the unit. Two orders of master-bedding surfaces are present within Unit 2. Large master-bedding planes dipping 1 to 3° to the south-south-west break the units into 1 to 3 m-thick packages within which smaller-scale master-bedding surfaces dip either to the north-west or the south-east at 1 to 10°. The cross-bedding within these smaller packages dips to the north-west at 30°.

Interpretation: Tidal currents maintained the same orientation as in Unit 1 with the locally dominant current directed toward the north-west. However, instead of sediment avalanching down a north-dipping lee side of the ridge, it accumulated on the obliquely up-current (southern/stoss) side that had been erosional before, during the deposition of Unit 1. The thickness of the cross-beds along the crest of the ridge indicates a possible water depth of 12 to 18 m. (See description and interpretation of Facies 2 above for depth reconstruction). The change in the ridge's accretion direction, from northward in Unit 1 to southward in Unit 2, is attributed to the ridge reaching a height (~25 to 30 m) where the shallow water depth at the ridge crest began to impede cross-ridge tidal flow. This caused deceleration of the dominant current as it flowed obliquely up the southern ridge flank and induced sediment deposition, thereby preserving the up-current accreting architecture of Facies 2 (cf. Reynaud *et al.*, 1999). Compound dunes migrating north-westward along the southern flank of the ridge deposited sediment and the troughs of the compound dunes formed the large-scale master-bedding planes which represent the general dip of the southern flank of the ridge in the same manner as seen in the Kaiser Bank (Reynaud *et al.*, 1999).

Unit 3

Description: Unit 3 is a relatively thin, 2 to 4 m-thick, sandy deposit composed of Facies 3. It overlies the northern flank of Unit 1 (Fig. 15). The base of Unit 3 is sharp, eroding into Unit 2. Mudstone percentage decreases upward as cross-bed size increases. Cross-beds are 50 to 100 cm thick and dip mostly to the north-west, approximately in the same direction as the gently dipping (1 to 4°) master-bedding planes within this unit, forming a forward-accretion architecture. Unit 3 is different from Unit 1 in that it is thinner, the erosive set boundaries are more gently dipping and the unit lacks Facies 1 and 2.

Interpretation: Palaeocurrent directions are the same as in Units 1 and 2, with a NW-SE orientation. Small simple dune-scale foresets dipping down the master-bedding planes indicate a downstream accretion direction that is along the ridge axis and is consistent with the interpretation of a compound dune (Dalrymple & Rhodes, 1995; Dalrymple, 2010) that migrated parallel to the ridge axis.

Summary of Ridge B evolution

Ridge B formed after the south-westward progradation of a river-dominated delta ceased and transgression had begun. The cessation of deltaic progradation coincided with the end of river influence. This is indicated by the increase in the intensity and diversity of bioturbation in the tidal sand ridge, relative to the underlying deltaic unit, as well as an increase in the influence of tidal currents. Tidal currents that impinged on the delta front reworked deltaic sediment into a shore-parallel elongate ridge constituted of compound dunes, simple dunes and current ripples. Bedform stability diagrams typically denote simple dunes as occurring within the 50 to 100 cm s^{-1} range, whereas compound dunes documented in the modern typically occur within the 50 to 200 cm s^{-1} range, the increase in velocity probably relating to the presence of deeper water and the requirement for slightly faster currents to form dunes in that situation (Southard & Boguchwal; 1990; Dalrymple & Rhodes, 1995). Thus, the north-west-directed current probably reached speeds of 50 to 200 cm s^{-1}, as evidenced by the presence of large compound dunes, whereas the south-easterly current reached speeds of only 50 to 100 cm s^{-1} as evidenced by the presence of only simple dunes migrating in this direction. The vast majority of inclined bedding dips to the NW on both the south and north sides of the ridge; that observation, combined with the reconstructed maximum current speeds, indicates that the north-westerly current was dominant over the south-easterly current at the scale of the entire ridge.

Initially the ridge grew by lateral accretion to the north-east as dunes migrated up and over the ridge crest and deposited sand along the northern flank of the ridge. This behaviour is suggested by the standard models for tidal-current ridges such as those in the Norfolk Bank area of the North Sea (Houbolt, 1968; Dalrymple, 2010). However, the ridge eventually reached a height where it began to impede cross-ridge currents. At this point, lateral accretion occurred on the southern side, preserving compound dunes climbing at a low angle up the stoss side of the ridge. A similar pattern of evolution is seen in the Kaiser Bank in the Celtic Sea (Reynaud *et al.*, 1999). Some mud was still being deposited in the trough adjacent to the ridge and the amount of mud within the ridge continued to be less in shallower water where higher current speeds inhibited mud deposition and preservation. While the ridge accreted to the south-west it may have also accreted on the northern margin, as it is not possible in the available outcrop to correlate the temporal overlap of Units 1 and 2 precisely. Unit 3 documents a north-westerly migrating compound dune on the northern flank of the ridge, which may have been coeval with Unit 2.

Ridge F

Description: Ridge F (Fig. 16) has a relatively simple internal architecture compared with Ridge B. Master-bedding surfaces dip toward the south-west (i.e. offshore), while ripple cross-lamination and dune cross-bedding dip mostly toward the WNW with rare dips to the ESE (Fig. 13F). There is a thin layer of Facies 4 on the top of the ridge which has relatively flat master-bedding surfaces. The ridge is composed predominantly of rippled Facies 4 (50%) and cross-bedded Facies 3 (35%), with lesser amounts of muddy Facies 5 (10%) and cross-bedded Facies 2 (5%) (Fig. 14).

Muddy Facies 5 typically occurs at the base of the ridge, which coarsens upward into cross-laminated fine sandstone of Facies 4 and then into 10 to 100 cm-thick cross-bedded sets of Facies 3. Facies 2 is found locally as metre-scale lenses surrounded by Facies 4. Along its north-eastern flank, the ridge erosively overlies deltaic sandstones. Here, the ridge typically lacks significant amounts of muddy Facies 5 at its base. Further to the south, the ridge overlies deeper-water marls and mud is more abundant at the toe of the accretional flank.

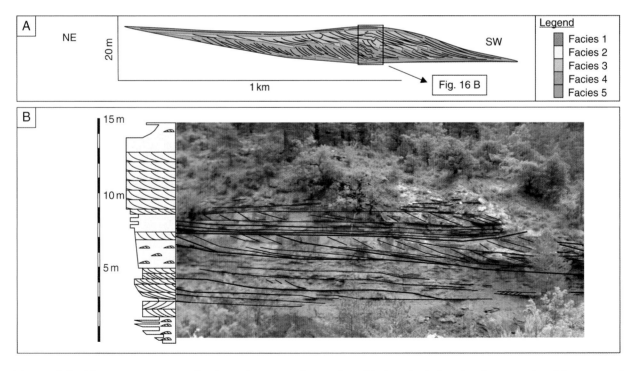

Fig. 16. (A) Oblique cross-section of Ridge F (see Fig. 3 for stratigraphic location) showing the general architecture and distribution of facies. (B) Panoramic photo with vertical log (measured 5 metres to the left of this photo) illustrate the facies and grain-size trend within the ridge.

Interpretation: Ridge F formed after the cessation of south-south-westward progradation of a river-dominated delta. The ridge is interpreted as transgressive because it erodes into underlying progradational deltaic deposits and it contains a fully marine trace-fossil assemblage that is distinct from the highly restricted trace-fossil suite in the underlying deltaic deposits.

The ridge is composed predominantly of ripples and small cross-beds with palaeocurrents indicating locally dominant WNW-directed and subordinate ESE-directed currents. Master-bedding planes indicate that the direction of ridge migration was towards the south-west, forming lateral-accretion deposits on the down-current side relative to the dominant current (at the scale of the ridge) that crossed the ridge crest obliquely.

Current ripples are present in sands ranging from fine to upper medium. Dunes are capable of forming within the coarser end of this grain-size range; thus, grain size was not a limiting factor on the distribution of dunes. Consequently, the tidal-current speed was probably less than $\sim 40\,\mathrm{cm\,s^{-1}}$ (Southard & Boguchwal, 1990) along the flanks of the ridge. The occurrence of dunes topographically higher than the ripples indicates that the currents were generally fastest at the crest of the ridge. Faster currents on the crest of the ridge probably indicate that this ridge had not aggraded to shallow friction-dominated depths (i.e. the ridge crest was not shallow enough to impede cross-ridge flow), hence the development of lateral-accretion bedding on its lee side.

RIDGE ORIGIN

Local flow patterns

At the scale of an individual ridge, irregularities of the coastline or on the sea floor influence the preferential pathway of tidal currents (Pingree, 1978; Williams *et al.*, 2000) and cause flood and ebb currents to become locally unequal in strength and duration. If sufficient sediment is present, tidal ridges can accumulate in the zone between the flood-dominant and ebb-dominant pathways created by these inequalities; thus, time-averaged maximum tidal currents (averaged over many tidal cycles) move in opposite directions on either side of the ridge, causing opposing directions of residual sediment movement on either side of the ridge crest (Pingree & Maddock, 1979; Huthnance, 1982;

McCave & Langhorne, 1982; Dyer & Huntley, 1999). Palaeocurrent observations from the ridges in the Roda Formation indicate that the currents moved in opposite directions on either side of the ridge crest: in the case of Ridge B, the residual sediment-transport direction was to the north-west on the south side and to the south-east on the north side (i.e. the residual sediment circulation was clockwise), whereas the opposite pattern existed for Ridge F, based on the assumption that deposition occurred on the side of the ridge dominated by the subordinate current (cf. Dalrymple, 2010). Such transport patterns match flow processes acting in modern examples, lending strong support for the ridge interpretation of these sandbodies.

Ridge type

Transgressive tidal ridges occur in 3 main settings: in the mouths of estuaries, on the open shelf away from shorelines, or adjacent to headlands (Dyer & Huntley, 1999). Tidal ridges in the Roda Formation are probably headland associated, as suggested by: 1) their intimate association with the progradational limit of deltaic bodies, with the ridges generally lying to the immediate west of the deltaic headland; 2) the sizes are consistent with modern headland analogues (Wood, 2003); 3) the ridges display a coast-parallel orientation; and 4) the character of the bioturbation is consistent with formation in a setting removed from significant river influence.

All documented modern headland-attached ridges are associated with bedrock headlands. However, in the Roda Formation the headland bulges to which the ridges are attached are deltaic in origin. As tidal currents approached the newly formed deltaic headland, they accelerated due to flow constriction. When the river was locally active, sediment was deposited rapidly and relatively continuously, such that tidal currents were unable to significantly rework the delta front. However, when the river was inactive and deltaic progradation had ceased, tidal currents reworked the deltaic sands into sandy ridges (Fig. 17). Tidal ridges are not present at the tip of each progradational tongue in the Roda Formation, which indicates that tidal currents were not always strong enough to erode and deposit significant quantities of sand. In the Roda Formation, tidal ridges only form on deltaic bodies that prograde at least 2 km into the basin (Fig. 3). This indicates that tidal currents were ineffective until the headland had

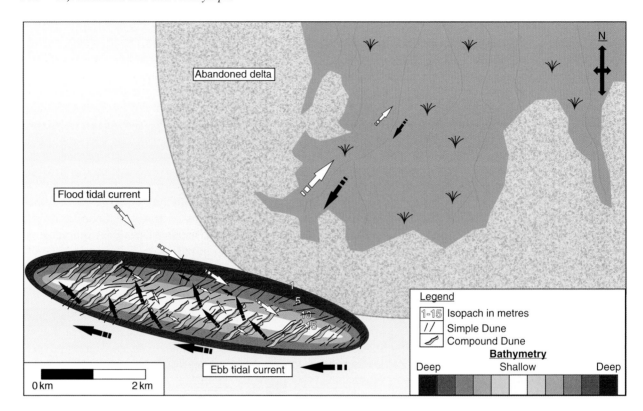

Fig. 17. Interpreted palaeogeographic reconstruction and flow conditions around the tidal ridges under investigation. The ridges are interpreted to represent headland-attached sandbodies that lay slightly to the west of the underlying deltaic headland because of the general dominance of the ebb (westerly flowing) currents.

grown large enough to cause significant local acceleration of the tidal currents. This palaeogeographic control led to the preferential stacking of several tidal sand ridges in a very narrow geographic corridor: all 6 ridges occur within a belt that is only 3 to 4 km wide (Fig. 3), which is small relative to the 25 km width of the entire basin. In the case of the Roda Formation, sufficient protrusion occurred primarily during accumulation of the LST in both sequences (Fig. 3), when the delta was at its most seaward location. Thus, the ridges are clustered both geographically and temporally. This is a result of the fact that there was an approximate balance between sedimentation and the creation of accommodation; in basins where this was not the case, greater geographic separation of subsequent ridges is to be expected.

COMPARISON BETWEEN THE RODA AND ESDOLOMADA RIDGES

An abundance of cross-bedded sandstone in which compound-dune deposits are common, the lack of mudstone and the scarcity of Facies 4

rippled sandstone imply that tidal currents were fast during the deposition of Ridges A to C in the Roda member (Fig. 14). Esdolomada member Ridges D to F, on the other hand, have a higher mudstone percentage, smaller cross-beds (Olariu *et al.*, 2012) and a higher proportion of Facies 4 rippled sandstone (Fig. 14), implying slower currents. Previous workers have attributed some of the current acceleration past the front of the deltas to the presence of the NNW-SSE-oriented, subaqueous Roda Anticline (Lopez-Blanco *et al.*, 2003) which lies immediately to the SW of the delta (Fig. 5). The growth of this anticline is thought to have been most active during Roda member time, allowing it to funnel the currents between the anticlinal crest and the delta front. However, the present authors note that the ridges in the Roda member and the lower part of the Esdolomada member are oriented WNW-ESE (Fig. 13 A to D) with their crest cutting across the trend of the Roda Anticline. By comparison, the ridges in the upper part of the Esdolomada member are oriented NNW-SSE, essentially parallel to the Roda Anticline (Fig. 5). This change in ridge orientation might suggest that the Roda Anticline was most prominent and had its strongest influence on the direction of the tidal currents later in the accumulation of the Roda Formation than suggested by previous workers.

Instead of a local control on the strength of the tidal currents, it is possible that larger-scale changes in basin geometry and bathymetry caused by ongoing tectonic movements and sediment accumulation might have altered the basin's natural tidal period which in turn influenced the degree to which the incoming tidal wave did or did not approach resonance. Thus, it is possible that the basin moved away from resonance during the transition from the Roda member to the Esdolomada member, leading to an overall decrease in the strength of the tidal currents.

Ridges in the Esdolomada member are capped by much thicker and more carbonate-rich flooding-surface deposits than those in the Roda member, implying a longer period of abandonment between episodes of progradation in the Esdolomada member. This is attributed to a more equal balance between sedimentation and the creation of accommodation in the Esdolomada, such that there was more vertical aggradation in more proximal areas and overall shorter distances of progradation (Fig. 3; Michaud, 2011), with longer periods between them, during which time the carbonate caps were developed.

CONTROLS ON THE VARIABILITY OF RIDGE FACIES

Sedimentary structures

As in all environments, the detailed sedimentary structures present within tidal sand ridges are dependent on the interplay of sediment grain size, water depth and current speed and/or wave energy. In the Roda Formation ridges, wave energy was universally low because of the small size of the basin, so the sedimentary structures are almost entirely current formed. This type of ridge would be found preferentially in small low-fetch basins or behind sheltered headlands. In larger basins, the proportion of wave-generated sedimentary structures, including hummocky cross stratification and graded storm beds would increase (cf. Rine *et al.*, 1991; Dalrymple & Hoogendoorn, 1997).

Where there is an abundance of medium-grained to very fine-grained sand and low current speeds (ca. 20 to 50 cm s^{-1}) current ripples will predominate (Fig. 18C and F), as in Ridge F. In other settings, where wave action is more pronounced, such fine sand can also include wave-generated sedimentary structures such as graded storm beds, hummocky cross stratification and wave ripples. Higher current velocities in medium-grained to coarse-grained sand will form simple dunes (Fig. 18A and B), as predominate in Ridge D (Olariu *et al.*, 2012) and/or compound dunes (Fig. 18D and E) as in Ridge B (50 to 200 cm s^{-1}; Southard & Boguchwal, 1990; Dalrymple & Rhodes, 1995). The degree of bioturbation can also vary widely and will be inversely correlated with current speed, the intensity of sediment movement and the length of time between periods of sea bed disturbance. The potential variability of these sedimentary structures within tidal ridges is summarized in Fig. 18.

Flow patterns associated with ridge accretion

Tidal ridges generally experience mutually evasive flow, with opposite directions of residual sediment movement on either side of the crest (Dyer & Huntley, 1999); however, there is always some amount of cross-ridge flow because the ridge is at a slight angle to the general direction of water movement (Huthnance, 1982). The importance of cross-ridge flow and whether the current is accelerated or decelerated at the ridge crest are dependent on the water depth at the crest. If the cross-ridge flow is not significantly impeded by the ridge (i.e. if the ridge crest is relatively deep), the currents will be accelerated as they approach the ridge crest, such that the crest experiences the fastest currents (Huthnance, 1982a; Suter, 2006). In this case, the bedforms on the ridge can pass upward from ripples lower on the flanks to dunes at the crest (Fig. 18D to E; Van De Meene *et al.*, 1996; Berné *et al.*, 1998) and the ridge will consist predominantly of lateral-accretion deposits that accumulated on the 'lee side' of the ridge where the locally dominant current is the regionally weaker one. Conversely, if the ridge crest is relatively shallow and impedes cross-ridge flow, then the currents can be slower at the ridge crest than in the adjacent troughs and the bedform distribution can be inverted relative to the preceding case, with ripples occurring mainly on the ridge crest (Fig. 18C; Schmitt *et al.*, 2007). Such ridges may also be prone to the development of swatchways (Fig. 18A; Dalrymple & Rhodes, 1995). The main deposits in these ridges will consist of lateral-accretion deposits that accumulated on

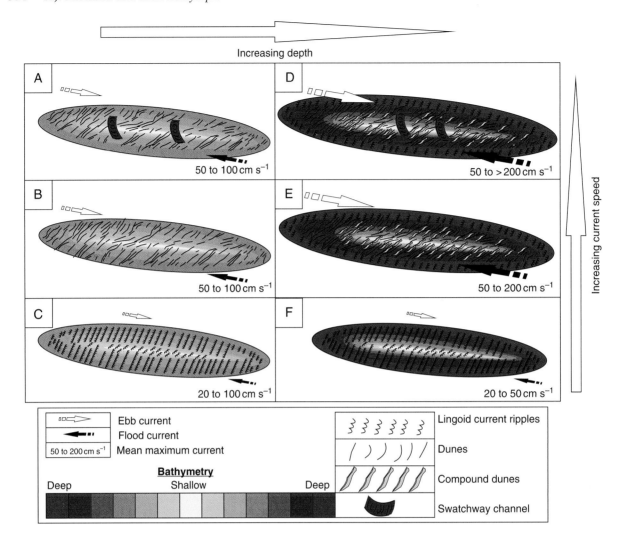

Fig. 18. Schematic diagrams showing the potential variation between tidal ridges due to changes in current velocity and water depth. The A to F sketches in this figure do not refer to ridges A to F in the Roda Formation but to ridge variation stages. This figure is based initially on the examples described from the Roda Formation, supplemented by a survey of the ridges documented in the literature cited in the text and theoretical considerations. Current speeds are displayed as mean maximum current speeds for the dominant and subordinate directions. Currents of 50 to 200 cm s⁻¹ are capable of forming dunes (A, B, D and E), whilst slower currents (20 to 50 cm s⁻¹) typically form ripples (C and F). If current speeds are periodically very high and the water depth over the ridge crest is shallow enough, obliquely cross-ridge swatchway channels can be incised into the top of the ridge. Deep ridges (D and E) have faster currents along the crest, so bedforms decrease in energy and grain size down the flanks, whereas shallow ridges (A and B) have slower current speeds at the crest because of frictional retardation of the cross-ridge flow, leading to the presence of smaller bedforms and finer grain sizes on the ridge crest. Fast currents are more effective at transporting sediment and, consequently, probably generate larger ridges than those formed by weak currents.

the up-current side where the regionally stronger flow predominates. The former situation appears to be most common on continental shelves (Houbolt, 1968; Berné *et al.*, 1994; Suter, 2006), whereas the later situation predominates in the inshore parts of estuaries and deltas and on shallow sand banks (Dalrymple & Rhodes, 1995; Chaumillon *et al.*, 2008; Dalrymple *et al.*, 2012);

although the Kaiser-I-Hind Bank on the outer continental shelf also appears to show stoss-side accretion and perched channels that have been interpreted as swatchways (Reynaud *et al.*, 1999). However, the precise water depth at which this transition occurs is probably a function of the general current speeds, with the transition occurring in deeper water as the general current speed increases (Fig. 18). Ridge B demonstrates that the ratio of water depth to current speed can change over the life of a single ridge, causing an evolution in the preserved architecture, with deposition changing from the lee to the stoss side as the ridge grew (Fig. 15).

Vertical Grain-size Changes

Vertical grain-size profiles through the ridges are variable because the cross-ridge flow can either accelerate or decelerate over the ridge. A coarsening-upward trend accompanied by an upward decrease in the amount of mud should be typical for ridges over which there is flow acceleration and that is what is observed in the Roda ridges (Figs 15C and 16B). This trend occurs because the currents generally accelerated toward the ridge crest: coarse sediment is moved up and onto the crest but cannot be removed because the currents decelerate away from the crest. More intense wave action at the crest, if present, would enhance this trend. These conditions should occur in the initial stages of ridge growth and continue until the ridge is buried or the ridge crest reaches a shallow water depth where friction becomes significant and current speeds at the crest begin to decrease (Huthnance, 1982).

Once the ridge builds high enough, currents can become slower at the crest due to friction, while the current in the surrounding lower areas will be faster. This situation would likely produce an up-current-lateral accretion characterized by a fining-upward trend with a sharp base, similar to what is seen in estuarine and deltaic channels (Olariu & Bhattacharya, 2006; Chaumillon *et al.*, 2008; Dalrymple *et al.*, 2012). Swatchway channels incising into ridge deposits may further complicate grain-size trends. No Roda ridges show this upward-fining pattern.

CONCLUSIONS

Tidal ridges typically accumulate in a zone between flood-dominant and ebb-dominant sediment-transport pathways. Where a headland protrudes into a basin that has sufficiently strong tidal currents, these pathways are formed in a lee-side eddy, thus creating a hospitable circulation pattern for the development of headland-attached tidal ridges. In modern times, these circulation patterns are documented only at rocky headlands but in this study we have shown that a deltaic headland can also alter the current pattern sufficiently to create tidal ridges.

Facies observations, palaeocurrent directions and isopach data, coupled with detailed tracing of stratigraphic contacts, led to the identification and description of six transgressive, headland-attached sand ridges in the Roda Formation. Observations indicate that the ridges formed on the western flank of deltaic headlands, within a small, rapidly subsiding basin. North-west-directed currents, related to basin-scale ebb tidal currents, were locally stronger than south-east-directed currents that are associated with basin-scale flood tidal currents. All ridges migrated obliquely downstream relative to the locally dominant (NW-directed) current, forming lateral-accretion deposits on the 'lee' side. However, some ridges grew large enough late in their evolution that currents were decelerated by bottom friction as they flowed over the ridge crest. At that point, ridges begin to accrete laterally on their upstream side.

In the Roda Formation, tidal ridges are not present at the tip of each progradational deltaic tongue. Instead, they are located only at the distal end of deltas that had prograded at least 2 km into the basin. This is because the tidal currents did not effectively winnow the deltaic headland until it had grown large enough to cause significant local acceleration of the tidal currents. Such long progradation distances occurred only in the lowstand systems tracts of the 3rd-order sequences. As a result, tidal-current ridges in the Roda Formation are preferentially stacked in relatively thin stratigraphic intervals and in a very narrow geographic corridor, thus representing an ideal exploration target for hydrocarbon reservoirs.

ACKNOWLEDGEMENTS

Thanks to Bryce Jablonski for his hard work in the field; Allard Martinius, Ron Steel and Cornel Olariu for an excellent introduction to the study area; and Jose Antonio Naval for his local experience and wisdom. The constructive comments provided by Allard Martinius, Jean-Yves Reynaud and an anonymous reviewer helped to improve the clarity of the presentation. This research was funded by a grant to RWD from the Natural Sciences and Engineering Research Council of Canada (NSERC).

REFERENCES

Bann, K.L., Fielding, C.R., MacEachern, J.A. and **Tye, S.C.** (2004) Differentiation of estuarine and offshore marine deposits using integrated ichnology and sedimentology: Permian Pebbley Beach Formation, Sydney Basin, Australia. In: *The Application of Ichnology to Palaeoenvironmental and Stratigraphic Analysis* (Ed. D. McIlroy). *Geol. Soc. London Spec. Publ.*, **228**, 179–211.

Berné, S., Lericolais, G., Marsset, T., Bourillet, J.F. and **De Batist, M.** (1998) Erosional offshore sand ridges and lowstand shorefaces: examples from tide- and wave-dominated environments of France. *J. Sed. Res.*, **68**, 540–555.

Berné, S., Trentesaux, A., Stolk, A., Missiaen, T. and **De Batist, M.** (1994) Architecture and long term evolution of a tidal sandbank: The Middelkerke Bank (southern North Sea). *Mar. Geol.*, **121**, 57–72.

Berthot, A. and **Pattiaratchi, C.** (2006) Field measurements of the three-dimensional current structure in the vicinity of a headland-associated linear sandbank. *Cont. Shelf Res.*, **26**, 295–317.

Bridge, J.S. and **Tye, R.S.** (2000) Interpreting the dimensions of ancient fluvial channel bars, channels and channel belts from wireline-logs and cores. *AAPG Bull.*, **84**, 1205–1228.

Burbank, D.W., Puigdefàbregas, C. and **Munoz, J.A.** (1992) The chronology of the Eocene tectonic and stratigraphic development of the Eastern Pyrenean Foreland Basin, Northeast Spain. *Geol. Soc. Am. Bull.*, **104**, 1101–1120.

Chaumillon, E., Bertin, X., Falchetto, H., Allard, J., Weber, N., Walker, P., Pouvreau, N. and **Woppelmann, G.** (2008) Multi time-scale evolution of a wide estuary linear sandbank, the Longe de Boyard, on the French Atlantic coast. *Mar. Geol.*, **251**, 209–223.

Dalrymple, R.W. (2010) Tidal depositional systems. In: *Facies Models 4* (Eds **N.P. James** and **R.W. Dalrymple**). St. John's, NF, Geological Association of Canada, 201–231.

Dalrymple, R.W. and **Hoogendoorn, E.L.** (1997) Erosion and deposition on migrating shoreface-attached ridges, Sable Island, Eastern Canada. *Geosci. Can.*, **24**, 25–36.

Dalrymple, R.W., Mackay, D.A., Ichaso, A.A. and **Choi, K.S.** (2012) Processes, morphodynamics and facies of tide-dominated estuaries. In: *Principles of Tidal Sedimentology* (Eds **R.A. Davis**, Jr. and **R.W. Dalrymple**). Dordrecht, Springer, pp. 79–107.

Dalrymple, R.W. and **Rhodes, R.N.** (1995) Estuarine dunes and bars. In: *Geomorphology and Sedimentology of Estuaries* (Ed. G.M.E. Perillo). *Dev. Sedimentol.*, **53**, 359–422.

Davis Jr., R.A. and **Balson, P.S.** (1992) Stratigraphy of a North Sea tidal sand ridge. *J. Sed. Petrol.*, **62**, 116–121.

Davis, R.A., Klay, J. and **Jewell, P.** (1993) Sedimentology and stratigraphy of tidal sand ridges Southwest Florida Inner Shelf. *J. Sed. Petrol.*, **63**, 91–104.

Dyer, K.R. and **Huntley, D.A.** (1999) The origin, classification and modelling of sand banks and ridges. *Cont. Shelf Res.*, **19**, 1285–1330.

Galloway, W.E. (1989) Genetic stratigraphic sequences in basin analysis; I, Architecture and genesis of flooding-surface bounded depositional units. *AAPG Bull.*, **73**, 125–142.

Horrillo-Caraballo, J.M. and **Reeve, D.E.** (2008) Morphodynamic behaviour of a nearshore sandbank system: The Great Yarmouth. *Mar. Geol.*, **254**, 91–106.

Houbolt, J.J.H.C. (1968) Recent sediments in the Southern Bight of the North Sea. *Geol. Mijnbouw*, **47**, 245–273.

Huthnance, J.M. (1982) On one mechanism forming linear sand banks. *Estuar. Coast. Shelf. Sci.*, **14**, 79–99.

Jin, H. and **Chough, K.** (2002) Erosional shelf ridges in the mid-eastern Yellow Sea. *Geo-Mar. Lett.*, **21**, 219–225.

Joseph, P., Hu, L.Y., Dubrole, O., Claude, D., Crumeyrolle, P., Lesueur, J.L. and **Soudet, H.J.** (1993) The Roda deltaic complex (Spain): from sedimentology to reservoir stochastic modelling. In: *Subsurface reservoir characterization from outcrop observations* (Eds **R. Eshard** and **B. Doligez**). Editions Technip, Paris, pp. 97–109.

Klein, G. de V. (1970) Depositional and dispersal dynamics of intertidal sand bars. *J. Sed. Res.*, **40**, 1095–1127.

Leclair, S.F. and **Bridge, J.S.** (2001) Quantitative interpretation of sedimentary structures formed by river dunes. *J. Sed. Res.*, **71**, 713–716.

Leren, B.L.S., Howell, J., Enge, H. and **Martinius, A.W.** (2010) Controls on stratigraphic architecture in contemporaneous delta systems from the Eocene Roda Sandstone, Tremp-Graus Basin, northern Spain. *Sed. Geol.*, **229**, 9–40.

Lopez-Blanco, M., Marzo, M. and **Munoz, J.A.** (2003) Low-amplitude, synsedimentary folding of a deltaic complex: Roda Sandstone (lower Eocene), South-Pyrenean Foreland Basin. *Basin Res.*, **15**, 73–95.

Martinius, A.W. (2012) Contrasting styles of siliciclastic tidal deposits in developing thrust-sheet-top basins-The Lower Eocene of the central Pyrenees (Spain). In: *Principles of Tidal Sedimentology* (Eds **R.A. Davis**, Jr. and **R.W. Dalrymple**). Dordrecht, Springer, pp. 473–506.

Martinius, A.W. (1995) Macrofauna associations and formation of shell concentrations in the early Eocene Roda Formation (southern Pyrenees, Spain). *Scripta Geol.*, **108**, 1–39.

Martinius, A.W. and **Molenaar, N.** (1991) A coral-mollusc (Goniaraea - Crassatella) dominated hardground community in a siliciclastic - carbonate sandstone (the Lower Eocene Roda Formation, southern Pyrenees, Spain). *Palaios*, **6**, 142–155.

Mathys, M. (2009) The Quaternary Geological Evolution of the Belgian Continental Shelf, Southern North Sea. Unpublished PhD. thesis. Ghent University, Ghent, p. 454.

McCave, I.N. and **Langhorne, D.N.** (1982) Sand waves and sediment transport around the end of a tidal sand bank. *Sedimentology*, **29**, 95–110.

Michaud, K.J. (2011) Facies Architecture and Stratigraphy of Tidal Ridges in the Eocene Roda Formation, Northern Spain. Queen's Theses & Dissertations Geological Sciences & Geological Engineering Graduate Theses, p. 128, http://hdl.handle.net/1974/6486.

Molenaar, N. and **Martinius, A.W.** (1996) Fossiliferous intervals and sequence boundaries in shallow marine, fan-deltaic deposits (early Eocene, Southern Pyrenees, Spain). *Palaeogeogr. Palaeoclimatol. Palaeoecol.*, **121**, 147–168.

Nio, S.D. (1976) Marine transgressions as a factor in the formation of sandwave complexes. *Geol. Mijnbouw*, **55**, 18–40.

Nio, **S.D.** and **Yang**, **C.S.** (1991) Sea-level fluctuations and the geometric variability of tide-dominated sandbodies. *Sed. Geol.*, **70**, 161–193.

Off, **T.** (1963) Rhythmic linear sand bodies caused by tidal currents. *AAPG Bull.*, **47**, 324–341.

Olariu, **C.** and **Bhattacharya**, **J.P.** (2006) Terminal distributary channels and delta front architecture of river-dominated delta systems. *J. Sed. Res.*, **76**, 212–233.

Olariu, **M.I.**, **Olariu**, **C. Steel**, **R.J.**, **Dalrymple**, **R.W.** and **Martinius**, **A.W.** (2012) Anatomy of a laterally migrating tidal bar in front of a delta system: Esdolomada Member, Roda Formation, Graus-Tremp Basin. *Sedimentology*, **59**, 356–378.

Pingree, **R.D.** (1978) The formation of the Shambles and other banks by tidal stirring of the seas. *J. Mar. Biol. Assoc. UK*, **58**, 211–226.

Pingree, **R.D.** and **Maddock**, **L.** (1979) The tidal physics of headland flows and offshore tidal bank formation. *Mar. Geol.*, **32**, 269–289.

Plaziat, **J.C.** (1981) Late Cretaceous to Late Eocene palaeogeographic evolution of southwest Europe. *Palaeogeogr. Palaeoclimatol. Palaeoecol.*, **36**, 263–320.

Puigdefàbregas, **C.**, **Muñoz**, **J.A.** and **Vergés**, **J.** (1992) Thrusting and foreland basin evolution in the southern Pyrenees. In: *Thrust Tectonics* (Ed. **K.R. McClay**). London, Chapman and Hall, pp. 247–254.

Reynaud, **J.Y.** and **Dalrymple**, **R.W.** (2012) Shallow-marine tidal deposits. In: *Principles of Tidal Sedimentology* (Eds **R.A. Davis**, Jr. and **R.W. Dalrymple**). Dordrecht, Springer, pp. 335–369.

Reynaud, **J.Y.**, **Tessier**, **B.**, **Proust**, **J.N.**, **Dalrymple**, **R.W.**, **Marsset**, **T.**, **De Batist**, **M.**, **Bourillet**, **J.F.** and **Lericolais**, **G.** (1999) Eustatic and hydrodynamic controls on the architecture of a deep shelf sand bank (Celtic Sea). *Sedimentology*, **46**, 703–721.

Rine, **J.M.**, **Tillman**, **R.W.**, **Culver**, **S.J.** and **Swift**, **D.J.P.** (1991) Generation of late Holocene sand ridges on the middle continental shelf of New Jersey, USA: Evidence for formation in a mid-shelf setting based on comparisons with a nearshore ridge. In: *Shelf Sand and Sandstone Bodies: Geometry, Facies and Sequence Stratigraphy* (Eds D.J.P. Swift, G.F. Oertel, R.W. Tillman and J.A. Thorne), *Int. Assoc. Sedimentol. Spec. Publ.*, **14**, 395–423.

Schmitt, **T.**, **Mitchell**, **N.C.** and **Ramsay**, **T.S.** (2007) Use of swath bathymetry in the investigation of sand dune geometry and migration around a near shore 'banner' tidal sandbank. In: *Coastal and Shelf Sediment Transport* (Eds P.S. Balson and M.B. Collins). *Geol. Soc. London Spec. Publ.*, **274**, 53–64.

Shaw, **J.**, **Duffy**, **G.**, **Taylor**, **R.B.**, **Chasse**, **J.** and **Frobel**, **D.** (2008) Role of a submarine bank in the long-term evolution of the northeast coast of Prince Edward Island, Canada. *Evolution*, **24**, 1249–1259.

Suter, **J.R.** (2006) Facies models revisited: clastic shelves. In: *Facies Models Revisited* (Eds H.W. Posamentier and R.G.Walker). *SEPM Spec. Publ.*, **84**, 339–397.

Southard, **J.B.** and **Boguchwal**, **L.A.** (1990) Bed configurations in steady unidirectional flows. Part 2. Synthesis of flume data. *J. Sed. Petrol.*, **60**, 658–679.

Tessier, **B.**, **Corbau**, **C.**, **Chamley**, **H.** and **Auffret**, **J.P.** (1999) Internal structure of shoreface banks revealed by high-resolution seismic reflection in a macrotidal environment (Dunkerque Area, northern France). *J. Coastal Res.*, **15**, 593–606.

Tinterri, **R.** (2007) The lower Eocene Roda Sandstone (south-central Pyrenees): An example of a flood-dominated river-delta system in a tectonically controlled basin. *Riv. Ital. Paleontol. Stratigr.*, **113**, 223–255.

Torricelli, **S.**, **Knezaurek**, **G.** and **Biffi**, **U.** (2006) Sequence biostratigraphy and paleoenvironmental reconstruction in the Early Eocene Figols Group of the Tremp-Graus Basin (south-central Pyrenees, Spain). *Palaeogeogr. Palaeoclimatol. Palaeoecol.*, **232**, 1–35.

Turner, **R.L.** and **Meyer**, **C.E.** (1980) Salinity tolerance of the brackish-water echinoderm Ophiophragmus filograneus (Ophiuroidea). *Mar. Ecol. Prog. Ser.*, **2**, 249–256.

Van Eden, **J.G.** (1970) A reconnaissance of deltaic environment in the Middle Eocene of the south-central Pyrenees, Spain. *Geol. Mijnbouw*, **49**, 145.

Van De Meene, **J.W.H.**, **Boersma**, **J.R.** and **Terwindt**, **J.H.J.** (1996) Sedimentary structures of combined flow deposits from the shoreface-connected ridges along the central Dutch coast. *Mar. Geol.*, **131**, 151–175.

Vergés, **J.** and **Burbank**, **D.W.** (1996) Eocene-Oligocene thrusting and basin configuration in the eastern and central Pyrenees (Spain). In: *Tertiary Basins of Spain* (Eds P.F. Friend and C.J. Dabrio). Cambridge, Cambridge University Press, pp. 120–133.

Vincent, **S.** (2001) The Sis palaeovalley: a record of proximal fluvial sedimentation and drainage basin development in response to Pyrenean mountain building. *Sedimentology*, **48**, 1235–1278.

Williams, **J.J.**, **MacDonald**, **N.J.**, **O'Connor**, **B.A.** and **Pan**, **S.** (2000) Offshore sandbank dynamics. *J. Mar. Syst.*, **24**, 153–173.

Wood, **L.J.** (2003) Predicting tidal sand reservoir architecture using data from modern and ancient depositional systems. *AAPG Mem.*, **80**, 1–22.

Yang, **C.-S.** (1989) Active, moribund and buried tidal sand ridges in the East China Sea and the southern Yellow Sea. *Mar. Geol.*, **88**, 97–116.

Index

Contributions to Modern and Ancient Tidal Sedimentology: Proceedings of the Tidalites 2012 Conference,
First Edition. Edited by Bernadette Tessier and Jean-Yves Reynaud.
© 2016 International Association of Sedimentologists. Published 2016 by John Wiley & Sons, Ltd.